Contents

Foreword v

1. *Elk Lake in Perspective* 1
   R. Y. Anderson, W. E. Dean, and J. P. Bradbury

2. *History of the Landscape in the Itasca Region* 7
   H. E. Wright, Jr.

3. *Climatic and Limnologic Setting of Elk Lake* 19
   R. O. Megard, J. P. Bradbury, and W. E. Dean

4. *Chronology of Elk Lake Sediments: Coring, Sampling, and Time-Series Construction* 37
   R. Y. Anderson, J. P. Bradbury, W. E. Dean, and M. Stuiver

5. *The Varve Chronometer in Elk Lake: Record of Climatic Variability and Evidence for Solar/Geomagnetic-^{14}C-Climate Connection* 45
   R. Y. Anderson

6. *On the Precision of the Elk Lake Varve Chronology* 69
   D. R. Sprowl

7. *Modern Sedimentation in Elk Lake, Clearwater County, Minnesota* 75
   E. B. Nuhfer, R. Y. Anderson, J. P. Bradbury, and W. E. Dean

8. *Environment of Deposition of $CaCO_3$ in Elk Lake, Minnesota* 97
   W. E. Dean and R. O. Megard

9. *Geochemistry of Surface Sediments of Minnesota Lakes* 115
   W. E. Dean, E. Gorham, and D. J. Swaine

10. *Physical Properties, Mineralogy, and Geochemistry of Holocene Varved Sediments from Elk Lake, Minnesota* 135
    W. E. Dean

11. **Geologic Implications of the Elk Lake Paleomagnetic Record** 159
   D. R. Sprowl and S. K. Banerjee

12. **Stable Carbon and Oxygen Isotope Studies of the Sediments
    of Elk Lake, Minnesota** 163
   W. E. Dean and M. Stuiver

13. **Fossil Pigments in Holocene Varved Sediments in Elk Lake,
    Minnesota** 181
   J. E. Sanger and R. J. Hay

14. **Surface Sample Analogues of Elk Lake Fossil Diatom Assemblages** 189
   R. B. Brugam

15. **Holocene Diatom Paleolimnology of Elk Lake, Minnesota** 215
   J. P. Bradbury and K. V. Dieterich-Rurup

16. **Postglacial Chrysophycean Cyst Record from Elk Lake, Minnesota** 239
   B. A. Zeeb and J. P. Smol

17. **Vegetation History of Elk Lake** 251
   C. Whitlock, P. J. Bartlein, and W. A. Watts

18. **Paleoclimatic Interpretation of the Elk Lake Pollen Record** 275
   P. J. Bartlein and C. Whitlock

19. **Fire, Climate Change, and Forest Processes During the Past
    2000 Years** 295
   J. S. Clark

20. **Holocene Climatic and Limnologic History of the North-Central
    United States as Recorded in the Varved Sediments of Elk Lake,
    Minnesota: A Synthesis** 309
   J. P. Bradbury, W. E. Dean, and R. Y. Anderson

Index 329

Elk Lake, Minnesota: Evidence for Rapid Climate Change in the North-Central United States

Edited by

J. Platt Bradbury
and
Walter E. Dean

U.S. Geological Survey
MS-919 (JPB), MS-939 (WED)
Box 25046, Federal Center
Denver, Colorado 80225

SPECIAL PAPER
276

1993

© 1993 The Geological Society of America, Inc.
All rights reserved.

All materials subject to this copyright and included
in this volume may be photocopied for the noncommercial
purpose of scientific or educational advancement.

Copyright is not claimed on any material prepared
wholly by government employees within the scope
of their employment.

Published by The Geological Society of America, Inc.
3300 Penrose Place, P.O. Box 9140, Boulder, Colorado 80301

Printed in U.S.A.

GSA Books Science Editor Richard A. Hoppin

Library of Congress Cataloging-in-Publication Data

Elk Lake, Minnesota : evidence for rapid climate change in the north
  -central United States / edited by J. Platt Bradbury and Walter E.
  Dean.
      p.   cm. — (Special paper ; 276)
    Includes bibliographical references and index.
    ISBN 0-8137-2276-4
    1. Climatic changes—Middle West.   2. Paleoclimatology—Middle
  West.   3. Elk Lake Region (Clearwater County, Minn.)—Climate.
  4. Varves—Minnesota—Elk Lake Region (Clearwater County)
  5. Sedimentology   I. Bradbury, J. Platt (John Platt)   II. Dean,
  Walter E.   III. Series: Special paper (Geological Society of
  America) ; 276.
  QC981.8.C5E55   1993
  551.6978—dc20   93-20393
                                                CIP

Cover photo: On the surface, Elk Lake with its forested shores resembles many of the other 10,000+ lakes in Minnesota (photograph), but the sediments underneath the lake contain annual layers (varves) that have marked off the seasons year after year for the last 10,400 yr (right edge of photograph). The thicknesses of these annual layers (to far right of photograph) are only one of the many indicators that tell of the time between 8,000 and 3,800 yr ago when Elk Lake was surrounded by prairie rather than forest and subjected to wind and dust that formed dune fields in Minnesota, Nebraska, and elsewhere in North America.

Elk Lake Data Set is available through:

National Geophysical Data Center   Phone: 303-497-6215
NOAA, E/GC1, Dept. 891   Fax: 303-497-6513
325 Broadway   Internet: Info@mail.ngdc.noaa.gov
Boulder, CO 80303, U.S.A.   Telex: 592811 NOAA MASC BDR

Data set includes varve thickness, pollen counts, diatom counts, isotopes, geochemistry, and mineralogy of samples from the Elk Lake core.

---

10   9   8   7   6   5   4   3   2   1

Foreword

We must now seek to understand our environment in terms of change on the scale of human generations. We need to learn about the sensitivity of our regional ecosystems in the face of changing climate. What are the natural ranges of variability? What are extreme rates of natural change? What controlled climate variability in the past? What are the thresholds and when were they crossed? It is a challenging but exciting task to extract multi-indicator information on complex biosphere-geosphere feedbacks from archives of the past. How is it done?

A lake is a system operating at the aqueous interface between atmosphere, geosphere, and biosphere. Each lake is a long-term climate sensitivity experiment, with multiple parameters recording different signals of changing climate at different rates, magnitude, and fidelity.

Minnesota, the land of thousands of lakes, sits in a special place near the center of the North American continent. It is uniquely characterized by a very steep climatic gradient at the junction of Arctic, Pacific, and Gulf of Mexico airmass systems. Tiny Elk Lake in north-central Minnesota is poised on the sharp, sensitive boundary between prairie and forest environments. The sediments of Elk Lake contain a chronicle of past climate dynamics written year by year in annual laminations. These annual rhythms are called varves. They do not derive from the annual melting of glaciers, but rather record seasonal biotic/abiotic interactions such as diatom blooms and the precipitation of carbonate minerals. Many Minnesota lake deposits contain intermittent varved sediment sections, but Elk Lake is special for its continuously varved Holocene record of climatic and environmental change.

Varves evoke rhythms with human dimensions, of years, of the sun. In lakes, the yearly solar cycle strikes the dominant beat, different from the ocean margins and their lunar tides. We respond to these seasonal rhythms that control the biological and economic cycles of our lives. Subconsciously, we are thus fascinated with rhythmically laminated sediments. Varved-lake archives provide a metronome for events in the past. Varve time!

This GSA special paper detailing the 10,400-year varved Holocene record in Elk Lake is a milestone. To deconvolve a climatic history from Elk Lake, the scientific team used a marine geological approach that has been highly successful for reconstructing past environments from ocean drilling cores. They document biotic and abiotic signatures of landscape history, including forest and prairie cover, wind transported dust, forest fires, soil, and magnetic particles. In addition, they detail the signatures of biotic and abiotic changes in the lake itself, such as precipitation of minerals, sedimentology, geochemical stratigraphy, stable isotope profiles from carbonate and organic components, and diatom and chrysophyte paleoecology. Studies of modern lake processes and particles collected in sediment traps have been used to calibrate the record. The whole is synthesized into a coherent paleoclimatic interpretation of the Holocene. In spite of the effort, the tone is preliminary, and posterity will determine if the new syntheses stand.

If we are to construct and compare decadal-scale paleoclimate signatures and rates of change across the globe, we need more such integrated, multidisciplinary paleoclimate studies, and we need the chronometry of varves.

Kerry Kelts, Director
Limnological Research Center
University of Minnesota
Minneapolis, Minnesota 55455

Elk Lake in perspective

Roger Y. Anderson
Department of Earth and Planetary Sciences, University of New Mexico, Albuquerque, New Mexico 87131
Walter E. Dean and J. Platt Bradbury
U.S. Geological Survey, Box 25046, Federal Center, Denver, Colorado 80225

ABSTRACT

Elk Lake is located in the forested region of north-central Minnesota at the headwaters of the Mississippi River and occupies one of countless basins left behind as the last great Pleistocene ice sheet retreated northward into Canada. In this respect it resembles many other moderately deep, dimictic, hard-water lakes in the north-central United States, the sediments of which contain a history of postglacial and Holocene climatic and environmental change. Elk Lake is different, however, because the Holocene sediments in the deeper part of the lake form an uninterrupted sequence of annual laminations or varves. The varves are a chronometer for timing precisely the biologic, geochemical, and sedimentological responses in the lake to cyclic and progressive changes in climate. The varves also, through profound changes in their composition, divide the history of Elk Lake into three, sharply defined episodes; a postglacial lake, a prairie lake, and a modern, mesic-forest lake. We use these episodes and the character of the varves as a framework to guide the reader to the chapters and discussions found in this volume.

CLIMATIC RECORD IN ELK LAKE

The Itasca region of north-central Minnesota has been recognized for many years as a sensitive location for paleoclimatic studies (McAndrews, 1966; Wright, 1976). The region is in a climatic transition zone at the southern margin of the polar front that separates the cold, dry arctic airstream to the north from a warmer and wetter tropical Atlantic airstream to the south (Wright, 1976) (Fig. 1). Dry Pacific air pushes into the Itasca area from the west, producing a sharp gradient in moisture that separates western prairie grassland from eastern forest. The three regional air masses join to form a climatic triple junction with sharp climatic gradients across Minnesota.

Elk Lake is situated in a pine-hardwood forest but is only 80 km east of the prairie-forest border (Fig. 2). During the more arid mid-Holocene, the present forest-prairie border (Fig. 2A) expanded eastward at least 100 km (Wright, 1976). As climate changed and the ice front withdrew from Minnesota, regional airstreams shifted and oscillated. The prairie-forest ecotone moved back and forth across Elk Lake, changing lake levels, altering the biota and geochemistry of the lake, and changing the vegetation and ground cover.

Sediments that accumulated on the bottom of Elk Lake (Fig. 3) recorded faithfully the oscillations in regional climate as changes in the abundance and composition of the biota and changes in the composition and texture of the sediments. Because the sediments are varved, the timing of key events can be determined precisely. Equally important, the natural chronometer provided by the varves allows us to examine climatic variability across a wide band of climatic frequencies and examine climatic cycles with periods of a few years to several thousand years (Fig. 3). How the varves formed, why they are preserved, and how changes in sediment components and properties measure changes in the lake are topics explored in chapters on modern sediment accumulation in Elk Lake (Nuhfer and others, Chapter 7). Using the varves as a natural chronometer (Anderson, Chapter 5) and the precision of the varve chronometer (Anderson and others, Chapter 4) are also explored elsewhere. Magnetic properties of the sediments permit correlation of the Elk Lake paleomagnetic record with those of other lakes, and help to validate the accuracy and precision of the varve chronometer (Sprowl and Banerjee, 1989 and Chapter 11; Sprowl, Chapter 6).

The pattern of climatic oscillation preserved in the varved sediments is complicated by successional changes in the lake and

---

Anderson, R. Y., Dean, W. E., and Bradbury, J. P., 1993, Elk Lake in perspective, *in* Bradbury, J. P., and Dean, W. E., eds., Elk Lake, Minnesota: Evidence for Rapid Climate Change in the North-Central United States: Boulder, Colorado, Geological Society of America Special Paper 276.

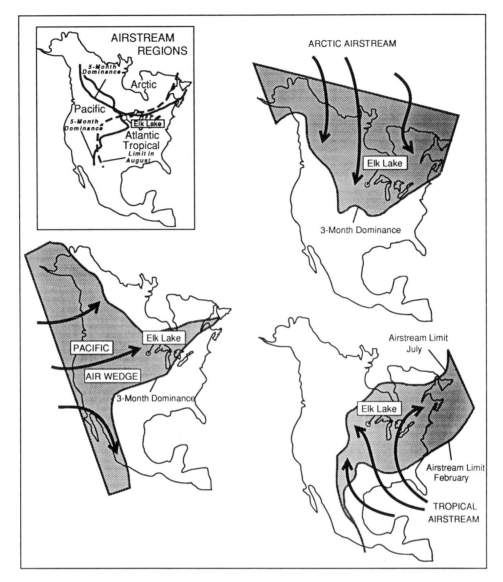

Figure 1. Regional airstreams of North America that affected the climatic environment of Elk Lake, the location of which is shown as a circle on each map. Modified from Bryson and Hare (1974).

its drainage basin. For example, the leaching of fresh till early in the history of the lake, at a time when the lake was surrounded by spruce and pine forest, supplied abundant manganese and calcium to the lake and its sediments (Fig. 4). After a dominance of prairie vegetation between 8 and 4 ka, mesic forest cover returned to the Elk Lake drainage. This forest was established on maturely weathered till that had most of its manganese content leached. However, anoxic soil conditions beneath the mesic forest allowed ground water to leach iron from the till and to subsequently deposit it in abundance in lake sediment (Fig. 4). These and other progressive changes recorded in lake sediment must be isolated from the climatic signal before the record of climatic variability can be identified and interpreted at local and regional scales.

GUIDE TO LONG-TERM CHANGES IN ELK LAKE

Elk Lake, when examined over its entire life span, is actually a succession of three geochemically and environmentally different lakes that occupied sequentially the same basin. The reasons for the succession are complex and reflect both a maturing of the lake and its drainage and a response to climatic forcing. Contributors to this volume have described and interpreted biotic, sedimentologic, geochemical, and magnetic parameters collected from the entire stratigraphic record; the following brief sketch of the history of the lake and its three lake phases may be helpful. Other chapters that are important to a general understanding of the Elk Lake sediment record include the glacial history of the Itasca

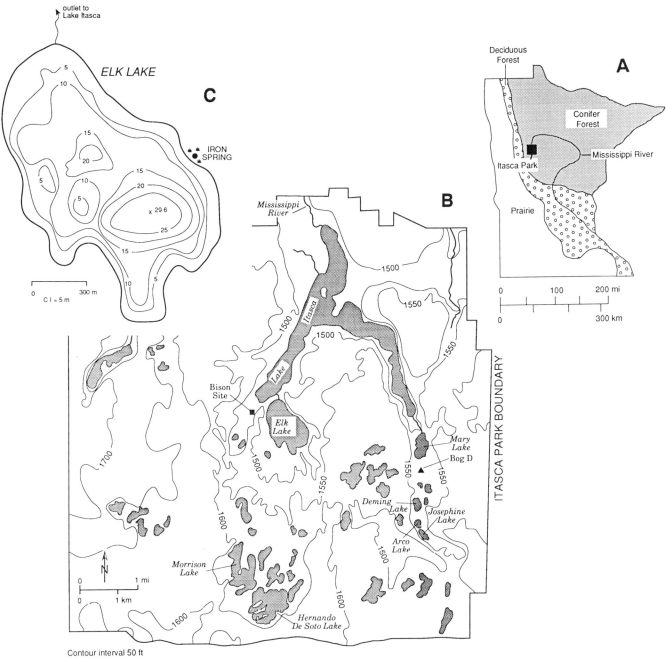

Figure 2. Maps of (A) Minnesota, (B) Itasca State Park, and (C) Elk Lake, showing the location of Elk Lake, general vegetation zones of Minnesota, topography and distribution of lakes in Itasca Park, bathymetry of Elk Lake, and location of the varved core (x) in the deepest part of the lake (29.6 m).

region (Wright, Chapter 2); fire, climate change, and forest changes in the Itasca region over the past 2000 years (Clark, Chapter 19); the climatic and limnologic setting of Elk Lake (Megard and others, Chapter 3); variations in diatom populations (Brugam, Chapter 14); chemical composition of surface sediments in lakes throughout Minnesota (Dean and others, Chapter 9); and the varve chronometer (Anderson, Chapter 5).

Postglacial Lake

The ice block that formed the deepest part of Elk Lake melted slowly and a small, temporary lake developed above the subterranean block of melting ice. Coarse sediment, plant debris, and abundant ostracodes of boreal forest aquatic habitats (R. M. Forester, 1987, personal commun.) accumulated in the shallow,

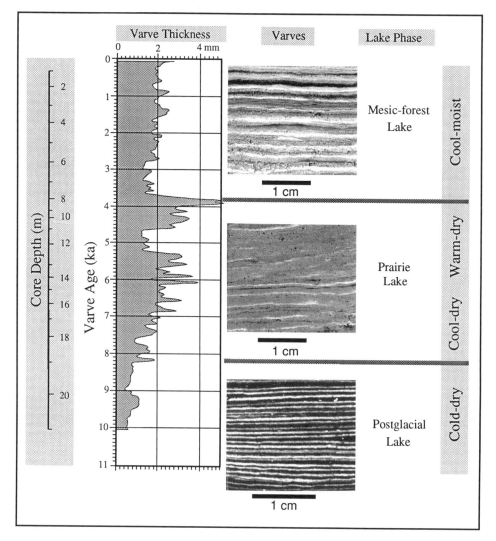

Figure 3. Smoothed plot of varve thickness as a function of time and photographs illustrating typical patterns of varves found in the three phases of Elk Lake.

transitory pond. When the ice block melted completely, it left a deep depression, and varved sediments began to accumulate beneath about 50 m of water (about 20 m deeper than at present because of sediment fill). By the time the glacier was several hundred kilometers to the northeast, the newly formed postglacial lake was surrounded by a forest of spruce and birch (see pollen zones in Fig. 4; Whitlock and others, Chapter 17) and receiving ground-water seepage that was rich in dissolved calcium, magnesium, manganese, and iron that had been leached from the fresh till (see Mn and Fe profiles in Fig. 4; Dean, Chapter 10). The lake was much deeper than at present and may have been anoxic the entire year. Dark-colored seasonal laminae, enriched in Fe, Mn, and diatoms, accumulated on the bottom, contrasting with light-colored, carbonate-rich summer laminae (Fig. 3). These sharply defined laminations persisted from about 10 to 8 ka, until oak savanna replaced the coniferous forest under drier mid-Holocene climate conditions.

Prairie Lake

The mid-Holocene time interval is widely recognized in Europe and the northeastern United States as a period that was warmer and drier than at present, and has been variously referred to as the altithermal, hypsithermal, or climatic optimum. This mid-Holocene climatic episode in Minnesota certainly was drier, but evidence for increased temperature throughout the interval is equivocal. We use Wright's (1976) term "prairie period" for the mid-Holocene event in northern Minnesota, marked simply by the presence of drier prairie-savanna vegetation with no connotation of temperature. Between 8.2 and 4.0 ka, oak savanna and prairie vegetation expanded eastward by as much as 100 km. Ostracode and diatom data (Forester and others, 1987; Bradbury and Dietrich-Rurup, Chapter 15) show that northwestern Minnesota was cooler than at present during much of the prairie period; conditions were similar to those found today in Canadian prairie

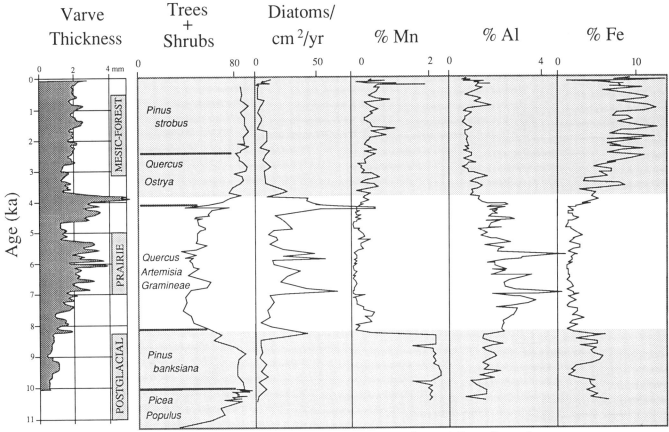

Figure 4. Plots of varve thickness, percentage of tree plus shrub pollen, total diatom flux, and percentages of manganese, aluminum, and iron as functions of time for the varved core from the deep basin of Elk Lake.

lakes. Cold, saline lake conditions are also suggested by a high $^{18}O:^{16}O$ ratio in precipitated $CaCO_3$ (Dean and Stuiver, Chapter 12).

The 4 ka of the mid-Holocene prairie period was a time of climatic transitions and oscillations and shifting dominance of the three airstreams that converge on Minnesota (Fig. 1; Bartlein and Whitlock, Chapter 18). With the advent of a drier climate, coniferous forests were replaced by oak savanna, a sparse distribution of sagebrush, and grass (see pollen zones in Fig. 4; Whitlock and others, Chapter 17).

Reduced vegetative ground cover in the Elk Lake drainage was accompanied by increased influx of detrital clastic material, mostly clay, into a somewhat shallower prairie lake. The increased influx of clastics, largely from eolian sources, is manifested as thicker varves (Fig. 3) and a higher proportion of elements such as aluminum (Fig. 4; Dean, Chapter 10). The fine-grained detrital material found in the varves was distributed in the lake by wind-driven turbulence. Primary productivity in the lake, reflected in the flux of diatoms (Fig. 4) and various geochemical parameters, increased significantly as nutrients from the drainage were recycled and released from lake sediments (Bradbury and Dieterich-Rurup, Chapter 15; Dean, Chapter 10; Sanger and Hay, Chapter 13; Dean and Stuiver, Chapter 12).

The laminations in lake sediment became obscure during parts of the prairie period as the amount of clay and silt in the deep part of the lake increased and as the thickness of the varves increased (Fig. 3). However, the seasonal production of organic materials and calcium carbonate was sufficient to define annual increments of sediment during most of the mid-Holocene; centennial pulses in the influx of clastic sediment appear to be associated with solar activity, and possibly with changes in the Earth's magnetic field (Anderson, Chapter 5).

During drought cycles, the elevation of the lake surface fell and the hardness of the water increased, producing a complex suite of carbonate minerals (Dean and Megard, Chapter 8). At times during the life of the prairie lake, large blooms of diatoms developed and became an important component of the varves.

Mesic-Forest Lake

By the time moist conditions returned at the end of the prairie period, about 4 ka, the glacial till in the drainage basin of the lake had become more completely weathered. After a mesic forest of pine, birch, and other hardwoods was established on weathered soils (see pollen zones in Fig. 4; Whitlock and others, Chapter 17), ground-water seepage carried a greater concentration of iron into the mesic-forest lake, and iron accumulated in increasing concentration, mainly as iron hydroxide and iron phosphate in the sediments of the modern lake (Fig. 4; Dean, Chapter 10). Magnetic properties of the sediments indicate that some of the increase in iron during the past 2 ka is due to addition of fine-grained magnetite that may be biogenic (Sprowl and Bannerjee, 1989 and Chapter 11).

The higher lake level and lower alkalinity of the mesic-forest lake were accompanied by changes in assemblages of ostracodes (Forester and others, 1987) and diatoms (Bradbury and Dieterich-Rurup, Chapter 15). The return of forest vegetation virtually shut off the supply of clastic sediment to the lake. Varves deposited in the modern lake are complex, and as many as five different components generated within the lake were deposited seasonally and organized into laminae rich in iron, manganese, diatoms, $CaCO_3$, and organic matter (Fig. 3; Anderson, Chapter 5).

CLIMATIC SUCCESSION VS. CLIMATIC OSCILLATION

The postglacial, prairie, and mesic-forest lake phases represent a successional change in the lacustrine environment that is difficult to separate from a directional shift in regional climate. The change in climate also appears as a succession because it represents only part of a major cycle in ice volume believed to be related to changes in insolation. The effects of different air masses on the lake environment are clearly shown in the response of ostracodes and diatoms (Forester and others, 1987; Bradbury and Dieterich-Rurup, Chapter 15). The pollen record (Whitlock and others, Chapter 17; Bartlein and Whitlock, Chapter 18) also reflects changes in vegetation that were induced by shifts in air masses rather than by ecologic succession.

The long-term (mid-Holocene) trend toward drier climate was interrupted by a strong, rapid reversal that lasted about 600 yr, between 4.6 and 5.2 ka (see varve thickness plot in Fig. 4). The climatic reversal is reflected in decreased sediment flux and is recorded by certain species of diatoms (Bradbury and Dieterich-Rurup, Chapter 15), assemblages of chrysophycean cysts (Zeeb and Smol, Chapter 16), the geochemistry of the sediments (Dean, Chapter 10), the magnetic susceptibility (Sprowl and Bannerjee, Chapter 11), and the character of the varves. The reversal is part of a pattern of climatic variability that cannot be attributed directly to local changes in the basin or to long-term regional trends in climate. This pattern is most strongly expressed on the time scale of decades to centuries during the prairie period, but also can be observed in sediments of the mesic-forest lake (Anderson, Chapter 5).

OBJECTIVES OF ELK LAKE STUDY

Our goal in assembling a detailed history of Elk Lake is to separate local and sequential changes in biota and geochemistry from changes that measure regional shifts and oscillations in climate. Another objective is to connect the paleoclimatic data directly to airstream patterns (Fig. 1; Bartlein and Whitlock, Chapter 18). A larger objective is to determine how effectively the observations of several investigators from several disciplines can be brought to bear on the study of a single locality. We believe that the sum of this volume is greater than its individual parts and the observations and interpretations in the synthesis (Chapter 20), and in separate chapters are more accurate for having used complementary data and interpretations.

REFERENCES CITED

Bryson, R. A., and Hare, F. K., 1974, Climates of North America, *in* Landsberg, H. E., ed., World survey of climatology, Volume 11: New York, Elsevier, 420 p.

Forester, R. M., DeLorme, L. D., and Bradbury, J. P., 1987, Mid-Holocene climate in northern Minnesota: Quaternary Research, v. 28, p. 263–273.

McAndrews, J. H., 1966, Postglacial history of prairie, savanna, and forest in northwestern Minnesota: Torrey Botanical Club Memoirs, v. 22, p. 1–72.

Sprowl, D. R., and Banerjee, S. K., 1989, The Holocene paleosecular variation record from Elk Lake, Minnesota: Journal of Geophysical Reseach, v. 94, p. 9369–9388.

Wright, H. E., 1976, The dynamic nature of Holocene vegetation—A problem in paleoclimatology, biogeography, and stratigraphic nomenclature: Quaternary Research, v. 6, p. 581–96.

Manuscript Accepted by the Society July 27, 1992

History of the landscape in the Itasca region

H. E. Wright, Jr.
Limnological Research Center, University of Minnesota, Minneapolis, Minnesota 55108

ABSTRACT

The natural landscape of Minnesota includes the readily visible aspects of the scenery—primarily the landforms, the vegetation, and the lakes. The landforms, including the basins in which the lakes and bogs are located, owe their origin to glaciation. The ice affected the area in one way or another for many thousands of years, up until about 11,000 years ago. Subsequently the lakes were filled with sediment as climate and vegetation gradually changed. The history of the glacial period is recorded by the topographic features of the region as well as by the composition and structure of the glacial drift. The postglacial environment history is recorded by the fossils in the lake sediments. Elk Lake is a typical glacial lake, except that it has the distinction of containing annually laminated sediments, which permit a precise chronology. Herein the sequence of glaciation and the development of the glacial landforms are considered first, then the regional climatic and vegetational history since the time of glaciation, and finally the characteristics of the lakes themselves, to provide a background for the detailed chapters on the stratigraphy of the various components of the Elk Lake sediments.

GLACIAL HISTORY

The most conspicuous topographic feature of the Elk Lake region in Minnesota is the prominent Itasca moraine, which trends east-west north of the town of Park Rapids (Fig. 1). The ice must have remained at the moraine for a long time in order to produce its great volume of till and outwash. The hilly topography of the moraine implies a heterogeneous distribution of glacial sediment in the terminal part of the ice sheet. For example, Elk Lake has 30 m of water and 20 m of sediment, and the hills to the east and west rise 25 m above the lake surface, producing a local original relief of 75 m. The depressions in this landscape represent parts of the ice that happened to contain relatively little rock debris—so-called ice blocks—and the hills are localized where the ice contained abundant such material. When all the ice melted away the resulting landscape was pockmarked with depressions between hills, generally with no particular pattern. The depressions now contain the lakes and wetlands that characterize the moraines.

The eastern part of the Itasca moraine contains a series of subparallel ridges slightly concave to the north-northeast. They may have resulted from upward thrusting of ice masses near the terminus (Mooers, 1990).

Elsewhere the Itasca moraine is transected by discontinuous north-south troughs or valleys, many of them identified by strings of lakes, such as the arms and extensions of Lake Itasca (Fig. 2). South of the Itasca moraine is the large Park Rapids sand plain, which laps southward onto the Wadena drumlin field. The drumlins are truncated on the east by the St. Croix moraine, which meets the Itasca moraine near Walker and is also fronted by outwash. North of Itasca Park are more moraines transected by north-south troughs. To the west are other moraines, which give way eventually to the plain of glacial Lake Agassiz.

Examination of these glacial features reveals differences in the color and composition of the materials, implying different source areas from which the ice came. The distinctive form of some of the features provides clues to the direction of ice movement and to the sequence of glaciation. Perhaps the easiest way to understand the origin of the glacial features of the area around Itasca Park is to trace the history of glaciation in northwestern Minnesota step by step.

Lobes of the Laurentide ice sheet moved into Minnesota

Wright, H. E., Jr., 1993, History of the landscape in the Itasca region, *in* Bradbury, J. P., and Dean, W. E., eds., Elk Lake, Minnesota: Evidence for Rapid Climate Change in the North-Central United States: Boulder, Colorado, Geological Society of America Special Paper 276.

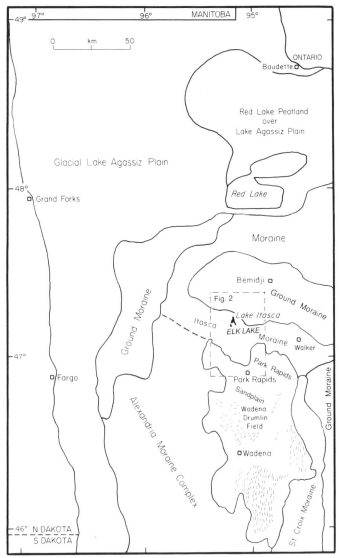

Figure 1. Map of northwestern Minnesota showing location of Elk Lake in the Itasca moraine.

from Ontario and Manitoba, and their positions and forms were determined largely by locations of lowlands in the bedrock. Thus on the east the great lowland now occupied by Lake Superior channeled the Superior lobe southwestward into the state, ultimately extending down to the Minneapolis area, where an extension of the Superior lowland existed (Fig. 3). The upland west of the Lake Superior basin was covered by the Rainy lobe. On the other side of the state was the lowland now occupied by the Red River, and this continues southeastward down the Minnesota Valley and then southward into central Iowa. This lowland channeled the Des Moines lobe. In north-central Minnesota, north of the Itasca region and centering near Red Lakes, was a shallower lowland connected northwestward to the Winnipeg area in Manitoba.

The oldest surficial glacial drift in the region is represented by the Wadena drumlin field, which is found in Wadena and Todd counties and adjacent areas (Fig. 1). The drumlins are oval hills generally about 2–7 km long, 0.5 km wide, and 5–15 m high. The drumlin field contains about 1200 individual hills in a fan-shaped pattern about 100 km long and equally broad at the outer edge of the fan. The drumlins are oriented parallel to the direction of ice flow. The axis of the fan trends southwest, but the drumlins on the two lateral margins shift around to west and to south. The field marks roughly the form of the Wadena ice lobe, which moved in a general direction to the southwest (Fig. 3). The outer arc of the fan terminates near the very large, curved Alexandria moraine complex, which may have marked the end of the Wadena lobe at the time the drumlins were formed. This entire moraine, as well as the outer fringe of the drumlin field, is covered with younger drift deposited by ice from the west and south.

The eastern margin of the Wadena drumlin field is also covered with younger drift, deposited as the St. Croix moraine of the Rainy lobe. The northern margin was buried by outwash from the Itasca moraine. The Wadena drumlin field therefore represents an enclave of older glacial drift completely surrounded by younger deposits.

The southwesterly orientation of the Wadena drumlins indicates that the ice that shaped them was moving from the northeast. The dominant rock types in the drift, especially in the southwestern part of the field, are dolomite and limestone apparently derived from Paleozoic rocks exposed to the northwest in southern Manitoba, because no such carbonate bedrock occurs in northern Minnesota or in western Ontario in the direction northeast from the Wadena drumlin field. It was therefore postulated that the Wadena lobe must have had its source northwest of the drumlin field, but that it was then diverted by the Rainy lobe (Wright, 1962, 1972).

According to this reconstruction, the ice of the Wadena lobe advanced southeastward into the Red Lakes lowland from the Winnipeg lowland in southern Manitoba. Meanwhile, on the east the Superior lobe filled the Lake Superior basin and the Rainy lobe advanced to the southwest across the upland of northwestern Ontario and adjacent northeastern Minnesota. The Wadena lobe, which was moving southeast across the Red Lakes lowland, was diverted to the southwest by the Rainy lobe, which was advancing to the southwest. Thus the Wadena lobe with its new trend reached Wadena County with a flow direction appropriate to mold the Wadena drumlin field, terminating at the Alexandria moraine complex. As it proceeded it became less and less contaminated with noncarbonate till acquired from its encounter with the Rainy lobe, so its carbonate content increased to the southwest.

The Alexandria moraine complex is a huge feature, and it stands out on any map of Minnesota as a great arcuate belt of lakes and hills. Although it was covered by a later advance of the Des Moines lobe from the west, the core of the moraine was probably formed by the Wadena lobe. The ice must have stood at this position for a very long time, during which the climate must have been relatively stable, so that the ice could maintain a steady

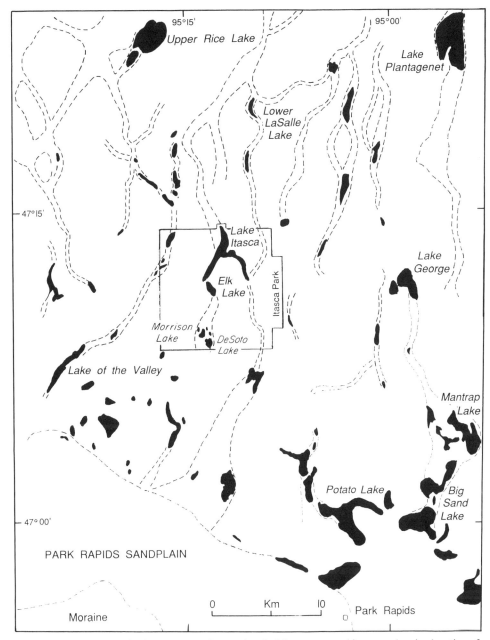

Figure 2. Tunnel valleys in the Itasca moraine, as sketched from topographic maps, showing location of Elk Lake.

terminus and mold such a large drumlin field and build such a large moraine.

Although this reconstruction of events explains adequately the major landforms and the general nature of the sediments, more recent research proposes a somewhat different explanation for the composition of the Wadena drumlin field (Goldstein, 1989). In this reconstruction the southwestward increase in carbonate content and other changes in lithologic components in the Wadena drumlins in the direction of ice flow are attributed not to *decreasing* contamination by the noncalcareous till of the obstructing Rainy lobe but rather to *increasing* contamination from an underlying strongly calcareous till being overridden by ice from the northeast, i.e., by the Rainy lobe itself.

In any case, a climatic change eventually caused the front of the Wadena lobe to retreat an unknown amount; the lobe then readvanced to an east-west front just south of Itasca Park, where it formed the Itasca moraine. The ice movement at this time may have been more from the north-northeast. The combined Rainy and Superior lobes, which covered all of the area to the east at this time, came to a terminus at the north-south St. Croix

Figure 3. Distribution of lakes in and near Lake Itasca State Park, illustrating the concentration in the tunnel valleys sketched in Figure 2.

moraine, which was joined to the east end of the Itasca moraine near Walker. The junction is complex, but the different drifts can be distinguished: that of the Wadena lobe is light brown and filled with carbonate fragments, and that of the Rainy lobe is darker brown and contains crystalline rock types derived from the northeast, and it does not contain carbonates.

The contemporaneity of the Itasca and St. Croix moraines is demonstrated by the form of their outwash plains. The great plain leading southward from the Itasca moraine at Park Rapids (Fig. 1) buries the northern part of the previously formed Wadena drumlin field, and the narrower plain leading westward from the St. Croix moraine buries the eastern edge of the drumlin field; the two plains are confluent, and the sand extends southward down the Crow River. The drainage must have continued farther south along the course of the Long Prairie River, eventually to the Minnesota Valley, because any possible course to the east was blocked by the St. Croix moraine. The course of the Long Prairie River has since been reversed to the northeast through a gap in the St. Croix moraine, because of blockage by the Grantsburg sublobe.

The Wadena lobe stood at the Itasca moraine for a long time, maybe hundreds of years. The time of this stillstand has been estimated at about 20,000 years ago, on the basis of a radiocarbon date on basal organic sediment from a lake deposit at a site in back of (east of) the St. Croix moraine, and therefore formed after retreat of the ice from that moraine (Birks, 1976). This date may be too old, however, because it was obtained from bulk fine-grained lake sediment that may have contained organic matter deficient in ^{14}C, either as the remains of algae that had incorporated CO_2 from ground water derived from carbonate-bearing tills, or as lignite particles derived from Cretaceous rocks. The moraine can only be shown to be older than 14,000 years, the date (based on several analyses of wood) for the advance of the Des Moines lobe, which overrode the St. Croix moraine in southern Minnesota at that time.

While the Wadena lobe stood at the Itasca moraine, the area affected by surface melting apparently extended for many kilometers back from the ice front. The meltwater produced found its way to the base of the ice through crevasses and other structures, and it flowed south under the ice to the actual terminus, discharging its load of sediment to form the Park Rapids outwash plain. These subglacial streams, which actually originated on the mountain of ice, had a very great hydrostatic head because of the ice thickness, and they flowed with tremendous velocity. They kept open the tunnels against the pressure of the still-active flowing ice. They had the power not only to carry the sand and stones that fell into the channel from the ice walls, but they also eroded the substratum deeply, creating great tunnel valleys 15–60 m deep in the drift beneath the ice sheet. These features are now represented by the linear troughs and strings of lakes that cut through the Itasca moraine and can be traced for several tens of kilometers (Fig. 2).

Eventually the stillstand of the Wadena lobe at the Itasca moraine ended, as did that of the Rainy and Superior lobes at the St. Croix moraine, because of another shift in climate. The ice began to thin at a greater rate in the terminal area, and its flow rate decreased. The hydrostatic head for the subglacial streams became reduced, and these streams no longer had the high velocities necessary to keep all the tunnels open or to carry coarse gravel and erode the substratum. Some of the tunnels were abandoned by the streams and were then filled with glacial ice collapsing from the walls and roof. Others may have remained occupied by smaller streams that deposited sand and gravel along the floors. As the ice walls eventually melted away, these gravel fillings were left as ridge-like eskers winding down the middle or along the sides of the old tunnel valleys. In the abandoned tunnel valleys the masses of collapsed ice became buried by rock debris, and when this ice eventually melted, series of depressions were left, now occupied by lakes and wetlands.

Most of the linear lakes and troughs in Itasca Park probably originated as depressions in such tunnel valleys. The main arms of Lake Itasca, along with their continuations to north and south, occupy such depressions. The deepest tunnel valley is partly occupied by Lower LaSalle Lake (Fig. 2), which is 65 m deep and has at least 8 m of sediment. Elk Lake has 20 m of sediment beneath 30 m of water and is in a depression almost as deep. The many-bayed De Soto and Morrison lakes (Fig. 3) may also lie in a complex tunnel valley, their long peninsulas possibly representing sections of eskers. Mantrap Lake (Fig. 2), just south of the park, may have the same origin. The Mississippi River, in its course northward from Lake Itasca, occupies first one tunnel valley and then another.

The Wadena, Rainy, and Superior lobes eventually wasted completely from their terminal moraines, although they left behind myriad stagnant ice blocks buried by a veneer of glacial drift. The ice then advanced, and the Des Moines lobe was particularly active. It formed a long tongue extending down the Red River valley and thence down the Minnesota Valley and southward across a low divide into the Des Moines River valley, reaching central Iowa about 14,000 years ago (Fig. 4). This tongue was held in on its west side by the Coteau des Prairies in South Dakota and the adjacent southwestern corner of Minnesota, and it formed a series of parallel lateral moraines on the steep eastern slopes of the coteau. The east side of the lobe was largely contained by the great Alexandria moraine complex, which had previously been formed by the Wadena lobe expanding in the opposite direction. The Des Moines lobe overtopped this moraine, however, and buried the western and southern edge of the Wadena drumlin field. It also broke completely across the St. Croix moraine and sent the long Grantsburg sublobe eastward across south-central Minnesota as far as Grantsburg, Wisconsin. This lobe was limited on its southern edge by the nose of the St. Croix moraine, the inner boundary of which extended from the Minneapolis area northeastward. It reached its terminus before 14,000 years ago, and as it retreated it left the vast Anoka sand-plain north of Minneapolis.

Another sublobe protruded from the Des Moines lobe in northwestern Minnesota, extending eastward as the St. Louis sublobe into the Red Lakes lowland, which had previously been occupied by the Wadena lobe. It must have branched from the Des Moines lobe south of the Canadian border rather than from the Winnipeg region, because its drift is characterized by fragments of Cretaceous shale, which underlies the Red River valley sediments in the United States but does not extend northward into the Winnipeg area.

The St. Louis sublobe moved eastward across the northern part of the state and almost reached the Lake Superior lowland. It sent a relative narrow tongue southward as far as Lake Mille Lacs, and on its northern flank it rode up over the Mesabi iron range, reaching its terminus about 12,000 years ago. It came very close to Itasca Park. It did not form a prominent moraine at its margin about 10 km north of the park; it merely placed a veneer of drift on top of the previously formed moraines of the Wadena lobe. The two glacial drifts are difficult to distinguish, because they are about the same color and are both very rich in carbonate fragments. The St. Louis sublobe drift, however, has fragments of Cretaceous shale as well as carbonate, and usually it is finer textured than the Wadena lobe drift, which is generally loose and sandy, as is easily seen in deep exposures within the park. North

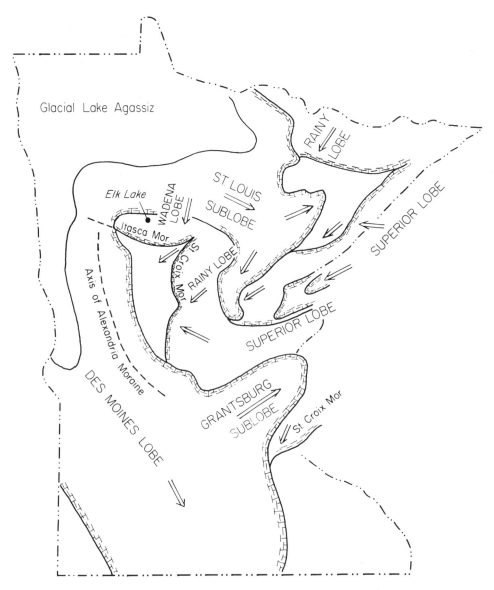

Figure 4. Map of Minnesota showing relations of different ice lobes that affected the area at different times during the Late Wisconsin glaciation.

of the Itasca moraine the St. Louis sublobe drift buried the northern extensions of the tunnel valleys that had formed during wastage of the Wadena lobe. Meltwater flowing south from the overlapping ice, however, apparently utilized these drainage channels, including the one occupied by Lake Itasca itself. The flat terrace of sand and gravel on which Lake Itasca post office is located at the northern edge of the park, for example, contains fragments of the Cretaceous shale, and shale was also found at the southern end of the west arm of the lake during archaeological excavations at the Itasca Bison Kill Site (Shay, 1971). Cretaceous pollen types were reported at the base of the Elk Lake sediment (Stark, 1976). The Itasca moraine and the trough in which Elk Lake is located were still partly filled with stagnant Wadena lobe ice, making it possible for St. Louis sublobe outwash to be carried along the tunnel valley, which as the ice melted became a series of unconnected lake depressions such as Elk Lake.

The St. Louis sublobe retreated soon after it reached its maximum, and in its wake it left several proglacial lakes, dammed by moraines on the south and by the ice front on the north or west. Glacial Lake Upham formed south of the Mesabi iron range. It drained eastward down the St. Louis River, which was blocked, however, by the contemporaneous Superior lobe and diverted southward to the St. Croix River. Glacial Lake Aitkin formed north of the Lake Mille Lacs area, and it drained southward through its moraine dam into the Mississippi River. As the ice withdrew into the Red Lakes lowland, glacial Lake Koochiching formed at its front (Hobbs, 1983), draining first southeastward into the Mississippi River system and then southwestward to the developing glacial Lake Agassiz, which at that time was confined to the southern end of the retreating Des Moines lobe in the Red River valley (Fig. 1). The first outlet of Lake Agassiz was at the south end, near present Lake Traverse, into the Minnesota River valley, and the outlet stream, called the glacial River Warren, carved a huge gorge or valley downstream to Minneapolis and beyond. The outlet stream was initially halted in its downward erosion by a pavement constructed of large boulders freed from the breached moraine dam and dropped on the floor of the spillway: thus, the lake level became stabilized (Matsch and Wright, 1967). A nearly continuous beach of sand and gravel, the Herman beach, was constructed around the entire lake at this stage, except for the northern side, which was still bordered by ice.

As the ice sheet continued to retreat into Canada, glacial Lake Agassiz became larger, spreading northward into Manitoba and western Ontario. The increased volume of water in the lake was reflected in an increased discharge over the spillway at the southern end, and the boulder pavement that had stabilized the lake at the Herman level was breached. This event occurred about 11,700 years ago. The outlet was then eroded to lower levels, becoming stabilized intermittently at various positions down to the Campbell beach. The ice sheet there retreated far enough north so that lower outlets to Lake Agassiz were uncovered to the east to Lake Superior, and the southern outlet was abandoned (Teller, 1987). Readvance of the ice temporarily closed these eastern outlets, and the lake rose once again to the Campbell level with renewed outflow to the south. With the final retreat of the ice, the Campbell level was abandoned for the last time about 9300 years ago, and eventually the ice retreated far enough so that Hudson Bay was opened; Lake Agassiz was drained completely about 8000 years ago.

Although the Lake Agassiz history did not directly affect the Itasca Park area, this large lake may have had some effect on the regional climate, especially when its eastern arm filled the Red Lakes lowland. At its maximum extent Lake Agassiz covered much of southern Manitoba and Saskatchewan, as well as the Red River Valley in Minnesota and North Dakota. It was as much as 100 deep in its central part, which is now a great clay plain.

The various strandlines of Lake Agassiz extend along the margin of the plain as low ridges of sand and gravel, and before the clay plain was drained for agriculture they provided the easiest routes of travel. The individual strandlines actually rise from south to north, as a result of rebound of the Earth's crust as a response to removal of the ice load. Thus the Campbell beach, which is 300 m above sea level at the southern outlet, rises gradually to about 315 m at the Canadian border, a distance of 200 km.

Shallow marginal parts of the lake plain are covered with sand, especially where tributary streams produced deltas. Some of these sand areas have since been modified to sand dunes by wind action. In some shallow areas the wave and current action was so strong that the sand was kept moving, and gravel lag or glacial till is exposed on the surface, only slightly modified by wave washing. Silt and clay were deposited only in deep water.

The history of glaciation in the Itasca region did not end with the retreat of the active glacial ice more than 14,000 years ago, because many blocks of stagnant ice had been left behind, buried by a veneer of glacial drift. These ice blocks were particularly prominent within moraines, which therefore may have had a greater total relief immediately after retreat of the active ice than they do now. But even in the outwash plains some ice blocks were localized, and the outwash plains that formed, such as the Park Rapids plain, have a few isolated ice-block lakes.

The course of melting of buried ice blocks is partly recorded by basal sediments of the lakes that take their place. In most of the lakes whose sediments have been studied in the Itasca region, as well as elsewhere in Minnesota, the bulk of the sediment is fine-grained organic ooze, representing relatively stable postglacial conditions of open water and relatively great depth. But in many cases (including Elk Lake) at the very base of the sediment section the ooze is underlain abruptly by a layer of terrestrial organic detritus, sometimes mixed with sand or silt. In addition to stems, twigs, leaves, cones, seeds, and other remains of forest plants, the material may contain fungal spores and hyphae, which are associated with decaying vegetation in soils. In some cases diatoms are found that do not live in water, but rather on soil, moss patches, and damp rocks (Florin and Wright, 1969). All these fossil remains suggest that the basal layer of sediment actually represents a superglacial forest floor, either in place or slightly transported into a shallow pool. The large lake itself could not have existed when this material was deposited; it probably only formed after the ice beneath melted out.

We thus can reconstruct a picture of incipient melting of buried ice blocks, forming small pools here and there similar to those observed today on stagnant ice of the Klutlan glacier in the Canadian Rocky Mountains (Wright, 1980). The localized breaching of the insulating mantle of glacial debris and vegetation—the area was forest-covered by this time—could have been accomplished by many natural events, such as erosion by small streams, wearing of animal trails, or slumping on slopes. Most of the superglacial pools were probably ephemeral, either rapidly drained or rapidly filled. But some became larger as the relatively warm water of the pool melted the ice on the banks and on the bottom. Such pools grew larger by slumping on the banks, and as soon as they were large enough and deep enough to keep from freezing to the bottom in winter the melting was accelerated, and perhaps in a few decades thereafter a full-fledged deep lake came into being.

The basal woody plant debris, recording the forest floor itself or a very shallow pool, became overlain abruptly by deep-water ooze. The total time involved in melting of buried ice blocks varied greatly, even locally, and probably depended on the thickness of the protecting mantle of glacial debris. In the Itasca lakes that have been studied, the basal woody debris is about 11,000 years old, yet we know from the general chronology of glacial events in Minnesota that the active Wadena lobe left the

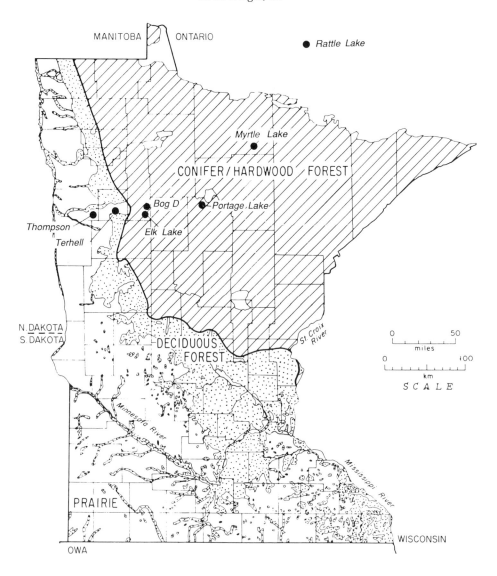
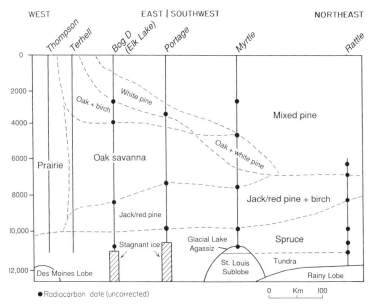

Itasca moraine more than 14,000 years ago. Therefore, the buried ice here might have survived for at least 3000 years. In a site on a sand plain in northeastern Minnesota, however, the basal wood trash dates from 13,500 years ago or before, whereas the sand plain itself had probably formed only a few hundred years before. Pollen studies of the sediment in this sand-plain lake show that tundra-type vegetation persisted in the region until at least 11,580 years ago. Tundra implies that the climate was still actually cold enough for ice to persist in the ground if it was adequately protected from the sun's direct heat in the summer. In this case the mantle of sand over the ice block must have been partly removed soon after the ice block was buried, allowing the heat of the sun to melt the exposed ice and create a pond large enough and deep enough to accelerate the melting. These two examples illustrate the differential survival of buried ice blocks in various areas.

VEGETATIONAL HISTORY

The history of vegetational succession in the Itasca region was first worked out through the pollen analysis of lake sediments in separate studies emphasizing different aspects of the subject. The vegetation history of this region is particularly amenable to detailed investigations, because the major vegetational belts are rather narrow (Fig. 5), and relatively slight changes in the climate might be expected to show significant changes in the vegetation. The conifer forest that dominates northwestern Ontario and northeastern Minnesota extends to the Itasca region and includes white and black spruce, balsam fir, tamarack, white, red, and jack pine, birch, and aspen, as well as a minor component of temperate deciduous trees. The western limit of this mixed coniferous-deciduous forest is about 30 km west of the park. From there westward for only about 10 km is the pure temperate deciduous forest: sugar maple, basswood, elm, bur oak, ironwood, and hazel dominate in various areas. Then comes a narrow belt, determined in part by favorable topographic situations, in which oak savanna and parkland prevailed before the prevention of prairie fires allowed oak shrubs to occupy the areas between the formerly open-grown oaks. Once established, the thick-barked bur oak resisted easily most prairie fires, which were regular events before the region was settled by farmers; the fires favored the development of prairie grasses in the open areas between the oaks. Groves of boxelder, aspen, or other deciduous trees also occurred in this transitional area, which is equivalent to the aspen parkland that occurs between the prairie and the boreal spruce forest of the Canadian prairie provinces. The vegetational belts in northern Minnesota are clearly controlled by the climatic gradient, which shows increasing total moisture, more snow, and lower summer temperatures to the northeast.

One of the first studies to demonstrate the relation between the vegetation and the pollen rain was made in the Itasca region. McAndrews (1966) collected a series of surface-sediment samples from lakes in an east-west transect through Itasca Park, from the dominantly conifer forest east of the park westward to the prairie. Then he studied the vegetation in this transect, to determine the relation between the vegetation and the pollen currently being deposited. This relation was difficult to express quantitatively, however, without making a very extensive quantitative survey of the vegetation from which the pollen was derived. Fortunately, in a sense this survey had already been made: just before the region was settled by homesteaders, about 1870–1890, the surveyors from the Federal Land Office laid out a grid of townships in the region, and they set a stake at every half-mile point. Each corner had to be identified by slashing the nearest "witness tree" in each of the four quadrants around the corner. The witness trees were identified and their distance from the corner noted. This provided a statistical sample of the tree vegetation of the time. Moreover, it was a measure of the vegetation before the disturbance introduced by agriculture, lumbering, and fire protection.

The pollen content of surface lake muds, however, reflects the modern disturbed vegetation, so McAndrews analyzed short cores of lake sediment that covered the time before settlement—the impact of disturbance could be identified in the pollen diagrams for these short cores by the sharp increase in the pollen of ragweed and other plants of agriculture. Thus he could compare directly and quantitatively the presettlement pollen rain with the presettlement vegetation, thereby providing a clearer basis for reconstructing past vegetation from the four pollen diagrams he completed.

In a study complementing that of McAndrews, Janssen (1966, 1967a) surveyed the vegetation bordering several different lakes and wetlands in the Itasca area and collected samples of surface sediment at increasing distances from the shore. Through pollen analysis of the samples he was able to distinguish what pollen was derived from local, extra-local, and regional sources. The results were then used to interpret the vegetation history of a small wetland in the park (Janssen, 1967b).

In the third study Shay (1971) excavated a prehistoric bison kill site in a small wetland at the southern end of the west arm of Lake Itasca near Elk Lake. He found from the pollen stratigraphy that the sediment had been eroded during the mid-Holocene, presumably because of a lower water level in Lake Itasca. Such an interpretation is consistent with the pollen evidence for dry climatic conditions during that time.

The pollen sequence in the Itasca region—and in fact throughout all but northeastern Minnesota—starts with high percentages of spruce (Fig. 5), indicating that the region was dominated by a spruce forest similar to that of Ontario and Manitoba to the north; however, pine, which today is a major component of

Figure 5. Vegetation map of Minnesota showing location of pollen sites used to infer the vegetation history along the cross section from Thompson Pond in the prairie to Terhell Pond in the deciduous forest to Bog D (near Elk Lake) to Rattle Lake in northwestern Ontario. Data for Thompson, Terhell, and Bog D are from McAndrews (1966), Portage from McAndrews (unpublished), Myrtle Lake from Janssen (1968), and Rattle Lake from Björck (1985). Dots show levels of radiocarbon dates.

the boreal forest, was absent from the late-glacial forest. In northeastern Minnesota, the spruce pollen dominance is preceded by an interval with pollen mostly of herbs. The inference of a tundra vegetation for this interval is supported by the discovery of fruits and seeds of tundra herbs and shrubs.

The spruce forest that was established in the Itasca area on the deglaciated landscape was succeeded by pine forest, then by deciduous forest, and finally by oak savanna or prairie about 8000 years ago, all as a result of steadily warming climate. At the culmination of this warming trend about 7200 years ago the prairie-forest border was perhaps 100 km east of its present position. Subsequently, with a climatic reversal, the forest returned to the west, and this time white pine became a major component (the earlier pine forest contained only red or jack pine). This climatic trend is best documented regionally by the development and westward expansion of the Red Lake peatlands in the lowland formerly occupied by the eastern arm of glacial Lake Agassiz (Wright and Glaser, 1983).

The postglacial vegetational history of the Itasca area history is epitomized by the biostratigraphic sequences in Elk Lake, as described in following chapters.

LAKES

Since the lakes of the Itasca region originated, late in the glacial period, their history has primarily involved the steady filling by organic sediment. The amount of mineral sediment washed into some lakes from the upland was fairly large in the beginning, before the land surfaces became fully stabilized by vegetation, and when soil erosion on the hill slopes may have been more prevalent than later. Subsequently most of the sediment that has collected on the bottom was produced within the lake, because it consists mostly of the products of growth of algae, rooted aquatic plants, and other organisms that live within the open water of the lake or in the marshy areas along the shore. Chemical components precipitated from the lake water through biologic processes are also present. Some organic material, such as seeds and twigs and leaves, was blown into the lake from the adjacent hillslope or washed in by streams after heavy rains. Seeds and certain other remains may occur in sufficient quantities in small lakes so that their stratigraphic analysis provides useful information about the vegetational history as supplements to the pollen sequence.

The sedimentary products of organic activity within the lake are not confined to remains of partially decomposed algae and other aquatic organisms. In some lakes, including most of the Itasca lakes, marl is the main type of sediment. Marl consists primarily of calcium carbonate, although it is usually mixed with organic or mineral detritus. Marl in the Itasca lakes is formed by the removal of dissolved carbon dioxide from the water during the process of photosynthesis by algae and other plants. This process causes calcium carbonate to precipitate as small crystals, which fall to the bottom of the lake or form on the leaves of submerged aquatic plants. Some of the carbonate is redissolved, but in lakes that have a large supply of calcium—such as those in the calcareous glacial drift of the Itasca area—much marl accumulates on the lake floor.

The thickness of organic sediments in lakes of the Itasca region is generally about 8–12 m. The rate of filling of lakes in forested regions is apparently fairly constant, according to close-interval radiocarbon dating of lake sediments in northeastern Minnesota as well as in other regions. This fact has been confirmed by the presence in the deepest part of Elk Lake of generally thin and even annual laminations (varves) throughout the entire section. The laminations can be complex, but basically they consist of a dark organic-clay winter layer and a light carbonate summer layer. These uniform laminations indicate a steady rate of sedimentation since shortly after the glacial period. They also point to the lack of subsequent disturbance of the sediment by bottom organisms.

Many lakes have been filled to such shallow depths that they have been overgrown by wetland vegetation from the side and now have no open water or perhaps only a small pond in the center. Bog D pond in Itasca Park, for which one of the major pollen studies was made by McAndrews (1966), is such a pond: borings across the entire breadth of the bog show that a lake originally filled the whole basin. In some cases the bog vegetation advances so rapidly that its edge floats over open water beneath.

Bogs that are encroaching on lakes in this manner frequently have a very clear zonation of vegetation. At the base of the hill slope that bounds the depression there may be a zone dominated by black ash, elm, balsam fir, or some other tree that can tolerate moist ground. In wetter situations alder may form a distinct zone, and here pools of water may actually stand among the alder thickets. Then comes the main shrub mat that may stretch far out into the basin, with trees of tamarack and black spruce. Shrubs like leather leaf and Labrador tea grow so thickly, commonly along with sphagnum moss, that the mat stands well above the water level, and the alder zone just mentioned, which has no shrub layer, then has the aspect of a moat around the bog. On the other edge of the bog next to the central pond, the trees are stunted or absent, and a mat of heath plants and sedges may form a floating fringe.

Studies of the sediments of Elk Lake for materials other than pollen grains reveal a history that is consistent with that worked out from the pollen sequence for the upland vegetation. The fossil organisms most useful for study of lake history are diatoms, which are a group of algae with intricately sculptured siliceous tests that are well preserved in most sediments. Diatom species are generally sensitive to the water chemistry, especially the acidic or alkaline nature or the salinity, so that the assemblage of diatoms in a particular lake reflects the ionic content of the water. Chemical analyses of the waters of many lakes throughout Minnesota have revealed a distinct gradient in the water chemistry from the relatively acid lakes with low ionic content in the conifer forest region in the northeast to relatively alkaline and comparatively saline lakes of the prairie region in the southwest (Gorham and others, 1983). If the upland vegetation in the Itasca region changed from boreal spruce forest to pine forest to deciduous forest to prairie and then partway back again, then the water chemistry should have changed in a comparable way, and these changes should be recorded stratigraphically in the sediments by changes in the assemblages of fossil diatoms. Investigations of pollen, diatoms, ostracods, and mollusks in several cores from Elk Lake were first undertaken by Stark (1976), and in this volume the results of more detailed analyses of some of these fossil groups as well as other components of the lake sediments are reported.

REFERENCES CITED

Birks, H.J.B., 1976, Late Wisconsin vegetational history at Wolf Creek, central Minnesota: Ecological Monographs, v. 46, p. 495–529.

Björck, S., 1985, Deglaciation chronology and revegetation in northwestern Ontario: Canadian Journal of Earth Sciences, v. 22, p. 850–871.

Florin, M. B., and Wright, H. E., Jr., 1969, Diatom evidence for the persistence of stagnant glacial ice in Minnesota: Geological Society of America Bulletin, v. 80, p. 695–704.

Goldstein, B. S., 1989, Sedimentology and genesis of the Wadena drumlin field, Minnesota, U.S.A., *in* Menzies, J., and Rose, J., eds., XII INQUA Congress Symposium on Subglacial Bedforms: Sedimentary Geology, v. 62, p. 241–277.

Gorham, E., Dean, W. E., and Sanger, J. E., 1983, The chemical composition of lakes in the north-central United States: Limnology and Oceanography, v. 28, p. 287–301.Hobbs, H. C., 1983, Drainage relationship of glacial Lake Aitkin and Upham and early Lake Agassiz in northeastern Minnesota, *in* Teller, J. T., and Clayton, L., eds., Glacial Lake Agassiz: Geological Association of Canada Special Paper 26, p. 245–260.

Janssen, C. R., 1966, Recent pollen spectra from the deciduous and coniferous-deciduous forest of northwestern Minnesota: A study in pollen dispersal: Ecology, v. 47, p. 804–825.

Janssen, C. R., 1967a, A floristic study of forests and bog vegetation mainly in the Itasca State Park area, northwestern Minnesota: Ecology, v. 48, p. 751–763.

Janssen, C. R., 1967b, A postglacial pollen diagram from a small *Typha* swamp in northwestern Minnesota, interpreted from pollen indicators of surface samples: Ecological Monographs, v. 37, p. 145–172.

Janssen, C. R., 1968, Myrtle Lake: A late and post-glacial pollen diagram from northern Minnesota: Canadian Journal of Botany, v. 46, p. 1397–1408.

Matsch, C. L., and Wright, H. E., Jr., 1967, The southern outlet of Lake Agassiz, *in* Mayer-Oakes, W. J., ed., Life, land, and water: Winnipeg, University of Manitoba, p. 121–140.

McAndrews, J. H., 1966, Postglacial history of prairie, savanna, and forest in northeastern Minnesota: Torrey Botanical Club Memoir 23, p. 1–72.

Mooers, H. D., 1990, Ice-marginal thrusting of drift and bedrock: Thermal regime, subglacial aquifers, and glacial surges: Canadian Journal of Earth Sciences, v. 27, p. 849–862.

Shay, C. T., 1971, The Itasca bison kill site. An ecological analysis: Minneapolis, Minnesota Historical Society, 133 p.

Stark, D. M., 1976, Paleolimnology of Elk Lake, Itasca State Park, northwestern Minnesota: Archiv für Hydrobiologie, Supplement 50, p. 208–274.

Teller, J. T., 1987, Proglacial lakes and the southern margin of the Laurentide ice sheet, *in* Ruddiman, W. F., and Wright, H. E., Jr., eds., North America and adjacent ice sheets during the last deglaciation: Geological Society of America, The Geology of North America, v. K-3, p. 39–70.

Wright, H. E., Jr., 1962, Role of the Wadena lobe in the Wisconsin glaciation of Minnesota: Geological Society of America Bulletin, v. 73, p. 73–100.

Wright, H. E., Jr., 1972, Quaternary history of Minnesota, *in* Sims, P. K., and Morey, G. B., eds., Geology of Minnesota: Minneapolis, Minnesota Geological Survey, p. 515–548.

Wright, H. E., Jr., 1980, Surge moraines of the Klutlan Glacier, Yukon Territory, Canada: Origin, wastage, vegetation succession, lake development, and application to the late-glacial of Minnesota: Quaternary Research, v. 14, p. 2–18.

Wright, H. E., Jr., and Glaser, P. H., 1983, Postglacial peatlands of the Lake Agassiz plain in northern Minnesota, *in* Teller, J. T., and Clayton, L., eds., Glacial Lake Agassiz: Geological Association of Canada Special Paper 26, p. 375–389.

CONTRIBUTION 389, LIMNOLOGICAL RESEARCH CENTER, UNIVERSITY OF MINNESOTA

MANUSCRIPT ACCEPTED BY THE SOCIETY JULY 27, 1992

Climatic and limnologic setting of Elk Lake

Robert O. Megard
Department of Ecology, Evolution, and Behavior, University of Minnesota, Minneapolis, Minnesota 55455
J. Platt Bradbury and Walter E. Dean
U.S. Geological Survey, Box 25046, Federal Center, Denver, Colorado 80225

ABSTRACT

Elk Lake is located on the Itasca moraine near the source of the Mississippi River in northwestern Minnesota. The basin is in calcareous glacial drift, and the lake water is a dilute solution of calcium and magnesium bicarbonate. Low-magnesian calcite formed by precipitation from the lake water has been a major component of the sediment throughout the lake's history. The sediment also is laminated with alternating light and dark, millimeter-thick layers containing diatoms, organic matter, $Fe(OH)_3$, and $CaCO_3$.

The sediment microstratigraphy has been preserved because the lake is unusually deep (maximum depth is 30 m) for its size (surface area is 1.01 km^2). Oxygen is present in low concentrations or absent in the deepest water during summer and winter. Water movements in the deepest part of the lake are insufficient some years for the complete aeration of the deepst water during spring and autumn circulation periods.

Phytoplankton photosynthesis, which occurs mostly in the surficial 6 m of water, typically removes 0.5 g C m^{-2} day^{-1} from the epilimnion, which becomes strongly oversaturated with calcite during late spring and summer as the pH increases above the equilibrium pH (7.73) for calcite saturation. Most of the $CaCO_3$ that makes up the light-colored layers of sediments probably is formed during the late summer, when concentrations of calcium in the epilimnion decrease most rapidly. The silica and organic matter that form the darker sediment laminae are deposited earlier in the year, during a period extending from April to late June, when silica decreases fastest in the epilimnion.

INTRODUCTION

Elk Lake (lat 47°12′N; long 95°15′W) is part of a complex drainage system that forms the headwaters of the Mississippi River in Itasca State Park, Clearwater County, northwestern Minnesota (Fig. 1). Elk Lake drains through a small stream, Chambers Creek, into Lake Itasca (Fig. 1), which has been designated as the headwaters of the Mississippi River.

The rolling topography of the Itasca Park region has many low hills, lakes, and wetlands and is developed on the Itasca moraine. The Itasca moraine formed when the Wadena ice lobe, with till containing Paleozoic carbonate debris acquired in southern Manitoba, arrived in the area about 20,000 years ago (Wright, Chapter 2; Wright and others, 1973). During the initial phase of glacier retreat, between 14,000 and 18,000 years ago, meltwater carved subglacial tunnel valleys into the landscape. Stagnant ice remained in deep depressions, protected from melting by a cover of drift from the receding glacier (Stark, 1976). A later ice advance between 12,000 and 14,000 years ago (the St. Louis sublobe of the Des Moines ice lobe) crossed the Red River Valley from west to east and deposited outwash with Cretaceous shale into some of the tunnel valleys that formed earlier (Wright, 1972).

Subsequent melting of the ice blocks during the final stages of the Wisconsin about 11,000 years ago produced a series of lakes and marshes aligned roughly north-south and complexly connected by stringers of coarse-clastic tunnel-valley sediments. Lake Itasca, Elk Lake, and numerous small basins mark the complex pattern of these ancient drainage channels (Wright, 1972; Stark, 1976). There are many marshes, bogs, and smaller lakes within the drainage basin of Elk Lake. The lake is fed by

Megard, R. O., Bradbury, J. P., and Dean, W. E., 1993, Climatic and limnologic setting of Elk Lake, *in* Bradbury, J. P., and Dean, W. E., eds., Elk Lake, Minnesota: Evidence for Rapid Climate Change in the North-Central United States: Boulder, Colorado, Geological Society of America Special Paper 276.

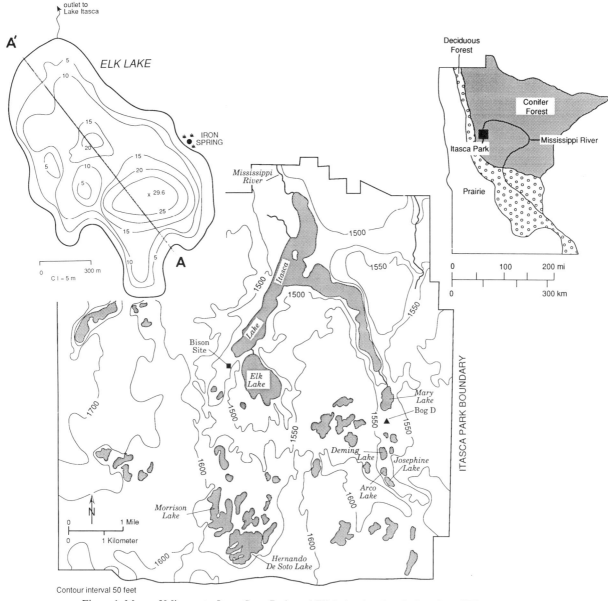

Figure 1. Maps of Minnesota, Itasca State Park, and Elk Lake showing the location of Elk Lake, general vegetation zones of Minnesota, bathymetry of Elk Lake, and location of the varved core in the deepest part of the lake. Line A–A′ is the echogram cross section shown in Figure 3.

numerous springs and four small streams that enter from the south and southwest.

CLIMATIC SETTING

Airstreams

Elk Lake lies near the mean annual junction of three dominant airstreams: (1) the arctic airstream that extends southward from the arctic regions and reaches the north-central United States in winter; (2) the dry, wedge-like Pacific airstream that follows the path of strongest westerlies as they enter western North America from the Pacific Ocean, having lost moisture by crossing the coastal and Rocky Mountain cordillera; and (3) the tropical airstream that brings warm, moist air northward from the Gulf of Mexico, especially during the summer months (Bryson and Hare, 1974).

Depending on the season, the presence and interaction of these different airstreams characterize the climate of northwestern Minnesota (Anderson and others, Chapter 1). The dry, warm Pacific airstream forms an eastward-pointing wedge that reaches the north-central United States in winter, and retreats northward into Canada in the summer. The main effect of the Pacific airstream on the climate of the Elk Lake region occurs when it entrains and lifts moist Gulf air to produce heavy snowfall. The arctic airstream also prevails in the Elk Lake region during the winter, although outbreaks of arctic air occur in the summer (Baldwin, 1973). Incursions of arctic air usually follow heavy snowfalls produced by the interaction of the westerlies and the Gulf Coast air, creating severe blizzard conditions.

The warm, moist tropical Gulf airstream does not usually invade northern Minnesota during the winter, but during the summer, circulation around low-pressure systems moving from west to east across continental United States causes substantial northward flow of Gulf air into the Elk Lake region. The contrast between the warm land in the southwestern United States and a cooler Pacific Ocean during the summer months (the southwest monsoon effect), and circulation around the west side of the Bermuda high-pressure system in the western North Atlantic enhances this northward flow of Gulf moisture. Interaction between this Gulf air with arctic air flowing almost straight south to meet it produces thunder storms that provide the principal source of moisture for the Elk Lake region in the spring and summer.

Temperature and precipitation

The Elk Lake region has a continental climate with cold winters and warm to hot summers. The seasonal extremes in temperature in the Itasca Park area for the 30 year period from 1951 to 1980 are listed in Table 1. The mean annual precipitation for this time interval was 66.6 cm. The variability in temperature and precipitation in the Itasca region for 1980 and 1984 is illustrated in Figure 2; they were comparatively dry and wet years with mean annual precipitations of 48.0 and 77.7 cm, respectively.

The seasonal progression of temperature and wind, generally associated with storms, is important in determining the span of the ice-free season, and the timing and amounts of nutrient fluxes into Elk Lake (Bradbury, 1988). Unfortunately, detailed observations (e.g., thaw dates, insolation, wind stress) have not been made routinely at Elk Lake or at the Itasca station, and have not been correlated with physical and biological limnological parameters such as lake temperature and seasonal algal productivity. Lack of such data reduces our understanding of

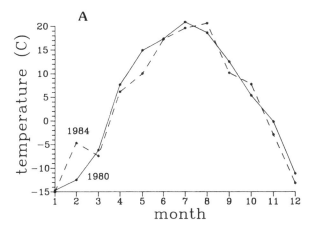

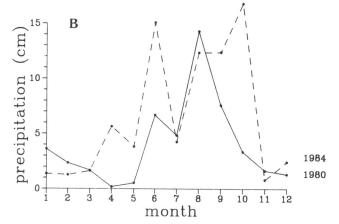

Figure 2. Monthly mean temperature (A) and precipitation (B) for Itasca State Park, Clearwater County, Minnesota during 1980, a dry year, and 1984, a wet year.

limnological-climatic relations and, consequently, our ability to make detailed climatic interpretations of paleolimnological changes. Nevertheless, climatic observations coupled with sediment-trap collections between 1979 and 1984 (Nuhfer and others, Chapter 7) provided important insights into the manner in which Elk Lake and its sedimentary components respond to seasonal climatic changes.

PHYSICAL FEATURES

Morphometry

Elk Lake is elliptical in shape with a surface area of 1.01 km^2, which is about the median surface area of a group of 1000 Minnesota lakes analyzed by Heiskary (1985). The median maximum depth of the lakes in Heiskary's survey is 10.4 m; Elk Lake, with a mean depth of 11.2 m and a maximum depth of 30 m (Fig. 1), is deeper than 95% of them. A mid-lake shoal at a depth of 12 m separates the 30 m depression from a second depression that is 21 m deep (Fig. 1). The lake volume is 11.3×10^6 m^3 (Table 2).

TABLE 1. AVERAGE SEASONAL TEMPERATURES AT ITASCA STATE PARK*

	°C
Mean annual temperature	3.0
Mean spring temperature	2.7
Mean summer temperature	17.9
Mean fall temperature	5.3
Mean winter temperature	-13.9

*Values are for the period 1951 to 1980 and have been corrected for 0800 hour observation (from Baker and others, 1985).

TABLE 2. MORPHOMETRY OF ELK LAKE AND CALCULATED FLUXES OF IRON AND MANGANESE INTO THE WATER FROM THE SEDIMENTS

Depth* (ft)	(m)	Area* (10⁶ m²)	Volume* (10⁶ m³)	Sed. Area* (10⁶ m²)	Fe Flux† (mmol m⁻² d⁻¹)	Mn Flux† (mmol m⁻² d⁻¹)
0	0.0	1.01				
			2.96	0.22	0.24	0.05
10	3.05	0.79				
			2.46	0.09	0.44	0.10
20	6.01	0.70				
			2.12	0.12	0.28	0.20
30	9.15	0.58				
			1.47	0.26	0.09	0.19
40	12.2	0.32				
			0.88	0.11	0.13	0.55
50	15.2	0.21				
			0.60	0.06	0.19	1.02
60	18.3	0.15				
			0.41	0.06	0.12	1.31
70	21.3	0.09				
			0.27	0.02	0.49	5.11
80	24.4	0.07				
			0.12	0.06	0.19	1.53
90	27.4	0.01				
			0.01	0.01	0.19	1.15
98.5	30.0					

Total volume = 11.3 x 10⁶ m³; Mean depth = 11.2 m; Maximum depth = 30 m; Relative depth = 2.6%.

*Based on a bathymetric map prepared by the Minnesota Department of Natural Resources, 1962.
†See text section on water chemistry for method of calculation.

Some relations between bottom topography of Elk Lake and layers within the water column when the lake is stratified during summer can be discerned on an echogram obtained with a high-frequency (192 kHz) echosounder along a midday transect across the lake from southeast to northwest (Fig. 3). The water layers can be discerned acoustically because zooplankton are more abundant in some layers than in others, and acoustic backscattering is proportional to the concentration of zooplankton (Clay and Medwin, 1977). Many comet-like reflections from fish occur near the top of the metalimnion.

The epilimnion includes depths to 6 m, where backscattering of high-frequency sound generally is low by comparison to that in deeper water. The metalimnion, between 6 and 15 m, is a layer in which a strong gradient of acoustic backscattering corresponds to a sharp temperature-related density gradient. The top of the metalimnion is marked by a thin layer of very low backscattering between 6 and 8 m; backscattering increases deeper in the metalimnion, where concentrations of Cladocera and Copepoda are highest. The highest acoustic backscattering is in the hypolimnion at depths greater than 15 m, due to high concentrations of aquatic midge larvae (*Chaoborus*), which are abundant in the nearly anoxic hypolimnetic water.

Ice cover

Elk Lake typically freezes in November or early December and thaws in April or early May. Although specific freeze and thaw dates are rarely recorded for Elk Lake, this information usually is available for Lake Itasca (Table 3) and documents the general freeze-thaw pattern for lakes in this region. Elk Lake usually thaws within one day of Lake Itasca, but it typically freezes a week later, or sometimes as much as two weeks later, because of its greater volume to surface-area ratio. When windy weather delays freezing of Lake Itasca, it can freeze on the same day as Elk Lake (Jon Ross, 1986, written commun.).

A half-century record of ice-out dates from Lake Itasca (Fig. 4) illustrates graphically the variability of spring climates in this region. Typically, extremely early or late ice-out dates are single-year events that are separated by several (3–7) years, a recurrence interval similar to that of El Niño events (Diaz and Pulwarty, 1992).

Ice cover on Elk Lake often reaches 1 m in thickness. The clarity of the ice and its ability to transmit light varies greatly from year to year and at different times throughout the winter. Snow commonly blankets the ice, reducing its ability to transmit light. Heavy snowfall will weigh the ice cover down, crack it, and

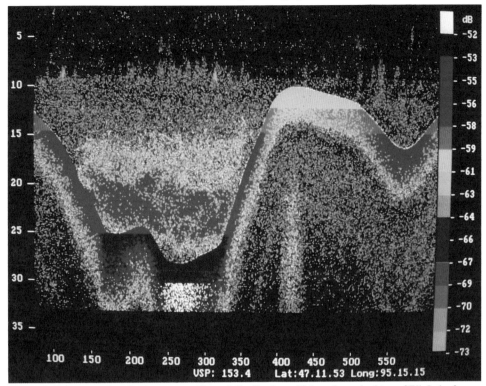

Figure 3. Echogram showing bottom profile and distribution of fish and zooplankton of Elk Lake from southeast to northwest (A–A′, Fig. 1). Echogram obtained during midday on 19 August, 1991, with a 192 kHz echosounder connected to a microcomputer. Left-hand vertical scale is depth in meters. Horizontal scale is a count of acoustic pulses emitted by the echosounder during transit across the lake and corresponds approximately to 1 m horizontal distance per pulse. Vertical color scale (right side) is a logarithmic scale, in decibels, for the strength of backscattered sound, which is proportional to zooplankton concentration. The scale ranges from –73 dB (light blue) to –53 dB (dark red). Comet-shaped echotraces near 8 m are due to fish.

allow lake water to flood the ice surface, where it can form frozen slush at the base of the overlying snow that further thickens the ice cover (Bradbury, 1988). Repeated fracturing of the primary ice layer may allow lenses of slush to build up beneath the snow in irregular patches.

Water Stratification

Elk Lake behaves as a more or less typical dimictic lake, circulating from ice-out in April or early May until stratification becomes fully developed in June (Figs. 5 and 6). Stratification weakens and deepens with cooler temperatures in September and October, and the lake circulates again in early November until the lake freezes. During periods of summer stagnation, hypolimnetic oxygen is depleted and anoxic conditions occur beneath the thermocline (Figs. 5, A and B, and 6, A and B). Oxygen depletion during winter stratification is less severe than in summer, and anoxic conditions may only exist at the sediment-water interface (A. L. Baker, unpublished data; Fig. 5B).

In 1980, the lake was thermally stratified by early May (Fig. 6A), and stratification persisted until late October. During stratification, the epilimnion included depths above 6 m, where temperatures were between 10 and 22 °C. The region of decreasing temperatures between 6 and 10 m is the metalimnion. Depths below 10 m, with temperatures less than 7 °C, composed the hypolimnion. The lake remained stratified until temperatures decreased to 5 °C at all depths in late October.

Oxygen concentrations in the epilimnion exceeded 12 mg/l during May 1980 (Fig. 6B), and fell below 8 mg/l during the summer as oxygen solubility decreased in response to increasing surface temperatures. Oxygen decreased more slowly in the upper metalimnion, where concentrations stayed above 10 mg/l until late July. High concentrations of dissolved oxygen were maintained in the metalimnion, in part because phytoplankton were more abundant than in the epilimnion. Light intensities are low in the metalimnion, but oxygenic photosynthesis nearly balances daily respiration, at least in the upper metalimnion. The overlying epilimnion insulates the metalimnion from the atmosphere and

TABLE 3. FREEZE AND THAW DATES FOR LAKE ITASCA AND ELK LAKE*

Year	Ice-out	Ice-in
LAKE ITASCA		
1978	27 April	10 November
1979	7 May	9 November
1980	21 April	15 November
1981	9 April	....
1982	27 April	....
1983	24 April	25 November
1984	19 April	17 November
1985	20 April	11 November
1986	2 April	13 November
1987	11 April	2 December (windy)
1988	17 April	18 December
1989	30 April	17 November
ELK LAKE		
1984	....	17 November
1985	....	17 November
1986	12 April	....
1987	....	2 December
1988	17 April	28 December
1989	29 April	28 December

*Data from Jon Ross, Itasca Biological Station (written commun., 1986; personal commun., 1989).

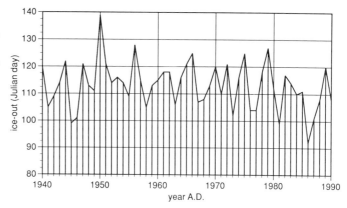

Figure 4. 50 year record of ice-out dates for Lake Itasca, Clearwater County, Minnesota. Data from Itasca State Park records.

prevents oxygen from escaping to the atmosphere. Oxygen concentrations in depths below 10 m were only 3 mg/l in May 1980, and decreased slowly during the summer until oxygen essentially disappeared in late August. Oxygen reappeared in the hypolimnion when the lake became isothermal in late October, but concentrations increased only to about 5 mg/l during autumn mixing. The autumn mixing period in 1980 was not long enough for complete areation, and concentrations were still low at the beginning of winter, when they began to decrease again.

The thermal stratification during summer leads to the stratification of many other substances in addition to oxygen, as indicated by the profiles of alkalinity, pH, specific conductance, and chlorophyll-a on July 21, 1986 (Fig. 7). Alkalinity and specific conductance both increase with depth in the metalimnion. The pH decreases throughout the metalimnion from values that may be as high as 9.0 at the surface to values of about 7.2 in the hypolimnion. The high concentrations of chlorophyll associated with the high concentrations of oxygen in the metalimnion indicate that the high oxygen concentration in the metalimnion during spring and early summer (Fig. 6B) is attributable, at least in part, to phytoplankton photosynthesis. Similar metalimnetic maxima of oxygen and chlorophyll are common in lakes of moderate productivity. The highest chlorophyll concentrations are in the same depth range as a maximum of optical turbidity that was first observed in Elk Lake by Baker and Brook (1971).

Minnesota lakes of this size usually are dimictic, i.e., the water mixes to all depths in spring and autumn; the deep water is aerated, and oxygen concentrations approach saturation at all depths. However, the 30 m depression in Elk Lake may be too deep to fully circulate, at least in some years. As discussed by Walker and Likens (1975) and Anderson and others (1985), a typical meromictic lake does not mix all the way to the bottom, so that a deep water mass, the monimolimnion, is perennially anoxic. Some lakes in temperate climatic regions are transitional, being dimictic some years but incompletely circulating in others, depending on the weather during spring and autumn mixing periods. Complete vertical mixing probably occurs in Elk Lake during both spring and autumn of many years, as is characteristic of dimictic lakes, but the 30 m depression has some properties of a meromictic lake because the deeper waters in this depression are not always completely aerated during mixing periods.

Meromixis is most likely to occur in lakes that are small, unusually deep, and protected from the wind by topography and/or surrounding forests. Walker and Likens (1975) used relative depth as an index of the potential for such morphometric meromixis. Relative depth is the maximum lake depth expressed as a percentage of the mean diameter. They found that the relative depth typically is between 2 and 10% in lakes that are meromictic due to morphometric factors. The topography and closed forest around Elk Lake provide some protection from the wind, and the morphology, with a relative depth of 2.5%, is marginally favorable for meromixis. Several lakes located in a narrow, protected valley south of the eastern arm of Lake Itasca (e.g., Deming, Arco, and Josephine lakes, Fig. 1) are more likely to be meromictic (Baker and Brook, 1971) because they are are much smaller than Elk Lake (surface areas are less than 0.05 km^2) and have larger relative depths (5–8%).

Meromixis may have been more persistent or more frequent earlier in Elk Lake's history when there was less sediment in the basin and the water was deeper. Under these conditions, anoxic conditions favorable for the formation and preservation of laminated sediments also may have been more persistent. However, meromixis would be less likely during times when water levels were lower than they are today.

Light transmission

Concentrations of phytoplankton chlorophyll in the water of Elk Lake typically are less than 5 mg/m^3 (e.g., Fig. 7), indicating low concentrations of planktic algae. The water is consequently more transparent than in many other lakes in the region; measurements of transparency with Secchi discs in Elk Lake typically are between 4 and 5 m. The median Secchi-depth transparency of 640 Minnesota lakes is 1.8 m (Heiskary, 1985), and Elk Lake is more transparent than about 75% of them.

The optical properties of Elk Lake water also can be described in terms of an attenuation coefficient for diffuse underwater light (K), which describes the exponential decrease of irradiance with respect to depth. As indicated by data from Elk and five other lakes (Fig. 8), the magnitude of the attenuation coefficient varies by a factor of 5 among lakes in this region, from 0.2/m in the most transparent lakes up to 1.0/m or more in lakes that are less transparent. The Secchi transparency is inversely proportional to the attenuation coefficient, as predicted by the Lambert-Bouger Law (Megard and Berman, 1989). Although Elk Lake is more transparent than nearby Lake Itasca, it is less transparent than two of the other lakes.

Surface sediments

Probably the most distinctive characteristic of the sediments in Elk Lake is that they contain millimeter-thick annual laminations of sediment components (varves; Anderson, Chapter 5; Anderson and others, Chapter 4). According to the classification of surface sediments of Minnesota lakes by Dean and Gorham (1976 and Chapter 9), the sediment in Elk Lake is high-carbonate group 3, typical of lakes in central and west-central Minnesota that contain calcium-magnesium bicarbonate waters of intermediate salinity in calcareous drift. The distinguishing characteristics of the sediments in these lakes is the high content of CaCO$_3$ (>30%), most of which is produced in the lake. The sediments of Elk Lake during the past 2000 years of sedimentation are unusual in that more than 95% of the inorganic and organic materials in the surface sediments were produced in the lake. Throughout its postglacial history, with the exception of the mid-Holocene, Elk Lake has received little allochthonous sediment (Dean and others, 1984). As shown in Table 4, most of the sediment is composed of autochthonous CaCO$_3$ (Dean and Megard, Chapter 8), organic matter, biogenic silica (diatom debris; Bradbury and Dieterich-Rurup, Chapter 15), and autochthonous iron, manganese, and phosphorus minerals (Dean, Chapter 10). The low content of terrigenous clastic debris is indicated by the extremely low aluminum concentration (0.8%; Table 4). Each of the autochthonous sediment components is deposited on a distinct seasonal schedule that can be detected in sediment traps and in the individual varve laminations (Anderson, Chapter 5; Nuhfer and others, Chapter 7).

LAKE CHEMISTRY

Water fluxes

Elk Lake, at an elevation of 450 m (1470 ft), is located near a major drainage divide on the crest of the Itasca moraine, broadly defined by the 1600 ft (485 m) topographic contour (Fig. 1). The effective catchment area for surface drainage to the lake probably is delineated by the 1550 ft (470 m) contour, which encloses an area only three or four times larger than the lake surface area. The catchment area contains many wetlands and several smaller lakes south of Elk Lake; four small streams enter the lake from the catchment. The amount of ground-water flow is difficult to evaluate quantitatively, but is probably large, because the unconsolidated glacial till surrounding Elk Lake is mostly sandy and porous. The chemical composition of water from the spring on the east shore of the lake (Fig. 1) probably is characteristic of the general ground water of the region. The salinity (conductivity) of the lake water is about half that of ground water from the spring (Table 5), which suggests that, on average, about half of the lake's water income is from direct precipitation and runoff from forest soils, sources that are more dilute than ground water.

Major ions

The concentrations of major ions in Elk Lake along with those in Elk Spring and several other carbonate lakes are given in Table 5: Lake Itasca and Lake Mendota, Wisconsin, Lawrence Lake, Michigan, and Green Lake, Fayetteville, New York. The waters of Elk Lake are dilute solutions of calcium and magnesium bicarbonate. Calcium and magnesium together compose about 90% of the cation equivalents in solution, whereas alkalinity, which is almost entirely bicarbonate, composes more than 95% of the dissolved anion equivalents. Concentrations of other major cations (sodium and potassium) and anions (sulfate and chloride) are very low. Carbonate minerals in the calcareous glacial drift that forms the lake basin apparently control the chemical composition of the lake water, as indicated by analyses of ground water from Elk Spring. Calcium and magnesium are the dominant cations in the spring water, as in the lake water, but the ground water is relatively rich in calcium; ionic ratios of calcium to magnesium are about twice as large as they are in the lake. The conductance and alkalinity of water from the spring also are about twice as large as they are in the lake water.

On the basis of this water chemistry, Elk Lake is classified as

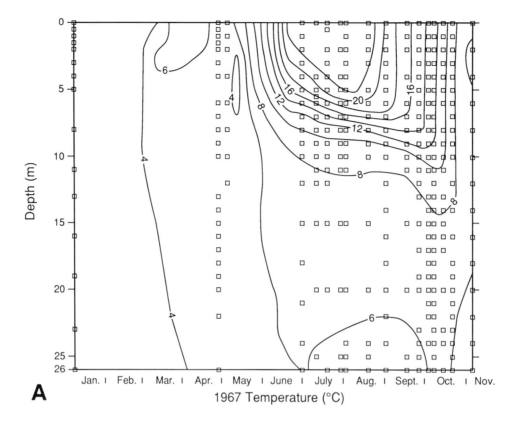

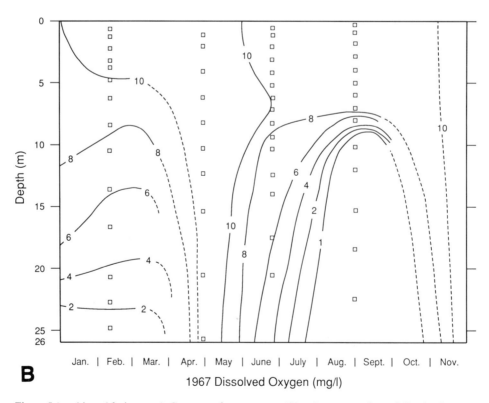

Figure 5 (on this and facing page). Contours of temperature (A) and concentrations of dissolved oxygen (B), phosphorus (C), and silicon (D) in Elk Lake in 1967. μM = micromoles.

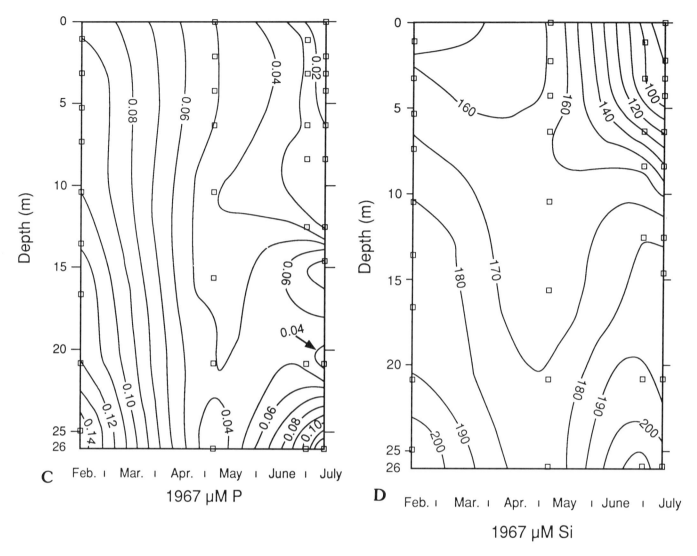

a group III lake (Gorham and others, 1983). Group III lakes in the north-central United States are most common in Minnesota, although some occur in Wisconsin and the North and South Dakota. They are on calcareous bedrock or glacial till, and all but the most dilute precipitate $CaCO_3$. The water chemistry of Elk Lake is extremely important in controlling the formation of carbonate minerals. For this reason, a detailed discussion of the water chemistry of Elk Lake is given in the chapter on the environment of deposition of $CaCO_3$ in Elk Lake (Dean and Megard, Chapter 8).

Ionic concentrations and the electrical conductance in Elk Lake are similar to those in nearby Lake Itasca and in Lake Mendota, Wisconsin, which is much larger (surface area = 39 km^2) than Elk Lake. Calcium, magnesium, and alkalinity are the dominant ions in both Elk and Mendota lakes, although magnesium exceeds calcium in Lake Mendota. Marl composes up to 60% of the profundal sediment deposited in Lake Mendota before the middle of the nineteenth century, when the watershed was cleared for agriculture (Murray, 1956; Brock, 1985). This caused a significant change in sedimentation. Sediments deposited more recently contain less carbonate and more organic matter, and a sharp boundary separates the black organic sediment from the older marl.

Concentrations of calcium, magnesium, and alkalinity in Lawrence Lake, Michigan, are about twice as large as they are in Elk Lake. Sulfate concentrations also are larger than in Elk Lake. Lawrence Lake is a much smaller lake (surface area = 0.05 km^2), and low-magnesian calcite has filled about 80% of the original basin (Otsuki and Wetzel, 1974).

Elk, Itasca, Lawrence, and Mendota lakes are different chemically from Green Lake, Fayetteville, New York. Green Lake is smaller (surface area = 0.26 km^2) than Elk Lake and is permanently meromictic. The modern sediments in Green Lake, like those in Elk Lake, are varved and calcareous (50 to 80% $CaCO_3$). However, the salinity of Green Lake is about ten times that of Elk Lake, and the dominant ions are calcium and sulfate. The high salinity and calcium sulfate composition derive from dissolution of gpysum ($CaSO_4 \cdot 2H_2O$) in Silurian evaporite beds that underlie the lake basin (Brunskill, 1969).

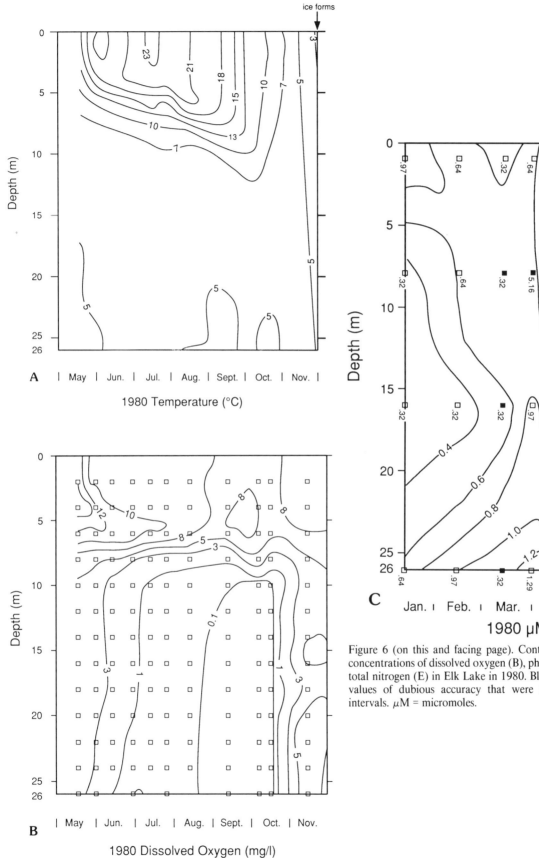

Figure 6 (on this and facing page). Contours of temperature (A) and concentrations of dissolved oxygen (B), phosphorus (C), silicon (D), and total nitrogen (E) in Elk Lake in 1980. Black squares in C and D mark values of dubious accuracy that were ignored for placing contour intervals. µM = micromoles.

Climatic and limnologic setting of Elk Lake

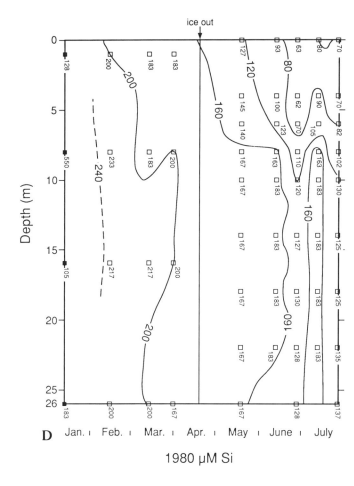

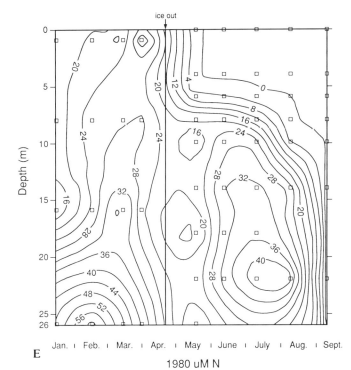

Iron and manganese

During periods of summer stagnation, concentrations of iron and manganese in the hypolimnion increase to more than 1.0 mg/l (Fig. 9). Iron and manganese enter Elk Lake predominantly in ground water, a principal source being Elk Spring, which is anoxic and has high concentrations of ferrous iron (860 μg/l). Iron and manganese are removed from the water column by precipitation during summer and winter stratification (Nuhfer and others, Chapter 7). Some of the ferric hydroxide is chemically reduced in the anoxic bottom water and sediments of the lake to produce the observed buildup of dissolved iron in the hypolimnion (Fig. 9B), but a considerable amount is buried in the sediments, as evidenced by the high iron concentrations in the sediments and the presence of distinct laminae of ferric hydroxide gel in the varves in the upper part of the core (Dean, Chapter 10; Anderson, Chapter 5). Manganese is much more sensitive to redox conditions than iron, and most of the manganese removed from the water by precipitation during overturn periods probably is recycled back to the hypolimnion during summer stagnation (Fig. 9A).

In order to obtain an estimate of the flux of iron and manganese from the sediments, we divided the lake into 2 m layers and determined the area of sediment exposed in a "bathtub ring" around the lake underlying each 2 m layer. We then calculated the amounts of iron and manganese in each 2 m layer of water for two dates: April 23, 1980, when the lake was isothermal and concentrations of iron and manganese were constant throughout the water column, and July 27, 1980, when the lake was stratified, most of the hypolimnion was anoxic, and high concentrations of iron and manganese had accumulated in the hypolimnion (Fig. 9). For this calculation we multiplied the volume of each layer (Table 2) by the average concentration of iron and manganese in each layer interpolated from measured concentrations of iron and manganese in the water column (Fig. 9). The difference between the amounts of iron and manganese in each 2 m layer between the two dates gives the amounts of iron and manganese that had accumulated in each layer over the 95 day period. Assuming that the daily rate of accumulation of iron and manganese was the same over this 95 day period, dividing the amount of increase by 95 yields the daily amount of iron and manganese added to each layer. If we also assume that all of the iron and manganese that accumulated in each layer each day came only from the sediments underlying each layer, we can normalize the rate of increase to a unit sediment surface area (m^2) and obtain flux rates of iron and manganese in mmol $\cdot$ m$^{-2} \cdot$ day^{-1} (Table 2). This is an oversimplification because there are other sources of iron and manganese, such as dissolution of precipitated iron and manganese minerals and mineraloids, and spring sources, but the calculation gives maximum flux rates of iron and manganese from the sediments. By summing the daily accumulations for each layer in the hypolimnion, we can obtain the total daily hypolimnetic accumulation rates of iron and manganese, which are 1.4 mmol $\cdot$ m$^{-2} \cdot$ day^{-1} and 15 mmol $\cdot$ m$^{-2} \cdot$ day^{-1}, respectively.

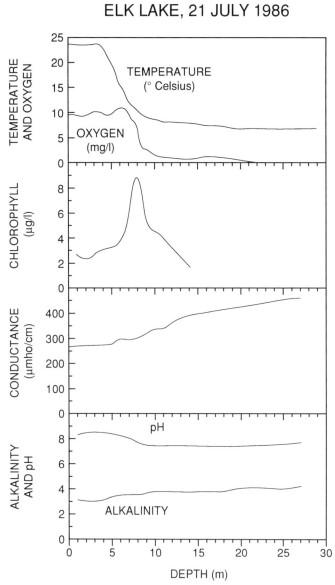

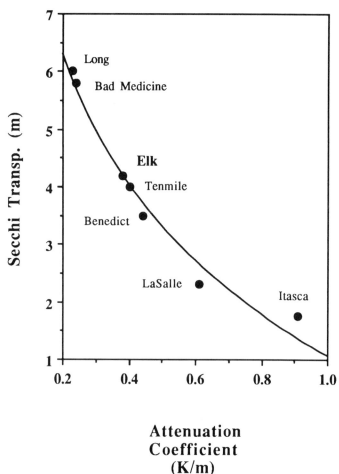

Figure 8. Relation between measurements of Secchi transparency and attenuation coefficients (K) for underwater light in Elk Lake and six other lakes in the vicinity of Elk Lake during July and August, 1984.

Figure 7. Stratification of alkalinity, pH, specific conductance, chlorophyll, temperature, and oxygen in Elk Lake for 21 July, 1986.

The same calculations for a 58 day interval between May 30, 1980, and July 27, 1980, give daily hypolimnetic accumulation rates of iron and manganese of 1.35 mmol · m^{-2} · day^{-1} and 5.4 mmol · m^{-2} · day^{-1}, respectively. A similar flux calculation by Carignan and Lean (1991) for a seasonally anoxic Canadian shield lake (Williams Bay, Jack's Lake, Ontario) gave daily hypolimnetic accumulation rates of iron and manganese of 0.026 mmol · m^{-2} · day^{-1} and 0.0030 mmol · m^{-2} · day^{-1}, respectively. The flux rates for Elk Lake, relative to those for Williams Bay calculated by Carignan and Lean (1991), are about 50 times higher for iron and about 500 times higher for manganese in Elk Lake. This difference is not too surprising, given the unusually high concentrations of iron and manganese in Elk Lake sediments (Dean, Chapter 10; Nuhfer and others, Chapter 7). The difference may also relate to unaccounted external sources of iron and manganese to Elk Lake that enter from Elk Spring and similar terrestrial and sublacustrine seeps. A. L. Baker (1992, written commun.) indicated that iron-rich water from Elk Spring enters the hypolimnion of Elk Lake as a density current during the summer because the cold (7 °C) and ionically more concentrated (2×) spring water is heavier than the epilimnetic water of the lake. Such sources probably contribute substantial amounts of iron to the lake and its sediments.

Sediments deposited over the past 2,000 years in the 30 m depression in Elk Lake contain an average of about 12% iron and 1% manganese (Table 3). Most of the iron precipitates as X-ray amorphous ferric hydroxide, identified by its bright orange color in sediment traps, particularly following fall overturn (Nuhfer and others, Chapter 7). Iron is also associated with sulfur and phosphorus in the sediments of Elk Lake (Dean, Chapter 10). Most of the manganese is deposited as X-ray amorphous manganese oxides and hydroxides, although rhodochrosite ($MnCO_3$) has been

TABLE 4. COMPOSITION OF ELK LAKE SEDIMENT DURING THE PAST 2,000 YEARS

Component	Percent	Comment
$CaCO_3$	40.0	Computed from total Ca
Organic matter	21.0	LOI 525 °C; C_{org} = 7.5%
SiO_2	18.0	Mostly biogenic as indicated by low Al_2O_3 concentration
Al_2O_3	0.8	
Fe_2O_3	16.0	11.5 percent as Fe
P_2O_5	2.3	1.0 percent as P
MnO	1.4	As hydrated manganese oxide and rhodochrosite ($MnCO_3$)
MgO	0.8	
Total S	0.3	
K_2O	0.2	
Na_2O	0.1	
Total	100.9	

Note: LOI is loss on ignition.

identified by X-ray diffraction analyses of profundal sediment and sediment-trap samples (Dean, Chapter 10; Nuhfer and others, Chapter 7). Mass accumulation rates for iron and manganese determined for the past 1,000 years in the deep basin of Elk Lake are about 7 mg · cm^{-2} · yr^{-1} and 0.5 mg · cm^{-2} · yr^{-1}, respectively (Dean, Chapter 10). These mass accumulation rates translate to yearly molar accumulation rates of 1250 mmol Fe · m^{-2} · yr^{-1} and 90 mmol Mn · m^{-2} · day^{-1}, or average daily molar accumulation rates of 3.4 mmol Fe · m^{-2} · day^{-1} and 0.25 mmol Mn · m^{-2} · day^{-1}. These rates represent average *net* accumulation rates after mineralization of iron and manganese across the sediment-water interface; the calculated daily hypolimnetic accumulation rates (Table 2) provide a measure of the amount of mineralization. For example, a comparison of the average net sediment accumulation rates in the deep basin (3.4 mmol Fe · m^{-2} · day^{-1} and 0.25 mmol Mn · m^{-2} · day^{-1}) with average hypolimnetic accumulation rates (1.4 mmol Fe · m^{-2} · day^{-1} and 15 mmol · m^{-2} · day^{-1}) shows that most of the iron that accumulates remains in the sediment, whereas most of the manganese is mineralized and lost to the hypolimnion during summer stratification.

Nutrients

Algal abundance and succession is controlled by seasonal changes in nutrients (chiefly silica, phosphorus, and nitrogen for diatoms and chrysophytes), light levels, and turbulence. Unfortunately, serial measurements of nutrients and other limnological parameters are rare and sporadic at Elk Lake and often miss the periods of circulation when rates of nutrient supply to the photic zone are particularly important. However, some data for 1967 and 1980 are available that permit a general view of nutrient fluxes relevant to the phytoplankton of Elk Lake.

Phosphorus and nitrogen, essential nutrients for phytoplankton growth, accumulate in the hypolimnion during periods of stratification (Figs. 5C and 6, C and E). Some phosphorus may be utilized in the formation of iron phosphate minerals in the water column, within the sediment, or at the sediment-water interface when anoxic conditions prevail during the summer and winter (e.g., Anthony, 1977; Nuhfer and others, Chapter 7; Dean, Chapter 10), and nitrate may be lost by denitrification at these times. The remaining dissolved nitrogen and phosphorus compounds become distributed to the epilimnion to be utilized by phytoplankton during periods of circulation. In addition, phosphate minerals formed in the water column or on the bottom may break down to release phosphorus when oxygenated water reaches the bottom sediments during circulation. During periods of circulation, nutrients are delivered from the hypolimnion, and heavy planktic algae such as diatoms are resuspended by turbulence in the photic zone where photosynthesis occurs. Increasing light levels during spring circulation often produces larger algal blooms than those in the fall.

TABLE 5. COMPOSITION OF ELK LAKE, ELK SPRING, AND SURFACE SAMPLES FROM OTHER LAKES

Sample	Date	Conductivities*	Ca^{++}	Mg^{++}	Na$^+$	K$^+$	Total Cations	Alk.	$SO_4^=$	Cl$^-$	Total anions
Elk Lake	3/14/80	334	1.80	1.44	0.36	0.05	3.69	3.40	0.01	0.04	3.45
	5/19/80	327	1.70	1.39	0.36	0.05	3.50	3.40	0.01	0.05	3.46
	7/12/80	299	1.25	1.31	0.36	0.04	2.96	3.20	0.01	0.04	3.25
Elk Spring	5/5/67		4.36	1.76				5.92			
	6/3/85	604	4.69	2.14	0.20	0.06	7.09	6.49	0.03	0.02	6.54
Lake Itasca	5/3/67		1.90	1.26	0.28	0.04	3.48	3.20	0.04	0.02	3.26
Lake Mendota		300	1.40	2.25	0.26	0.09	4.00	3.00	0.17	0.34	3.51
Lawrence Lake			3.48	2.04				4.51	0.29	0.25	5.05
Green Lake		2,470	20.96	5.92	0.74	0.07	27.69	3.3	23.32	0.96	27.56

Note: Concentrations of all cations and anions are in meq L^{-1}. Data for three other North American lakes with sediments rich in calcium carbonate are from Brock (1985; Lake Mendota), Otsuki and Wetzel (1974; Lawrence Lake), and Brunskill and Ludlam (1969; Fayetteville Green Lake).
*Conductivities are in μmho cm^{-1} at 25 °C.

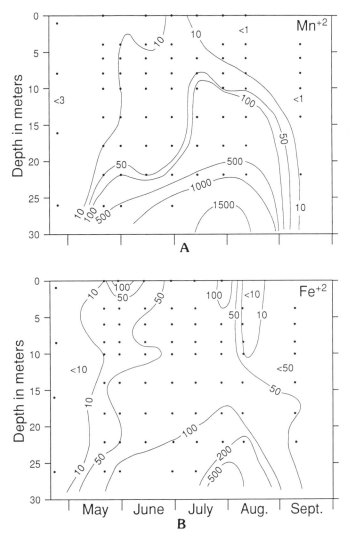

Figure 9. Contours of concentrations of dissolved Mn (A) and Fe (B) in Elk Lake from April to September, 1980. Values are in µg l.

of the silica entering the lake. For example, at nearby Willams Lake, Hubbard County, Minnesota, silica in ground water flowing into the lake is between 16 and 22 mg/l (James Labaugh, 1989, personal commun.). A similar situation exists in Crystal Lake, Wisconsin, where ground-water–borne silica (6 mg/l) entering the lake in the spring as a result of snow melt dominates the lake's silica budget and supplies this necessary nutrient for the spring diatom bloom (Hurley and others, 1985). Analogous lakes in the north-central United States are characterized by comparatively short hydraulic residence times and ground-water inflows with 5–7 mg/l silica (Stauffer, 1987).

The depth distribution of silicon in Elk Lake from winter to summer, 1967 (Fig. 5D), shows a moderate amount of silicon dissolution from underlying sediments in the hypolimnion, in contrast to this period in 1980, when silicon concentrations were not elevated in hypolimnetic waters (Fig. 6D). Both profiles show uptake of silicon in the epilimnion during the summer by diatom and chrysophyte algae.

BIOLOGICAL PROPERTIES

Phytoplankton successional dynamics

Although the kinds and successional dynamics of phytoplankton in Elk Lake have received no systematic study, two reasonably complete analyses by A. L. Baker in the fall of 1967 and the late spring of 1968 indicate that cyanobacteria (bluegreen algae) and Bacillariophyta (diatoms) dominate in the fall and spring, respectively (Fig. 10, Table 6). This suggests that Elk Lake conforms in a general way to the character of phytoplankton succession typical of many temperate fresh-water lakes (e.g., Hutchinson, 1967, p. 400). Sediment-trap studies (Nuhfer and others, Chapter 7; Bradbury, 1988) provide additional information about the variability of diatom succession and productivity in Elk Lake between 1979 and 1984. In particular, *Fragilaria crotonensis* dominates the spring and summer diatom populations during years when the period of spring circulation is short (late ice-out and early summer stratification), whereas *Stephanodiscus minutulus* dominates when spring circulation is early and long. The dominance of these two diatom species is probably related to rates of supply of silicon and phosphorus as a consequence of the strength and length of spring circulation (Bradbury, 1988). Less is known about diatom dominance during summer stagnation and fall circulation periods, although *Fragilaria crotonensis* and *Rhizosolenia eriensis* were comparatively common in Elk Lake in the fall of 1967 (Table 6) and in fall accumulations in sediment traps in 1979 and 1983 (Bradbury, 1988; Nuhfer and others, Chapter 7).

Phytoplankton and Photosynthesis

Phytoplankton analyses throughout the water column of Elk Lake in late spring of 1968 (May 1, 1968; Fig. 11) indicate that most of the phytoplankton was located in the epilimnion, al-

Although some 1980 phosphorus analyses are anomalously and possibly erroneous (Fig. 6C), overall 1980 winter and spring phosphorus concentrations are about 10 times higher than 1967 values, suggesting considerable variation in this nutrient from year to year. However, the phosphorus data for 1980 are not comparable with those for 1967, because the 1980 samples were stored for several weeks before analysis.

Silicon, an essential nutrient for diatom growth, enters the lake both from the hypolimnion, where sedimented diatom frustules partly dissolve, and from the ground water entering Elk Lake. Although the magnitude of ground-water flow into Elk Lake is unknown, the silica concentration (21 mg/L = 750 µmol) in Elk Spring on the east shore of Elk Lake (Fig. 1) is comparatively high, and if similar values characterize the general ground water in the area, it is probable that ground water supplies most

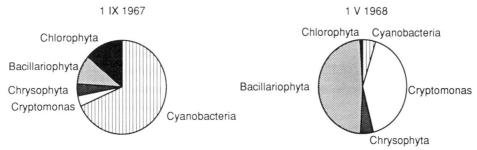

Figure 10. Proportional representation of phytoplankton groups present in Elk Lake epilimnion during the fall of 1967 (1 IX 1967) and throughout the water column during the spring of 1968 (1 V 1968). Data from A. L. Baker (1987, written commun.).

though diatoms were fairly common at all depths. Cyanobacteria appear to be stratified at a depth of 5 m, which may have been near the thermocline at that time because algal stratification typically accompanies thermal stratification in Elk Lake and other nearby lakes (Baker and Brook, 1971). The highest chlorophyll concentrations in Elk Lake are usually in the thermocline (Fig. 7).

The photosynthetic activity of the phytoplankton (Fig. 12) is related to sedimentary processes in Elk Lake, because the phytoplankton have two important but fundamentally different effects on the sediment composition. The varves are composed of couplets of alternating dark olive and light gray laminae (Anderson, Chapter 5). The dark layers contain siliceous cells (frustules) of diatoms, and therefore the thickness of these layers is related to the abundance of these algae in the lake water. However, the light layers are composed primarily of $CaCO_3$, which is formed in surface water oversaturated with calcite due to phytoplankton photosynthesis (Dean and Megard, Chapter 8).

The photosynthetic activity of the phytoplankton was measured on six dates during 1966 and 1967. Samples of water were collected at depth increments, and 300 ml subsamples were transferred to transparent and opaque biological oxygen demand (BOD) bottles. The subsamples were suspended in the lake at the depths from which they were collected for incubation periods of 6 hours, beginning at noon; rates of oxygenic photosynthesis and respiration were calculated from the changes of oxygen concentration in the bottles. As indicated by the profiles in Figure 12, oxygenic photosynthesis often was measurable down to depths of 8 or 9 m. The maximum activity was typically in the epilimnion at 1.5 m depth, where light was saturating, but there was often a secondary maximum in the metalimnion, usually near 6 m, where concentrations of oxygen are highest during spring and early summer (Figs. 5B and 6B) and chlorophyll concentrations exceed 8 mg/m^3 (Fig. 7).

Integral oxygenic photosynthesis (π), computed from the areas enclosed by the profiles, varied from 0.5 g $O_2 \cdot m^{-2} \cdot 6\ h^{-1}$ (during a cloudy day) to 1.7 g $O_2 \cdot m^{-2} \cdot 6\ h^{-1}$. The mean was 1.3 g $O_2 \cdot m^{-2}\ 6\ h^{-1}$. The mean daily rate was 2.6 g $O_2 \cdot m^{-2}$ day^{-1}, given the daily rate—twice the rate measured during 6 h. The mean volumetric rate of respiration in the epilimnion for a 24 h day, assumed to be 4 times the oxygen decrease during 6 h in opaque bottles, was 0.21 g $O_2 \cdot m^{-3} \cdot$ day^{-1}. This corresponds to a daily integral respiration rate of 1.3 g $O_2 \cdot m^{-2} \cdot$ day^{-1} in a 6 m epilimnetic mixed layer. Because this was about half as much as gross integral photosynthesis, the net daily integral rate of oxygen production also was 1.3 g O_2/m^2 (= 0.04 mol/m^2). The corresponding net integral rate of carbon assimilation would be 0.5 g C $\cdot m^{-2} \cdot$ day^{-1} if the phytoplankton used all of the photoreductant produced by oxygenic photosynthesis for carbon dioxide reduction.

The productivity of phytoplankton is lower in Elk Lake than in Lake Mendota, Wisconsin, where concentrations of chlorophyll in the mixed layer typically are between 10 and 25 mg/m^3 and mean daily integral carbon fixation vary from 0.6 g C/m^2 during autumn to 1.8 g C/m^2 during spring and summer (Brock, 1985). The comparison is only approximate, however, because different methods were used. Phytoplankton productivity in Lake Mendota was calculated from measurements of ^{14}C uptake, and the estimates of daily carbon fixation are not corrected for losses due to respiration.

Zooplankton

The zooplankton assemblage in Elk Lake is similar to the assemblages in other lakes of the region. The abundances of zooplankton at depth increments in mid August, 1990, are shown in Table 7. Vertical distribution is closely related to the stratified environment, as can be seen by comparing the zooplankton concentrations at various depths with the concentrations of dissolved oxygen (Table 7). Zooplankton depth distribution is also shown in the echogram from an acoustic transect (Fig. 3).

Two groups of microcrustacea, Copepoda and Cladocera, were most abundant in the epilimnion and metalimnion. The aquatic larvae of an insect, *Chaoborus*, were most abundant in the hypolimnion at depths greater than 15 m, where oxygen concentrations were less than 0.5 mg/l. Two genera, *Diaptomus* and *Cyclops*, were the most abundant copepods. Adults of *Diaptomus* were most abundant in the epilimnion, whereas adult *Cyclops* were more abundant in the metalimnion (7.5–10.5 m). However, immature copepods (nauplii and copepodids) were more abundant than adults at almost all depths, and much more abundant than adults in the metalimnion.

TABLE 6. MEAN CONCENTRATIONS OF PHYTOPLANKTON IN THE UPPER METER OF ELK LAKE*

	September 1, 1967	May 1, 1968
Cryptophyta		
Cryptomonas spp.	4	258
Bacillariophyta		
Synedra acus		114
Tabellaria flocculosa		114
Fragilaria crotonensis	42	25
Asterionella formosa	4	37
Cyclotella sp.	1	
Navicula		30
Aulacoseira italica		29
Stephanodiscus niagarae		2
Euglenophyta		
Trachelomonas sp.	2	1
Cyanobacteria		
Anabaena sp.	205	
Aphanocapsa sp.	18	
Microsystis aeruginosa	1	
Oscillatoria sp.	19	1
Gomphosphaeria sp.	11	
Aphanizomenon flos-squae		9
Chroococcus sp.	7	
Oscillatoria trichoides	1	
Oscillatoria agardhii		1
Oscillatoria redeckii		1
Arthrospira sp.	1	
Chrysophyta		
Mallomonas pseudocoronata	1	25
Dinobryon sociale	13	4
Dinobryon stetuleria	1	
Chlorophyta		
Dictyosphaerium pulchellum	43	
Crucigenia	5	
Ulothrix		2
Staurastrum pingue	1	
Pyrrhophyta		
Ceratium hirundinella	1	
Hemidinium sp.		1

*Values in units (naturally occurring unicells, filaments, or colonies) per milliliter, from A. L. Baker (written commun., 1987).

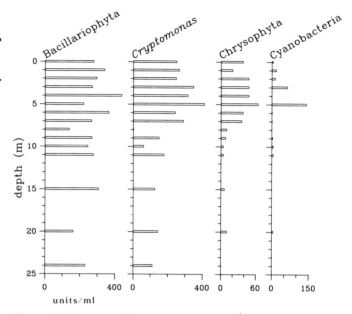

Figure 11. Abundance of algal groups (units · ml^{-1}) at different depths in Elk Lake on May, 1, 1968. The units represent unicells, filaments, or colonies depending upon the species. Data from A. L. Baker (1987, written commun.).

The Cladocera also were stratified. *Diaphanosoma* was present only in the epilimnion, whereas *Daphnia galeata* was more abundant in the metalimnion than in the epilimnion; *Daphnia catawba* was restricted to the metalimnion.

These samples were collected during the daytime. The distributions would be substantially different at night because many of the microcrustacea and *Chaoborus* migrate toward the surface after sunset. The zooplankton are probably most abundant during the daytime in the metalimnion and hypolimnion because they are invisible to predators at the low light intensities in deep water.

Bosmina longirostris, a small cladoceran, is abundant in the littoral vegetation although it occurs sparsely in the plankton. Remains of *Bosmina longirostris* and *Daphnia longispina* (= *D. galeata*) dominate the profundal sediments in Elk Lake throughout its 10,000 year limnological history (Boucherle, 1982), because the former cladoceran preserves more easily than the latter.

Benthic invertebrates

The most common and paleolimnologically relevant benthic invertebrates in Elk Lake are cladocerans, ostracodes, mollusks, and chironomids. Preliminary studies of the distribution and ecology of these groups in Elk Lake (e.g., Stark, 1976) generally conform with distributional patterns of other temperate, hardwater dimictic lakes.

Cladocerans. *Chydorus sphaericus* dominates the chydorid cladoceran fauna living among *Chara* beds in shoals at the northern end of Elk Lake (Whiteside, 1974). Relative to other species, *Chydorus sphaericus* is the most capable of swimming short distances (10–20 cm) above its preferred *Chara* habitat. Perhaps because of this, and because *Chydorus sphaericus* also associates with planktic cyanobacteria (e.g., *Anabaena*), either as a food resource or as a platform for feeding on other microorganisms, *C. sphaericus* commonly occurs in the plankton and consequently in the profundal sediments of dimictic lakes (Frey, 1960; Bradbury and Megard, 1972). *Alonella excisa, Pleuroxus denticulatus, P. procurvus, Camptocercus rectirostis, Acroperus harpae,* and *Pseudochydorus globosus* are subdominant in *Chara* habitats at Elk Lake, but mud substrates are dominated by *Alona*

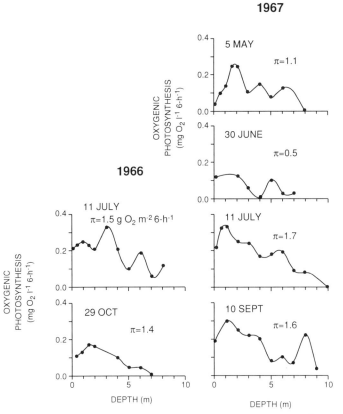

Figure 12. Depth profiles of oxygenic photosynthesis measured during 6 hr incubation periods on 6 dates in 1966 and 1967.

Ostracods. The ostracod fauna of Elk Lake is dominated by the genera *Candona, Cypridopsis, Cyclocypris, Cytherissa, Limnocythere, Physocypria,* and *Darwinula.* Stark (1976) found that *Cypridopsis vidua* and *Cyclocypris* sp. dominated in samples of bagged vegetation collected more than 30 cm above the lake floor. Their presence in the upper parts of submerged plants and also occasionally in the plankton of the lake is presumably due to their swimming ability. *Physocypria pustulosa, Darwinula stevensoni,* and *Candona ohioensis* characterize the littoral (to 6 m depth) benthic ostracod community. However, *Candona ohioensis* and *C. decora* have been found living on the sublacustrine ridge in the center of Elk Lake (Fig. 1) at a depth of about 12 m (Stark, 1976), which typically lies below well-oxygenated water during much of the summer (Figs. 5B and 6B). These ostracods may have a greater tolerance for seasonally low oxygen conditions than species confined to the littoral zone proper. Stark (1976) reviewed many of the factors that might control ostracode distribution, including temperature, oxygen, and substrate conditions, but more extensive trapping and collecting is required before the habitat preferences of these crustaceans can be seen approximately known.

Mollusks. *Gyralus, Amnicola, Marstonia, Valvata,* and *Physa* characterize the gastropods of Elk Lake. Their distribution is closely related to the shallow water of the littoral zone, although water currents may have redeposited mollusks in deeper water earlier in the history of the lake (Stark, 1976; Birks and others, 1976). The pelecypod *Pisidium* apparently prefers somewhat deeper water and finer substrates in Elk Lake.

Chironomids. Stark's (1976) survey of the distribution and composition of benthic organisms in Elk Lake identified a number of chironomid genera. As in many temperate dimictic lakes, *Chironomus plumosus* has the widest depth distribution, dominating in water between 10 and 15 m deep because of its ability to withstand low oxygen tension seasonally. Other species of *Chironomus, Procladius, Pentaneura, Tanytarsus,* and *Endochironomus* are restricted to the littoral zone because they require water with a higher oxygen content.

quadrangulata (Whiteside, 1974), and *Sida crystallina, Pleuroxus procurvus,* and *Graptolebris testudinaria* characterize floating and submerged aquatic vegetation (*Potamogeton* species and *Ceratophyllum demersum*) (Quade, 1969).

In general, the chydorids are most abundant in spring and fall; populations are lower in mid summer (Whiteside, 1974), perhaps as a result of predation (e.g., Goulden, 1971).

TABLE 7. CONCENTRATIONS OF ZOOPLANKTON AND INSECT LARVAE IN ELK LAKE*

Depth (m)	Copepoda			Cladocera			Insecta	Oxygen
	Diaptomus	*Cyclops*	Immature Copepoda	*Diaphanosoma*	*Daphnia galeata*	*Daphnia catawba*	*Chaoborus*	(mg l^{-1})
2.5	9,900	3,600	21,200	1,100	80	....	....	10.5
4.5	11,800	3.700	16,800	800	170	....	....	10.2
5.5	5,900	4,500	15,500	1,100	1,200	....	....	9.7
7.5	7,100	15,000	6,200	80	32,800	300	....	7.1
9.5	5,200	11,700	25,300	....	29,500	2,800	....	1.6
10.5	300	15,800	127,700	....	2,700	7,000	....	0.5
15.5	900	500	11,100	....	300	100	500	0.4
20.5	100	100	100	....	....	....	250	0.3
25.5	....	....	200	....	....	....	40	0.2

*Concentration values are in number of organisms per m³. Samples were collected at midday on August 16, 1990, with 1 m vertical tows that filtered 24 l of lake water.

CONCLUSION

This limnological overview of Elk Lake lacks the detail required to characterize fully the biological and physical processes operating in the lake today. Long-term monitoring of the physical and biological limnology of Elk Lake will be needed to fill out this preliminary information and to document the interannual variability of the lake so that the modern limnological basements can be used for paleolimnological and paleoclimatic reconstructions. Modern limnological baselines can be used for paleolimnological and paleoclimatic reconstructions. Modern limnological studies that include long-term monitoring must become a prerequisite for future paleolimnological studies.

ACKNOWLEDGMENTS

We gratefully acknowledge the interest and help of A. L. Baker and Jon Ross in providing data and insights about the limnology and biology of Elk and Itasca lakes.

REFERENCES CITED

Anderson, R. Y., Dean, W. E., Bradbury, J. P., and Love, D., 1985, Meromictic lakes and varved lake sediments in North America: U.S. Geological Survey Bulletin 1607, 19 p.

Anthony, R. S., 1977, Iron-rich, rhythmically laminated sediments in Lake of the Clouds, northeastern Minnesota: Limnology and Oceanography, v. 22, p. 45–54.

Baker, A. L., and Brook, A. J., 1971, Optical density profiles as an aid to the study of microstratified phytoplankton populations in lakes: Archiv für Hydrobiologie, v. 69, p. 214–233.

Baker, D. G., Kuehnast, E. L., and Zandlo, J. A., 1985, Normal temperatures (1951–80) and their application, in Climate of Minnesota, Part XV: St. Paul, Minnesota Agricultural Experimental Station, Part XV: 66 p.

Baldwin, J. L., 1973, Climates of the United States: U.S. Department of Commerce, National Oceanographic and Atmospheric Administration, 113 p.

Birks, H. H., Whiteside, M. C., Stark, D. M., and Bright, R. C., 1976, Recent paleolimnology of three lakes in northwestern Minnesota: Quaternary Research, v. 6, p. 249–272.

Boucherle, M. M., 1982, An ecological history of Elk Lake, Clearwater County, Minnesota, based on Cladocera remains [Ph.D. thesis]: Bloomington, University of Indiana, 135 p.

Bradbury, J. P., 1988, A climatic-limnologic model of diatom succession for paleolimnological interpretation of varved sediments at Elk Lake, Minnesota: Journal of Paleolimnology, v. 1, p. 115–131.

Bradbury, J. P., and Megard, R. O., 1972, A stratigraphic record of pollution in Shagawa Lake, northeastern Minnesota: Geological Society of America Bulletin, v. 83, p. 2639–2648.

Brock, T. D., 1985, A eutrophic lake. Lake Mendota, Wisconsin: New York, Springer-Verlag, 308 p.

Brunskill, G. J., 1969, Fayetteville Green Lake, New York. II. Precipitation and sedimentation of calcite in a meromictic lake with varved sediments: Limnology and Oceanography, v. 14, p. 830–847.

Brunskill, G. J., and Ludlam, S. D., 1969, Fayetteville Green Lake, New York. I. Physical and chemical limnology: Limnology and Oceanography, v. 14, p. 817–829.

Bryson, R. A., and Hare, F. K., 1974, Climates of North America, in Bryson, R. A., and Hare, F. K., eds., Climates of North America (World Survey of Climatology, Volume 11): Elsevier, New York, p. 1–47.

Carignan, R., and Lean, D.R.S., 1991, Regeneration of dissolved substances in a seasonally anoxic lake: The relative importance of processes occurring in the water column and in the sediments: Limnology and Oceanography, v. 36, p. 683–707.

Clay, C. S., and Medwin, H., 1977, Acoustical oceanography: New York, John Wiley and Sons, 544 p.

Dean, W. E., and Gorham, E., 1976, Major chemical and mineral components of profundal surface sediments in Minnesota lakes: Limnology and Oceanography, v. 21, p. 259–284.

Dean, W. E., Bradbury, J. P., Anderson, R. Y., and Barnosky, C. W., 1984, The variability of Holocene climate change: Evidence from varved lake sediments: Science, v. 226, p. 1191–1194.

Diaz, H. F., and Pulwarty, R. S., 1992, A comparison of Southern Oscillation and El Niño signals in the tropics, in Diaz, H. F., and Markgraf, V., eds., El Niño: Historical and paleoclimatic aspects of the Southern Oscillation: Cambridge, Cambridge University Press (in press).

Frey, D. G., 1960, The ecological significance of cladoceran remains in lake sediments: Ecology, v. 41, p. 684–699.

Gorham, E., Dean, W. E., and Sanger, J. E., 1983, The chemical composition of lakes in the north-central United States: Limnology and Oceanography, v. 28, p. 287–301.

Goulden, C. E., 1971, Environmental control of the abundance and distribution of the chydorid Cladocera: Limnology and Oceanography, v. 16, p. 320–331.

Heiskary, S. A., 1985, Trophic status of Minnesota lakes: St. Paul, Minnesota Pollution Control Agency, 39 p.

Hurley, J. P., Armstrong, D. E., and Bowser, C. J., 1985, Groundwater as a silica source for diatom production in a precipitation-dominated lake: Science, v. 227, p. 1576–1578.

Hutchinson, G. E., 1967, A treatise on limnology, Volume 2: Introduction to lake biology and the limnoplankton: New York, John Wiley and Sons, 1115 p.

Megard, R. O., and Berman, T., 1989, Effects of algae on the transparency of the southeastern Mediterranean Sea: Limnology and Oceanography, v. 34, p. 1640–1655.

Murray, R. C., 1956, Recent sediments of three Wisconsin lakes: Geological Society of America Bulletin, v. 67, p. 883–910.

Otsuki, A., and Wetzel, R. E., 1974, Calcium and total alkalinity budgets of a small hard-water lake: Archiv für Hydrobiologie, v. 73, p. 14–30.

Quade, H. W., 1969, Cladoceran faunas associated with aquatic macrophytes in some lakes in northwestern Minnesota: Ecology, v. 50, p. 170–179.

Stark, D., 1976, Paleolimnology of Elk Lake, Itasca State Park, northwestern Minnesota: Archiv für Hydrobiologie, Supplement 50, p. 208–274.

Stauffer, R. E., 1987, A comparative analysis of iron, manganese, silica, phosphorus, and sulfur in the hypolimnia of calcareous lakes: Water Research, v. 21, p. 1009–1022.

Walker, K. F., and Likens, G. E., 1975, Meromixis and a reconsidered typology of lake circulation patterns: Verhandlungen Inernationale Vereinigung für Limnologie, v. 19, p. 442–458.

Whiteside, M. C., 1974, Chydorid (Cladocera) ecology: Seasonal patterns and abundance of populations in Elk Lake, Minnesota: Ecology, v. 55, p. 538–550.

Wright, H. E., Jr., 1972, Quaternary history of Minnesota, in Sims, P. K., and Mory, G. B., eds., Geology of Minnesota: Minneapolis Minnesota Geological Survey, p. 515–548.

Wright, H. E., Jr., Matsch, C. L., and Cushing, F. J., 1973, Superior and Des Moines lobes, in Black, R. F., Goldthwait, R. P., and Willman, H. B., eds., The Wisconsin Stage: Geological Society of America Memoir 136, p. 153–185.

MANUSCRIPT ACCEPTED BY THE SOCIETY JULY 27, 1992

Geological Society of America
Special Paper 276
1993

Chronology of Elk Lake sediments: Coring, sampling, and time-series construction

Roger Y. Anderson
Department of Earth and Planetary Sciences, University of New Mexico, Albuquerque, New Mexico 87131
J. Platt Bradbury and Walter E. Dean
U.S. Geological Survey, Box 25046, Federal Center, Denver, Colorado 80225
Minze Stuiver
Quaternary Research Center, University of Washington, Seattle, Washington 98105

ABSTRACT

A 22 m series of cores from a continuously laminated sequence of postglacial sediment was recovered from 29.6 m of water from the deepest part of Elk Lake, Clearwater County, Minnesota, by piston and freeze-coring methods during the winters of 1978 and 1982. A varve time series constructed and used as a basis for subsampling the cores and samples, based on the varve chronology, allows precise determination of fluxes of geochemical and biological sediment components. Chronological and petrographic studies have shown that the laminations are varves and their measurement and enumeration has produced a 10,400 year time series that estimates the rates and timing of paleolimnologic and paleoenvironmental changes in Elk Lake and its drainage. A radiocarbon date from surface sediment is 850 years. The difference between radiocarbon and varve dates continues down core; varve dates are older than radiocarbon dates, probably because of systematic incorporation of dead carbon (as bicarbonate) in organic matter in the sediment. Varve-dated boundaries of pollen zones in the Elk Lake cores compare closely with the ages of the same zones in cores from nearby lakes that have been radiocarbon dated.

INTRODUCTION

In 1967 a core was collected from the 30 m profundal basin in Elk Lake as part of a sampling program to determine the geochemistry of surface sediments of Minnesota lakes (Dean and Gorham, 1976; Dean and others, Chapter 9). At this time it was noted that the entire 1 m core contained varve-like laminations. Longer piston cores were collected in 1968, 1969, and 1970 with maximum recovery of about 9 m. These cores showed that all of the sediment recovered was laminated, but a complete Holocene section was not obtained. A 22 m core of sediment was collected through the ice in December, 1978, as part of a U.S. Geological Survey Climate Change Program initiative to obtain high-resolution records of Holocene environments. This core recovered 20.4 m of laminated sediment and a short (2.2 m) section of unlaminated sediment at the base. Petrographic studies of the laminations (Anderson, Chapter 5) and investigations of modern sediment fluxes in Elk Lake (Nuhfer and others, Chapter 7) indicate that the laminations are produced seasonally. The seasonal laminae were grouped into units that represented one year of accumulation (varves), and the base of the laminated sequence was dated by varve enumeration as 10,400 calendar (varve) years and by ^{14}C dating as 11,380 yr B.P. (see discussion below). The base of the short unlaminated section was ^{14}C dated as 17,000 yr B.P. The 1978 core was supplemented with another core in 1982 to ensure that the entire Holocene varve sequence was available. Varve interpretation and the counting and measuring of varves was performed on the 1978 and 1982 cores by R. Y. Anderson, and a complete varve time series was constructed from these data. The assembled varve chronology is shown in Anderson (Chapter 5, Fig. 18), and in Figure 1 as a smoothed time-series profile of varve thickness.

Anderson, R. Y., Bradbury, J. P., Dean, W. E., and Stuiver, M., 1993, Chronology of Elk Lake sediments: Coring, sampling, and time-series construction, *in* Bradbury, J. P., and Dean, W. E., eds., Elk Lake, Minnesota: Evidence for Rapid Climate Change in the North-Central United States: Boulder, Colorado, Geological Society of America Special Paper 276.

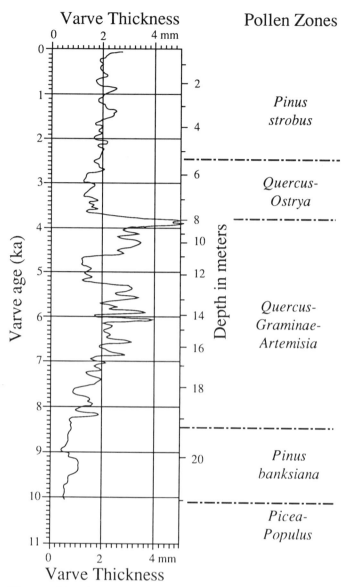

Figure 1. Plot of varve thickness vs. time in varve years before A.D. 1927 (ka) and depth of sediment in the 30 m hole of Elk Lake.

The petrology of sediment components and the complex lamination of the varves were examined before the varve time series was constructed. A discussion of varve petrology (Anderson, Chapter 5) outlines criteria used in varve interpretation, counting, and measuring. The validity of grouping seasonal laminae into annual units and the accuracy of the Elk Lake varve time series have been checked by comparing the varve chronology with radiocarbon dates on the same core, and by comparing events in the palynological record to radiocarbon-dated events in sediment core obtained from nearby localities. As evaluated below, the varve chronology parallels closely estimates of age obtained by less-direct methods. Additional confirmation of the varve chronology comes from an independent varve study of oriented cores collected from Elk Lake for paleomagnetic studies (Sprowl Chapter 6).

METHODS

Coring

The Elk Lake varve chronology was constructed from core segments (Fig. 2) collected through the ice in December, 1978, with a modified Livingstone piston corer (Wright, 1967). The 29.6 m water column was cased from the ice to several meters into the bottom sediment to ensure that the corer would reenter the same hole and to restrain the coring extension rods. Below the end of the casing, however, the corer does not necessarily follow the same path into the sediment. The top 12.7 m of varved sediment was collected in four drives using a 3.5 m core barrel. Immediately following extrusion from the core barrel, the cores were cut into 1 m lengths, wrapped in plastic wrap and aluminum foil, and placed on the ice to freeze. A 1 m core barrel was used to collect sediment between 12.7 and 22.4 m subbottom, when it became physically impossible to push the 3.5 m core barrel any further into the sediment by hand.

A freeze-coring technique (Wright, 1980) was used through a separate hole in the ice to obtain an undisturbed section of the sediment-water interface and the top part of the sediment section. The technique involved filling a weighted, hollow, rectangular, stainless-steel box with a mixture of alcohol and dry ice, lowering the tube about 1 m into the sediment, leaving it for about 0.5 h, recovering the tube, and peeling the frozen slab of sediment away from the face of the box.

Core splicing

After an initial compilation it was discovered that the third 3.5 m coring drive in the 1978 core had duplicated the second drive, and there was an uncored gap between 6.5 and 8.5 m. This part of the section was collected in 1982 by an oriented core taken for paleomagnetic studies and was spliced into the 1978 core. A prominent 1-cm-thick clay layer containing a 0.4-cm-thick, waxy, nearly pure layer of the diatom *Rhizosolenia* occurs at a depth of ~39 m near the top of a thick sequence of clastic-rich varves (Fig. 2). Another distinctive layer composed of *Stephanodiscus minutulus* occurs at a depth of ~36 m. These layers, as well as changes in varve character, were used to correlate the core segments and splice the 1982 core segment into the 1978 core.

The 30 cm level in the frozen box core that recovered the sediment-water interface marks the lower boundary of a thick, distinctive layer of nonlaminated or weakly laminated sediment referred to as the "disturbed layer." This layer apparently represents disturbance due to logging activities and flooding of Elk Lake in A.D. 1903, and thereby dates that level in the frozen box core. The *Ambrosia* rise (A.D. 1890) begins at a depth of about 35 cm in the frozen box core as determined by correlation with

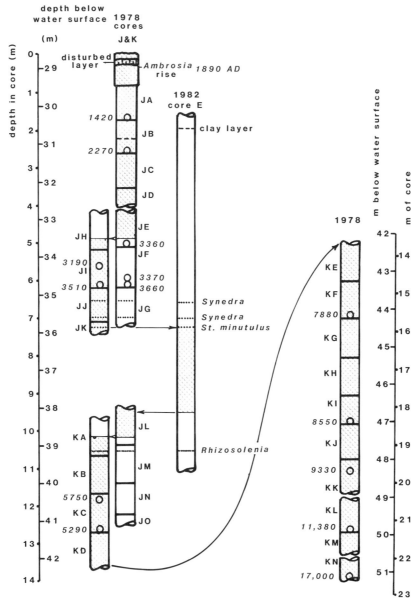

Figure 2. Stratigraphic positions of core segments for the 1978 (J-K) core and the 1982 (FL-82) core. Key marker layers used for correlation between the 1978 and 1982 cores are shown. Numbers are uncalibrated ^{14}C ages (see Table 1). Pattern indicates actual core segments used to construct the varve chronology.

other frozen box cores from Elk Lake. The number of varves that have accumulated above the (1903) disturbed layer is variable in cores collected since 1978, and, on the basis of probable varves above the disturbed layer, the top of the frozen box core is estimated to be at about A.D. 1927, and 53 years of sediment are missing from the top of the time series in the 1978 core. It is suspected that coring activity since 1967 has disrupted the upper sediment layers and the most reliable datum in the varve chronology is the (1903) base of the disturbed layer. A prominent 1-cm-thick clay layer occurs about 2.2 m below the 1978 sediment-water interface and permits correlation of the 1978 core with subsequent cores taken from Elk Lake. Approximately 720 varves occur between the clay layer and the base of the (1903) logging disturbance, placing the clay layer at about A.D. 1180.

Construction of varve time series

Frozen core sections were cut in half longitudinally with a band saw. The exposed surface was scraped clean and sprayed with water to form a clear, ice-glazed surface. The still-frozen core was then photographed beside a metric scale and appropriate labels. The photographs were enlarged 2.5 times for counting and measuring of varves. Interpretations of annual groupings of laminae were made on frozen core surfaces with the aid of a magnifying lens and/or a stereomicroscope. The varve boundaries were then transferred to the photographs for thickness measurements. A plastic marker was inserted in the core every 50th varve to aid in sampling. Varve measurements and counts were assembled into a master time series (Anderson, Chapter 5, Fig. 18) by tabulating the thickness measurements from the photographs.

A small amount of disturbed material was present at the top of each of the 13 coring drives. The thicknesses of varves on each side of the disturbed intervals were determined, and the average of these varve thicknesses was used to extrapolate the number of varves in the disturbed interval. A total of 609 estimated varves in coring breaks was patched into the master varve time series.

For all chapters in this volume, the following conventions are used for time designations: ka (for kilo annum) = thousands of varve years ago, T_0 = A.D. 1927 (the beginning of Elk Lake varve chronology); ^{14}C yr B.P. = radiocarbon years before A.D. 1950; cal. yr B.P. = calibrated radiocarbon years.

Sampling

Samples of sediment about 1 cm wide that included 50 varves were collected for analyses of diatoms, geochemistry, and mineralogy. Duplicate samples of 50 varves were used for plant-pigment analyses. A 1.0 cm^3 sample was collected every 50th varve for measurement of physical properties, including water content and bulk density, using a constant-volume sampler with a rectangular cross section. One to about five varves were sampled depending upon varve thickness. A 0.5 cm^3 sample by pollen analysis was collected adjacent to the 1.0 cm^3 physical properties sample using a constant-volume sampler. A 5.0 cm^3 sample for paleomagnetic studies was collected with a constant-volume sampler between successive 50 year pollen and physical properties samples. After paleomagnetic measurements were made, the 5 cm^3 samples were used for cladoceran studies.

The frozen box core was sampled only for diatom and geochemical analyses, although the top of the box core did not contain enough material for geochemical analyses. The uppermost sample for diatom stratigraphy is A.D. 1900, based on its position relative to the 1903 base of the disturbed layer. The uppermost geochemical sample is at A.D. 1864, and the uppermost sample for all other studies is at the top of core segment JA, about A.D. 1795.

EVALUATION OF ELK LAKE VARVE CHRONOLOGY

Extrapolations between core drives, distortion of the core during freezing, and problems of interpreting annual groups of laminae have contributed to errors in the varve chronology. The amount of error can be evaluated by examining the internal consistency of the Elk Lake chronology [see discussions in Anderson, Chapter 5, and Sprowel, Chapter 6)] and by comparing radiocarbon dates obtained from Elk Lake with dates for key events at other localities.

Comparison with radiocarbon chronology

Fourteen radiocarbon dates were obtained from varved Elk Lake sediments (Table 1). In addition, one date was obtained from surface sediment and another from the stratigraphic horizon of the increase in abundance of ragweed pollen ("*Ambrosia* rise") dated at A.D. 1890 (Birks and others, 1976). A plot comparing the varve chronology with calibrated radiocarbon dates shows essential agreement between the two methods (Fig. 3). If we assume that varve years are calendar years, then the calibrated ^{14}C age for the midpoint of each sample (Table 1), based on tree-ring data, can be determined using the calibration curve of Stuiver and Reimer (1986), and Figure 3 can be used to convert varve and radiocarbon years. (A more detailed plot for each varve year (Fig. 4) can be used to convert varve years to uncalibrated ^{14}C years.)

Varve dates are consistently younger than radiocarbon dates, suggesting the presence of old carbon in the lake's carbon reservoir. The difference (Δ) between uncalibrated ^{14}C age and the calibrated ^{14}C age is a measure of the amount of old carbon in bicarbonate ions in Elk Lake water derived from leaching of calcareous glacial drift. In general, values of Δ decrease with decreasing age (Fig. 5), which is expected if the concentration of dead carbon from leaching of the calcareous drift was large initially and if the production of dead carbon decreased with time.

Additional evidence for the presence of old carbon in the carbon cycle of Elk Lake comes from radiocarbon dates of samples from the unlaminated sublittoral ostracode core. Organic carbon in sediment from the base of the sublittoral core gave a ^{14}C age of 10,800 ± 60 yr B.P., and wood from the same horizon gave a ^{14}C age of 10,170 ± 150 yr B.P., an expected relation if wood recorded the uncalibrated age from atmospheric CO_2 and if organic carbon in lake sediments recorded the age of carbon in the lacustrine reservoir.

Comparison with vegetation history

The accuracy of the varve chronology of the Elk Lake core also can be evaluated by referencing the pollen stratigraphy of the core (Whitlock and others, Chapter 17) to radiocarbon-dated pollen stratigraphies in cores from other sites. Pollen records from north-central Minnesota document several vegetation changes over the past 11,000 years, and some changes are rapid and distinctive enough to be correlated with those in Elk Lake. The most fully analyzed site closest to Elk Lake is Bog D, 4.8 km southeast of Elk Lake (McAndrews, 1966). The following dated vegetation changes from Bog D are the most useful for correlation.

1. *Picea* fall. A rapid decrease in percentage of *Picea* (spruce) pollen from 30% to 5% at 11,000 ± 90 yr B.P.

TABLE 1. RADIOCARBON DATES ON SAMPLES FROM THE 1978 ELK LAKE VARVED CORE*

Depth in core (m)	Varve age (yr)	Cal. yr B.P.	^{14}C yr B.P.	Δ	Lab number
0.0	0		420 ± 060		QL4018
x.x	88		1,160 ± 065		QL4017
1.7	648	720	1,420 ± 080	700	QL1560
2.6	1,100	1,100	2,270 ± 080	1,160	QL1561
5.0	2,216	2,216	3,360 ± 070	1,144	QL1562
5.6	2,317	2,400	3,190 ± 100	790	QL1493
6.0	2,634	2,450	3,370 ± 070	920	QL1563
6.1	2,666	2,400	3,660 ± 130	1,260	QL1492
6.2	2,731	2,650	3,510 ± 090	860	QL1564
11.8	5,084	4,540	5,750 ± 120	1,210	QL1565
12.6	5,654	4,950	5,290 ± 100	340	QL1494
15.5	6,694	5,850	7,880 ± 050	1,950	QL1566
18.4	7,983	7,150	8,550 ± 140	1,400	QL1495
19.7	9,061	8,150	9,830 ± 150	1,680	QL1496
21.3	10,500		11,380 ± 180		QL1497
22.4	No varves		17,000 ± 800		QL1498

*Cal. yr B.P. for a varve year is based on tree-ring chronology. Reservoir effect (Δ) is difference between uncalibrated ^{14}C yr B.P. age and Cal. yr B.P. age of sample.

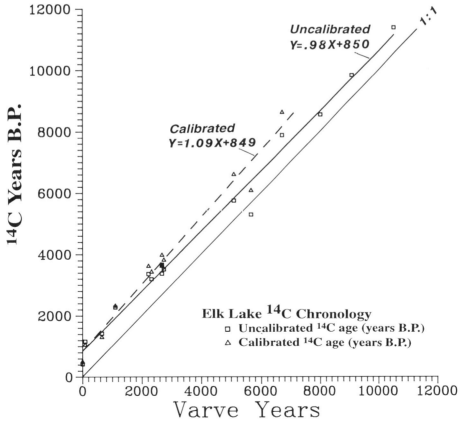

Figure 3. Plot of uncalibrated and calibrated radiocarbon dates from Elk Lake sediments against calendar (varve) years.

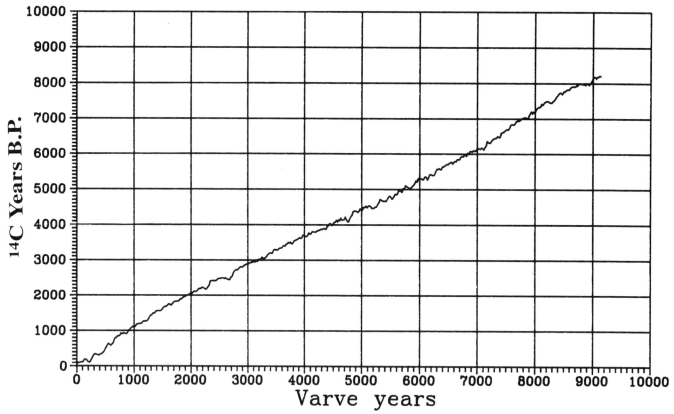

Figure 4. Plot of calendar years (varve years) beginning at 1927 (the youngest varve recovered in Elk Lake) against uncalibrated radiocarbon ages (data from Stuiver and Reimer, 1986).

2. *Artemisia* rise. A steep increase in percentage of *Artemisia* (sagebrush) pollen from 5% to 25% at 8560 ± 120 yr B.P.

3. *Betula* rise. An increase in percentage of *Betula* (birch) pollen from 10% to 25% at 3930 ± 100 yr B.P.

4. *Pinus strobus* rise. An increase in percentage of *Pinus strobus* (white pine) from about 10% to greater than 40% beginning 2730 ± 75 yr B.P.

When these pollen-zone boundaries are compared to the equivalent varve-dated pollen-zone boundaries in the Elk Lake core (Fig. 1; Whitlock and others, Chapter 17), the following correlations result.

Pollen zone	Bog D (^{14}C yr B.P.)		Elk Lake (Varve years)
	Atmospheric	Calibrated	
Picea fall	11,000 ± 090	12,200	10,200
Artemisia rise	8,560 ± 120	9,200–9,700	8,500
Betula rise	3,930 ± 100	4,400	4,000
Pinus strobus rise	2,730 ± 075	2,800	2,700

There is good agreement between the younger varve-dated pollen zones at Elk Lake and calibrated radiocarbon dates for the same zones at Bog D. Dates for the *Picea* fall at Wolf Creek, in central Minnesota but within the same ecotone between forest and prairie vegetation as Elk Lake, range from 10,500 to 10,100 yr B.P. (Birks, 1976). Sites to the northeast of Elk Lake—Kylen Lake and Weber Lake, 260 km northeast, and Lake of the Clouds, 325 km northeast—record *Picea* declines at dates between 9,100 and 9,300 yr B.P. (Fries, 1962; Birks, 1981; Craig, 1972); the declines suggest a time-transgressive demise of *Picea* from southwest to northeast Minnesota (Wright, 1968; Amundson and Wright, 1979). At the Itasca bison kill site on Nicollet Creek, less than 1 km northwest of Elk Lake, radiocarbon dates on *Larix* wood within sandy detritus containing 60% *Picea* pollen are between 9,360 and 9,690 yr B.P. (Shay, 1971). Webb and others (1983) have accordingly used a date of 10,000 yr B.P. for the decline of *Picea* in this area. The ^{14}C dates of 10,170 ± 150 yr B.P. and 10,800 ± 60 yr B.P. for wood and sediment, respectively, from the sublittoral core from Elk Lake are from within the *Picea* decline, according to pollen analyses. In view of dates obtained elsewhere for the *Picea* fall, the uncalibrated 11,000 yr B.P. date at Bog D seems anomalously old, as does the calibrated date for the rise in *Artemisia*. The assumed date of 10,000 yr B.P. for the *Picea* fall in the Elk Lake region (Webb and others, 1983) is consistent with wood dates in or near Elk Lake, and brings the dated pollen-zone sequence from Bog D into agreement with the Elk Lake varve chronology.

REFERENCES CITED

Amundson, D. C., and Wright, H. E., Jr., 1979, Forest changes in Minnesota at the end of the Pleistocene: Ecological Monographs, v. 49, p. 1–16.
Birks, H.J.B., 1976, Late-Wisconsin vegetation history at Wolf Creek, central Minnesota: Ecological Monographs, v. 46, p. 395–429.
——, 1981, Late Wisconsin vegetational and climatic history at Kylen Lake, northeastern Minnesota: Quaternary Research, v. 16, p. 322–355.
Birks, H. H., Whiteside, M. C., Stark, D. M., and Bright, R. C., 1976, Recent paleolimnology of three lakes in northwestern Minnesota: Quaternary Research, v. 6, p. 249–272.
Craig, A. J., 1972, Pollen influx to laminated sediments: A pollen diagram from northeastern Minnesota: Ecology, v. 53, p. 46–57.
Dean, W. E., and Gorham, E., 1976, Major chemical and mineral components of profundal surface sediments in Minnesota lakes: Limnology and Oceanography, v. 21, p. 259–284.
Fries, M., 1962, Pollen profiles of late Pleistocene and recent sediments from Weber Lake, Minnesota: Ecology, v. 43, p. 295–308.
McAndrews, J. H., 1966, Postglacial history of prairie, savanna, and forest in northwestern Minnesota: Torrey Botanical Club Memoirs, v. 22, p. 1–72.
Shay, C. T., 1971, The Itasca bison kill site: An ecological analysis: St. Paul, Minnesota Historical Society, Minnesota Prehistoric Archeology Series, 133 p.
Stuiver, M., and Reimer, P. J., 1986, A computer program for radiocarbon age calibration: Radiocarbon, v. 28, p. 1022–1030.
Webb, T. B., Cushing, E. J., and Wright, H. E., Jr., 1983, Holocene changes in vegetation of the midwest, *in* Wright, H. E., ed., Late Quaternary environments of the United States, Volume 2, The Holocene: Minneapolis, University of Minnesota Press, p. 142–165.
Wright, H. E., Jr., 1967, A square-rod piston sampler for lake sediments: Journal of Sedimentary Petrology, v. 37, p. 975–976.
——, 1968, The roles of pine and spruce in the forest history of Minnesota and adjacent areas: Ecology, v. 49, p. 937–955.
Wright, H. E., Jr., 1980, Cores of soft lake sediments: Boreas, v. 9, p. 107–114.

MANUSCRIPT ACCEPTED BY THE SOCIETY JULY 27, 1992

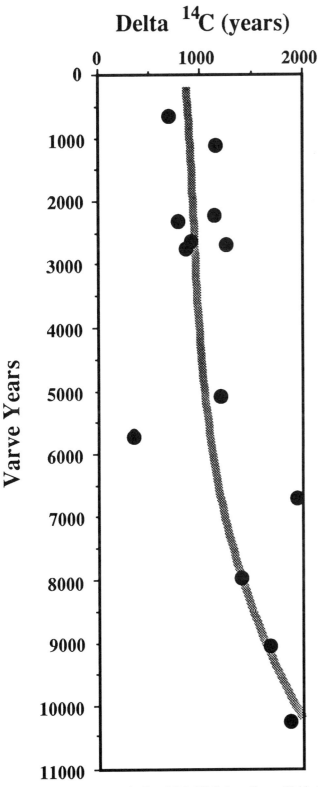

Figure 5. Plot of the reservoir effect (Δ) in Elk Lake sediment (Table 1). Difference (Δ) between uncalibrated ^{14}C age in sediment samples and calibrated ^{14}C age for each dated sample is plotted against varve years.

Geological Society of America
Special Paper 276
1993

The varve chronometer in Elk Lake: Record of climatic variability and evidence for solar-geomagnetic-^{14}C-climate connection

Roger Y. Anderson
Department of Earth and Planetary Sciences, University of New Mexico, Albuquerque, New Mexico 87131

ABSTRACT

Varves in Elk Lake are composed of seasonally deposited laminations of diatoms, calcite, aragonite, and layers enriched in Mn, Fe, organic matter, and clay and/or silt. The proportions of these components and the character of varve laminations changed systematically over the past 10,000+ years and define a post-glacial lake rich in calcium and manganese, a mid-Holocene prairie lake with sediments enriched in clay and silt, and a modern lake rich in Fe. Sequential changes in varve composition during the three phases of lake development are the result of a maturing lake and drainage basin and a systematic shift in airstream movements, accompanied by changes in precipitation, vegetation, and sedimentational responses to climatic forcing.

Changes in varve thickness between 3.8 and 8.0 ka are attributed to eolian processes and are believed to reflect changes in regional surface winds. Within this ~4000 year time window is a 2000 year interval when time series for Δ^{14}C (from tree rings) and varve thickness cross correlate and when both express periodicity at ~200 years, 40–50 years, and 20–25 years, supporting meteorological evidence that solar-geomagnetic events lead to changes in both cosmic particle flux and tropospheric winds. The Earth's magnetic dipole moment reached its lowest value in the Holocene during the same 2000 year interval, suggesting that solar-geomagnetic events had a greater effect on the wind field when the Earth's magnetic field was weak. A period of ~100+ years is weakly expressed in iron-rich varves during the past 3800 years, with a shift from cyclicity of ~100 years to 40–50 years between the Medieval Warm Epoch and the Little Ice Age.

INTRODUCTION

Today, the basin that contains Elk Lake (Fig. 1a; Megard and others, Chapter 3, Fig. 1) has a deep hole that reduces the vigor of seasonal circulation and lowers the concentration of dissolved oxygen. Newly formed layers of sediment are protected from benthonic, burrowing organisms. Similar conditions prevailed through the Holocene, preserving laminated sediments. Seasonally generated laminations, organized into annual groupings or varves have been assembled into a continuous 10,400 yr time series of varve thickness. The varves serve as a chronometer and also provide a compositional framework for examining the history of Elk Lake and its record of climatic variability. This history includes three distinct stages that are reflected in the composition of varved sediment (see Anderson and others, Chapter 1).

Post-glacial lake

About 14 ka, withdrawal of the ice sheet stranded a large block of ice that had melted by about 3000 yr later. The ice block formed a deep lake basin in a terrain of unweathered till, clothed by coniferous forest. This early lake and its environs was dominated by cold, anticyclonic winds from the ice margin to the northeast (see discussion of air masses and Fig. 3 in Anderson and

Anderson, R. Y., 1993, The varve chronometer in Elk Lake: Record of climatic variability and evidence for solar-geomagnetic-^{14}C-climate connection, *in* Bradbury, J. P., and Dean, W. E., eds., Elk Lake, Minnesota: Evidence for Rapid Climate Change in the North-Central United States: Boulder, Colorado, Geological Society of America Special Paper 276.

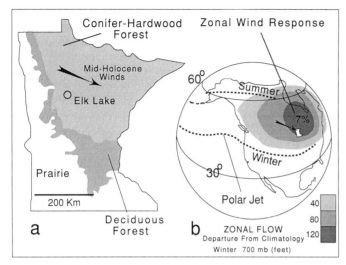

Figure 1a. Location of Elk Lake with respect to forest and prairie. b: Location of Elk Lake with respect to region of solar-geomagnetic meteorologic response (Stolov and Shapiro, 1974) and core of polar front jet in winter and summer at 6.0 ka (Kutzbach, 1987).

others, Chapter 1). Thin, dark-colored layers of sediment, enriched in Mn and diatoms, accumulated on the bottom, and during warm seasons, the dark materials were set off by thin, light-colored layers enriched in calcium carbonate.

Prairie lake

About 8.0 ka, after further decay of the ice sheet, cold anticyclonic winds were replaced by incursions of relatively dry Pacific air that reached progressively farther into the continental interior. In response to these incursions, the prairie shifted northeastward and there ensued a 4000 yr period of drought in north-central Minnesota. The more saline mid-Holocene lake (Forester and others, 1987) was surrounded by scattered stands of oak, sparse grass, *Artemisia,* and some open, bare ground (Whitlock and others, Chapter 17; Grigal and others, 1976). The varves that accumulated in Elk Lake more than doubled in thickness as a result of the influx of eolian clay and silt. The loess was suspended from the region to the west and was carried to Elk Lake by dry westerly winds. Occasionally massive blooms of diatoms, or precipitates composed of aragonite, added exceptionally pure laminae to the varved sediment.

Modern lake

Within a few centuries, at about 3.8 ka, the tropical airstream moved northward, bringing additional moisture, and a new balance was struck between Arctic, Pacific, and Tropical airstreams. The new expression of moisture and seasonality brought pine, hardwoods, and forest soils to the Elk Lake drainage. Subsequent leaching of till ensured that greater quantities of iron would be carried into the lake by ground water (Dean, Chapter 10). As deciduous forest replaced prairie, laminations enriched in iron hydroxide replaced laminae enriched in clay and silt.

Changes in three dominant components (Mn-Ca → clay-silt → Fe) define the three stages in lake history. In addition, sequential changes in the character of varves acted as a natural filter of climatic variability, allowing only certain frequencies of climatic change to be recorded. For example, climatic change over several thousand years is expressed mainly by one-way changes in the character and composition of varves. Climatic oscillations on the scale of one or more millennia certainly occurred, but were effectively filtered out by changes in varve composition that accompanied the weathering of glacial till and the regional effects that were associated with the retreat of the ice sheet. Hence, changes in the composition and thickness of varves in Elk Lake do not adequately define millennial oscillations in climate. The varve chronometer and the record preserved at Elk Lake are most useful for examining climatic variability in the range of decades to centuries.

Although the varved record at Elk Lake has the potential for annual resolution, in this chapter and other accounts of investigations at Elk Lake, results are based on sampling intervals of 50 years or greater because of demands on time. One exception is the parameter of varve thickness, which has annual resolution. One segment of the varve time series, from 3.8 to ~8.0 ka (prairie lake stage), has a dominant (eolian) component. Within this segment, changes in varve thickness have spectral patterns that provide information about winds and climatic variability.

In this chapter varved sediments are described in sufficient detail to establish the varve time series as a chronometer and to support interpretations of climatic variability. Descriptions of varves are organized according to the three stages of lake history, and components within varves are defined with reference to seasonal and lacustrine processes that are believed to be responsible for their generation. A discussion of climatic change is organized according to the three stages of lake history, followed by descriptions of climatic variability, as interpreted from the record of varve thickness. The section on climatic variability during the prairie lake stage describes a potentially important association between varve thickness (regional winds) and secular variation of ^{14}C (cosmic particle flux). This association suggests that significant changes in climate may be linked to solar activity through effects related to Earth's magnetic field (Anderson, 1992a).

ELK LAKE VARVES

Origin and types of varves

In a temperate, mid-continent setting such as at Elk Lake, the annual cycle has enough variance (Mitchell, 1976) to affect the reproduction and growth of organisms, as well as physical and chemical processes that occur within and around the lake. Many of the components of sediment that accumulated on the bottom of Elk Lake, as for most lakes, were produced seasonally in the overlying water column. Other components were delivered by winds, storms, runoff, and wind-driven overturn and circulation in the lake, processes that also have strong seasonal associations (Nuhfer and others, Chapter 7). Because dominant components were produced or transported at different times of the year, they accumulated on the bottom as concentrations of materials that later became organized into discrete, identifiable laminae. Pairs of dominant components (e.g., carbonate and clay), if generated every year, or almost every year, form couplets of laminae organized into annual groupings or varves, provided the sediment remains undisturbed (Anderson and Dean, 1988). At times in Elk Lake, more than two dominant components accumulated on the bottom in a single year to form an annual triplet of laminations. At times in the modern lake stage, as many as four discrete types of laminae can be recognized within a single varve.

Because Elk Lake underwent profound changes in geochemistry and sedimentation as a result of the progressive effects of weathering, and from major shifts in climate and vegetation, character and composition of laminae also changed markedly. As many as ten recognizable types of laminations accumulated during the past 10,400 yr (Table 1) and the proportion of materials incorporated into the laminations changed during the three stages of the lake's development (Table 2).

Seasonal associations of major components collected in sed-

TABLE 1. CHARACTER AND COMPOSITION OF SEASONALLY GENERATED LAMINATIONS PRODUCED IN ELK LAKE

Laminae Type Composition	Components	Generating Processes	Season of Accumulation
Organic	Disseminated dark organic matter	Life-form organization and fragmentation	Spring through Fall
Diatomaceous			
Pure	Pure bloom, e.g., *Rhizosolenia Stephanodiscus*	Phytoplankton blooms	Spring, ~Fall
Concentrated	Mostly monospecific diatoms, admixture of other species	Phytoplankton blooms	Spring, ~Fall
Impure	Mixed diatom species and terrigenous/organic debris	Resuspension of shallow sediment	Various, Summer?
Carbonate			
Calcitic	Carbonate aggregates, masses coatings (with Fe and Mn) *Phacotus*, calcified cysts, rare ostracodes	Chemical and organic precipitation	Summer, Late Summer
Dolomitic	Yellow-tan-white dolomitic calcite	Chemical precipitation	Summer
Aragonitic	Aragonite rods	Chemical precipitation	Summer
Ferric iron	Brown, iron-rich aggregates, films	Precipitation and flocculation?	Late Fall, Early Winter
Iron-manganese	Mn-rich layers and rinds, large porportion of iron	Precipitation	Fall, Winter?
Clay/silt	Clay and silt-size quartz and feldspar, fragmented diatoms, organic matter	Wind transport Advection of bottom/ shallow sediments	Various, Summer?

TABLE 2. TYPES OF LAMINATIONS PRODUCED DURING THREE SEQUENTIAL PHASES OF LAKE DEVELOPMENT

Lake Stage	Lamination Type						
	Organic	Diatomaceous		Carbonate			Ferric	Clay/Silt
		Concentrated	Impure	Dolomitic	Calcitic	Aragonitic		
Post-glacial Lake	X	X		X			X	
Prairie Lake	X	X	X	X	X	X	X
Modern Lake	X	X			X		X	

iment traps (Nuhfer and others, Chapter 7), and the close agreement between the varve and the radiocarbon chronology (Anderson and others, Chapter 1; Sprowl, Chapter 6), confirm that the groupings identified in Elk Lake sediments represent the annual cycle and are varves. Although annual groups of laminae can be complex, and some types of laminae may be added or omitted from the annual cycle, annual bundles can usually be recognized from the occurrence of distinctive and persistent laminations, such as light-colored layers of carbonate or dark-colored layers enriched in organic matter (see Tables 1 and 2, and criteria in the discussion of *Varves as a Chronometer*).

Preservation of varves

The deep hole or depression in the wide, southeastern part of the lake (Fig. 2) is the largest of several holes left by stranded blocks of ice in the Elk Lake basin. Varves began to form and to be preserved shortly after the initial melting of the ice block, which was accompanied by the influx of sand, silt, and clay into the depression above the ice. The first sediment to accumulate in the basin was bioturbated and contains coarse detritus (Fig. 3). As the ice block continued to melt, the basin deepened, depth of water in the deep hole reached ~50 m, slow circulation reduced the concentration of dissolved oxygen, and laminated sediments (Fig. 4) were preserved. Occasional introduction of turbidites into the lower part of the sequence of varved

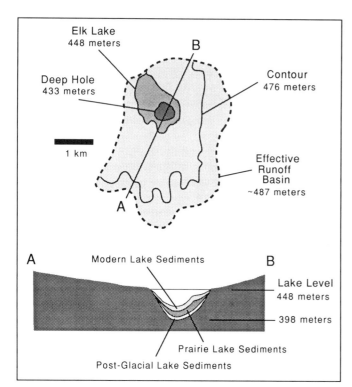

Figure 2. Cross section and map of Elk Lake showing depth and sediment thickness for the early Holocene lake, the prairie lake, and the modern lake, and relief and runoff area of drainage basin.

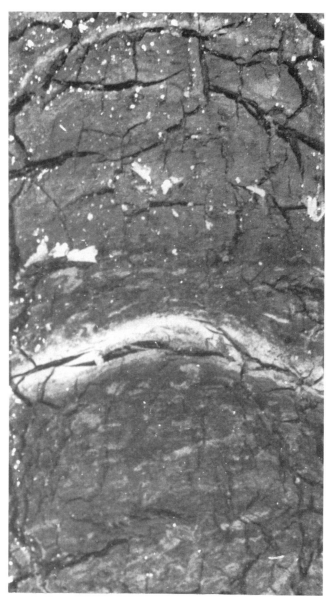

Figure 3. Layered, silty clay at base of varved sediments in the deep hole of Elk Lake. Depth is 50.3 m below water surface; age is 11.0+ ka. Note bioturbation below thick, light-colored layer. Bar scale = 1 cm.

sediments (Fig. 4) suggests that some slumping occurred along the margin of the basin. Elk Lake today is not a typical chemically stratified or meromictic lake. However, even with a depth of ~30 m, circulation within the deep hole is sufficiently weak to reduce the concentration of dissolved oxygen and exclude a population of sediment-disturbing organisms.

Diagenesis of varved sediments

Although laminations are generally well preserved, some reorganization of soluble materials has occurred after burial. For

Figure 5. SEM photograph of material in Fe-Mn-rich lamina and vivianite twin. EDS relative peaks = P 1216; Fe 773 and 105; Mn 334. Age is ~10.14 ka.

Figure 4. Core slab of alternating calcite and Mn-Fe-rich laminae near base of varved sequence. Note thick turbidite layer. Depth is 49.55 m below water surface; age is ~10.4 ka. Bar scale = 1 cm.

example, light blue aggregates of a mineral rich in iron and phosphorous (vivianite, Fig. 5) have grown at the interface between carbonate laminae and laminae enriched in organic matter. Vivianite grains are subspherical, but often have angular corners. Within laminae these grains develop a graded size profile, with coarser crystallites adjacent to the interface between laminae. In addition, larger twin crystals of vivianite (Fig. 5) are scattered through the core and bright blue masses can be seen on exposed surfaces. Vivianite is concentrated in the high-manganese sediment in the post-glacial lake stage, but also occurs in the iron-rich modern lake sediments. In addition to vivianite, masses of white, pure aragonite, occurring as radiating needles, have grown diagenetically and have displaced brown, organic-enriched laminae in parts of the core. Pyrite framboids also were observed in the clastic-rich, mid-Holocene part of the sequence.

Manganese oxides can be seen filling or staining cellular organic structures and as coatings or rinds on other materials in manganese-rich parts of the core. The rinds and fillings probably were formed during early diagenesis. Soluble components were mobilized during reduction in the sediment, but the volume of dissolved material and distance of transport within sediment are probably small. Some unknown proportion of Mn and Fe precipitated within the water mass returns to solution before reaching the bottom (Nuhfer and others, Chapter 7). Sediment-trap investigations in other lakes (Anderson and Dean, 1988) indicate that dissolution takes place as particles of calcium carbonate settle through the water column, and this phenomenon may partly account for a variable association between changes in calcium carbonate and changes in varve thickness. X-ray diffraction shows that manganese carbonate (rhodochrosite) is a common constituent of the sediment, especially in the manganese-rich parts of the sequence. Rhodochrosite was not observed in a crystalline form in a scanning electron microscope (SEM) or thin sections, and it is apparently finely divided and incorporated into laminae with high concentrations of manganese.

VARVES AS A CHRONOMETER

The varve time series reconstructed from Elk Lake sediments is the basis for sample collections and analyses presented in other chapters, and it is important to define its limitations. The annual time series for Elk Lake was constructed by interpreting, counting, and measuring varves in several cores obtained from Elk Lake (Anderson and others, Chapter 4). Most of the record was obtained from a single core that had zones of disturbed sediment between core drives, and for which the varve count was extrapolated. Such zones introduce errors that can be corrected with study of additional cores. In addition to errors related to the completeness and condition of the cores, inaccurate varve counts may originate in errorneous interpretation of annual groups of laminae. For example, exceptionally thick or thin laminae may not be recognized as part of an annual group, and types of laminae that normally are persistent from year to year may be missing. The brief summary of varve components and types of laminae described previously (Table 1) is a guide to associations that were used to place laminae into annual groupings. These seasonal associations have been confirmed by investigations using sediment traps (Nuhfer and others, Chapter 7), and they show that carbonate laminae represent the warm season, extending into the fall; ferric laminae represent seasonal circulation, mainly during late fall and early winter overturn; and diatom-enriched laminae generally accumulate in spring through summer. Dark, organic-rich laminae probably represent a background accumulation of organic debris and organisms that are concentrated from spring to fall, and at times when accumulation of other components is reduced.

No single lamina type or component was used to define the annual cycle throughout the core. The most persistent and reliable component proved to be dark-colored, organic laminae, and these were used when other components were missing or erratically developed. For example, the introduction of thick layers of clay and silt, and impure diatomaceous layers during part of the prairie lake stage nearly obliterated evidence for annual grouping. However, even where the thickness of clastic layers exceeds 1 cm, dark organic laminae are generally recognizable and were used to identify annual units of accumulation. Only a few zones were found, at times of greatest clastic influx during the mid-Holocene, where poor definition of laminae was responsible for an inaccurate varve count. Within these zones, about 200 varve couplets are in zones where varve thickness was extrapolated from adjacent varves.

In carbonate laminae, unusually thin or thick layers and possibly missing layers may be the result of dissolution of carbonate in the water column. For complex groupings of laminae, as in the modern lake stage, it was often necessary to switch criteria from one type of laminae to another, based on persistent patterns observed elsewhere in the core. Concentrated diatom laminae, alternating with organic laminae, constitute the principal seasonal cycle in the modern lake stage, and this pair was usually the basis for annual grouping. In the post-glacial lake stage, carbonate and Fe-Mn alternations are regular and clearly defined (Fig. 3), and, accordingly, varve counts are more accurate.

I emphasize that the varve chronostratigraphy incorporates subjective interpretation of the annual groupings based on associations observed throughout the core. Hence, the varve time series contains interpretive and splicing errors that contribute to inaccuracy. Interpretation of annual groupings was completed before varve counts were compiled into the time series. This precaution, combined with the slow and difficult task of interpreting 10,400 varves, assured freedom from conscious bias.

Estimates of interpretive error in the lower part of the modern lake stage, where two cores were available, indicated an error of about 15%. This part of the Elk Lake sequence contained complex varves that had been subjected to refreezing and ice-crystal wedging, and the estimate probably represents a maximum error. The error for the thick clay-silt laminations in the prairie lake stage, in spite of loss of definition in parts of the core, probably is less than 10%, mainly because relatively few zones lost their laminated character. Interpretive error in the well-defined varves of the post-glacial lake is estimated to be low, probably less than 5%.

Another estimate of error was obtained by comparing the varve chronostratigraphy with radiocarbon age values throughout the 10,000+ years of record (Dean and Stuiver, Chapter 12). Comparison with uncorrected radiocarbon values gives a varve chronology that is too old by ~2%. Comparison with corrected (calibrated) radiocarbon years for the past 7000 years of record gives a varve chronology that is about 9% too young. The use of multiple and overlapping cores will eventually lead to improved interpretation and a more accurate varve time series. Although the Elk Lake varve time series can never have the accuracy of a tree-ring record based on many trees, consistent agreement between 15 radiocarbon dates and the varve chronology shows the general validity of the varve interpretation and increases confidence in results and interpretations reported in this and other chapters.

SEQUENTIAL CHANGES IN ELK LAKE VARVES

Post-glacial lake (~8 to 11 ka)

Bathymetry and environs. Elk Lake basin had an area of about 1 km^2 and a maximum depth of 50 m shortly after the ice block that formed the deep hole had melted (Fig. 2). Today maximum water depth is 29.5 m in the deep hole and the average depth outside the deep hole is ~12 m (Megard and others, Chapter 3. If one assumes a proportional rate of accumulation outside the deep hole throughout the history of the lake, average depth around the deep hole shortly after the basin formed was about 35 m.

A piston core taken in the deep hole penetrated about 0.8 m of banded, coarse and fine clayey silt beneath the varved sequence (Fig. 3). Organic content is low in this sediment but increases in the upper 30 cm. Some zones in the banded silt and clay contain discontinuous, organic-rich layers. The environment, however, is lacustrine rather than fluvial. Organic-rich sediment in the upper 30 cm of the pre-varve sequence is disturbed by simple animal borings that show the presence of a bottom fauna, indicating that the earliest lake was oxic. The concentration of dissolved oxygen must have decreased as the lake deepened or as more organic material was produced in the lake, until the bottom water was sufficiently disaerobic to exclude burrowing organisms and permit the preservation of laminations.

The poorly drained catchment area around the post-glacial lake supported a forest of spruce, with some birch and larch near the ice margin. As the ice sheet continued to melt back, this vegetation was replaced by pine, birch, and elm, which occupied the drainage until xeric conditions became evident at ~8.0 ka. The provenance for sediment and material through which ground water seeped was unweathered till that contained freshly ground limestone and other unaltered rock. With little or no overland flow in the drainage during the post-glacial lake stage, seepage of ground water through unweathered till delivered dissolved Ca, Mn, and Fe to the lake to be precipitated as $CaCO_3$ and Mn-Fe oxyhydroxides, constituents that characterize the sediments of the post-glacial lake (Dean, Chapter 10).

Varves in the post-glacial lake. The earliest-formed laminations appear in the slabbed core as sharply defined laminae of dark gray to black sediment, alternating with thinner, almost white laminae that contain mostly calcium carbonate (Fig. 4). Thin sections of the early-formed varves reveal that dark laminae consist of reddish-brown aggregates and thick, reddish-brown rinds that coat organic matter and fine-grained, lighter-colored sediment (Fig. 6). SEM-EDS analysis reveals that this darker material is enriched in compounds of manganese and iron (Fig. 7). Laminae enriched in diatoms contain abundant Mn (Fig. 8) and other, dark-colored laminae rich in organic matter also are enriched in Fe and Mn. Some laminae contain concentrations of phosphorous in the form of diagenetic vivianite. Thin, light-colored laminae appear in thin section as layers of fine-grained, yellow-tan carbonate, generally with a well-defined base, changing upward to whitish, even finer-grained sediment with a poorly defined upper boundary. Geochemical and mineralogical analyses suggest that this material is dolomitic (Dean, Chapter 10). Some light-colored laminae are coated by dark-colored rinds of Mn and Fe oxyhydroxides that shrink and pull apart upon drying, fragmenting the laminae.

Clay and silt was introduced to the deepest part of the lake as early as 9.5 ka, but the first significant episode of clastic influx did not occur until 8.3 ka. The thickness of varves increased to more than 1 cm during this episode, which lasted for ~500 yr and is interpreted as a period of dryness (see discussion of prairie lake stage). About 9.0 ka, and before the onset of dry conditions, varves lose their reddish brown color in thin section, even though chemical analyses indicate that concentrations of Mn and Fe are high. Diatoms are not so abundant in varves formed before 9.0 ka that they form discrete laminae. Later in the post-glacial lake stage, seasonal accumulations of diatoms are visible as discrete layers and varves consist of well-defined laminae of calcium carbonate that alternate with thick, relatively pure blooms of *Stephanodiscus*. Within this time interval, varves that contain few clastics may have diatomaceous laminae with sharp lower contacts (Fig. 9).

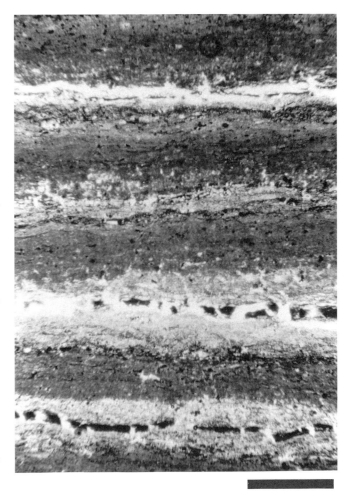

Figure 6. Thin section of core slab in Figure 4. Fragmented dark layers, enriched in Fe and Mn, set off by light-colored diatom-rich laminae, overlain by carbonate layers and segregations of dark organic matter. Bar scale = 1 mm.

Prairie Lake (~8.0 to 3.8 ka)

Bathymetry, environs, and processes of transport. By ~8 ka, slightly less than 4 m of varved sediment had accumulated in the deep basin, leaving maximum water depth at about 46 m (Fig. 2). Drawdown during the following dry prairie lake stage was about 6 m (Stark, 1976). Assuming that the rest of the lake also was filling slowly, water depth surrounding the deep hole during the prairie lake stage was probably never less than 30 m. By the end of the prairie lake stage, at about 3.8 ka, 9 m of additional sediment rich in clay and silt had accumulated in the deepest part of the lake and the lake bottom approached its present configuration (Fig. 2).

Much of the overland flow in the Elk Lake catchment is drained into small depressions and swampy areas. Gentle slopes with gradients of less than 20 m/km lie to the south, southwest, and to the northeast of the lake and rise about 30 m above the lake surface (Fig. 2). During the prairie lake stage, these slopes

Figure 7. SEM photograph of material in fine-grained, light-colored lamina showing subrhombic calcite grains and aggregated material enriched in Ca. EDS relative peaks = Ca 5484 and 535; Mn 265; and Si 263. Age is ~10.14 ka.

Figure 8. SEM photograph of material in mixed, diatom-Mn lamina. EDS relative peaks = Ca 1572 and 152; Mn 1018 and 272; Si 292; S 160. Age is ~10.38 ka.

were occupied by sagebrush, grass, and herbs (Whitlock and others, Chapter 17), and this vegetative association is accompanied by open areas and some bare ground. Overland flow from the low slopes, with an effective runoff area of ~10 km^2, must have carried some sediment to the lake. However, with no large streams, it is unlikely that the volume of clay and silt that reached the deepest part of the basin during the prairie lake stage was transported by surface flows or underflows in the lake. Transport within the lake of the fine to coarse silt that is significant fraction of the clastic component is unlikely because depth of water over most of the lake was greater than ~30 m, and far in excess of wave base, which for Elk Lake would have been less than 3 m for a wind of 40 knots (Sly, 1978).

A smoothed profile of the thickness of Elk Lake varves (Fig. 10b) shows multiple oscillations in thick varves that reflect pulse-like incursions of clay and silt (Fig. 10a) during the prairie lake stage. Silt in the size range of 20 μm to 40 μm, is abundant in thick varves, with some clastic grains of sand size. The Elk Lake varve-thickness profile resembles closely a profile for magnetic susceptibility obtained from Lake Ann (Keen and Shane, 1990), which is to the southeast of Elk Lake and within the Anoka sand plain. Magnetic susceptibility in sediments from Lake Ann measures the proportion of eolian sand that was carried into the lake from nearby dunes, and is unquestionably a record of eolian activity. In Lake Ann, within the resolution of its chronostratigraphy, onset and termination of clastic influx occurred at the same time as in Elk Lake. Both lakes contain an episode, centered on 5.0 ka, when the supply of clastic material was interrupted. These similarities, especially the interruption in the influx of clastic material, suggest that the regional surface winds that carried sand into Lake Ann also brought clay and silt to Elk Lake.

Grigal and others (1976) described strong, eolian and dune activity in north-central Minnesota between 5.0 and 8.0 ka. Because eolian sediment transport began and ended at approximately the same time over a wide area of Minnesota, a reasonable inference is that clastic sediment added to Elk Lake varves during this same time interval is the fine-grained eolian fraction (loess) carried by regional winds. Elk Lake lies along the eastern margin of a major region of dust generation (Changery, 1983), and today, in winter, layers of dust are observed to accumulate on the ice of Elk Lake after periods of storminess. Suspension of dust in modern times is related to agriculture, but in the dry mid-Holocene this region had sparse vegetation (Grigal and others, 1976). Threshold wind velocity for entrainment of dust from a

Figure 9. Core slab in transition into prairie lake stage. In upper part, note thick white laminae with sharper lower contact produced by diatom blooms alternating with thicker layers also enriched in diatoms, organic matter, and clastic material. Age is ~8.25 to 8.28 ka. Bar scale = 1 cm.

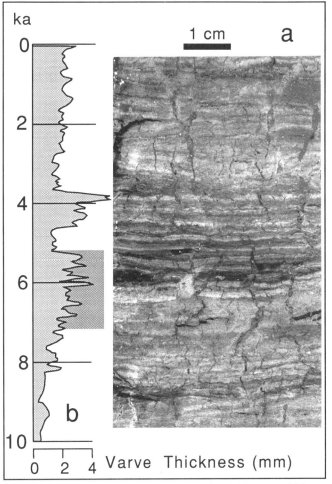

Figure 10. a: Varved sediment from zone of variable varve thickness (shaded area in b). Lightest shade laminae are carbonate and diatoms. Thicker, medium-shade laminae are largely clay and silt of eolian origin. Note that ~10 thinner couplets occur between zones of thicker, clay-rich layers. Cycle period is ~20–25 yr. Bar scale = 1 cm. b: Filtered time series for varve thickness. Note that ~200 yr oscillations occur in the shaded time window when varve thickness correlates with changes in atmospheric (tree ring) ^{14}C.

partly vegetated surface is comparable to threshold velocity in areas of agriculture (Brazel, 1989), and in the mid-Holocene, fine-grained eolian material probably was carried eastward by strong winds to Elk Lake and its drainage.

Several features suggest that transport of clastics also occurred as a result of circulation in the lake. For example, clastic layers containing mixed diatom assemblages and broken diatoms indicate that some of the sediment that reached the deep hole was resuspended from a shallower depth. Pure blooms of diatoms occur within thick, clastic laminae that also contain diatoms (Fig. 11), suggesting that the diatom frustules accumulated rapidly within a prolonged interval when clay and silt was suspended in the lake. Sediment-trap investigations (Nuhfer and others, Chapter 7) show that episodes of circulation during seasonal overturn may last for several months. Even though the lake bottom was well below wave base for orbital wave motion, weaker currents within the lake during prolonged episodes of seasonal circulation probably redistributed clay-size material and some material of silt size, adding an undetermined volume of clastics to sediment that was accumulating in the deepest part of the lake.

Varves in the prairie lake. Seven distinctive types of laminae are recognizable in the varved sediments that accumulated during the prairie lake stage (Table 2). In addition, rare blooms of individual diatoms such as *Rhizosolenia* occur as brittle or waxy-appearing layers, 2–6 mm thick (only about six such layers occur in the core, and they are confined to the prairie lake stage). The most informative components, in terms of changes in climate and vegetation, are contained in laminae that were unique to the

Figure 11. Thin section of thick laminae from prairie lake stage. Alternation of concentrated diatom and organic enriched laminae. Note thick, impure layer of diatoms mixed with clastics beneath dark organic lamina. Age is ~5.69 ka. Bar scale = 1 mm.

Figure 12. SEM photograph of material in diatomaceous lamina below carbonate-rich lamina in moderately thick varves from the prairie lake stage. Note fragmented character of some diatoms. EDS relative peaks = Ca 2193 and 255; Si 646; Fe 79. Age is ~5.64 ka.

mid-Holocene, including clay-silt laminae, impure diatomaceous laminae, and laminae composed of calcite and aragonite.

The average thickness of varves increased from less than 1 mm before the onset of dry climate to nearly 4 mm during full development of the prairie lake stage. Increased thickness is the result of blooms of diatoms and/or increased influx of fine, detrital clastic material. Clastic sediments are recognizable as relatively pure layers of clay and fine to coarse silt, and as impure mixtures of diatoms and clastics. Diatoms of various types are mixed together in the thick, gray layers that contain a clay matrix. Diatoms in some of the thicker layers appear to be segregated and graded upward, but most layers have a uniform distribution in size of frustules. Some diatoms in the thick laminae are fragmented (Fig. 12). Increased clay content in the thick, impure diatomaceous layers makes these layers partly opaque in thin section. Diatoms usually do not form separate laminae in materials with high clastic content and the thick, medium-gray, clastic-rich layers often are set off only by thin, dark, organic-rich laminae. Annual groups of thick, clastic laminae often lack clear definition and are sometimes difficult to recognize in thin sections and core slabs.

Thin, light-colored laminae of fine-grained calcium carbonate generally can be observed between diatomaceous laminae, and they clearly define the annual cycle. In addition, some clay-rich laminae and concentrated diatom laminae are set off by tan-colored laminae that stand out sharply as thin layers in the gray background of the slabbed core (Fig. 13). SEM-EDS examination reveals that the tan layers consist mainly of small rods of aragonite (Fig. 14). Thin sections show that diatom-rich layers may be separated both by calcite and aragonite laminae (Fig. 15). Precipitation of aragonite is favored both by increased salinity and higher water temperature (Dean and Megard, Chapter 8), and aragonite laminae only occur between 7.0 and 6.3 ka (Dean, Chapter 10), at a time when salinity in Elk Lake was highest (Forester and others, 1987). Rarely, however, does the seasonal production of pure aragonite laminae continue for more than a few years without interruption, suggesting that formation of aragonite remained close to a threshold.

Figure 13. Slab of core from early in prairie lake stage showing thick, diatomaceous and clastic-enriched varves interrupted intermittently by thin, light-colored aragonite laminae. Age is ~7.49 ka. Bar scale = 1 cm.

Modern lake (Present to 3.8 ka)

Bathymetry and environs. When humid conditions and forest returned to Elk Lake, depth of water in the deepest part of the lake was about 37 m (Fig. 2). Water depth is now 29.5 m and about 8 m of partly compacted iron-rich sediment has accumulated in the past ~3800 yr. Today, the margin of the deep hole that preserves varves lies about 15 m beneath the lake surface. This depth was about 25 m, with an average depth for the lake of about 20 m at the beginning of the modern lake stage.

Retreat of the prairie to the west and return of the forest effectively shut off the supply of clastic sediment to the lake and reduced the supply of nutrients. The combination of mesic forest,

Figure 14. SEM photograph of material in aragonite laminae shown in Figure 3. Note rod-like shape of aragonite crystals. EDS relative peaks = Ca 2370 and 245; Si 632.

more deeply weathered till, plus increased precipitation, as indicated by changes in vegetation (Bartlein and Whitlock, Chapter 18), increased the importance of ground-water seepage as a source of dissolved chemical species and altered the chemical balance in the lake. Calcium continued to enter the lake through ground water, to be precipitated as low-magnesium calcite. The concentration of dissolved iron progressively increased in the lake and accumulated in the sediment. Sediments deposited at the end of the prairie lake stage contain 2%–3% Fe, whereas sediments deposited during the past 2000 yr contain an average of 15% Fe. Concentrations of Mn and P also increased, but to a lesser extent. The main residences for Fe and Mn are amorphous Fe and Mn oxyhydroxides, Fe and/or Mn phosphate minerals, Mn carbonate (rhodochrosite), and minor amounts of pyrite (Dean, Chapter 10). Materials related to intense redox cycling of iron and manganese (Nuhfer and others, Chapter 7), together with organic matter and biogenic silica as diatoms, compose more than 98% of the modern lake sediment. These compositional changes, as compared to the prairie lake and post-glacial lake stages, are reflected in the complex, annual groups of laminae that accumulated in the modern lake.

Varves in the modern lake. Core slabs from the upper 8 m of the varve sequence contain ferric hydroxide laminae, dark-

Figure 15. Thin section of core slab in Figure 13. Thick, diatom-enriched laminae are set off by organic segregations and fine-grained calcium carbonate. Note medium-shade lamina of pure aragonite. Bar scale = 1 cm.

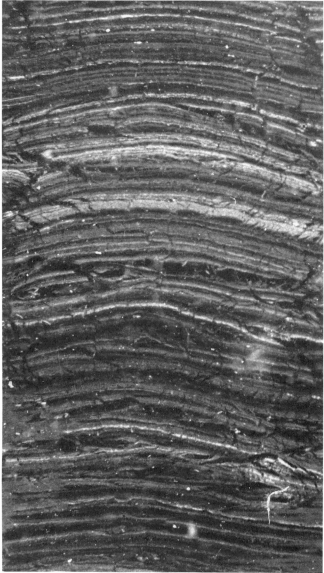

Figure 16. Slab of core from modern lake stage. Dark shades are organic-rich and iron-rich laminae. Light shades are diatom-rich and carbonate-rich laminae. Age is ~2.5 ka. Bar scale = 1 cm.

colored organic laminae, light-colored concentrations of diatoms, and light-colored layers of carbonate (Tables 1 and 2; Fig. 16). These components have been illustrated previously for other parts of the varved sequence and can be characterized further in modern lake sediments as the following.

Ferric hydroxide laminae. Laminae enriched in iron appear dark brown to reddish-brown in slabs and thin sections, and commonly are gel-like in appearance. The bulk of the material in these laminae appears to be a mixture of the remains of zooplankton, diatoms, pollen, algal and plant debris, and aggregated masses and grains containing Mn, Ca, P, and S, as determined by EDS analysis. However, Fe is usually the main component in EDS analyses and oxyhydroxides of iron probably are responsible for the brown to reddish-brown color.

Organic laminae. Organic laminae are black to dark gray in core slabs, amber in thin section, and contain diatoms and concentrations of brown amorphous organic matter, often organized into thin, discontinuous layers. Black to dark gray organic laminae are persistent in all phases of lake development.

Concentrated diatom laminae. The color of diatom layers in slabs is generally light gray to tan to white, depending on the admixture of organic material and carbonate. Diatom frustules are sometimes abundant within associated ferric hydroxide, organic, and carbonate laminae, but are not concentrated into discrete layers. In most diatom laminae, several diatom species are represented, although diatoms often occur as blooms and concentrations of single species, especially species of *Stephanodiscus*. Contacts between laminae produced by diatom blooms and adjacent laminae that have lower concentrations of diatoms are moderately to poorly defined.

Carbonate laminae. Aggregates and masses in laminae that EDS analyses indicate contain only Ca, generally show subcrystalline development, with occasional grains approaching a rhombic form. In addition, Ca is incorporated into aggregates of various shapes, along with admixtures of Fe and Mn. X-ray diffraction analyses indicate that low-magnesium calcite is the dominant carbonate mineral deposited during the past 4000 yr. Carbonate laminae also contain calcispheres and the algae *Phacotus*. Ostracode carapaces are rare in the deep part of the lake. Laminae with high concentrations of calcium carbonate are tan to white and have well-defined margins, often with a sharper lower contact.

Although only four of the ten types of laminations found in Elk Lake varved sediments occur in the modern lake stage (Table 2), three and sometimes four of these types are found together in annual groupings that are difficult to interpret as annual accumulations. In core slabs, reddish-brown, ferric hydroxide laminae alternate with dark gray, organic laminae and light-colored laminae, and this triplet is the most persistent indicator of the annual cycle.

EXPRESSION OF CLIMATIC VARIABILITY IN ELK LAKE VARVES

Resolution

Elk Lake and its sediments have been studied from the perspective of several disciplines, and the sampling intervals used for most of these investigations (50 yr or greater) is most effective for resolving the sequential changes that occurred in and around Elk Lake in response to changes in climate and weathering over several millennia. Because the thickness and composition of the varves have been affected profoundly by sequential changes in the three stages of the lake's history, climatic variability on the scale of one or two millennia is hard to isolate in the varve record. However, interannual to century-scale patterns of variability are expressed as changes in varve thickness and can be observed within each of the three stages of lake development. To the extent that the thickness of varves can be linked to processes that were responsive to climatic forcing, the varve thickness time-series provides insight into patterns of climatic variability that are rarely resolved in paleoclimatic records.

Climatic variability recorded in post-glacial lake

A smoothed plot of varve thickness for the interval from ~10.2 to ~7.7 ka reveals two ~500 yr episodes (~7.8 to ~8.3 ka), ~9.1 to ~9.6 ka) when thickness increased significantly (Fig. 17). Each zone of thick varves is separated by about 800 yr of relatively thin varves, with about 1400 yr between long-term maxima and minima. Geochemical analyses (Dean, Chapter 10) reveal that the thicker varves in both zones are the result of an increase in the rate of accumulation of a clastic-sediment fraction, and the long-term oscillation during the post-glacial lake stage can be attributed to changes in the quantity of clay and silt reaching the deepest part of the lake.

The older zone of thicker varves (~9.1 to ~9.6 ka) occurs after the shift to a pine and fern pollen assemblage (Whitlock and others, Chapter 17). Winter and summer may have been effectively drier and there may have been an increased incidence of fire. The weaker oscillations in varve thickness prior to ~8.3 ka, including the older zone of less variant varve thickness, have a period of about 80 to 90 yr (Fig. 17). Changes in thickness related to the 80 to 90 yr oscillation occur largely in thicker laminae that are enriched in diatoms, organic matter, and Mn, rather than in the thin laminae enriched in carbonate.

The younger zone of thick varves (~7.8 to ~8.3 ka) has a greater clastic content and higher variance, and the beginning of this zone is coincident with a significant percent increase in *Artemisia*, grass, and oak pollen (Whitlock and others, Chapter 17), signaling an increased flux of clastics that was related to a change to xeric vegetation. Oscillations in the thickness of varves within this younger zone are principally the result of changes in the rate of accumulation of clay and silt. Spectral density for varve thickness shows that the period for these oscillations is ~200 yr, with increased spectral density also at ~40 to 50 yr and ~22 yr (Fig. 17). Other increases in spectral density in this segment (not illustrated) occur at 12.5 yr and 5 yr. Periods of ~200, 40–50, and 22 yr reflect a pattern of oscillation that accompanies thick varves enriched in clay and silt, and that pattern finds even clearer and stronger expression in the prairie lake stage.

Climatic variability recorded in the prairie lake

The abundance of clay and silt and the thickness of laminations continued to increase after ~8.0 ka, reflecting the eastward advance of the prairie and increased eolian activity in north-central Minnesota (Grigal and others, 1976). The regional distribution of plants (Whitlock and others, Chapter 17) and abundance of ostracodes in the lake (Forester and others, 1987), indicate that the climate was cool and dry during the early part of the prairie lake stage. During this interval, clastic material reached the deep part of the lake in persistent pulses that lasted for decades, and the influx of clay and silt greatly increased the thickness and variance of the laminations. The amplitude of these oscillations intensified gradually until about 5.4 ka, when, inexplicably, the transport of clastic material was interrupted for ~600 yr, only to resume in equally strong pulses after 4.7 ka. A second and final interruption in the transport of clastic sediments occurred at about 3.8 ka, in a transition that took place in few centuries (Fig. 18), marking an end to the prairie lake stage and a return of moisture that continues to the present day.

The shutting off and turning on of clastic-sediment influx at 5.4 ka and 4.7 ka, and the final shut off at 3.8 ka, raises the question of a mechanism to account for such rapid responses.

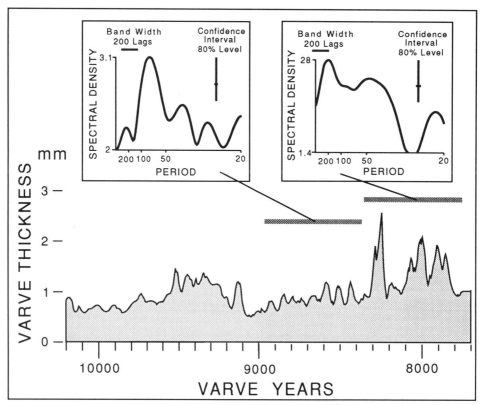

Figure 17. Time series of varve thickness in post-glacial lake stage (19 yr weighted filter). The 500 yr intervals of thicker varves are enriched in clastic material. Power spectra for 600 yr segments show thickness cycles with a period of ~80–90 yr shifting to a period of ~200 yr in thick varves enriched in clay and silt.

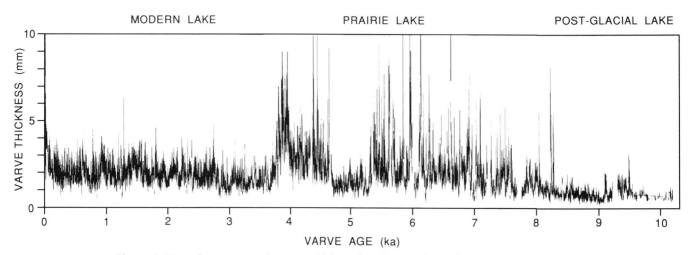

Figure 18. Plots of raw, unsmoothed varve thickness for Elk Lake time series. Variability is expressed chiefly at interannual to century time scale. Note abrupt changes in thickness at 3.8, 4.7, and 5.4 ka.

Abrupt transitions in sedimentation are characteristic of mechanical (shear) thresholds (Anderson, 1986) in which transport occurs after a critical value is exceeded, either by increasing energy to overcome friction, or by reducing shear strength. A mechanical threshold related to wind and soil moisture is implicated by eolian-sized material, and by evidence for drought and eolian activity in the region (Grigal and others, 1976; Keen and Shane, 1990). If such a threshold was the controlling mechanism, then long-term changes in regional winds could explain abrupt transitions through thresholds for dust suspension that are related to shear strength of exposed soils, and affected by drought. An inference drawn from the abrupt response in Elk Lake is that the oscillation in moisture was longer than the ~600 yr interval during which deposition of clastics was interrupted. Although a long-term change in the supply of moisture may have been more gradual, the shift to conditions that must have resembled those of the dust bowl era occurred in a few decades (Fig. 18).

The Elk Lake varve time series is most informative about variability for oscillations of less than a thousand years. In order to expedite visual examination of the temporal patterns of oscillation in varve thickness, an evolutionary spectrum was constructed for the past 9000 yr. The evolutionary spectrum (Fig. 19a) was constructed from values for power spectral density obtained from 600 yr segments of the time series, using a 300 yr overlap with adjacent segments. Spectral methods are those used by Anderson and Koopmans (1963) and Jenkins and Watts (1968), and calculations were done with a spectral program modified at Brown University. Relative changes in the variance of varve thickness, as depicted in Figures 10 and 18, appear in the evolutionary spectrum as a general increase in spectral density between 3.8 and ~8.5 ka that corresponds to an increase in the clastic sediment fraction during the prairie lake stage.

The structure of the evolutionary spectrum is most consistent when varves are thickest and most variable, between 3.8 and

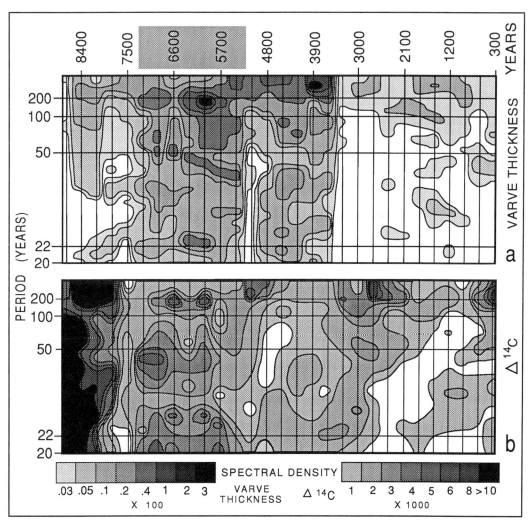

Figure 19. a: Evolutionary spectrum for the Elk Lake varve-thickness time series. b: Evolutionary spectrum for decadal values of $\Delta^{14}C$ in tree rings. Linear plots (200 yr lag) of overlapping 600 yr segments. Note high spectral densities at ~200, 40–50, and 22 yr expressed between 5.7 and 6.6 ka. Shaded area on time scale is time window when spectral densities are high at ~200 yr in both records.

7.3 ka. Within this time interval, increased power spectral density is expressed in three frequency bands, corresponding to periods of ~200, 40–50, and 22 yr. High spectral density in the frequency range of 40 to 50 yr appears to change systematically over time, with peak spectral density of ~50 yr at 6.9 ka shifting to ~40 yr at 5.4 ka and continuing to about 4.0 ka (Fig. 19a). Spectral peaks within the three frequency bands are significant at the 80% confidence level and are recognizable in individual spectra (e.g., Fig. 20). The three periods expressed in the segment between 5.3 and 6.3 ka (Fig. 20), and the three frequency bands that have higher spectral density between 3.8 and 7.3 ka, repeat, with greater clarity, the three-period spectral structure found in the segment from 7.8 to 8.4 ka, in which thick varves are associated with clastic sediment (cf. Figs. 17, 19a, and 20).

The three-period spectral structure is a feature associated with the influx of the clastic-sediment fraction. Because this influx occurred in strong pulses at a time scale that includes Maunder and other solar deviations, a similar evolutionary spectrum was constructed for the >9000 yr record of secular variation in ^{14}C, which reflects changes in the flux of cosmic particles and is a proxy for solar activity. Comparison of the two evolutionary spectra revealed a possible link between changes in varve thickness, ^{14}C (solar activity), and the strength of the geomagnetic field.

Varve thickness–solar–geomagnetic field connection: Evidence in varves from the prairie lake

Comparison of varve thickness with secular variation of ^{14}C. The evolutionary spectrum for $\Delta^{14}C$ (Fig. 19b) was constructed, using methods previously described, from a decadal time series derived from tree rings, detrended by a fifth-order polynomial, and provided by C. P. Sonett, University of Arizona.

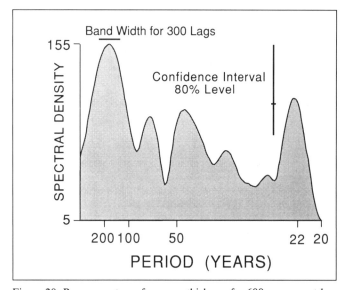

Figure 20. Power spectrum for varve thickness for 600 yr segment between 6.0 and 6.6 ka, showing periodicity at ~200 and 22 yr. See Figure 19a for position in evolutionary spectrum.

The evolutionary spectrum for ^{14}C, within the interval between 5.3 and 7.3 ka, resembles the evolutionary spectrum obtained from varve thickness, with increased spectral density at ~200 yr, between 40 and 50 yr, and between 20 and 25 yr (cf. a and b in Fig. 19). Only that part of the ^{14}C evolutionary spectrum between 3.8 and 8.0 ka is applicable to identifying associations with eolian activity because varve thickness is directly related to the clay and silt fraction only within this interval.

The three-band spectral structure is best developed in a relatively narrow, 2000 yr window of time between 5.3 and 7.3 ka. An overlap of the two evolutionary spectra for this time window (Fig. 21a) shows the congruence in time of high spectral density at a period of ~200 yr. This congruence suggested that a direct visual correlation and a statistical cross correlation between the two time series might be found between 5.3 and 7.3 ka. Plots of the two time series were aligned according to varve and tree-ring chronologies. The segment with the best visual correlation, from 6.3 to 7.3 ka, yielded the highest correlation coefficient at a lag of 50 yr. Realignment of the two time series to remove the lag yielded a visual correlation that is clearest between 6.3 and 7.3 ka (Fig. 22a). The alignment for the segment between 6.3 and 7.3 ka is supported by a cross correlation significant at the 95% confidence level (Fig. 22b). The long-term trend in varve thickness and $\Delta^{14}C$ (Fig. 22a) is part of a longer, 2100–2400 yr cycle in ^{14}C (Damon et al., 1989; Damon, 1988; Sonnet and Finney, 1990). Removal of this trend by first difference preserved ~200 yr oscillations and correlation remained above the 95% confidence level.

Accuracy of the varve thickness–^{14}C correlation. Correlation between varve thickness and secular changes in ^{14}C is less certain from 5.3 to 6.3 ka, even though the evolutionary spectrum (Fig. 19a) has the three-band spectral structure. Laminations are thickest in this zone, thick layers are more highly enriched in clay and silt, organic-rich laminae often are hard to identify, and annual groups are more poorly defined. This poorer definition of laminae may account for a weaker agreement in this part of the correlated segment. Gaps in the varve chronology may also be responsible for lack of agreement between the two records. Overlapping cores were not available for reconstructing the Elk Lake varve time series, and a disturbed zone of extrapolated varves occurs at core tops between each core section (Fig. 23). The correlation depicted for the second half of the 2000 yr segment represents the visual correlation obtained by removing 60 years at the top of a core near the middle of the segment. The weaker association between ^{14}C and varve thickness between 5.3 and 6.3 ka, for which the cross correlation has a lag of –170 yr, may be the result of a less-accurate chronology in zones where there were strong pulses in the influx of clay and silt. If so, collection of overlapping cores may improve correlation for the segment from 5.3 to 6.3 ka.

The tree-ring chronology is probably accurate to within a few decades (Stuiver and others, 1986). Although accuracy for the varve time series is much lower, radiocarbon age control (Anderson and others, Chapter 4) indicates that the varve chronol-

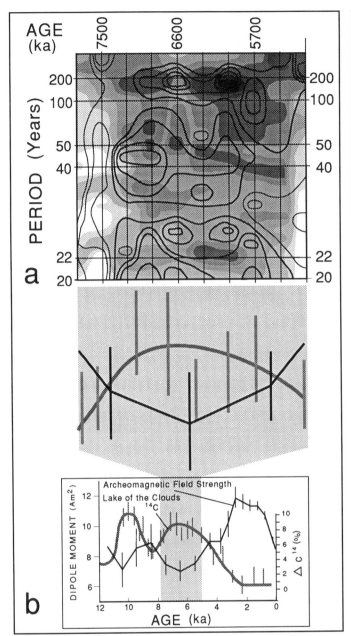

Figure 21. a: Evolutionary spectrum for varve thickness (Fig. 19a) overlain by spectrum for $\Delta^{14}C$ (solid lines, Fig. 19b). Note that high spectral density near 200 yr coincides in both records. b: Plot of magnetic dipole moment and $\Delta^{14}C$ in sediments from Lake of the Clouds. Note that high positive values of $\Delta^{14}C$ in Lake of the Clouds and ~200 yr periodicity in varve thickness (Elk Lake) and $\Delta^{14}C$ coincide with a weak dipole moment. Lake of the Clouds data from Stuiver (1970). Dipole data from McElhinny and Senanayake (1982). See Fig. 19 for scale of spectral density.

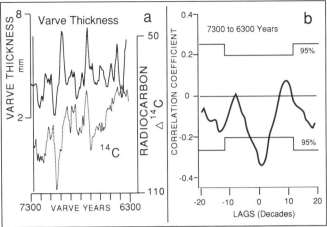

Figure 22. a: Time series between 6.3 and 7.3 ka for Elk Lake varve thickness (19 yr weighted filter) and decadal values for $\Delta^{14}C$ in tree rings. Radiocarbon time series is plotted inversely. b: Cross correlation for the above time series. r = –0.33 at 0 lag.

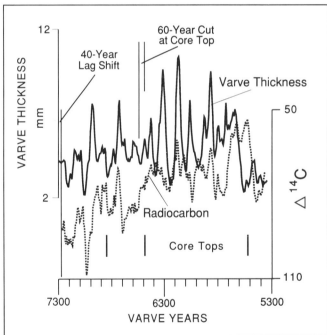

Figure 23. Time series for varve thickness and decadal values of $\Delta^{14}C$ (19 yr weighted filter) for segment from 5.3 to 7.3 ka showing position of core tops. Radiocarbon time series is plotted inversely.

ogy is probably accurate to within a few centuries. Hence, the position of the two time series relative to each other, as illustrated in Figures 22a and 23, cannot be far from true alignment; the available age control supports the visual agreement and the relatively high coefficient for cross correlation obtained for the segment from 6.3 to 7.3 ka. A significant cross correlation, however, does not necessarily mean that a real temporal correlation has been identified.

Even though a direct correlation between varve thickness and ^{14}C appears to be demonstrated, this does not mean that a high coefficient of correlation is needed to confirm an association between the two variables. Expression of the three-period spectral structure in both time series, and especially the restriction of this structure to the same relatively narrow window in time, also suggest a valid association. The ~200 yr period in this association is best developed in both varve thickness and $\Delta^{14}C$ near 6.0 ka, at a time when the >9000 yr record of secular variation of ^{14}C contains especially prominent deviations similar to the set of cycles developed during Wolf, Sporer, Maunder, and Dalton deviations (see Fig. 5 in Damon and others, 1989). Taken together, the expression of periodicity at the time scale of Maunder deviations in solar activity and the similarity between changes in varve thickness and a proxy for solar activity invite an examination of evidence and mechanisms that might account for these associations.

Associations between ^{14}C and varve thickness. Secular variation is the result of changes in the rate of production of ^{14}C, which is regulated largely by cosmic rays and secondary neutron flux. Changes in ^{14}C are in close agreement with changes in solar activity, as measured by sunspots and other solar indices, and the rate of production is greater when solar activity is relatively weak in ~200 yr and other solar cycles (Damon and others, 1989; Damon, 1988). Secular variation of ^{14}C is also affected by oceanographic and geologic processes, but isotopic variation attributed to these sources (Damon and Sternberg, 1989) takes place over a longer time scale than the interannual to century scale oscillations in the Elk Lake material. At the time scale of interest, spectral analysis of the long record of secular variation in ^{14}C, obtained from tree rings, has identified several periodicities that appear to be related to cycles of solar activity. Most prominent among these oscillations is a cycle with a period of ~200 yr that is related to Maudner-scale deviations in solar activity (Stuiver and others, 1991; Stuiver and Braziunus, 1989; Damon, 1988).

High spectral density at ~200 yr in the evolutionary spectrum for ^{14}C (Fig. 19b) is in agreement with these earlier results. The evolutionary spectrum for varve thickness also has its highest spectral density at a period of ~200 yr (Fig. 19a). The maximum in spectral density in both time series occurs at virtually the same time in both records, suggesting the possibility of a causal connection between changes in solar activity and varve thickness for solar cycles of ~200 yr (Anderson, 1992a). Because the association at a period of ~200 yr appears to be part of a coherent, three-period spectral structure, expressed in both evolutionary spectra, periodicity at 40–50 yr and 20–25 yr may also be part of a solar–varve-thickness association. The negative association between varve thickness and ^{14}C shown by the cross correlation requires confirmation by additional investigation. If confirmed, it will mean that thick varves are associated with lower rates of production of ^{14}C and with stronger solar activity.

As discussed in the section on varves in the prairie lake stage, changes in varve thickness during the time window from 5.3 to 7.3 ka, when ^{14}C and varve thickness appear to be associated, are the result of changes in rates of influx of clay and silt. Because these materials probably are eolian, a link to solar activity through wind fields in the troposphere may be indicated. This possibility is reinforced by evidence that solar-geomagnetic processes, in addition to affecting the rate of production of ^{14}C, also affects the strength of zonal winds.

Solar-geomagnetic events and surface winds at Elk Lake. Solar flares and geomagnetic disturbances are accompanied by changes in cosmic ray flux. McDonald and Roberts (1960) examined changes in cyclonic activity after solar-geomagnetic events and found that strong events in winter were followed, several days later, by changes in the depth of subpolar lows and in associated cyclones over the north Pacific. Stolov and Shapiro (1974) observed that the response included changes of as much as 7% in zonal winds at 700 mb and that these changes occurred over North America between lat 40° and 60°N (Fig. 1b). Subsequently, statistical examination of meteorological data following several hundred similar disturbances, over more than 30 years, has confirmed the atmospheric response (Tinsley and Deen, 1991). Tinsley and Deen (1991) have also shown that the strength of regional winds over the Atlantic, at a similar latitude, change in association with the 11 yr cycle of solar activity, with weaker winds during maxima of solar activity. Elk Lake (lat 47°13'N, long 95°W) lies within the region between lat 40° and 60°N (Fig. 1) where short-term solar-geomagnetic events have been followed by a wind-field response, and was in a favorable geographic location for recording changes in winds that might be related to these observations.

Strong solar-geomagnetic events can be expected to accompany maxima in solar activity at all frequencies of change, including ~200 yr Maunder-type deviations. Hence, there is reason to expect that the cumulative effects of strong events would be expressed as long-term changes in zonal winds. A general circulation model (GCM) for the mid-Holocene (Kutzbach, 1987) places the core of the polar jet to the north of Elk Lake in the summer and to the south in the winter (Fig. 1b). Seasonally strong winds associated with the core of the jet must have moved back and forth across Elk Lake and the dry prairie and bare ground that lay to the west, providing a means for transporting eolian materials. Given evidence for a short-term effect of geomagnetic events on zonal winds, the association found at Elk Lake between varve thickness (regional surface winds) and ^{14}C supports a case for a long-term wind-field response to changes in solar activity and the geomagnetic field.

The relation between varve thickness and ^{14}C, after the sense of the association is confirmed, may be a source of information about the response of the tropospheric wind field to long-term changes in solar activity. For example, transport of dust and the number and magnitude of dust events are closely related to antecedent precipitation (drought) (Brazel, 1989), and reduced moisture appears to be accompanied by weaker winds. In the dust-bowl era, the velocity of the wind field was generally reduced in the continental interior (J. O. Fletcher, 1991, personal commun.) and the GCM for 6.0 ka (Kutzbach, 1987) also shows a weakening of surface winds at the same time that dunes were active during the dry mid-Holocene in northern Minnesota (Grigal and others, 1976). If, as suggested by these observations, drought resulted in thicker varves, then thicker varves are associated with weaker surface winds. Because the association between ^{14}C and varve thickness observed at Elk Lake appears to be a negative one, a reduction in cosmic particle flux (reduced rate of ^{14}C production and stronger solar activity) would be accompanied by a weaker wind field and century-scale episodes of drought.

A wind-field response to changes in solar activity is suggested by the evidence cited. However, a causal pathway between solar activity and changes in regional winds, as recorded in meteorologic data, remains to be demonstrated. Although a response by zonal winds may be indicated, it is not yet clear how tropospheric winds above and at the Earth's surface are affected by changes in cosmic particle flux. A mechanism related to ionization in the atmosphere by MeV-GeV cosmic particle flux (Tinsley and Dean, 1991) is consistent with the associations found at Elk Lake, but other phenomena also may explain or contribute to associations between changes in cosmic ray flux (solar activity) and winds (Dickinson, 1975).

Associations with the geomagnetic field. The associations observed for cycles of ~200 yr and their strongest expression in the time window between 5.3 and 7.3 ka have implications for both longer and shorter oscillations in solar activity and the geomagnetic field. For example, the time window centered around 6.0 ka corresponds to a prolonged interval when archeomagnetic data show that the geomagnetic dipole moment reached its lowest value in the past 10,000 yr (Fig. 21b) (Merrill and McElhinny, 1983; McElhinny and Senanayake, 1982). Archeomagnetic evidence for a weak dipole moment during the 2000 yr time window is supported by a high rate of production of ^{14}C, as recorded in Lake of the Clouds, Minnesota (Stuiver, 1970) (Fig. 21b), because higher production of ^{14}C at this time scale reflects an increase in cosmic ray flux that can be attributed to a weakening of the dipole moment (Damon, 1988). The inference drawn from ~200 yr oscillations in ^{14}C, during a time interval with a low dipole moment, is that cosmic rays that entered the atmosphere at middle latitudes were more susceptible to modulation by Maunder-scale deviations in solar activity when the Earth's magnetic field was weak. A related inference is that ~200 yr oscillations in tropospheric winds were also stronger when the Earth had a weak magnetic field (Anderson, 1992a).

The varve and tree-ring ^{14}C records are ~10,000 yr long. Within this interval there was a 4000 yr window of time in the mid-Holocene when the thickness of varves reflected changes in wind activity and when strong ~200 yr oscillations could have been recorded (before and after this 4000 yr interval, changes in varve thickness are the result of changes in nutrient supply, diatom abundance, water chemistry, and lake circulation [Dean and others, 1984] and do not reflect a direct response to changes in surface winds). Because ~200 yr oscillations are expressed during only part of the available 4000 yr interval, their strongest expression during the minimum in dipole moment strengthens the inference that changes in solar activity had a greater effect on the wind field when the Earth's magnetic field was weak.

Solar-Climate Connection? Congruent changes in ^{14}C and varve thickness, as documented at Elk Lake, may reflect a link between solar activity the geomagnetic field, and changes in climate, the connection being the most firm when the Earth has a weak magnetic field. Another explanation is that these associations are a result of coincidence. The following brief summary of evidence supports a search for causal relations.

An ~200 yr oscillation appears in evolutionary spectra of both ^{14}C and varve thickness, and an ~200 yr period is the strongest one expressed in both spectra. A period of ~200 yr also is the strongest one found by other investigators in the >9000 yr record of secular variation in ^{14}C (Stuiver and others, 1991; Damon and others, 1989; Damon, 1988), suggesting that the evolutionary spectra are reasonable representations of patterns of variability. Another example of a long-term association of ^{14}C with climatic variables has been documented in studies of the width of annual rings in bristlecone pine (Sonnet and Suess, 1984; Suess and Linick, 1990; Thomson, 1990). In that example the association is for cycles with a period of ~200 yr. The approximate period of Maunder and Sporer deviations in sunspot number and solar activity is ~200 yr and historical and instrumental evidence for an association between solar activity and regional and global temperature, and other climatic events, is for these deviations (Eddy, 1976). In the few ancient varved deposits of sufficient length to measure long-term variability, and in a geologic (evaporite) setting believed to be highly responsive to climatic forcing, the strongest period expressed is ~200 yr (Anderson, 1992b). Collectively, these examples document an ~200 yr cycle in solar activity. Where there is a well-documented association of ^{14}C with climatic variables, or with proxies for climatic variables, the period expressed is also ~200 yr.

The 22 yr period seen in varve thickness is strongest during ~200 yr oscillations (Fig. 19a), and is associated with drought in the region around Elk Lake. A 22 yr cycle has been reported for historical episodes of drought in the continental interior and the correlation of drought with the Hale solar cycle is one of the few documented and accepted statistical associations with solar activity (Mitchell, 1979). A response to drought in cycles of ~200 yr and 22 yr, periods that are related to the Maunder and Hale solar cycles, reinforces the association between ^{14}C and varve thickness, as expressed in the three-period spectral structure.

In summary, Elk Lake, located at a middle latitude and in a region of drought, was in a setting where changes in tropospheric winds, if affected by solar activity, could be recorded. The record recovered from Elk Lake shows that changes in varve thickness (tropospheric winds) are aligned with changes in ^{14}C (cosmic particle flux), and most clearly expressed, within a relatively narrow window of time in the mid-Holocene when the geomagnetic field was weak. If one accepts evidence that solar-geomagnetic events lead to changes in both cosmic particle flux (Damon, 1988) and tropospheric winds (Tinsley and Deen, 1991), then the associations found at Elk Lake are reasonable.

Climatic variability recorded in the modern lake

Controls on varve thickness. Variability in the thickness of varves that accumulated in the modern, iron-enriched lake originates with major components that contributed to varve thickness in four types of laminae (ferric, organic, concentrated diatom, and carbonate). Because no one of these components was a dominant contributor, characterizing and interpreting variability depends upon establishing associations between varve thickness and geochemical and biologic parameters. The analyses include %FE, %organic matter, %loss on ignition (LOI at 525 °C), %CaCO$_3$, and diatom species abundance. The geochemical and biological sampling interval is generally 50 yr per sample; analytical procedures are described in Dean (Chapter 10).

Aluminum, an element of the minor clastic fraction, shows no relation to varve thickness (Fig. 24). Calcium carbonate, an important component by volume, is negatively associated with % organic matter, and negatively and weakly associated with varve thickness (Fig. 24). CaCO$_3$ ranges from about 25 to 50 wt% and generally occurs as thin laminations that appear to have little contribution to overall changes in varve thickness, a relation that is confirmed in a time-series plot of geochemical parameters (Fig. 25). Together, organic matter and iron compose by weight about 10% to 20% of sediment samples, and most of the remainder, other than carbonate, is biogenic silica. Biogenic silica is an important constituent by volume, but the most abundant diatom species show little relation to varve thickness.

Ferric hydroxide and organic matter can be considered as a single component contributing to changes in varve thickness in the modern lake, even though laminae in which these materials are concentrated often can be identified separately. Both organic matter and iron are positively, but not strongly, associated with varve thickness (Fig. 24). Laminae of ferric hydroxide and organic matter contain flocculated material and appear to be less dense than carbonate laminae. Although iron and organic matter constitute a lower percent by weight than carbonate, they make an important contribution to changes in varve thickness. Except for the past few centuries, the association between Fe and varve thickness is positive, with strong, somewhat regular fluctuations in %Fe and less-regular changes in varve thickness (Fig. 25).

Sediment-trap investigations (Nuhfer and others, Chapter 7) indicate that settling of materials containing ferric hydroxide and organic matter is greatest during seasonal circulation, especially in the fall and early winter, but also in spring. Sediments collected in the trap in the deep hole also include some fraction of resuspended materials from outside the deep hole. Periods of circulation and resuspension may last for several months, and these processes contribute to the formation and thickness of laminae enriched in ferric hydroxide and organic matter. Hence, the positive association between varve thickness and the ferric-organic

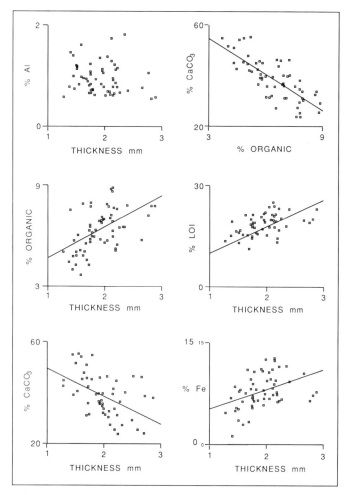

Figure 24. Scatter diagrams for components contributing to varve thickness in modern lake stage. Note that Fe and organic matter are positively associated with varve thickness.

component, although not strong, suggest that thicker varves reflect greater accumulation of ferric hydroxide during stronger circulation in the lake. Seasonal circulation was the result of several factors related to climate, including strength of regional winds, rate of spring warming, and the duration of circulation (Nuhfer and others, Chapter 7).

Expression of climatic variability. The evolutionary spectrum for varve thickness (Fig. 19a), in addition to illustrating the temporal pattern of variability in the clastic sediment fraction before 4.8 ka, also shows changes in variability during the modern lake stage. The shutting off of the supply of clastics at 3.8 ka is reflected in reduced variance in varve thickness and lower spectral density. For six to eight centuries thereafter, there appears to have been a continuation of high spectral density at 70 to 90 yr that was established at about 4.5 ka. This pattern was followed by a prolonged episode of stronger cyclicity and increased spectral density between ~1.4 and ~2.5 ka, with higher spectral density from 100–200 yr, shifting to ~90 yr. These changes also can be observed in a smoothed plot of the time series (Fig. 26). From

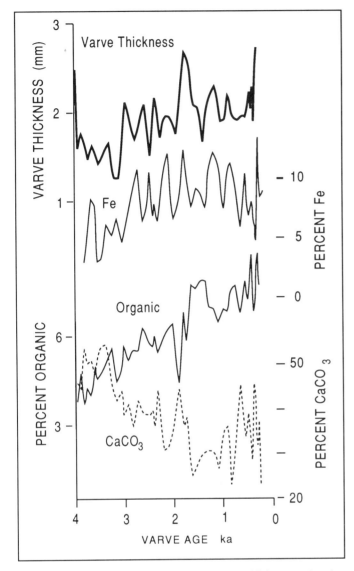

Figure 25. Comparison of time series for varve thickness and major components contributing to thickness. ~50 yr sample intervals.

~1.0 to ~1.8 ka, spectral density was stronger at a period of ~35 yr.

Changes in periodicity over time appear to coincide with major changes in distribution of vegetation, as interpreted from pollen data (Whitlock and others, Chapter 17; Bartlein and Whitlock Chapter 18). For example, irregular cycles in varve thickness, accompanied by weak spectral density, were recorded after termination of the prairie lake stage, from about 3.8 to 3.0 ka, and during the relatively dry interval when oak and other hardwoods replaced herbs and grass. This vegetation response was followed by migration of *Pinus strobus,* and an assemblage that reflects a cooler and moister climate. The wetter climate between ~1.4 and 2.5 ka (Fig. 26) was associated with stronger 100 yr cycles in varve thickness (Figs. 19a and 26) and, presumably, with similar oscillations in winds, storms, and lake circulation.

The 40–50 year period that was part of the three-period spectral structure during the prairie lake stage, returns at about the time of the Little Ice Age (Figs. 19a and 27). The transition from strong, regular cycles of ~100 yr to 40–50 yr occurs at about the end of the Medieval Warm Epoch, when several segments of the time series have thicker varves (Fig. 27). The first segment with thick varves in the Medieval Warm Epoch contains a distinctive, 1-cm-thick clay layer that is found in all cores recovered from the lake. After about A.D. 1800, thickness of varves increased markedly, owing to lack of compaction. Although cycles in varve thickness during the past 1000 yr suggest the potential for an informative paleoclimatic record, subdecadal sampling and analysis of major components such as iron, aluminum, biogenic silica, and diatom species abundance, will be required before this information can be extracted and interpreted.

CONCLUSIONS

On a time scale of several millennia, the varves in Elk Lake are most useful for characterizing progressive stages in the development of the lake and its drainage. On a shorter time scale, the varves provide a higher resolution than available from radiocarbon dating alone, and time series for varve thickness define patterns of decadal to centennial climatic variability. One such pattern, developed during the dry mid-Holocene, may reflect changes in regional surface winds. These cycles have periods of ~200, 40–50, and 22 yr, and two of these periods are recognized in historical solar activity. All three periods expressed in the varve record are found in a proxy for solar activity (^{14}C). Most important, the three-period spectral structure that is common to both records occurred during the same relatively narrow window of time in the mid-Holocene. This 2000 yr window corresponds to the time interval when the geomagnetic dipole moment reached its lowest value in the Holocene.

There is other, independent evidence that zonal winds in the troposphere are affected by geomagnetic events linked to solar activity. By accepting this evidence, the spectral associations between varve thickness and ^{14}C become associations between solar activity and the tropospheric wind field. This association is strongest when the Earth has a weak magnetic field. In the context of Elk Lake, thicker varves appear to go with weaker winds, stronger solar activity, and conditions of drought. If so, the record at Elk Lake reveals century-scale episodes of drought in the dry mid-Holocene.

In spite of the apparent detail of investigations at Elk Lake, the reconstruction, sampling, and analysis of Elk Lake varved sediments are incomplete. A reconstruction based on several correlative and overlapping cores will be required to remove extrapolated zones from the time series and refine initial varve counts. For sediments accumulated in the modern lake stage, the sampling interval must be reduced from 50 to ~5 yr in order to examine variability of components in independent time series that reflect changes in climate. A comprehensive program to observe climatic variables, circulation in the lake, and processes that

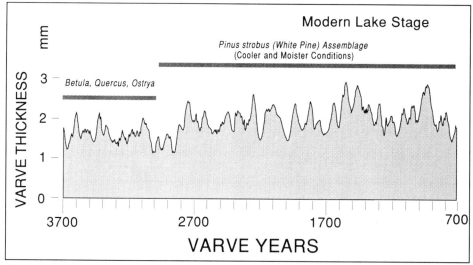

Figure 26. Smoothed time series for varve thickness in modern lake stage (19 yr weighted filter). Note ~100 yr cycles in varve thickness and increased regularity between ~2.7 and 1.5 ka (see Fig. 19a).

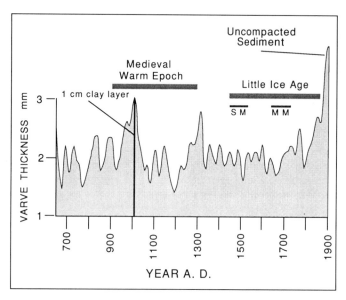

Figure 27. Smoothed time series for the past 1000+ yr (19 yr weighted filter). Note weak but regular cycles of ~40–50 yr during Little Ice Age (see Fig. 19a). SM is sporer minimum; MM is Maunder Minimum.

generate varves will have to be maintained for several years in order to interpret the paleoclimatic record of the last few thousand years.

Limitations demonstrated by this investigation are the time and experience required to assemble and interpret such records. Clearly, information recovered must be worth so much effort. Even with an incomplete chronostratigraphy, with most sampling done at a low resolution, and with minimal obervation of limnology, the results from this investigation show promise, and they reveal the kinds of information to be expected from more detailed studies. For example, results from the time series for varve thickness show that it is possible to reconstruct interannual to century-scale patterns of climatic variability, and that these patterns, measured in a single lake, are regional in extent. Furthermore, if regional winds are indeed linked to solar activity, then varved records preserved at other sites will be sources of information about local and regional responses to physical and atmospheric phenomena that have an element of regularity. I hope that the picture of climatic variability that has emerged from Elk Lake will teach patience for research that requires years of effort before fruition.

ACKNOWLEDGMENTS

I thank Walter E. Dean and J. Platt Bradbury, U.S. Geological Survey, Denver, Colorado, for assistance in collecting and compiling Elk Lake varve data. I thank Charles P. Sonett, University of Arizona, Tucson, for ^{14}C data and the following for comments on the manuscript: P. E. Damon, University of Arizona, Tucson; Joseph O. Fletcher, NOAA-ERL, Boulder, Colorado; Donald C. Herzog, U.S. Geological Survey, Denver; Douglas W. Kirkland, Mobil Research, Dallas; Minze Stuiver, University of Washington, Seattle; and Brian A. Tinsley, University of Texas, Dallas. This work was completed under a National Science Foundation Grant.

REFERENCES CITED

Anderson, R. Y., 1986, The varve microcosm: Propagator of cyclic bedding: Paleoceanography, v. 1, p. 373–382.
——, 1992a, Possible connection between surface winds, solar activity and the Earth's magnetic field: Nature, v. 358, p. 51–53.

——, 1992b, Solar variability captured in climatic and high-resolution paleoclimatic records: A geologic perspective, in Sonett, C. P., and Giampapa, M. S., eds., The sun in time: Tucson, University of Arizona Press, Space Science Series, p. 543–561.

Anderson, R. Y., and Dean, W. E., 1988, Lacustrine varve formation through time: Palaeogeography, Palaeoclimatology, Palaeoecology, v. 62, p. 215–235.

Anderson, R. Y., and Koopmans, L. H., 1963, Harmonic analysis of varve time series: Journal of Geophysical Research, v. 68, p. 877–893.

Brazel, A. J., 1989, Dust and climate in the American Southwest, in Leinen, M., and Sarnthein, M., eds., Paleoclimatology and paleometeorology: Modern and past patterns of global atmospheric transport: Hingham, Maine, Kluwer Academic Publishers, p. 65–96.

Changery, M. J., 1983, A dust climatology of the western United States: Asheville, North Carolina, National Oceanographic and Atmospheric Administration, National Climate Data Center, NUREG/CR-3211, RB, p. A1–B6.

Damon, P. E., 1988, Production and decay of radiocarbon and its modulation by geomagnetic field-solar activity changes with possible implications for global environment, in Stephenson, F. R., and Wolfendale, A. W., eds., Secular, solar, and geomagnetic variations in the last 10,000 years: Hingham Maine, Kluwer Academic Publishers, p. 267–285.

Damon, P. E., and Sternberg, R. E., 1989, Global production and decay of radiocarbon: Radiocarbon, v. 31 (3), 697–703.

Damon, P. E., Cheng, S., and Linick, T. W., 1989, Fine and hyperfine structure in the spectrum of secular variations in atmospheric ^{14}C: Radiocarbon, v. 31, p. 697–703.

Dean, W. E., Bradbury, J. P., Anderson, R. Y., and Barnosky, C. J., 1984, Variability of Holocene climatic change: Evidence from varved lake sediments: Science, v. 226, p. 1191–1194.

Dickinson, R. E., 1975, Solar variability and the lower atmosphere: American Meteorological Society Bulletin, v. 56, p. 1240–1248.

Eddy, J. A., 1976, The Maunder Minimum: Science, v. 192, p. 1189–1192.

Forester, R. M., Delorme, L. D., and Bradbury, J. P., 1987, Mid-Holocene climate in northern Minnesota: Quaternary Research, v. 28, p. 263–273.

Grigal, D. F., Severson, R. C., and Goltz, G. E., 1976, Evidence of eolian activity in north-central Minnesota 8,000 to 5000 yr ago: Geological Society of America Bulletin, v. 87, p. 1251–1254.

Jenkins, G. M., and Watts, D. G., 1968, Spectral analysis and its applications: San Francisco, Holden-Day, p. 171–208.

Keen, K. L., and Shane, L.C.K., 1990, A continuous record of Holocene eolian activity and vegetation change at Lake Ann, east-central Minnesota: Geological Society of America Bulletin, v. 102, p. 1646–1657.

Kutzbach, J. E., 1987, Model simulations of climatic patterns during the deglaciation of North America, in Ruddiman, W. F., and Wright, H. E., Jr., eds., North America and adjacent oceans during the last deglaciation: Geological Society of America, Decade of North American Geology, The Geology of North America, v. K3, p. 425–446.

McDonald, N. J., and Roberts, W. O., 1960, Further evidence of a solar corpuscular influence on large-scale circulation at 300 Mb: Journal of Geophysical Research, v. 65, p. 529–534.

McElhinny, M. W., and Senanayake, W. E., 1982, Variations in the geomagnetic dipole 1: The past 50,000 years: Journal of Geomagnetism and Geoelectricity, v. 34, p. 39–51.

Merrill, R. T., and McElhinny, M. W., 1983, The Earth's magnetic field, its history, origin, and planetary perspective: London, Academic Press, 401 p.

Mitchell, J. M., Jr., 1976, An overview of climatic variability and its causal mechanisms: Quaternary Research, v. 6, p. 481–493.

Mitchell, J. M., Jr., 1979, Evidence of a 22-year rhythm in the western United States related to the Hale solar cycle since the 17th century, in McCormac, B. M., and Seliga, T. A., eds., Solar-terrestrial influences on weather and climate: Dordrecht, Holland, D. Reidel Publishing Company, p. 125–143.

Sly, P. G., 1978, Sedimentary processes in lakes, in Lehrman, A., ed., Lakes, chemistry, geology, and physics: New York, Springer-Verlag, 363 p.

Sonett, C. P., and Finney, S. A., 1990, The spectrum of radiocarbon, in Pecker, J. C., and Runcorn, S. K., eds., The Earth's climate and variability of the Sun over recent millennia: Geophysical, astronomical, and archeological aspects: Royal Society of London Philosophical Transactions, ser. A, v. 330, p. 413–426.

Sonett, C. P., and Suess, H. E., 1984, Correlation of bristlecone pine ring widths with atmospheric carbon variations: A climate-sun relation: Nature, v. 307, p. 141–143.

Stark, D. M., 1976, Paleolimnology of Elk Lake, Itasca State Park, northwestern Minnesota: Archiv für Hydrobiologie, Supplement 50 (Monographische Beitrage), p. 208–274.

Stolov, H. L., and Shapiro, R., 1974, Investigation of the responses of the general circulation at 700 Mbar to solar-geomagnetic disturbance: Journal of Geophysical Research, v. 79, p. 2161–2170.

Stuiver, M., 1970, Long-term C^{14} variations, in Olsson, I. U., ed., Radiocarbon variations and absolute chronology (Proceedings of the 12th Nobel Symposium, Uppsala, Sweden): New York, John Wiley & Sons, 625 p.

Stuiver, M., and Braziunus, T. F., 1989, Atmospheric ^{14}C and century-scale solar oscillations: Nature, v. 338, p. 405–408.

Stuiver, M., Kromer, B., Becker, B., and Ferguson, C. W., 1986, Radiocarbon age calibration back to 13,300 years B.P. and the age matching of the German oak and U.S. bristlecone pine chronologies: Radiocarbon, v. 28, p. 969–979.

Stuiver, M., Braziunus, T. F., Becker, B., and Kromer, B., 1991, Climatic, solar, oceanic, and geomagnetic influences on late-glacial and Holocene atmospheric $^{14}C/^{12}C$ change: Quaternary Research, v. 35, p. 1–24.

Suess, H. E., and Linick, T. W., 1990, The ^{14}C record in bristlecone pine of the past 8000 years based on the dendrochronology of the late C. W. Ferguson, in Pecker, J. C., and Runcorn, S. K., eds., The Earth's climate and variability of the Sun over recent millennia: Geophysical, astronomical, and archeological aspects: Royal Society of London Philosophical Transactions, ser. A, p. 403–412.

Thomson, D. J., 1990, Time-series analysis of Holocene climatic data: Royal Society of London Philosophical Transactions, Royal Society of London, ser. A, v. 330, p. 601–616.

Tinsley, B. A., and Deen, G. W., 1991, Apparent tropospheric response to MeV-GeV particle flux variations: A connection via electrofreezing of supercooled water in high-level clouds: Journal of Geophysical Research, v. 96, p. 22283–22296.

MANUSCRIPT ACCEPTED BY THE SOCIETY JULY 27, 1992

On the precision of the Elk Lake varve chronology

Donald R. Sprowl*
Department of Geology, 120 Lindley Hall, University of Kansas, Lawrence, Kansas 66045-2124

ABSTRACT

Multiple varve counts over common intervals in four parallel cores permit estimation of the precision of the Elk Lake varve chronology. At the 95% confidence level, the imprecision of the counts averages 12%. External evidence and comparison of independent varve counts suggest that the accuracy of the varve chronology is well within this limit. Magnetic susceptibility is shown to be an exceptional tool for determining stratigraphic correlations, allowing unambiguous matching of all cores used.

INTRODUCTION

Anderson and others (Chapter 4) have discussed an Elk Lake stratigraphy and varve chronology derived in large part from the long core taken by the U.S. Geological Survey in 1978. Their comparison of the varve chronology with independent estimates of sediment age argues strongly for the overall accuracy of the varve chronology. Paleomagnetic correlations with distant sites provide additional evidence for the accuracy of the varve chronology (see Sprowl and Banerjee, Chapter 11). A separate question relates to the precision of the varve counts, which cannot be evaluated from a single record of the varves. This paper details an independent counting of the varves utilizing cores taken in 1982 and 1983 for paleomagnetic work. Because multiple cores are available from the same intervals, precision of varve counting can be estimated.

SEDIMENT STRATIGRAPHY

Figure 1 plots magnetic susceptibility (χ) versus depth below ice surface for the major sediment cores used in this study: 78j, 78k, 83a, 83b, 82c, and 82e. The horizontal scale is relative to core 78j, with the other cores shifted progressively 5 units to the right for visual clarity. The vertical scale is relative to the 1983 cores. To first order, magnetic susceptibility is a measure of the concentration of magnetite present in a sample. Corresponding features in χ mark points of common depositional age. The χ logs, combined with prominent sedimentologic marker horizons, permit unambiguous time of deposition correlation between the cores. The uncertainty in the χ correlations is 2–4 cm.

In the upper 4 m of the section, χ variations are not very distinctive and correlations are ambiguous. Through this region, correlation can be made using sedimentologic marker horizons and ARM (anhysteretic remanent magnetization), which shows a marked change in character at 31 m adjusted (1983) depth (Fig. 2).

In Figures 1 and 2, depths are adjusted to a reference depth scale based on the stratigraphy from the 1983 field notes (Sprowl and Banerjee, 1985, 1989). This 1983 reference depth scale is self consistent but is not identical to the 1978 depth reference of Anderson and others (Chapter 4). For example, the 1978 depth reference does not require any change in depth for the bottom of the 78j core, while the 1983 depth reference does (Fig. 2). Both are equally valid and the critical requirements are met, namely, all cores are properly correlated with regard to time of deposition and both references give exactly the same relative stratigraphy of core overlaps and gaps in coring.

VARVE CHRONOLOGY

Counting the Varves

The varves were counted from the 1982 and 1983 cores in a fashion very similar to that of Anderson and others (Chapter 4), with the exception that most of the core segments were preserved and counted in an unfrozen state. Varve recognition is hindered by freezing in the upper part of the section. However, freezing appears to be advantageous in the lower section because it prevents loss of varve distinction due to oxidation.

The entire sedimentary section was divided into 35 counting

---

*Present address: Department of Chemistry, Louisiana College, Pineville, Louisiana 71359.

Sprowl, D. R., 1993, On the precision of the Elk Lake varve chronology, *in* Bradbury, J. P., and Dean, W. E., eds., Elk Lake, Minnesota: Evidence for Rapid Climate Change in the North-Central United States: Boulder, Colorado, Geological Society of America Special Paper 276.

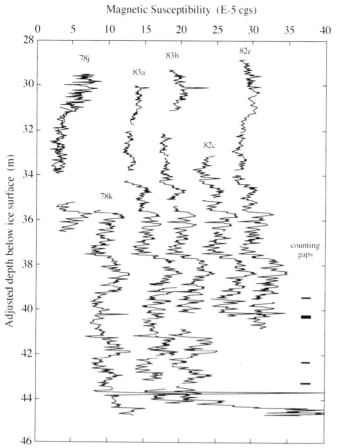

Figure 1. Correlated χ. The magnetic susceptibility logs are plotted for cores 78j, 78k, 83a, 83b, 82c, and 82e, with depths adjusted to match time of deposition between the cores. The χ scale is with regard to core 78j, with the other cores shifted progressively 5 units to the right for visual clarity. The solid bars on the right of the graph indicate gaps in the stratigraphy across which varve counts were interpolated.

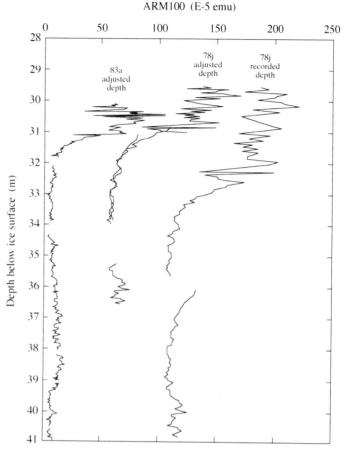

Figure 2. Correlation of core 78j with core 83a on the basis of ARM. Horizontal scale is relative to the left most curve, with remaining cores shifted progressively 50 units to the right for visual clarity. The depth scale is that of the 1983 stratigraphy.

intervals with interval boundaries occurring at prominent sedimentary horizons. Paleomagnetic sampling left two halves from each core that could be counted independently. Multiple cores covering given intervals provided additional redundancy. Two-fold redundancy from core 83b is the norm throughout the section. As much as nine-fold redundancy is available in the upper section. Table 1 lists the sedimentologic markers that divide the section into counting intervals. Table 2 lists the core segments that were used in the counting. Table 3 uses the codes given in Tables 1 and 2 to detail the varve counts from each of the counting intervals. Most of the gaps in the 83b section are covered by sections from cores 78k, 82c, 82e, and 83a. Four gaps could not be covered in this manner. The widths of these gaps, which represent either missing material or material in which the varves were thoroughly obscured, are unambiguously defined in Figure 1, and the sedimentation rates were interpolated from sedimentation rates on either side of the gap. The gaps and their counts are as follows: between horizons LL and MM, 4 cm, 38 yr; between horizons NN and OO, 16 cm, 80 yr; between horizons SS and YY, 5 cm, 40 yr; between horizons AAA and BBB, 9 cm, 73 yr. The total number of uncounted years in the chronology is thus 231. The stratigraphy of the gaps is plotted in Figure 1.

Estimating counting precision

The redundancy in the varve counting detailed in Table 3 permits evaluation of the precision of the varve counts. The variance among the multiple counts from common intervals is taken as a measure of counting imprecision. Thus, two standard deviations from the mean estimates the limits of 95% confidence on the precision of the counts. These limits are listed in Table 3. Intervals with less than two-fold redundancy are assigned the highest variance from a bordering interval. The overall uncertainty in the varve counts is about 12% at the 95% confidence level. This is taken as a conservative estimate of the precision of the counts and a conservative limit to the accuracy of the varve chronology. External evidence (Anderson and others, Chapter 4; Sprowl and Banerjee, Chapter 11) suggests that the varve chronology is accurate to well within these limits.

TABLE 1. DESCRIPTION OF STRATIGRAPHIC MARKER HORIZONS

Horizon code	Adjusted depth (cm)	Horizon description	Horizon code	Adjusted depth (cm)	Horizon description
Ambrosia	2,888	Ambrosia rise	BB	3,696	Thick black, middle of 83b04, 36-37
A	2,898	Double white	CC	3,723	Single white, bottom of 83b04, 36-37
B	2,941	Double white	DD	3,734	Single black, top of 83b05, 37-38
C	2,968	Double white	EE	3,766	Thick white, middle of 83b05, 37-38
D	2,985	Diatom bloom	FF	3,790	Thick white, bottom of 83b05, 37-38
E	3,018	Prominant silt layer	GG	3,809	Diatom bloom, top of 83b05, 38-39
F	3,032	Double white	HH	3,842	Thick black, middle of 83b05, 38-39
G	3,040	Double white	II	3,877	Single white, bottom of 83b05, 38-39
H	3,074	Diatom bloom	JJ	3,885	Thick black, top of 83b06, 39-40
I	3,083	Diatom bloom	KK	3,918	Single white, middle of 83b06, 39-40
J	3,094	Double white	LL	3,966	Thick white, bottom of 83b06, 39-40
K	3,103	Double white	MM	3,970	Single white, top of 83b06, 40-41
L	3,122	Single white, bottom of 83b01, 30-31	NN	4,012	Single white, middle of 83b06, 40-41
M	3,151	Single white, top of 82b03	OO	4,089	Single white, bottom of 83b08
N	3,239	Double white	PP	4,132	Single white, top of 83b08
O	3,255	Diatom bloom	QQ	4,167	Single white, top of 83b09
P	3,283	Double white	RR	4,200	Single white, middle of 83b09
Q	3,308	Diatom bloom, bottom of 83b02, 32-33	SS	4,236	Single white, bottom of 83b09
R	3,330	Single white, top of 83b03, 33-34	YY	4,261	Single white, top of 83b12
S	3,383	Thick black, middle of 83b03, 33-34	ZZ	4,303	Thick white, middle of 83b12
T	3,417	Single white, bottom of 83b03, 33-34	AAA	4,335	Single white, bottom of 83b12
U	3,429	Single white, top of 83b03, 34-35	BBB	4,363	Single white, top of 83b13
V	3,467	Single white, middle of 83b03, 34-35	CCC	4,372	Single white
W	3,508	Single black, bottom of 83b03, 34-35	DDD	4,380	Single white
X	3,554	Single white, top of 83b04, 35-36	EEE	4,405	Single white
Y	3,600	Single white, middle of 83b04, 35-36	FFF	4,420	Single white, bottom of 83b13
Z	3,620	Thick black, bottom of 83b04, 35-36	GGG	4,423	Single white, top of 83b14
AA	3,641	Thick black, top of 83b04, 36-37	HHH	4,430	Transition to nonvarved, late glacial

TABLE 2. CORE SEGMENT IDENTIFICATION CODES

Code	Core segment	Code	Core segment	Code	Core segment	Code	Core segment
a	82b1 (Surface)	t	83a08a	mm	83b04a, 35-36	fff	83b12a
b	84aa	u	83a09a	nn	83b04b, 35-36	ggg	83b12b
c	84ab	v	83a11a	oo	83b04a, 36-37	hhh	83b13a
d	84ba	w	82c01	pp	83b04b, 36-37	iii	83b13b
e	84bb	x	82c03	qq	83b05a, 37-38	jjj	83b14a
f	84ca	y	82c04	rr	83b05b, 37-38	kkk	83b14b
g	84cb	z	82e05a	ss	83b05a, 38-39	lll	83a01b
h	82e01a, 30-30	aa	78ki	tt	83b05b, 38-39	mmm	83a02b
i	82e01b, 30-30	bb	78kk	uu	83b06a, 39-40	nnn	83a03b
j	82a02a, 29-31	cc	78kl	vv	83b06b, 39-40	ooo	83a04b
k	82a02b, 29-31	dd	83b01b, 29-30	ww	83b06a, 40-41	ppp	83a08b
l	82b03a, 31-33	ee	83b01a, 30-31	xx	83b06b, 40-41	qqq	83a09b
m	83a01a	ff	83b01b, 30-31	yy	83b08a	rrr	83a11b
n	83a02a	gg	83b02a, 32-33	zz	83b09a	sss	82e05b
o	83a03a	hh	83b02b, 32-33	aaa	83b09b	ttt	78kf
p	83a04a	ii	83b03a, 33-34	bbb	83b10a	uuu	78kg
q	83a05	jj	83b03b, 33-34	ccc	83b10b	vvv	78kh
r	83a06	kk	83b03a, 34-35	ddd	83b11a	www	83a12
s	83a07	ll	83b03b, 34-35	eee	83b11b		

TABLE 3. VARVE COUNTS*

Elk Varves Horizon	Adjusted Depth (cm)	C1	C2	C3	Replicate Counts C4	C5	C6	C7	C8	C9	Max.	Average	n	Std Dev.	% Dev.	± at 95%	Cumm Age	Cumm Dev. at 95%
Ambrosia	2,888	88									88	88.0	1	...			88	4
A	2,898	129 d ?									129	129.0	1	...	2.3	4.1	217	10
B	2,941	220 b	215 c	215 d	212 e	205 f	212 g				220	213.2	6	5.0	2.3	6.0	437	20
C	2,968	148 b	149c	151 d	148 e	151 f	148 g	147 j	153 k		153	148.9	9	2.4	1.6	9.9	590	25
D	2,985	104 d	103 e	106 f	115 g	104 j	114 k	92 dd		145 dd	115	105.4	7	7.7	7.3	15.4	705	40
E	3,018	179 h	176 i	181 j	197 k	172 dd					197	181.0	5	9.6	5.3	19.1	902	59
F	3,032	74 h	77 i	71 j	75 k						77	74.3	4	2.5	3.4	5.0	979	64
G	3,040	50 j	57 k								57	53.5	2	4.9	9.3	9.9	1,036	74
H	3,074	166 j	191 k	201 ee	196 ff						201	188.5	4	15.5	8.2	31.1	1,237	105
I	3,083	58 j	65 k	67 ee	72 ff						72	65.5	4	5.8	8.9	11.6	1,309	117
J	3,094	59 k	77 ee	74 ff							77	70.0	3	9.6	13.8	19.3	1,386	136
K	3,103	68 ee	66 ff								68	67.0	2	1.4	2.1	2.8	1,454	139
L	3,122	125 ee	125 ff								125	125.0	2	0.0	0.0	0.0	1,579	139
M	3,151	143 m	157 lll								157	150.0	2	9.9	6.6	19.8	1,736	159
N	3,239	483 l									483	483.0	1	...	7.7	74.0	2,219	233
O	3,255	97 l	104 gg	113 hh							113	104.7	3	8.0	7.7	16.0	2,332	249
P	3,283	161 l	189 gg	183 hh							189	177.7	3	14.7	8.3	29.5	2,521	278
Q	3,308	150 l	161 ff	147 hh							161	152.7	3	7.4	4.8	14.7	2,682	293
R	3,330	127 n	119 mmm								127	123.0	2	5.7	4.6	11.3	2,809	304
S	3,383	373 ii	372 jj								373	372.5	2	0.7	0.2	1.4	3,182	306
T	3,417	240 ii	241 jj								241	240.5	2	0.7	0.3	1.4	3,423	307
U	3,429	68 w									68	68.0	1	...	1.1	1.5	3,491	309
V	3,467	195 kk	198 ll								198	196.5	2	2.1	1.1	4.2	3,689	313
W	3,508	161 kk	192 ll								192	176.5	2	21.9	12.4	43.8	3,881	357
X	3,554	191 o	210 nnn								210	200.5	2	13.4	6.7	26.9	4,091	384
Y	3,600	215 mm	218 nn								218	216.5	2	2.1	1.0	4.2	4,309	388
Z	3,620	95 mm	93 nn								95	94.0	2	1.4	1.5	2.8	4,404	391
AA	3,641	89 p ?	80 ooo ?	59 z	53 sss						89	70.3	4	17.0	24.3	34.1	4,493	425
BB	3,696	286 oo	294 pp								294	290.0	2	5.7	2.0	11.3	4,787	436
CC	3,723	215 oo	213 pp								215	214.0	2	1.4	0.7	2.8	5,002	439
DD	3,734	83 p	85 ooo								85	84.0	2	1.4	1.7	2.8	5,087	442
EE	3,766	231 qq	251 rr								251	141.0	2	14.1	5.9	28.3	5,338	470
FF	3,790	127 qq	163 rr								163	145.0	2	25.5	17.6	50.9	5,501	521
GG	3,809	72 x									72	72.0	1	...	17.6	25.3	5,573	546
HH	3,842	164 ss	171 tt								171	167.5	2	4.9	3.0	9.9	5,744	556
II	3,877	184 ss	179 tt								184	181.5	2	3.5	1.9	7.1	5,928	563
JJ	3,885	57 q	50 y	51 y							57	52.7	3	3.8	7.2	7.6	5,985	571

TABLE 3. VARVE COUNTS* (continued)

Horizon	Elk Varves	Adjusted Depth (cm)	C1	C2	C3	Replicate Counts C4	C5	C6	C7	C8	C9	Max.	Average	n	Std Dev.	% Dev.	± at 95%	Cumm Age	Cumm Dev. at 95%
KK		3,918	192 uu	180 vv								192	186.0	2	8.5	4.6	17.0	6,177	588
LL		3,966	387 uu	313 vv								387	350.0	2	52.3	15.0	104.7	6,564	693
MM		3,970	38 @									38	38.0	1	...	15.0	11.4	6,602	704
NN		4,012	280 ww	276 xx								280	278.0	2	2.8	1.0	5.7	6,882	710
OO		4,089	491 r,s	(@ 80)	463 r,s	(@ 80)	425 ttt,uuu	(@ 55)				491	459.7	3	33.1	7.2	66.3	7,373	776
PP		4,132	180 yy									180	180.0	1	...	15.1	54.4	7,553	830
QQ		4,167	152 t	156 ppp	198 vvv							198	168.7	3	25.5	15.1	51.0	7,751	881
RR		4,200	204 zz	171 aaa								204	187.5	2	23.3	12.4	46.7	7,955	928
SS		4,236	228 aa	128 aaa								238	233.0	2	7.1	3.0	14.1	8,193	942
YY		4,261	198 aa	(@ 40)								198	198.0	3	16.6	8.4	33.2	8,391	975
ZZ		4,303	259 fff	230 ggg								259	244.5	2	20.5	8.4	41.0	8,650	1,016
AAA		4,335	284 fff	276 ggg								284	280.0	2	5.7	2.0	11.3	8,934	1,027
BBB		4,363	284 www	(@ 73)								284	284.0	1	...	5.4	30.4	9,218	1,058
CCC		4,372	127 hhh	137 iii								137	132.0	2	7.1	5.4	14.1	9,355	1,072
DDD		4,380	64 hhh	73 iii								73	68.5	2	6.4	9.3	12.7	9,428	1,085
EEE		4,405	268 hhh	299 iii								299	283.5	2	21.9	7.7	43.8	9,727	1,129
FFF		4,420	250 hhh	256 iii								256	253.0	2	4.2	1.7	8.5	9,983	1,137
GGG		4,423	58 hhh									58	58.0	1	...	19.3	22.4	10,041	1,160
HHH		4,430	60 jjj	79 kkk								79	69.5	2	13.4	19.3	26.9	10,120	1,186

*? = Identification of marker horizon is uncertain; @ = Number of varves interpolated through a gap in the core.

The above analysis assumes that the counting errors are normally distributed with zero mean, but this is only true if the probability of counting too many varves is the same as that of counting too few. Too many varves can be counted when subannual sets of laminations appear to be annual. However, varves can be obscured or left undistinguished sedimentologically or by the cleaning or polishing technique used. It is my (subjective) judgement that counting too few varves is more likely than counting too many, and I expect the errors to be biased on the low side. In the limiting case, varve counting would be a binomial process where each of the varves (n) has a probability, p, of being counted. The expected value of the count, k, is then np, and it is clear that $k < n$ for any $p < 1$. However, estimating n from k when p is unknown is an unsolved problem. Because of this expected negative bias in the varve counting process, the highest count from a given interval was used as the estimator of the actual number of varves present. Presumably, this still underestimates the actual number of varves present.

Comparison of varve chronologies

The varve chronology developed from the 1982 and 1983 cores can now be compared with the chronology developed from the 1978 core by Anderson and others (Chapter 4). The samples for which Anderson and others list varve and radiocarbon ages have been correlated to the 1982–1983 chronology through Figure 1. There is some uncertainty (~10 cm, ~50 yr) in these correlations. The comparison of the two chronologies is given in Table 4 and in Figure 3. For each sample, the difference in the two varve ages is less than the precision limits given in Table 3. The age of the bottom of the section is given as 10,400 yr B.P. from the 1978 chronology and 10,120 varve yr from the 1982 and 1983 chronology. These values difer by only 3%.

The numerical estimates of counting precision derived here give an indication of the degree of preservation and recovery of the varve record by the combined geologic and experimental processes. A separate question is that of varve recognition and interpretation. The uncertainties in this process are estimated by the difference between the independent chronologies of this paper and the work of Anderson and others. Comparison of the two chronologies suggests that at Elk Lake, varve identification is not a significant difficulty relative to varve preservation during the experimental process.

REFERENCES CITED

Sprowl, D. R., and Banerjee, S. K., 1985, High-resolution paleomagnetic record of geomagnetic field fluctuations from the varved sediments of Elk Lake, Minnesota: Geology, v. 13, p. 531–533.

Sprowl, D. R., and Banerjee, S. K., 1989, The Holocene paleosecular variation record from Elk Lake, Minnesota: Journal of Geophysical Research, v. 94, p. 9369–9388.

MANUSCRIPT ACCEPTED BY THE SOCIETY JULY 27, 1992

TABLE 4. ELK LAKE RADIOCARBON DATES

U.W. reference	Listed depth (cm)	Adjusted depth (cm)	C¹⁴ age	1982–1983 Varve age	1978 Varve age	Difference
QL4018	2,865	2,880	420	0	0	0
QL4017	2,873	2,888	1,160	88	88	0
QL1560	3,035	2,988	1,420	723	648	75
QL1651	3,125	3,038	2,270	1,022	1,100	-78
QL1562	3,365	3,185	3,360	1,923	2,216	-293
QL1493	3,425	3,223	3,190	2,131	2,317	-186
QL1563	3,465	3,277	3,370	2,481	2,634	-154
QL1492	3,475	3,295	3,660	2,598	2,666	-68
QL1564	3,485	3,276	3,510	2,474	2,731	-257
QL1565	4,045	3,736	5,750	5,103	5,084	19
QL1494	4,125	3,802	5,290	5,546	5,654	-108
QL1566	4,415	4,023	7,880	6,952	6,694	258
QL1495	4,705	4,252	8,550	8,320	7.983	337
QL1496	4,835	4,358	9,830	9,167	9,061	106
QL1497	4,995	4,430	11,380	10,120	10,400	-280
QL1498	5,105	....	17,000	....	....	

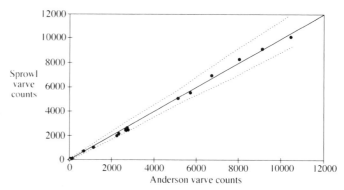

Figure 3. Comparison of 1978 varve chronology with 1982–1983 varve chronology. Cumulative ages are compared at horizons for which radiocarbon dates are available (filled circles). The solid line represents 1:1 correlation. The dashed lines indicate the estimated limits of 95% confidence in the precision of the 1982–1983 chronology.

Geological Society of America
Special Paper 276
1993

Modern sedimentation in Elk Lake, Clearwater County, Minnesota

Edward B. Nuhfer
Department of Geography and Geology, University of Colorado at Denver, Denver, Colorado 80217-3364
Roger Y. Anderson
Department of Earth and Planetary Sciences, University of New Mexico, Albuquerque, New Mexico 87131
J. Platt Bradbury, Walter E. Dean
U.S. Geological Survey, Box 25046, Federal Center, Denver, Colorado 80225

ABSTRACT

The varved sediments of Elk Lake, Clearwater County, Minnesota, contain a 10,000 year record of climatic and limnologic events. Sediment traps deployed in the lake's water column from 1979 to 1981 and from 1983 to 1984 collected samples that permitted us to identify materials, to see the timing of sedimentation events, and to deduce processes that form the microlaminae within varves. Fall and spring microlaminae consist mainly of sequential accumulations of biogenic silica and resuspended calcitic and siliceous materials. Precipitates of iron, manganese, and organic detritus dominate the thin winter microlaminae. Calcitic microlaminae are deposited in summer. Concentrated iron and manganese precipitates form when the onset of seasonal circulation (especially in autumn) oxygenates the lower water column, but precipitation of these metals also continues throughout periods of seasonal stratification, when these dissolved elements migrate upward and are converted to particles that rain back to the bottom. Mineraloids dominate the sediment; minerals compose only a minor part and include quartz, calcite, rhodochrosite, and rockbridgeite (iron phosphate).

The bulk of the bottom accumulation occurs during the longer, calmer summer and winter periods, but important contributions are also made during spring and autumn overturn events. Sediment resuspended from the shallows accumulates together with newly formed endogenic sediment, and can even briefly dominate the seston in autumn and spring. Vigor and duration of seasonal circulations in the upper water column dictate the amount of resuspended sediment contributed annually to a varve. When abrupt warming within days after ice-out stratifies the lake, sedimentation in that year is diminished by resultant suppression of plankton blooms and lack of vernal resuspension that would normally move sediment from the littoral areas into the deep parts of Elk Lake. Thin sections of varves confirm that resuspension during autumn and spring is a varve-forming process that has probably varied in importance as a function of climate and changing morphometry due to infilling. Through an entire year, sediment traps catch a greater proportion of material from spring and autumn overturns than accumulates on the bottom. Lake morphometry is the most important factor governing sediment resuspension and associated annual accumulation rates in traps.

Nuhfer, E. B., Anderson, R. Y., Bradbury, J. P., and Dean, W. E., 1993, Modern sedimentation in Elk Lake, Clearwater County, Minnesota, *in* Bradbury, J. P., and Dean, W. E., eds., Elk Lake, Minnesota: Evidence for Rapid Climate Change in the North-Central United States: Boulder, Colorado, Geological Society of America Special Paper 276.

INTRODUCTION

Elk Lake (Fig. 1) provided a unique opportunity to study a continuous (10,000 years) varved record of paleoenvironmental change. Because the lake is extant and still producing varves, the opportunity also was available to study contemporary sedimentation and deduce from direct observation the seasonal processes that produced the varves.

We captured modern sediments in special traps (Fig. 2) continuously from 1979 through 1984. The records from the traps provide valuable details about lacustrine varve-forming processes and their occurrence in time that could not be deduced from study of bottom sediments or water analyses.

Definitions of terms used

Apparent sedimentation rate is the rate of sediment accumulation, expressed in terms of thickness per unit time or as mass accumulation rate, seen in sampling devices such as sediment traps. Apparent sedimentation rates may have little to do with the true rates of accumulation on the bottom.

Lake seasons are the four periods of the year that correspond to winter ice-cover, spring circulation, summer stratification, and autumn circulation. Lengths of these respective lake seasons differ from year to year. Lake seasons do not coincide precisely with the conventional calendar definitions of winter, spring, summer, and fall. Unless otherwise specified, season names in this paper refer to lake seasons.

Sediment flux is the rate of sediment flow into a sampling device or onto the bottom without special implications about whether the units are thickness or weight per unit time.

Sedimentation rate is the rate of actual accumulation expressed in terms of thickness per unit time or as mass accumulation rate on the lake bottom.

Mass accumulation rate is the rate of accumulation expressed in units of weight per area per time; in this case $mg/cm^2/day$ or $mg/cm^2/yr$.

This chapter tells how the modern seston was sampled, how the timing of events was deduced, the details about events and their magnitude over the sampling period, differences between sedimentary processes in the upper and lower water column, and the relation of seston collected in traps to bottom accumulations and varve-forming processes.

METHODS

Sampling

Elk Lake regularly deposits only a few millimeters of laminated sediment each year. Time-marking sediment traps (Fig. 2) similar to those used by Anderson (1977), Nuhfer (1979), and Nuhfer and Anderson (1985) provided the continuous, high-resolution sampling record required to resolve events under those conditions. Traps were deployed from December 17, 1978, to June 20, 1984, but suitable samples were collected only between July 20, 1979, and June 10, 1981, and between August 12, 1983, and June 20, 1984. In some years, traps were hung on a single line at two different depths so that the trap mouths sampled from depths of 12 and 26 m. Dilute formalin buffered with sodium borate was added to some accumulation tubes (Fig. 2) at the time of deployment in order to prevent biogenic decay and to kill any burrowing organisms that might enter the tube and disturb the sequence of accumulation. Our results are based primarily on records of accumulation from two deep traps for the periods July 20, 1979, to June 24, 1980, and August 12, 1983, to June 30, 1984, and from two shallow traps for the periods July 20, 1979, to June 24, 1980, and from June 25, 1980, through June 10, 1981. The shallow trap sample collected in 1979–1980 was sparse, and permitted analyses of only six subsamples. Data from this trap were used for mineralogy and to yield an average composition that confirmed differences between trapped shallow seston and trapped deep seston. The samples from the other traps were used to develop high-resolution profiles based upon detailed chemical analyses and diatom studies (see also Bradbury, 1988, and Bradbury and Dieterich-Rurup, Chapter 15). Records from deep traps from the sampling period 1979–1980 were also utilized to study mineralogy of Elk Lake sediments by X-ray diffraction.

Calculated sedimentation rates and daily temperature and precipitation records for the Itasca region for this 5.5 yr period show distinct cycles as well as unique events that varied in duration from several days to several weeks. The timing of sedimentation events in the traps was deduced primarily from the positions of the Teflon time markers (usually at 10, 15, or 30 day; intervals) in the transparent accumulation tube (Fig. 2), as recorded in field notes and photos made after trap removal.

Most trap subsample divisions were made at time markers or at abrupt changes in color or fabric between Teflon time-marker layers (Fig. 2). Exceptions occurred where an unmarked zone or zone of uniform appearance exceeded 2 cm in thickness. Then subsamples were taken in 2 cm increments. Dates were estimated by interpolation in samples not bounded above and below by Teflon markers. These estimated dates were checked against episodes of circulation and stratification as inferred from climate records and seasonal sedimentary associations (e.g., diatom blooms and ferric hydroxide floc at overturns). Dates deduced even in the most dubious intervals are probably accurate to within 5 days. During some intervals the amount of sediment that accumulated between Teflon time markers was too small to analyze. Such small samples were combined with adjacent samples in the profile until material sufficient for analysis was obtained.

Figure 1. Maps of Minnesota, Itasca State Park, and Elk Lake showing the location of Elk Lake, general vegetation zones of Minnesota, bathymetry of Elk Lake, and location of the trap and varved core in the deepest part of the lake.

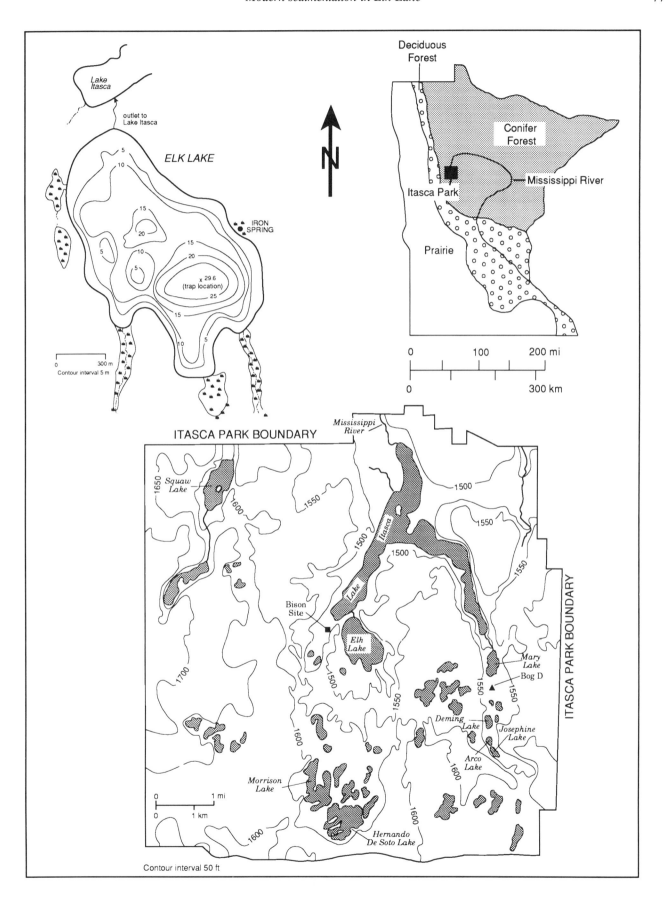

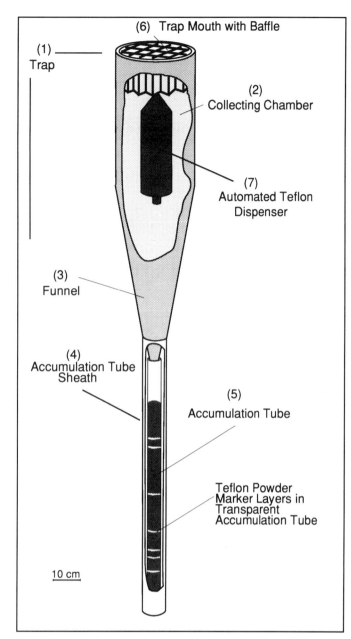

Figure 2. Schematic of time-marking sediment trap (modified from Anderson, 1977). (1) Entire trap unit, (2) cylindrical collecting chamber, (3) tapered funnel at 20° angle, (4) accumulation tube sheath, (5) transparent accumulation tube, (6) mouth of trap with baffles, and (7) automated Teflon granule dispenser.

The apparent mass-accumulation rates of constituents by dry weight that might be expected to accumulate on 1 cm^2 of the lake bottom in a day were calculated by dividing the thickness of accumulation of each sample in centimeters in the trap by the number of days in the sample interval and multiplying this quantity times 1 cm^2 times a mean density of 1060 mg/cm^3 times the mean volume fraction of dry material (0.15 based on an average of 85% water-filled porosity) and finally dividing by the trap magnification as determined by the ratio of the cross-sectional area of the funnel mouth to the area of the accumulation tube. This gives mass-accumulation rates in units of mg/cm^2/day. These apparent rates have little relation to actual sedimentation rates on the bottom.

Analyses

Samples for chemical analyses were dried at 100 °C, then ashed. Weight loss was recorded at 550 °C and again at 1000 °C (Dean, 1974). Organic matter is taken as equivalent to loss on ignition (LOI) at 550 °C. The ashed residues left after LOI at 1000 °C were dissolved for atomic absorption (AA), inductively coupled plasma (ICP), and colorimetric analyses by fusion with $LiBO_2$ and subsequent digestion in a 1M HCl-2% HF solution as described by Nuhfer and Romanosky (1979). Methods of sample preparation for diatom studies are given by Bradbury (1988) and Bradbury and Dieterich-Rurup (Chapter 15).

Samples for X-ray diffraction (XRD) mineralogy were prepared by removal of organic matter by low temperature (<100 °C) ashing for one week in a low-temperature asher. A 20 mg aliquot of each ashed sample was then dispersed in water and drawn onto a Millipore filter (method of Nuhfer and Renton, 1979) prior to scanning with an X-ray diffractometer.

Relations between air temperatures and seasonal lake circulation events

Seasonal stratification and circulation events determine the cycling of nutrients and the duration of productivity, which are important factors in governing seasonal sedimentation at Elk Lake. The timing and magnitude of these events are dictated by a heat budget that is controlled dominantly by changing air temperatures. Temperature and rainfall data (Fig. 3, A and B) contemporary with the sediment-trap collections came from the weather station at the University of Minnesota's Forestry and Biological Station at Lake Itasca (Fig. 1). The station recorded maximum and minimum daily temperatures and daily precipitation. These temperatures served as the basis for determining dates of ice-up and ice-out on Elk Lake over the period of study.

We used the known water temperatures (limnological data collected in 1967 by A. L. Baker from Elk Lake and a single recorded observation on ice-out date of April 21, 1980) and daily records of weather from these same years to calibrate the relation between air temperature and water temperatures. Thereafter, dates of thermal events (changing water temperatures, development and duration of circulations and freeze-thaw) in any year could be approximated from daily records of air temperature. We calculated the dates of ice-cover and ice-out because we did not have continuous measurements of water temperatures, directly observed dates of ice cover, or known duration of overturn events during our sampling period.

We began by calculating a weighted daily mean air temper-

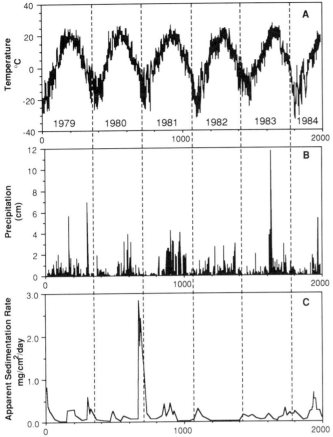

Figure 3. Average daily temperature (A) and precipitation (B) for the Itasca region of northwestern Minnesota for the period 1979 to 1984, and (C) calculated mass-accumulation rates of sediment (in mg/cm²/day) from sediment traps deployed at Elk Lake from January 1, 1979, (day 1) to June 20, 1984. All data in C are from deep traps except from June 24, 1980 (day 580) through June 10, 1981 (day 891), when it was necessary to use data from that shallow trap to achieve a complete record. Highest peaks in C occur in days 667–727 (October 29, 1980, through December 28, 1980), where the shallow trap captured the autumn overturn. There was no clear relation between rainfall and trapping rates, thus indicating that lake circulation is more important than basin erosion for placing sediment into these traps. Dates given correspond to the top of each sample.

ature in degrees Celsius (T_{ma}) for each day from the sum obtained after multiplying each daily maximum temperature by the representative fraction of each 24 hr day occupied by daylight hours (D_f) and the minimum-temperature by ($1-D_f$). Lengths of daylight hours at the latitude of Elk Lake were obtained from sunrise and sunset times recorded for each day in the *Astronomical Almanac* published by the U.S. Naval Observatory Nautical Almanac Office (1985). Simpler mean temperatures calculated by averaging the maximum and minimum temperatures recorded for each day take into account only this locality's characteristically wide diurnal temperature fluctuation and neglect the influence of time imposed by relative durations of day (warmest temperatures) and night (coolest temperatures) through most of the year. By taking into account the duration of a given temperature as well as its magnitude, it was possible to improve slightly on the estimate of lake water conditions based on a heat budget.

Baker's notes gave water column temperatures on 12 known dates between July 29, 1967, and November 27, 1967, and recorded the erosion of the hypolimnion, development of holomictic circulation, and eventual achievement of ice cover. The notes contained a detailed record of the erosion of the hypolimnion during fall of 1967. In August, 1967, the thermocline was established at 7 m depth with a temperature of the epilimnion (T_{epi}) of 24 °C, but by September 1 the thermocline declined to 8 m with T_{epi} of 19 °C, and by September 26 to 8.5 m at 15 °C—an erosion of the hypolimnion of 0.5 m in 25 days. Thereafter rapid erosion followed, and by October 14 the zone of circulation averaged about 10 °C and extended to a depth of 14 m. By October 21 mixing extended to 22 m and by October 27 to 26 m. By October 28, the entire lake circulated at 7 °C. The fastest rates of growth of the epilimnion (up to 4 m/day) occurred when the zone reached about 9 °C.

We averaged the T_{ma} for several intervals of days in 1967 until we achieved the best possible correlation between Baker's observations on the temperature of the epilimnion (T_{epi}) and the moving averages of weighted atmospheric T_{ma}. We selected the 25 day interval as the basis to calculate moving averages. The relation between average epilimnion temperature (T_{epi}) and 25 day averages of weighted mean daily air temperatures (T_{ma25}), both in degrees Celsius, on any given day was linear according to:

$$T_{epi} = 3.09 + 0.872\,(T_{ma25}); \ (r = 0.99). \tag{1}$$

Ice-out dates for each year are approximated when $T_{epi} = 0$, as calculated from the 25 day moving averages of weighted mean air temperature. The usefulness of this model for estimating the time of autumn overturn is corroborated by direct temperature measurement for 1979 by Megard and others (Chapter 3).

Estimation of dates of ice-out required a different model because the lake was protected from wind-driven circulation by the overlying ice and responded more slowly to changes in air temperature. Further, when water varies above and below 4 °C and when it converts back and forth to ice at 0 °C, heat-capacity changes accompany changes in molecular bonding and ordering that occur at these temperatures. We had only a single precise date of observation of April 21, 1980, for an ice-out date at Elk Lake plus a series of approximate dates based on nearby Lake Itasca. We experimented with T_{ma} averaged over time and found that a moving average using 45 day intervals (T_{ma45}) predicted ice-out on April 21, 1980, the date corresponding to the moving average of T_{ma45}, yielding the first spring temperatures above 0 °C. Results from these two models to predict ice-up and ice-out dates agree with recorded observations at nearby Lake Itasca, where ice-up dates at Elk Lake generally occur on the same day to a week after Lake Itasca ice-up, and ice-out dates are about the same in both lakes (Jon Ross, University of Minnesota Forestry and Biological Station, 1986, written commun.).

After ice-out, the circulating lake can be modeled by equation 1 to estimate the date of onset of summer stratification. This date was established as the first date that T_{epi} reaches 10 °C. Table 1 summarizes the calculated dates and duration of major thermal events.

RESULTS

Temporal variations in apparent mass-accumulation rates and relative proportions of organic and inorganic matter (Figs. 3C and 4) reflect changes in events and processes that occurred within the lake. Variations in composition of the inorganic fraction are useful for deducing the actual processes (Fig. 5).

Blooms of major silica-contributing diatom genera were an aid to pinpointing times of events. Diatoms and chrysophytes were important contributors of biogenic silica to the sediment. Most contributions of biogenic silica occur during spring or fall overturns, or both (Fig. 6, A, B, and C). One diatom, *Asterionella formosa,* may bloom under ice cover. Nuhfer (1979) noted that this same diatom also blooms beneath ice at East Twin Lake, Ohio.

1979–1980 Sediment-trap results

The sediment trap for the period 1979–1980 was placed in Elk Lake on July 20, 1979, following one of the coldest and wettest springs on record. At this time, the lake was well stratified, with the epilimnion probably at about 22 to 24 °C and extending to a depth of about 6 m. Cooling began in mid-September of 1979, and by October 11 the base of the epilimnion had deepened from 7 to 14 m. Rapid erosion of the hypolimnion occurred thereafter, and the lake was in full circulation by October 18. Fall circulation was an event of about 30 days duration. Ice-up occurred about November 26, 1979, and persisted to April 21, 1980. The lake stratified thermally almost immediately after ice-out, and spring overturn persisted for less than a week.

The mass-accumulation rate in mg/cm^2/day (dry weight) for each of 14 chemical constituents in each subsample of the trap was calculated from its concentration, the known thickness of each trap subsample (corrected for trap magnification), an approximate density of 1060 mg/cm^3, and a water content of 85%. The mass of each element in every subsample was totaled, and then adjusted by the total days represented in the trap to 365 days to give an annual mass-accumulation rate in mg/cm^2/yr (Table 2).

Trapped sediment was dominated by silica, iron, calcium, manganese, and organic matter (Table 2; Fig. 4A). Calcium reached maximum concentration in late August (sample beginning 8/20/79) and remained high through September 20, 1979 (Fig. 5). Chrysophyte scales and a diatom assemblage dominated by *Fragilaria crotonensis* and *Tabbelaria flocculosa* (Fig. 6A) contributed significant silica through October 31, 1979.

Sediment fluxes were low through summer (Fig. 4A), until increases in early October resulted from circulation-induced resuspension and plankton blooms (including diatom production). Temperatures decreased abruptly at the beginning of October, and significant deepening of the epilimnion was occurring by October 11. Sediments that accumulated after that date became increasingly enriched in iron and manganese (Fig. 5).

On October 31, a rainfall of 6.93 cm coincided with the peak of a diatom bloom (Fig. 6A). The bloom was likely caused when nutrients were delivered to the photic zone from below by wind-driven circulation. The storm that brought the wind and rain coincided fortuitously with the time of thermal destratification. Surprisingly, this time of peak rainfall produced no other significant sediment response in the trap. Instead, the compositional record reflects only slight increases in silica, alumina, and manganese (Fig. 5), all in accord with a sediment composition dictated by circulation, mixing, and gradual dilution of summer calcite and full diatoms with endogenic oxides and resuspended sediment. An apparent mass-accumulation rate of 0.4 mg/cm^2/day (sample for the period ending November 10, 1979; Fig. 4A) corresponded to the peak of autumn overturn when the entire lake was in circulation.

Rates of sedimentation declined through the fall to a winter minimum of 0.02 mg/cm^2/day beneath ice cover in February (Fig. 4A). Ice cover was established by November 12, but little snowfall occurred until early January, 1980. Perhaps this led to an ice cover that transmitted light for two months and permitted the late bloom of *Asterionella formosa* under ice noted in late November and December samples. Samples of sediment that accumulated under ice are dominated by iron and manganese oxides (Fig. 5) with some silica contributed by the diatom bloom under ice in November and December. Increases in sediment accumulation in the trap under ice in March through early April are attributed to entrapment of migrating copepods. Abundant copepods are manifest in winter samples that are enriched in organic matter and phosphorus (Fig. 5).

Abrupt ice-out at about April 20, 1980, was accompanied by weather that remained warm and dry. These conditions quickly established stratification which kept apparent sedimenta-

TABLE 1. CALCULATED TIMES OF ELK LAKE THERMAL EVENTS

	Year				
	1979	1980	1981	1983	1984
Ice-out	n.c.	April 21	April 3	n.c.	April 22
Ice-up	Nov. 26	Nov. 29	n.c.	Nov. 28	n.c.
Onset of Summer Stratification	n.c.	April 27	May 1	n.c.	May 15
Duration of Spring Circulation	n.c.	6 days	27 days	n.c.	23 days
Duration of Fall Circulation	30 days	46 days	n.c.	60 days	n.c

Note: n.c. = not calculated; date falls outside collection period of samples available for analysis.

tion rates below 0.2 mg/cm^2/day. Rates declined thereafter under summer stratification between May 21 and June 24, 1980, to 0.05 mg/cm^2/day. As the result of a very brief and incomplete vernal circulation, the spring diatom bloom was small and was dominated by *Fragilaria crotonensis*, chrysophyte scales, and *Cyclotella stelligera* (Fig. 6A). High manganese concentrations in samples that accumulated after April 10 also are an indicator of incomplete vernal overturn. Lack of recycled or contributed dissolved nutrients apparently attenuated subsequent summer green algal or blue-green cyanobacterial blooms.

1980–1981 Shallow sediment trap results

The trap was installed at a depth of 12 m on June 25, 1980, after the hot, dry spring which had stratified the lake less than a

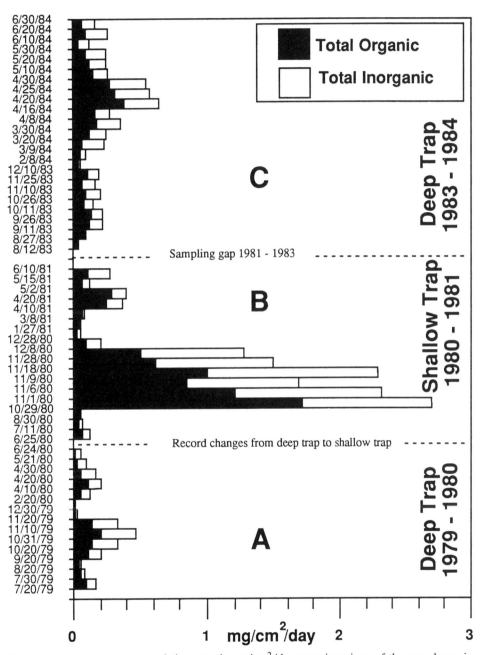

Figure 4. Apparent mass-accumulation rates in mg/cm^2/day at various times of the year shown in context of compositions of the organic (LOI 550 °C) and inorganic fractions. Dates plotted correspond to the top of the sample. Number of days in any sample interval may be found by subtracting the dates that bound the sample. A: Trap record from 28 m depth over the period July 20, 1979, to June 24, 1980; B: Trap record from 12 m depth over the period June 25, 1980, to June 10, 1981; C: Trap record from 28 m depth for the period August 12, 1983, to June 30, 1984. Compositions of the inorganic fractions are plotted in Figure 5.

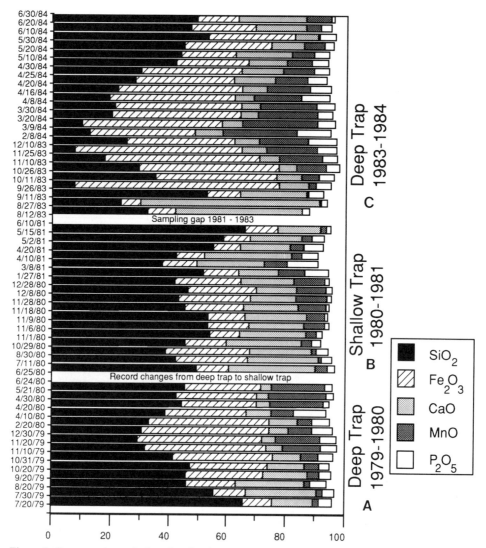

Figure 5. Concentrations of selected major elements reported as weight percent oxides in the inorganic fraction of Elk Lake sediment during the three periods of trap sampling. Deficit summations to 100% are the sums of measured Al_2O_3, Na_2O, K_2O, MgO, and TiO_2 (not plotted). Dates plotted correspond to the top of the sample. Number of days in any sample interval may be found by subtracting the dates that bound the samples. A: 1979–1980 deep trap; B: 1980–1981 shallow trap; C: 1983–1984 deep trap.

week after ice-out. Sedimentation rates began low (Fig. 4B), remained low until overturn, and summer rains had no effects on apparent sedimentation rates. Summer samples were rich in silica, iron, and phosphorus with low organic matter and little calcium (Figs. 4B and 5). Although more than 25.5 cm of rainfall occurred during summer, one on August 20, 1980, of 3.92 cm, any nutrients supplied by runoff were not reflected in increased apparent sedimentation rates or plankton blooms (Figs. 4B and 6B).

In fall of 1980, cooling began about September 13, and by October 11 the base of the hypolimnion extended to a depth of 18 m. By October 15 the lake was in full circulation which persisted until ice-up on November 29. When fall holomictic circulation commenced, the sediments were very enriched in iron and manganese (Fig. 5), but as circulation progressed, rates increased (Fig. 4B) and compositions changed toward enrichment in silica and calcium (Fig. 5). The gradual change in composition occurred as circulation added resuspended material from the lake shallows, and available nutrients promoted diatom blooms dominated by *Fragilaria crotonensis* and chryosphyte scales (Fig. 6B). Cooling kept the lake in holomictic circulation for more than six weeks during a fall season characterized by vacillations in temperature before freezing. Resuspension through these weeks kept apparent sedimentation rates high from early November through early December (Fig. 4B). *Asterionella formosa* bloomed again under ice cover (Fig. 6B) when apparent sedimentation rates remained low. Ice-out occurred early, around April 3, with-

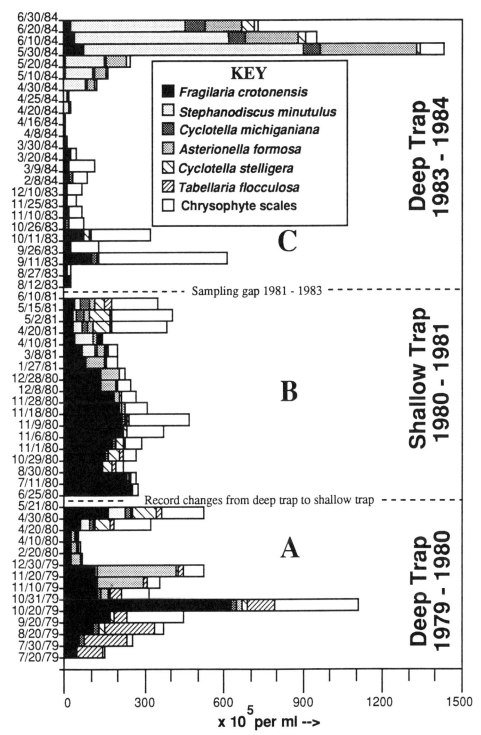

Figure 6. A: Abundances of dominant diatoms in samples collected at depth of 28 m from the 1979–1980 sediment trap. B: Abundances of dominant diatoms in samples collected at depth of 12 m from the 1980–1981 sediment trap. C: Abundances of dominant diatoms in samples collected at depth of 28 m from the 1983–1984 sediment trap.

TABLE 2. COMPOSITION AND ACCUMULATION RATES OF SEDIMENT TRAP CONSTITUENTS*

	Average Compositions			Mass Accumulation Rates		
	1979-80 (wt.%)	1980-81 (wt.%)	1983-84 (wt.%)	1979-80 (mg/cm^2/yr)	1980-81 (mg/cm^2/yr)	1983-84 (mg/cm^2/yr)
SiO_2	42.63	48.33	29.15	11.33	36.13	11.63
Al_2O_3	2.44	2.87	1.59	0.68	2.17	0.56
Fe_2O_3	28.76	15.76	34.5	8.35	11.66	14.65
CaO	10.39	20.3	14.74	2.84	14.55	5.08
Na_2O	0.73	0.68	0.83	0.19	0.46	0.35
K_2O	0.84	0.85	0.91	0.23	0.49	0.36
MgO	0.86	1.71	1.12	0.24	1.20	0.42
TiO_2	0.09	0.19	0.15	0.027	0.128	0.056
MnO	8.95	5.67	10.80	2.13	4.58	4.61
P_2O_5	4.87	3.47	5.48	1.31	1.67	2.29
Sr	0.27	0.02	0.71	0.005	0.015	0.013
Ba	0.019	0.14	0.034	0.072	0.110	0.300
Organic matter	47.2	57.5	53.5	12.40	36.85	21.26
Carbonate-CO_2	3.27	7.12	7	1.18	5.39	3.10
Total	153	164	161	40.97	115.41	54.68

*Compositions are given for elemental oxides as wt.% in high temperature ash, and for organic matter and carbonate-CO_2 as wt.% in whole dried sample. Fluxes of Fe, Mn, and P are higher in 1983-84 than in 1979-80 because the brief spring overturn in 1980 prevented oxygenation and circulation of the water column in spring of that year.

out producing either a striking increase in sedimentation rate or a responsive diatom bloom (Figs. 4 and 6). A 27 day period of spring circulation lasted until May 1, when thermal stratification was established again. The plankton bloom developed in early May and persisted through the end of the sampling period in June (Fig. 6B). This gradually raised the apparent sedimentation rate and silica content of the trap sample (Fig. 5).

The shallow trap of 1980–1981, in conjunction with a shallow trap from 1979–1980, allowed us to contrast shallow-water and deep-water processes in Elk Lake. Trapping rates (Fig. 3C; cf. Fig. 4, A, B, and C) in the shallow trap were, for brief periods, several times higher than in deep traps. Over a year, the mass accumulations collected in the shallow trap were about two to three times that of the deep traps (Table 2). Relative to sediment accumulated in the deep trap, the sediment in the upper water column was enriched in calcium carbonate, strontium, silica (mostly diatomaceous), alumina, and organic matter, whereas sediment from the lower water column was enriched in iron, manganese, phosphorus, and barium.

The shallow trap shows greater variations in sediment mass-accumulation rates between its subsamples (Fig. 4B vs. Fig. 4, A and C), but deeper traps show greater variations in both chemistry (Fig. 5) and plankton populations (Fig. 6). Diatom counts are lower in the trap samples from the upper water column where the seston is enriched in resuspended material. Events are discernible in the shallow trap, but they are not manifested by as large or as abrupt changes in chemistry or plankton in comparison with deeper traps.

1983–1984 Sediment-trap results

The 1983–1984 deep trap recorded higher apparent mass accumulations than the 1979–1980 deep trap (Fig. 4, A and C; Table 2), the major reason being the stronger spring overturn in 1984 as compared with that of 1980. In the 1983–1984 sampling period, sediment flux was low in summer. Cooling began just before mid-September and the epilimnion declined from 8 m on September 14, to 14 m by September 25. The August and September samples were very rich in calcium (Fig. 5) as result of calcite precipitation from the epilimnion in late summer by algae and blue-green cyanobacteria (Dean and Megard, Chapter 8). The sample with the highest calcium concentrations (August 27 to September 11, 1983) had a high organic content, primarily in the form of filamentous algae or cyanobacteria.

In September, moderate increases in apparent sedimentation rates (Fig. 4C) accompanied the bloom of diatoms and chrysophytes (Fig. 6C). The fall bloom (Fig. 6C) was dominated by chrysophyte scales, with lesser amounts of *Fragilaria crotonensis* and *Cyclotella michigianiana*. On October 1, 1983, a cold front entered the area, bringing more than 4 cm of rainfall, and an abrupt temperature decrease by October 4 (Fig. 3B). Lake circulation expanded into the deep depression, and a layer of ferric hydroxide was deposited in the sample by October 11, 1983 (Fig. 5).

Warm temperatures prevented further rapid circulation until about October 15, when the thermocline was at about 18 m. The lake was in full circulation by October 20; diatom blooms persisted and mottled sediment composed of a mixture of newly formed ferric hydroxide mixed with resuspended calcite-rich summer sediment and newly produced diatoms entered the trap through November 25, 1983 (Figs. 5 and 6).

Ice cover was established about November 28, thus making the autumn circulation about a 60-day-long event with full mixing during the last 30 days. Minimum rates of sedimentation occurred from January through early March (Fig. 4C), and diatoms practically disappeared (Fig. 6C). Sediment flux under ice cover in March through mid-April resulted from entrapped copepod migrations, which contributed organic matter and phosphorus (Fig. 5) to the trap. Maximum sedimentation in this study period corresponded to the spring overturn after ice-out about April 22, 1984 (Fig. 4C). Ice-out was followed by significant rains on April 27 and 28. Thermal stratification did not begin until about May 14, thus making the spring overturn a 25 to 30 day event preceding summer stratification. Spring storms in 1984 promoted a long, full circulation period which produced high rates of sedimentation and a large plankton bloom dominated by chrysophytes, *Stephanodiscus minutulus,* and *Asterionella formosa* that gradually increased to a peak by June 10 (Fig. 6C). Resuspension of littoral materials after April 22 increased alumina slightly and also resuspended calcite and diatoms from the shallows of the lake and transported minor amounts of these to the trap. Chironomids were abundant in these spring samples, their presence in accord with the observations of Stark (1976), who noted that they were found in all water depths during spring at Elk Lake.

Seston mineralogy

The study of XRD sediment mineralogy is based on samples from three traps collected during the period July 20, 1979, through June 24, 1980—one shallow trap with formalin hung at 12 m, a deep trap (28 m) with formalin, and a deep trap without formalin (Fig. 7). The presence of formalin in the traps as a preservative was a concern because of the possibility of dissolution of delicate minerals or inhibition of early mineral-forming reactions.

Production of organic matter in Elk Lake accounts for about 50% of total annual sediment weight and consists of settling phytoplankton, zooplankton, and zooplankton pellets. Most of the remaining seston consists of mineraloids, i.e., X-ray amorphous inorganic material. The most abundant mineraloid is biogenic opal, and most of this is from the diatoms *Fragilaria crotonensis, Asterionella formosa, Stephanodiscus minutulus, Tabbelaria flocculosa,* and *Cyclotella* species and from chrysophyte scales and cysts (Bradbury, 1988; Bradbury and Dieterich-Rurup, Chapter 15). Although diatoms and chrysophyte scales are a part of the trapped sediment at all times of the year, the major blooms (Fig. 6, A, B, and C) occur in association with the cycling of nutrients during the spring and fall overturns. Biogenic opal makes up 30%–40% of the inorganic fraction of material that accumulates annually in sediment traps.

The mineral fraction of the sediment (that which can be detected by XRD) is minor and most minerals are endogenic. XRD on small samples with low concentrations of minerals is only semiquantitative, and all of the sediment in the traps is too fine to permit adequate study with a light microscope.

Clastic materials (deduced by looking at concentrations of aluminum, sodium, titanium, and potassium in addition to XRD peaks) compose only a minor part of the sediment trapped (about 10 to 15% of the inorganic fraction and only about 5 to 7% of the whole, dried sediment). Only quartz was revealed clearly on diffractograms, and it was found in all samples from all traps. Some clay-sized quartz is present in the seston of other lakes we have studied that lie within glacial till (Nuhfer, 1979), and its presence here is expected. Increased amounts of clastic material in the shallow traps indicate that concentrations of clastic minerals are enriched in the littoral zone and are cycled in the shallows by wind-driven circulation. Periods of high rainfall did not produce discernible pulses of clastic influx into the traps.

About 10 to 20% of the inorganic fraction of seston consists of calcite. Calcite is far more abundant in the shallow trap (Fig. 8), but also is present in both deep traps. Elk Lake waters contain high concentrations of dissolved calcium and bicarbonate, and the epilimnion is oversaturated with calcite during summer stratification (Dean and Megard, Chapter 8). Calcite precipitated in summer has been detected in trap samples from many lakes, including those not nearly as high in dissolved calcium as Elk Lake. Precipitation of calcite in Elk Lake probably results from blooms of green algae and blue-green cyanobacteria (mostly filamentous types) rather than diatoms, because the peak blooms of diatoms seldom coincide with the late summer peak of calcite precipitation. Coprecipitation of magnesium and some strontium occurs with calcium carbonate in Elk Lake as revealed by positive correlations between MgO and Sr with CaO in the chemical analyses of the sediment trap samples.

Two additional minerals were detected in samples from all three traps (Fig. 7, A, B, and C). These were rhodochrosite ($MnCO_3$), and an iron-phosphate mineral, probably lipscombite [$Fe^{2+}Fe_2^{3+}$ $(PO_4)_2(OH)$] and/or rockbridgeite [(Fe, Mn)Fe_4 $(PO_4)_3(OH)_5$], based on XRD peak occurrences at 4.85 A, 3.18 A, and 1.96 A. The latter mineral will be assumed for discussion purposes. Rhodochrosite is most abundant in the deep trap without preservative and some is found in all but one of the shallow trap samples. In the deep trap with preservative, rhodochrosite was detected in only four samples, three associated with overturns and the fourth being the uppermost sample. Rockbridgeite is found in all three traps but is most prevalent in the deep trap with preservative.

Rhodochrosite is rare in lacustrine sediments. Jones and Bowser (1978) noted that the Eh-pH stability region for rhodochrosite is reached in many lakes where dissolved manganese is sufficiently abundant that rhodochrosite should be found. Because rhodochrosite is abundant in the deep trap without preservative, moderately abundant in the shallow trap, and not abundant in the deep trap with formalin, it is apparent that the

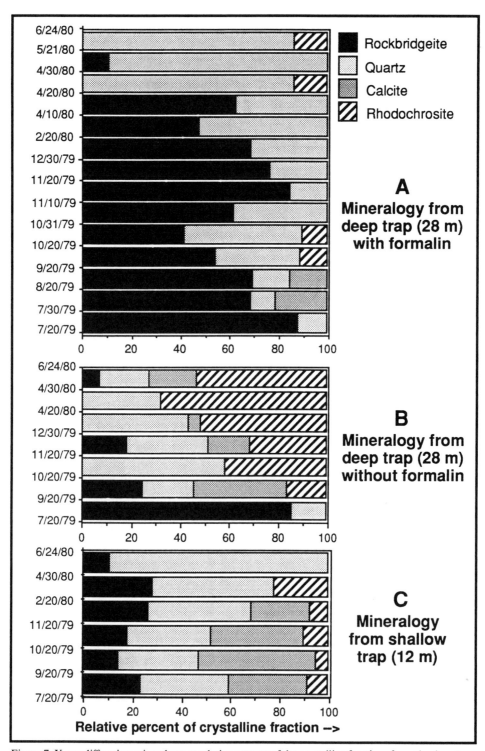

Figure 7. X-ray diffraction mineralogy as relative percent of the crystalline fractions from the deep trap with formalin (A), deep trap without formalin (B), and shallow trap (C). The crystalline fraction is only a small part of the sediment samples.

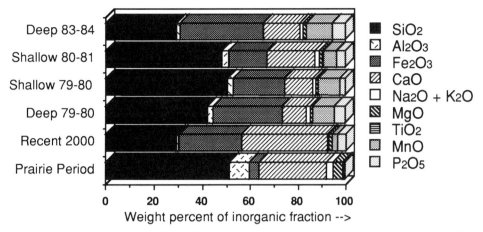

Figure 8. Comparison of compositions of inorganic component of sediment from two deep traps at 28 m (Deep 83-84 and Deep 79-80); the shallow trap (Shallow 80-81); a shallow trap at 12 m (not further detailed in text) that collected enough material (6 samples) to allow use here as an average (Shallow 79-80); the most recent 2000 yr of varve (Recent 2000); and the mean value of the thick clastic-rich varves of 3700–5400 varve yr (Prairie Period).

formation occurs quickly in either the trap or within the hypolimnion, and that the formation of rhodochrosite is inhibited by the formalin if it forms in the trap, or is dissolved by formalin if it forms in the hypolimnion. The general upward increase of rhodochrosite in the deep trap without formalin suggests possible migration of manganese in the accumulation tube and conversion to rhodochrosite within the trap.

Dean and Megard (Chapter 8) postulate that rhodochrosite may form at times of overturn when bicarbonate-rich bottom waters circulate to shallow areas. The trap profiles do not show restrictions of rhodochrosite abundances to periods of overturn, but there is a general tendency toward maximum amounts in spring and late fall (Fig. 7). The rhodochrosite profile in Figure 7B hints that the mineral could form near the sediment-water interface, as MnO_2 in the sediments dissolves, migrates upward, and encounters bicarbonate waters near the interface. Resuspension that placed the upper layer of bottom sediment into traps would then produce the distribution of rhodochrosite observed in the trap in Figure 7B. CO_2 produced by decomposition is suppressed by formalin and possibly accounts for less rhodochrosite in the deep trap with preservative (even though manganese solids are abundant) and in the shallow trap. Further decreases in rhodochrosite in the shallow trap probably resulted from less-abundant total manganese being available to enter the trap in any form at that depth.

Iron phosphate minerals, such as rockbridgeite, are thought to form during early diagenesis, but their abundance in the traps suggests that they could form in the water column. Rockbridgeite has not been previously reported from lakes. McConnell (1943) noted that the mineral is found where bat excrement contacts iron-rich rocks, and Fisher (1966) predicted that it should occur in iron-rich sediments. The occurrence of rockbridgeite apparently has been strongly influenced by preservative (Fig. 7); rockbridgeite dominates the crystalline phases in the deep trap with preservative, but is restricted to the lowermost samples of the trap without preservative. It is found in all subsamples from the shallow trap as a minor constituent.

DISCUSSION

Apparent sedimentation rates and compositions in traps

Varves are annual accumulations of sediment on the lake bottom. In order to compare the sediment composition and accumulation rates of trap samples with those of bottom sediments, an average annual trap sample composition was calculated as well as an average annual mass-accumulation rate (Table 2). From the compacted varve density (1360 mg/cm^3), moisture (70%), and thickness (average about 2 mm), we estimate that "true" average annual mass-accumulation rates to the lake bottom are about 82 mg/cm^2/yr (0.22 mg/cm^2/day) of dry solid material, including 20% combustible organic matter. The annual mass-accumulation rates in the 1979–1980 and 1983–1984 deep traps (Table 2) are less than the "true" rate, and the mass-accumulation rate in the shallow trap is greater than the "true" rate of fill as noted from varves. Traps seem to catch sediment during winter (Fig. 4) at rates that correspond most closely to true rates of fill.

In small lakes, apparent rates of sediment accumulation measured by traps (like our shallow ones that are hung in the water column) are usually greater than actual rates on the lake bottom. Seasonal circulation resuspends large amounts of materials from the littoral zone, and, through time, sediment traps hung in the circulating epilimnion receive and retain a larger dose of resuspended material than does the lake floor, especially the floor of a deep hole such as exists in Elk Lake. Resuspension is such an important process that annual trap accumulations in small shallow lakes are composed of far more resuspended material than new exogenic material (Nuhfer, 1979). In lakes, like Elk Lake, that have well-vegetated drainage basins, periods of peak sediment trap accumulation rates correspond to periods of dimictic circulation. In such lakes, a heavy rainfall does not necessarily produce a clear sediment response in traps. Rainfalls above certain thresholds do produce clear sediment pulses in traps in lakes located in poorly vegetated basins (Nuhfer, 1979; Anderson and others, 1985b).

Mass accumulations recorded in traps through the different lake seasons at Elk Lake are given in Table 3. The amounts in Table 3 are heavily dependent upon the length of the lake's seasons. For example, the lake's spring in 1980 lasted only six days. Even though the mass-accumulation rate was a little more than 0.2 mg/cm^2/day, total mass accumulation over that season was minor.

Disparity between bottom sediment and trapped sediment compositions

Seasonal variations in compositions and rates of accumulation of seston are great (Figs. 4 and 5). The seasons for most bottom accumulation are winter and summer, but in the trap, amounts of sample caught during autumn and spring are often lower than coeval amounts that accumulate on the bottom. For this reason the average composition of a year's trap record may differ from the composition of a varve deposited during the same year.

Figure 8 reveals the disparity in compositions between average annual trap samples and bottom samples at Elk Lake. The composition of sediment that accumulated in the past 2000 years differs in composition from contemporary seston (trap sample) by being far richer in calcium, depleted in silica, and very depleted in manganese. Iron and phosphorus in these bottom sediments appear to be close to the content of the average seston in the lake.

Sediment samples from the mid-Holocene prairie period are high in clastic materials (as reflected in high alumina, Na, and K, Fig. 8; Dean, Chapter 10), and no contemporary seston sample in any traps has comparable compositions. However, all contemporary seston is a little higher in clastic materials than the varves deposited during the last 2000 years (Fig. 8).

Effects of trap location. Sediment traps collect seston from the water column at the depth at which they are suspended. The sediment in traps thus reflects what is in the water column, not what accumulates finally on the lake bottom. A trap sampling more shallow depths is closer to the zone of biogenic production and is more exposed to circulation, with resultant littoral sediment resuspension. Therefore, silica (in clay-sized clastic particles and diatom debris), aluminum, sodium, potassium, and titanium (in clastic particles) are enriched in shallow Elk Lake traps relative to the deeper traps.

At Elk Lake, the mobile elements that precipitated in traps were retained there, and their concentrations in the trap sample are higher than the concentrations in the bottom sediment. Figure

TABLE 3. APPARENT MASS ACCUMULATIONS (mg/cm^2)*

Source of Trap Data	Summer Stratification	Autumn Circulation	Winter Ice Cover	Spring Circulation
1983-84 deep	6.1	8.5	29.2	11.1
1980-81 shallow	5.6	92.8	5.7	11.2
1979-80 deep	13.6	9.9	17.3	0.1

*The seasons labeled here correspond to lake's seasons as defined by thermal events rather than to the standard seasons as defined by calendar dates. At Elk Lake, summer ranges between 150 and 180 days, autumn from 30 to 60 days, winter from 120 to 150 days, and spring from 5 to 30 days in approximate duration.

8 shows that a significant portion of iron is retained in bottom sediments, but most manganese is recycled back to the water column. The high iron content of the bottom samples over the past 2000 yr is much higher than that expected for a typical lake in glacial till. A continuously rich source of iron is required that cannot be accounted for by detrital contributions of hematite and limonite. Local springs such as the "iron spring" (Fig. 1) contain 800 ppm iron, and ground water may account for the iron enrichment of the recent 2000 yr of sediment (Megard and others, Chapter 3).

Trap samples typically have porosity values >85%, typical of very recently deposited uncompacted sediment. Important early diagenetic processes such as settling, compaction, formation of pyrite, microbial decomposition of organic matter, and resolution of mobile constituents don't have time to progress very far in traps that are in place for only a year (Nuhfer, 1990).

The organic content (Table 2) is slightly higher in shallow traps than deep traps, but bottom sediments by comparison are depleted in organic matter. The water confined within the small-diameter accumulation tube does not circulate and contains only a limited supply of oxygen that is quickly used up after a few days of sediment accumulation and decomposition. The sample found in traps is enriched in organic matter relative to bottom sample because the sample is withdrawn from oxygen and grazing plankton that otherwise tend to break down organic solids if they are allowed to persist for longer periods in the lake (Nuhfer, 1979). The normal processes of aerobic and anaerobic decomposition of organic matter that would occur at the bottom are therefore abated, particularly if formalin is used in traps (see also Knauer, 1984).

Iron, silica, and manganese are the dominant inorganic constituents at Elk Lake. Precipitation of iron and manganese oxyhydroxides occurs below about 6 m (the normal depth of the base of the epilimnion in summer) when holomictic circulation mixes bottom waters that are enriched in dissolved iron and manganese with epilimnion waters enriched in oxygen and resuspended sediment. Sediment that is very high in iron oxides can be produced as circulation begins in autumn, when a deepening epilimnion encroaches gently on the deeper water and oxidizes large amounts of iron. This early stage lasts only a few days, before the more vigorous wind-driven circulation churns sediment from the lake shallows and mixes it with the iron-rich endogenic seston.

Precipitation of oxidized particles also takes place when upward migration of iron and manganese in the hypolimnion eventually encounters oxygenated waters above. This produces rains of iron and manganese oxyhydroxide particles that persist for months at Elk Lake during the periods of stratification. Summer and especially winter rains of oxyhydroxides in Elk Lake represent a pattern of sedimentation that is unique among the lakes which we have studied by sediment traps. Upward movement of dissolved iron and manganese in the hypolimnion of Elk Lake through summer is confirmed by water analyses performed in an unpublished study by A. L. Baker in 1967 and by Dean and Megard (Chapter 8).

Under summer stratification, a rain of manganese oxyhydroxides rather than iron hydroxide is more likely, because manganese is more easily reduced than iron. Consequently manganese leaves the sediment in greater quantities and reaches higher concentrations that extend to more shallow depths sooner than those of iron (see Stumm and Morgan, 1981). Calculations by Megard and others (Chapter 3) show that most of the iron that accumulates remains in the sediment, whereas most of the manganese is remineralized and lost to the hypolimnion during summer stratification. Manganese oxide rains occurred in the shallow trap in summer of 1980, after a brief ineffective spring overturn, thus revealing that migration during that summer eventually brought significant dissolved manganese even to depths of less than 12 m.

Fine oxyhydroxide particles have large, chemically active surface areas and can adsorb numerous cations and anions (Tessier and others, 1985; Sigg and others, 1987), including phosphorus and barium. Such rains can probably account for fluxes of ferromanganese oxides and trace metals being greater at Lake Zurich in deeper traps than shallow traps (Sigg and others, 1987). Concentrations of iron, manganese, and, to a lesser extent, phosphorus, are at their peak in the Elk Lake seston after winter stratification (Fig. 5). During this period a large part of the sediment accumulation is endogenic, and up to 80% of the inorganic fraction in deep traps and up to 35% in shallow traps, is attributable to minerals and mineraloids containing these three elements. The slow but consistent rain of oxides into the deep trap during the long periods of summer and winter stratification eventually brings appreciable mass to traps over the course of an entire year.

Effects of events in time. The recent 2000-yr-old bottom sediments in Elk Lake are enriched several times in calcium (as endogenic calcite) and are somewhat poorer in silica relative to either the deep or shallow traps (Fig. 8). Most bottom-sediment accumulation in Elk Lake occurs during seasons of stratification (summer and winter) and is compositionally influenced by the sediment typical of these seasons. In contrast, much sediment-trap accumulation occurs during the seasons of circulation (autumn and spring) when the water column contains more resuspended materials. Thus shallow trap samples are influenced compositionally by a greater proportion of resuspended material. One might explain the enrichment of calcium in bottom samples (Fig. 8) by stating that calcite is more stable than other mobile constituents which are also sedimented but subsequently are recycled back to the water column. If this were indeed the dominant explanation, then bottom sediments should be also enriched in silica. They are not. Instead, the trap samples are enriched in silica because silica forms during the periods of circulation in plankton blooms when sediment is captured in traps at maximum rates.

Disregarding the organic content, the recent 2000 yr of bottom sediment most closely resemble in composition the deep trap samples of early September, 1983, when most collected sediment was endogenic. At that time, rates of accumulation were moderate (~ 0.2 mg/cm^2/day), resuspension was minimal, the seston was high in calcite, diatoms were sparse, some iron was contributed by oxide rains, and manganese had been largely remobilized into the water column. No sample collected in traps resembles the highly clastic varves of the mid-Holocene prairie period, which are characterized by detrital quartz and feldspars and, consequently, higher concentrations of SiO_2, Al_2O_3, Na_2O, and K_2O (Fig. 8; Dean, Chapter 10).

Disparity between bottom sediment and trapped sediment accumulation rates

A number of workers have tried to relate apparent sedimentation rates in traps to true lacustrine sedimentation rates, but the literature shows no clear relation. Because shallow traps sample the zone of circulation, and cores sample the zone of accumulation, the samples from these two zones are not equivalent. Only through fortuitous circumstances should mass-accumulation rates on the lake bottom agree with rates measured in the sediment traps. Representative studies that attempted to relate trap accumulation to bottom accumulation used sediment traps of various designs deployed above and below thermoclines. Most conclusions were based only on variations in bulk mass between shallow and deep sediment-trap catches (see Hilton and others, 1986, for a review). Some workers (e.g., Pennington, 1974, in Blelham Tarn; Sigg and others, 1987, in Lake Zurich) noted greater mass accumulations in deeper traps. This was attributed to sediment focusing (Likens and Davis, 1975; Serruya, 1977). Although sediment focusing is a valid concept (at least for most lakes) that describes a preferred accumulation of sediment on a lake bottom in deeper, sheltered areas, sediment focusing is not consistently reflected by traps capturing higher mass accumulations in these deeper, sheltered areas. Elk Lake is not the first situation in which we have seen higher accumulation rates recorded in shallow traps. In fact, Elk Lake is more the rule than the exception in this regard. Other lakes we have studied (e.g., Soap Lake, Washington, from 1978 to 1983; Williams Lake, Washington, from 1980 to 1984) produced much higher rates of sediment accumulation in the upper traps, even though coring at Williams Lake confirmed sediment focusing (Anderson and others, 1985a). A grid study using traps at Morgan Lake, New Mexico (Nuhfer, 1979; Nuhfer and others, 1985), showed no significant difference between shallow and deep trap accumulation rates. Apparently the actual differences in bottom accumulation that exist between deep sheltered areas and shallow areas are obscured wherever traps capture sediment more efficiently than does the lake bottom.

Influence of lake morphometry on trapping rates. No property of small dimictic lakes has greater influence on trapping rates than lake morphometry. The upper 15 m or so of Elk Lake behaves much like a shallow dimictic lake, and resuspended sediment from the lake shallows constitutes much of the trap catch during seasonal mixing. Gasith (1975) and Nuhfer and Anderson (1985) demonstrated how seasonal resuspension contributes most of the sediment captured annually by traps in shallow dimictic lakes. Seasonal circulation-induced influx is the main cause of trap rates that overestimate true sedimentation rates (Nuhfer, 1990). During circulation, resuspended sediment moves in the water column and enters the trap at high rates coincident with the time when no accumulation is taking place on the lake bottom. Baffles (Fig. 2) do not protect successfully against sediment being introduced to traps during seasonal circulation at rates much higher than true sedimentation rates. During circulation, sediment is captured more efficiently by the trap than by the lake bottom. Fine sediment accumulation is better pictured as the capture of moving particles rather than by the accumulation of particles that simply fall freely like raindrops into a rain gauge. Internal turbulence within the lake can move large amounts of sediment over large areas (Nuhfer, 1988) and can affect the apparent sedimentation rate registered by a sediment trap placed in the water column.

Apparent sedimentation rates noted in traps are highly correlative with lake morphometry (Fig. 9). Lake morphometry should exert a strong influence on sediment-trap accumulations because seasonal circulation is wind driven and a larger surface area provides a greater wind fetch and working surface, whereas a smaller mean depth dictates that a larger fraction of the lake bottom will be scoured during circulation. Håkansson (1982) established a quantitative link between lake morphometry and internal lake sediment transport (Håkansson and Jansson, 1983) by a study of nine Swedish lakes with areas larger than 1 km². Håkansson (1982) noted that the percentage of a lake floor dominated by erosion and transportation was strongly correlated to the "dynamic ratio" (DR), a value obtained by dividing the square root of the surface area (a) in km² by the mean depth (D_m) (in m) (thus DR = $a^{1/2}/D_m$). Because very small lakes often have 100% of their bottom areas involved in erosion and transportation, a plot of DR vs. this percentage is of little use to see differences between small lakes ($a < 1$ km²). However, a plot using sediment-trapping rates instead of percentages of bottom areas can show differences in the degree of resuspension and sediment recycling between small lakes.

Nuhfer (1990) tested the influence of morphometry on trap accumulations at small lakes. Morphometric parameters considered included surface area, mean depth, dynamic ratio, length, orientation of longest axis with dominant regional wind direction, and a "shelter factor," wherein relief at 36 points around the lake was averaged. Of these, multiple regression analyses showed that surface area, mean depth, and length were, in that order, the important predictors of mass accumulations in traps. As a lake's surface area is elongated, the ratio of area of the likely sources of resuspended sediment (littoral zone) to the total lake area increases. Length may be more important in small lakes because no point is very far from shore, and increased lengths in lakes of small area ensure that the trap is close to sources of resuspension from the littoral and shore areas. A "morphometric factor" (MF) devised by multiplying the area in km² (a) by the length (L) in km and dividing by the mean depth (D_m) (in meters) predicts sediment accumulations in traps (Y) in units of mg/cm²/yr (dry sediment weight):

$$Y = 11.7 + 202 \text{ MF (with } r = 0.95). \tag{2}$$

For small lakes studied (Fig. 9), this morphometric factor gave a better prediction of trapping rates than Håkansson's dynamic ratio, although his dynamic ratios were also strongly correlated. In interpreting the varve profile, Figure 9 is useful because it indicates that a dry period with a drop in lake level should result in a group of thicker varves as a greater proportion of the lake bottom is exposed and more vigorous wind-driven resuspension contributes more material to the deep depression during fall and spring. A rise in lake level should result in thinner varves with couplets that are more purely new endogenic material, because of less contribution during spring and autumn from wind-driven resuspension.

Figure 9 also can be used to obtain a rough estimate for maximum currents likely to be achieved in the bottom of Elk

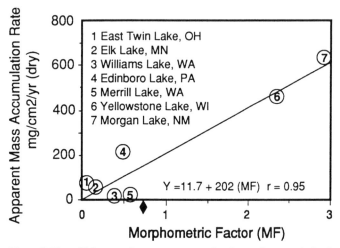

Figure 9. Plot of lake morphometry vs. annual sediment flux rate derived from time-marking sediment trap of type shown in Figure 2. Areas of lakes in this figure range in size from 0.3 to 5.8 km². The mathematical relation between trapping rates and morphometry documents the very strong positive relation between the two. The actual equation may change as more trap data are gathered from other lakes and will likely vary with other kinds of traps.

Lake. Lemmin and Imboden (1987) used photographs of deflections of a tethered submerged float to measure bottom current rates of about 1 cm/s with maxima of 2.5 cm/s at Baldeggersee. Because the MF of Baldeggersee is about 0.7 (diamond on abscissa of Fig. 9), it can be inferred from Figure 9 that the current flows in the bottom of the deep hole of Elk Lake average below 1 cm/s, a velocity sufficient to transport fine particles, but not to erode them once they are deposited (Hjulstrom, 1935).

The deep depression in Elk Lake has accumulated more than 21 m of varved sediments, even though the lake currently lacks a monimolimnion. With 21 m of sediment accumulation, Elk Lake has slowly decreased its mean depth and may be nearing the end of maintaining favorable morphometry for the preservation of varved sediments.

Importance of the design of the sediment trap. Gardner (1980) demonstrated the importance of sediment trap shape on apparent sedimentation rates, and we noted the effects of incorporating baffles in trap design (Nuhfer and Anderson, 1985). There have been many sediment trap studies done in lakes, but the traps varied widely in design and almost no two workers used the same kind. Most traps used in lakes were simple bottles, jugs, or funnels, nearly all of them unbaffled, and few authors gave consideration to effects of trap shape or design on their reported results. This weakness confounds efforts to tie together sediment trap studies from the literature (e.g., Hilton, 1985) in order to make quantitative generalizations about sediment redistribution in lakes. The high correlation in Figure 9 is certainly helped by using data from just one type of trap.

Inflated rates produced by mobile zooplankton. Traps have been calibrated and tested in tanks or flumes (e.g., Gardner, 1980) to see how traps catch sediment. Yet, tank experiments are inadequate to predict trap response in the field because sediments in tanks respond only to the whims of currents. Zooplankton are different because they are mobile and they compose a large amount of the seston caught by traps.

In times of stratification, many zooplankton, particularly copepods, establish themselves in dense populations at discrete depths in response to stratification of phyptoplankton populations on which they graze. As vertical migration occurs, copepods pass over the mouth of the sediment trap, each pass literally pumping thousands of individuals into the trap, where they die and enter the accumulation tube. The resultant accumulation seen in the trap record is thus a synthetic event produced by the trap.

Pumping of copepods is important at Elk Lake in spring. This was particularly true in one 1978–1979 trap when the accumulation tube was almost completely filled with copepods and little use could be made of the sample. Copepod grazing results in pelletization and sedimentation of diatoms and other phytoplankton (Haberyan, 1985) and probably other solids. In Williams Lake, Washington, we noted that thick accumulations of copepods (introduced by plankton pumping) in traps in spring and winter of 1981 were associated with increased deposition of fine tephra from the May 18, 1980, eruption of Mount St. Helens (Anderson and others, 1985b). In Elk Lake, elevated silica values in samples from March 30 to April 16, 1980, are concurrent with increases in sedimentation rate, organic matter, and P_2O_5. This may result from copepod migration and pelletization of detrital material along with food.

Use of Trap Results in Varve Interpretation

The traps provide a means for indirect examination of processes that contribute to bottom sedimentation and allow an investigator to resolve effects of processes that occur only briefly in time (Fig. 10). When Elk Lake is stratified, sedimentation occurs within a few days between shallow production of materials and bottom deposition. When seasonal mixing occurs, suspended sediment is generally homogenized throughout the circulating zone; then sediment is advected to the lake depths at an even faster rate. In spite of complications noted in the previous section of this chapter, an "ideal modern Elk Lake varve" can be constructed (Fig. 11) from trap records alone.

In thin sections from the upper part of the stratigraphy of Elk Lake, most varves generally have three to four microlaminae (Fig. 12). The most distinctive microlaminae are the summer carbonate layers, and these tend to set off the other laminae and help to define the annual package of sediment (Anderson, Chapter 5). Summer layers may contain smaller centric diatoms from the spring bloom that settle after stratification is established. Above the summer layer is the fall overturn deposit, characterized in varying thickness and distinctiveness (depending on the year), by a mottled fabric with a mixture of diatoms, ferric hydroxide, and organic matter, as well as some carbonates. This autumn layer is overlain by a thick, rather uniform winter layer of fine, brown organic matter which probably derives additional color from colloidal ferric hydroxide and some manganese oxides. The upper and lower boundaries of this winter organic layer appear darker, perhaps as an artifact of thin-section preparation, but possibly because these are more manganese-rich parts of the lamination. Above the brown winter laminae is another mottled zone, the spring layer, that is similar in appearance and distinctiveness to the autumn microlaminae. The thickness of the spring layer depends on the vigor and duration of overturn. Because the spring overturn can be either of long duration if produced by a cool, windy spring, or of short duration if spring is characterized by abrupt warming and stratification (as occurred in spring of 1980), the spring microlaminae are variable and may even be missing in some years. The end of spring is consistently shown by the deposition of the next summer carbonate layer.

Sometimes the overturns produce such abundant diatom blooms that the events are marked by very pure layers of one or two species (Fig. 13). With burial, compaction occurs most obviously in the winter organic layers; below 15 m the organic layers may become mere wisps.

The effects of lake processes on sediment traps and bottom accumulation in Elk Lake are summarized in Figure 14. Additional descriptions of varves are found in Anderson (Chapter 5). The epilimnion (Fig. 14) is a zone of resuspension from lake

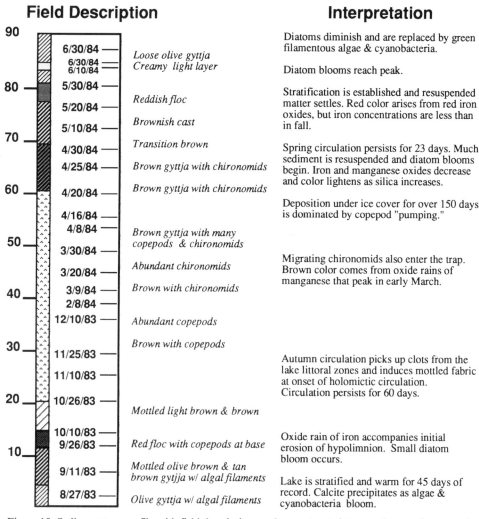

Figure 10. Sediment-trap profile with field descriptions and reconstructed temporal events interpreted for the 1983–1984 sediment-trap record. Vertical scale is in centimeters.

shallows for most of the year. The zone of circulation reaches to the bottom during overturns and oxygenates the lake column, but does not reentrain sediment from the deep hole. Shallow traps spend a greater part of the year in the zone of resuspension, and sediments in shallow traps have more littoral materials and higher rates from capturing of resuspended material during fall and spring. In the hypolimnion at Elk Lake, large amounts of manganese, iron, and phosphorus are dissolved and migrate upward. When they encounter oxygen at shallower depths (Interface in Fig. 14), the metals oxidize and rain downward into the deep trap. Relative to shallow traps, deep traps have slower apparent sedimentation rates and the samples have higher concentrations of manganese, iron, and phosphorus that accumulate mostly in summer and winter. Less resuspended material reaches the deep traps and lake bottom. Even deep trap samples differ from bottom sediments because diagenesis occurs in the varved bottom sediments.

Anderson and others (Chapter 1) note that the history of Elk Lake includes three periods. During the early lake period between 10,300 and 8300 varve years, rhythmic sedimentation was dominated by a series of thin (<1 mm) varves consisting of light carbonate layers and dark iron- and manganese-rich layers precipitated within a lake surrounded by coniferous forest. The lake would have been much deeper at that time with a lower morphometric factor and contribution from resuspended sediments would have been very small. Thus these oldest varves reflect almost pure summer-winter deposits. A monimolimnion made more dense by dissolved iron and manganese could conceivably have existed then in the deep depression.

The second, or prairie lake, period (about 8400 to 3700 varve yr) was characterized by thick (2 to 4 mm) varves composed of detrital clastic materials separated by thin endogenic carbonate- and organic-rich layers. We know that traps in lakes inside drainage basins that are sparsely vegetated reveal sediment

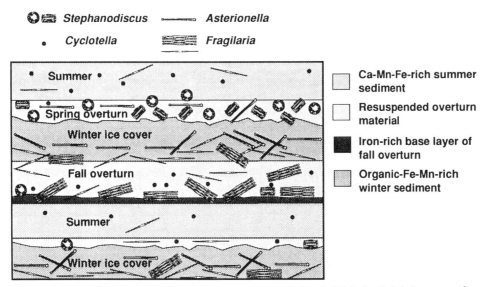

Figure 11. An idealized varve based on sediment trap accumulations in Elk Lake. Labeled seasons refer to lake seasons, not calendar seasons. Thickness of spring laminae are particularly variable from year to year. Today, resuspended material consisting of organic detritus, calcite, and older diatoms is a common constituent of fall and spring microlaminae. Older lake varves show overturns with more pure diatomaceous material and less contamination from littoral zones.

pulses in response to rainfall (Nuhfer and others, 1985; Anderson and others, 1985b). A dry climate with sparse ground cover would favor sheetwash and wind transport of clastic material to enrich the lake sediment in detrital sediment. A dry period also indicates a shallower lake that is more susceptible to wind-driven resuspension and mobilization of littoral sediments into the deeper parts of the lake. Thus the prairie period sediments have more contribution from the littoral area, and sample compositions are more influenced by autumn and spring contributions than are the varves of the early lake period. Within the prairie period was a 500 yr period (4800 to 5300 varve yr) characterized by thin, uniform (<1.5 mm) laminae. Our data from traps lead us to interpret that 500 year period as a return to wet climatic conditions with a rise in lake level, a reduced seasonal resuspension contribution to the deep hole, and a return to sediment dominated by winter-summer laminae.

The modern lake period is represented by ~2-mm-thick varves that consist of iron-rich laminae separated by organic-rich and carbonate-rich laminae. These have been deposited since the lake was surrounded by mesic forest about 3700 varve years ago. These modern varves, depicted in Figures 11 and 12, consist of laminae dominantly contributed in winter and summer but with varying, significant contributions from the lake shallows during autumn and spring. More autumn and spring contribution now occurs because more than 20 m of sediment deposited over the past 10,000 yr have made the lake more shallow, and more of the bottom is vulnerable to resuspension. In this latter period, most littoral sediment consists of endogenic material. The well-vegetated drainage basin now prevents runoff from contributing much clastic material to the littoral zones, where sediment is dominantly recycled endogenic silica and calcite.

Most of the sediment characteristics displayed by the three periods of the lake are recapitulated somewhere within the modern annual record of sedimentation. Only a well-established clastic influx is missing from modern (trap) records. On the basis of trap studies we have made at other lakes with well-vegetated drainage basins, this is expected.

CONCLUSIONS

Accumulations in sediment traps show that modern laminations in Elk Lake are created by distinct seasonal processes. Sediments typical of modern Elk Lake consist of endogenic material: organic matter, Fe-Mn oxyhydroxides, biogenic silica, endogenic calcite, and either endogenic or very early diagenetic rhodochrosite and iron phosphate.

Deposition of Mn and Fe oxyhydroxides on the lake bottom occurs during both summer stratification and winter stagnation. These seasons are long in the lake, about 150 days each, and although sedimentation rates are slow, the longer proportion of the year dedicated to precipitation during these seasons makes them the dominant periods of modern varve building. Most of the manganese deposited is redissolved back into the water column, but most iron remains in the bottom sediment. Continuous ground-water contributions of dissolved iron to the lake may have enabled the lake to continue to deposit iron-rich sediments over the past 2000 years.

Resuspension during spring and autumn transports solid materials to, but not from, the deep depression. Seasonal resuspensions are important to the varve-forming process in Elk Lake, and resuspended material is mostly recycled endogenic material from the lake shallows. Factors that would favor thicker varves

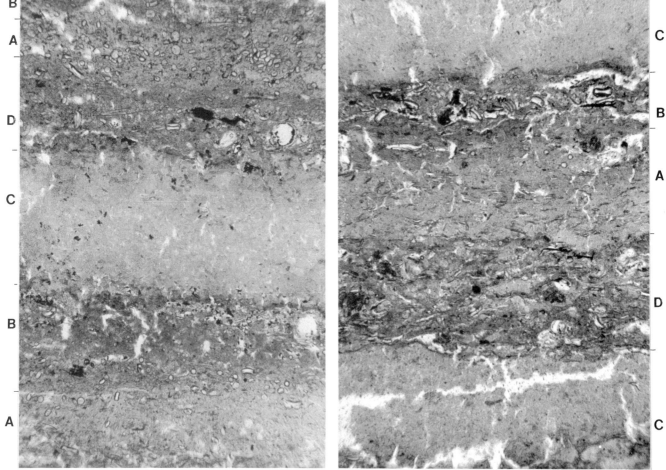

Figure 12. Thin section of laminated sample (JD 26-28) from Elk Lake core about 1500 varve yr. Left (140 ×) is annual varve centered on winter lamina (C) revealing: A—Light brown calcitic summer sample with small centrales diatoms; B—darker red-brown autumn material of consisting of Fe-Mn oxides, resuspended material and diatoms; C—golden-brown Fe-Mn organic-rich lamina of winter; D—moderate brown mottled spring resuspended material with various diatoms including larger genera. Right (140 ×) is annual varve centered on summer sample (A). Seasonal laminae have same letter designations as at left. Mottling in spring sample (D) shows well. Varves from the past 2000 yr have significant contributions during all seasons.

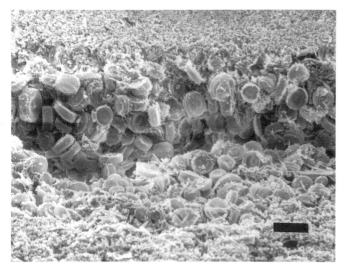

Figure 13. Scanning electron micrograph of a relatively pure microlamina of the diatom *Stephanodiscus niagare*. Bar scale = 100 μm.

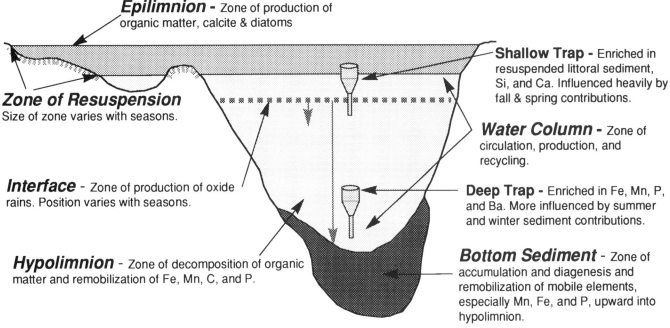

Figure 14. Schematic (not to scale) depicting composite materials and processes at Elk Lake that affect trap and bottom sediment compositions and fluxes.

include (1) climate conditions that prolong the periods of circulation and shorten the periods of winter ice cover, (2) climate changes that cause lowering of the mean depth of the lake and produce sparser vegetation in the lake basin, (3) morphometric changes that result from infilling of the lake basin with sediments, and (4) conditions that enrich nutrients in the water sufficient to cause heavy plankton blooms and increase of biogenic contributions to sediment.

Trap records provide insight into processes that created the Elk Lake sediment profile. However, sedimentation rates and sediment compositions deduced from sediment traps should not be presumed as being equivocal to those of bottom sediments.

ACKNOWLEDGMENTS

Edward Nuhfer thanks the U.S. Geological Survey for the intermittent support between 1980 and 1992 as a W.A.E.-status employee. That support permitted lab analyses. He also thanks his former employer, the University of Wisconsin at Platteville, for support of research at Yellowstone Lake, Wisconsin, between 1985 and 1991 that was critical to the derivation of Figure 9, and for the sabbatical leave to Colorado in 1988–1989 that permitted him to devote time to this manuscript as well as to a number of other projects. He also thanks his present employer, the University of Colorado at Denver, for their financial contribution in support of this publication through the Fund for University Scholarly Publications. All of us appreciate the efforts of David Bosanko and Jon Ross of the Itasca Biological Station for their help in field work at Elk Lake and those of our manuscript reviewers Al Swain and John King, who provided both encouraging comments and useful suggestions.

REFERENCES CITED

Anderson, R. Y., 1977, Short term sedimentation response in lakes in western United States as measured by automated sampling: Limnology and Oceanography, v. 22, p. 423–433.

Anderson, R. Y., Nuhfer, E. B., and Dean, W. E., 1985a, Sinking of volcanic ash in uncompacted sediment in Williams Lake, Washington: Science, v. 225, p. 505–508.

—— , 1985b, Sedimentation in a blast-zone lake at Mount St. Helens, Washington—Implications for varve formation: Geology, v. 13, p. 348–352.

Bradbury, J. P., 1988, A climatic-limnologic model of diatom succession for paleolimnological interpretation of varved sediments at Elk Lake, Minnesota: Journal of Paleolimnology, v. 1, p. 115–131.

Dean, W. E., 1974, Determination of carbonate and organic matter in calcareous sediments and sedimentary rocks by loss on ignition: Comparison with other methods: Journal of Sedimentary Petrology, v. 44, p. 242–248.

Fisher, D. J., 1966, Cacoxenite from Arkansas: American Mineralogist, v. 51, p. 1811–1814.

Gardner, W. D., 1980, Sediment trap dynamics and calibration: A laboratory calibration: Journal of Marine Research, v. 38, p. 17–39.

Gasith, A., 1975, Tripton sedimentation in eutrophic lakes—Simple correction for the resuspended matter: Internationale Vereinigung fuer Theoretische und Angewandte Limnologie, Verhandlungen, v. 19, p. 116–122.

Haberyan, K. A., 1985, The role of copepod fecal pellets in the deposition of diatoms in Lake Tanganyika: Limnology and Oceanography, v. 30, p. 1010–1023.

Håkansson, L., 1982, Bottom dynamics and morphometry—The dynamic ratio: Water Resources Research, v. 18, p. 1444–1450.

Håkansson, L., and Jansson, M. P., 1983, Principles of lake sedimentology: Berlin, Springer-Verlag, 316 p.

Hilton, J., 1985, A conceptual framework for predicting the occurrence of sediment focusing and sediment redistribution in small lakes: Limnology and Oceanography, v. 30, p. 1131–1143.

Hilton, J., Lishman, J. P., and Allen, P. V., 1986, The dominant processes of sediment distribution and focusing in a small monomictic lake: Limnology and Oceanography, v. 31, p. 125–133.

Hjulstrom, F., 1935, Studies of the morphological activity of rivers as illustrated by the River Fyris: University of Upsala Geology Institute, Bulletin, v. 25, p. 221–527.

Jones, B. F., and Bowser, C. J., 1978, The mineralogy and related chemistry of lake sediments, *in* Lerman, A., ed., Lakes—Chemistry, geology, physics: Berlin, Springer-Verlag, p. 179–236.

Knauer, G. A., 1984, In-situ effects of selected preservatives on total carbon, nitrogen and metals collected in sediment traps: Journal of Marine Research, v. 42, p. 445–462.

Lemmin, U., and Imboden, D. M., 1987, Dynamics of bottom currents in a small lake: Limnology and Oceanography, v. 32, p. 62–75.

Likens, G. E., and Davis, M. B., 1975, Post-glacial history of Mirror Lake and its watershed in New Hampshire, U.S.A.—An initial report: International Vereinigung fuer Theoretische und Angewandte Limnologie, Verhandlungen, v. 19, p. 982–993.

McConnell, D., 1943, Phosphatization at Malpelo Island, Colombia: Geological Society of America Bulletin, v. 54, p. 707–716.

Nuhfer, E. B., 1979, Temporal and lateral variations in the geochemistry, mineralogy and microscopy of seston collected in automated samplers from selected lakes in Ohio, Pennsylvania and New Mexico [Ph.D. thesis]: Albuquerque, University of New Mexico, 396 p.

——, 1988, Comment *on* "Sediment yield history of a small basin in southern Utah, 1937–1976: Implications for land management and geomorphology": Geology, v. 16, p. 759–760.

——, 1990, Lacustrine sediment trap catches and bottom sediments—The reasons for disparity in fluxes and composition: Geological Society of America Abstracts with Programs, v. 23, no. 7, p. A318.

Nuhfer, E. B., and Anderson, R. Y., 1985, Changes in sediment composition during seasonal resuspension in small shallow dimictic inland lakes: Sedimentary Geology, v. 31, p. 131–158.

Nuhfer, E. B., and Renton, J. J., 1979, Preparation of membrane filter mounts for X-ray diffraction, *in* Nuhfer, E. B., and others, eds., Procedures for petrophysical, mineralogical and geochemical characterization of fine-grained clastic rocks and sediments: National Technical Information Service/Morgantown Energy Technology Center/CR79126, 39 p.

Nuhfer, E. B., and Romanosky, R. R., 1979, Chemical analysis of earth materials from $LiBO_2$-HCl-HF dissolutions, *in* Nuhfer, E. B., and others, eds., Procedures for petrophysical, mineralogical and geochemical characterization of fine-grained clastic rocks and sediments: National Technical Information Service/Morgantown Energy Technology Center/CR79126, 39 p.

Nuhfer, E. B., Anderson, R. Y., and Dean, W. E., 1985, Some applications of time-marking sediment traps and their implications for monitoring of sediment discharge from disturbed areas: Lexington, Kentucky, Proceedings of the National Symposium on Surface Mining, Sedimentation, Hydrology, and Reclamation, p. 9–18.

Pennington, W., 1974, Seston and sediment in five Lake District lakes: Journal of Ecology, v. 62, p. 215–251.

Serruya, C., 1977, Rates of resuspension at Lake Kinneret, *in* Golterman, H. L., ed., Interactions between sediments and fresh water: The Hague, W. Junk B. V. Publishers, p. 48–56.

Sigg, L., Sturm, M., and Kistler, D., 1987, Vertical transport of heavy metals by settling particles at Lake Zurich: Limnology and Oceanography, v. 32, p. 112–130.

Stark, D. M., 1976, Paleolimnology of Elk Lake, Itasca State Park, northwestern Minnesota: Archiv für Hydrobiologie Supplement 50, p. 208–274.

Stumm, W., and Morgan, J. J., 1981, Aquatic chemistry (second edition): New York, Wiley, 780 p.

Tessier, A., Rapin, F., and Carignan, R., 1985, Trace metals in oxic lake sediments: Possible adsorption onto iron oxyhydroxides: Geochimica et Cosmochimica Acta, v. 49, p. 183–194.

U.S. Naval Observatory Nautical Almanac Office, 1985, The astronomical almanac: Washington, D.C., Government Printing Office, p. A14–A21.

Manuscript Accepted by the Society July 27, 1992

Environment of deposition of CaCO$_3$ in Elk Lake, Minnesota

Walter E. Dean
U.S. Geological Survey, MS 939, Box 25046, Federal Center, Denver, Colorado 80225
Robert O. Megard
Department of Ecology, Evolution, and Behavior, University of Minnesota, Minneapolis, Minnesota 55455

ABSTRACT

Elk Lake is near the present forest-prairie border in northwestern Minnesota, and is also located on the boundary between hard-water lakes that are typical of once-glaciated parts of the north-central United States and more saline prairie lakes of western Minnesota and the Dakotas. The sediments of the prairie lakes just west of Elk Lake are unusual in that they commonly contain high-Mg calcite and dolomite in addition to low-Mg calcite, which is the dominant carbonate mineral in most marl lakes. During the mid-Holocene dry period, prairie conditions expanded eastward into the forested regions of Minnesota. Variations in types and abundances of carbonate minerals in the Holocene sediments of Elk Lake recorded this climatic change.

Studies of primary productivity, carbonate saturation, water chemistry, and sediment-trap samples show that low-Mg calcite precipitates during the summer, triggered by algal photosynthesis. The epilimnion of Elk Lake is always oversaturated with calcite, and the degree of oversaturation increases progressively during the summer. The pH of the epilimnion increases from <8.0 after spring overturn to almost 9.0 in late summer in response to photosynthetic removal of CO$_2$ during the summer months. The rate of calcium depletion from the epilimnion is proportional to the increase in pH and the rate of photosynthetic carbon fixation.

Today the only carbonate minerals that are accumulating in the sediments of Elk Lake are low-Mg calcite and manganese carbonate (rhodochrosite). Rhodochrosite, and probably manganese oxyhydroxide, precipitates when manganese-rich anoxic bottom waters come in contact with carbonate-rich oxic surface waters. During the arid mid-Holocene prairie period, however, low-Mg calcite, dolomite, aragonite, and rhodochrosite all accumulated in the sediments of Elk Lake. Dolomite formed in Elk Lake during this period in response to a higher Mg:Ca ratio in the water, just as it is forming today in lakes of the prairie regions of western Minnesota. The coincident occurrence of aragonite and biological indicators of high salinity suggests that the salinity of Elk Lake and the Mg:Ca ratio were higher than in any of the present prairie lakes of western Minnesota.

INTRODUCTION

Temperate-zone lakes that have drainage basins in carbonate bedrock or calcareous glacial drift commonly have high concentrations (usually more than 1.0 milliequivalent per liter, or epm) of total dissolved alkaline-earth cations, mostly calcium and magnesium, and are saturated with CaCO$_3$ as low-Mg (<4 mole% Mg) calcite at least during the late summer. The sediment in these lakes typically contains more than 30% CaCO$_3$. Such CaCO$_3$-rich sediment is called marl, and lakes that contain marl are often referred to as marl lakes. The major sources of CaCO$_3$ in marl lakes are (1) inorganically precipitated CaCO$_3$, (2) photosynthesis-induced, inorganically precipitated (bio-induced; Dean, 1981) CaCO$_3$, (3) biogenic CaCO$_3$ in the form of remains of calcareous plants and animals, and (4) detrital carbonate minerals derived from carbonate rocks in the drainage basin. Most of this $CaCO_3$ is in the form of low-Mg calcite, although minor amounts of high-Mg (>4 mole% Mg) calcite, dolomite, and/or aragonite may occur (Müller and others, 1972; Jones and Bowser, 1978; Kelts and Hsü, 1978; Dean, 1981; Dean and Fouch, 1983).

Biogenic $CaCO_3$, mainly detritus from molluscs and macroscopic calcareous algae (e.g., *Chara*), may be the dominant source of $CaCO_3$ in the shallow-water (littoral) part of a marl lake, and thick deposits of $CaCO_3$ commonly form a platform or marl bench out from shore to the depth limit of rooted aquatic vegetation. Some of this biogenic $CaCO_3$ may be transported to the deeper (profundal) part of the lake by slumping and turbidity currents, but most $CaCO_3$ in profundal sediments is produced in the surface waters (pelagic zone) by inorganic or bioinduced precipitation (Dean, 1981). Pelagic, microscopic calcareous organisms are rare in lakes and consist of only a few species of calcareous algae. This is the most important difference between pelagic lacustrine and marine carbonates; pelagic marine carbonates consist entirely of the remains of calcareous planktonic plants and animals.

The bottom waters of Elk Lake are anoxic for enough of the year to limit burrowing organisms and permit the preservation of annual sediment laminations (varves; Dean and others, 1984; Anderson, Chapter 5). Varved sediments are common in temperate lakes (Anderson and others, 1985), but varved sediments containing many climatically sensitive chemical, mineral, and biologic components, as are found in the sediments of Elk Lake (Dean and others, 1984), are rare. Elk Lake also is near the present forest-prairie border (Fig. 1) and near the boundary between hard-water lakes that are common in areas underlain by calcareous glacial till between western Minnesota and New York and more saline prairie lakes of western Minnesota and the Dakotas (Gorham and others, 1982, 1983) (Fig. 2). The prairie lakes just west of the forest-prairie border in Minnesota are unusual in that their sediments commonly contain high-Mg calcite and dolomite (Dean and Gorham, 1976a) in addition to low-Mg calcite, which is the dominant mineral in most marl lakes. Therefore, Elk Lake and other lakes in northwestern Minnesota are ideally located for studying processes involved in the deposition and diagenesis of carbonate minerals in marl lakes, and particularly in changes in carbonate deposition through time. Elk Lake is especially well suited for these studies because the varved sediments provide precise calibration for studying the rates and timing of changes as the forest-prairie border shifted back and forth in response to changes in the balance between evaporation and precipitation. Elk Lake and other lakes in Itasca State Park (Fig. 1), Minnesota, also are ideal because they have been exposed to a minimum of anthropogenic disturbance.

Our investigations into the deposition of $CaCO_3$ in Minnesota lakes in general and in Elk Lake in particular date back to the mid-1960s. At that time, Megard measured the photosynthetic activity of phytoplankton in a number of lakes throughout Minnesota. In 1965, Kenneth Deffeyes suggested that carbonate saturation measurements be made routinely during the primary productivty experiments. The suggested measurements were made for several years and the results were reported by Megard (1967, 1968). In 1967, Dean began an investigation of the mineral and chemical composition of surface sediments throughout Minnesota (Dean and Gorham, 1976a, 1976b). A regional investigation of water chemistry of Minnesota lakes also was conducted at that time, and this investigation was later expanded to include lakes in adjacent states (Gorham and others, 1983). During these investigations the complex carbonate mineralogy of the prairie lakes was realized, and the varved character of Elk Lake sediments discovered. In 1978, the U.S. Geological Survey Program on Climate Change initiated several investigations to provide high-resolution paleoclimatic records of the Holocene. The varved character of the Elk Lake sediments, its location near the climatically sensitive line of zero net balance between precipitation and evaporation, and the presence of several sediment components (e.g., carbonate minerals) sensitive to this climatic boundary suggested that Elk Lake had the potential of providing a very high resolution (annual or even seasonal) paleoclimatic record. Our initial studies indicated that the carbonate minerals in Elk Lake sediments probably would be useful climate proxy variables. In order to better interpret changes in composition of Elk Lake sediments with time, additional studies on carbonate deposition in Elk Lake were needed. To this end, water chemistry investigations were conducted in 1979 and 1980.

In this chapter we examine the environment of deposition of carbonate minerals in Elk Lake from three different perspectives. First, we examine carbonate deposition in Elk Lake within the context of carbonate deposition in lakes across the steep climatic gradient from forested regions of eastern Minnesota to prairie regions of western Minnesota. We document this regional framework in terms of water chemistry as well as sediment chemistry and mineralogy. Second, we discuss the results of studies on water chemistry, carbonate saturation, and primary productivity as they relate to carbonate deposition. Finally, we apply what we have learned about the formation of carbonate minerals in modern lake sediments to interpret the carbonate mineralogy of the Elk Lake core.

CARBONATE DEPOSITION IN MINNESOTA LAKES

Water chemistry

An empirical approach to the precipitation of $CaCO_3$ is based on observations of water chemistry and amount of $CaCO_3$ in profundal sediments in lakes of the north-central United States (Fig. 3). Precipitation of $CaCO_3$, as indicated by the presence of measurable amounts of $CaCO_3$ in profundal sediments, does not occur until a log cation value of about 0.3 (2.0 epm) is reached. In lakes having total cation concentrations above this threshold, the amount of $CaCO_3$ that actually does precipitate is highly variable, as indicated by the large range of $CaCO_3$ content in sediments from lakes within a narrow range of total cation con-

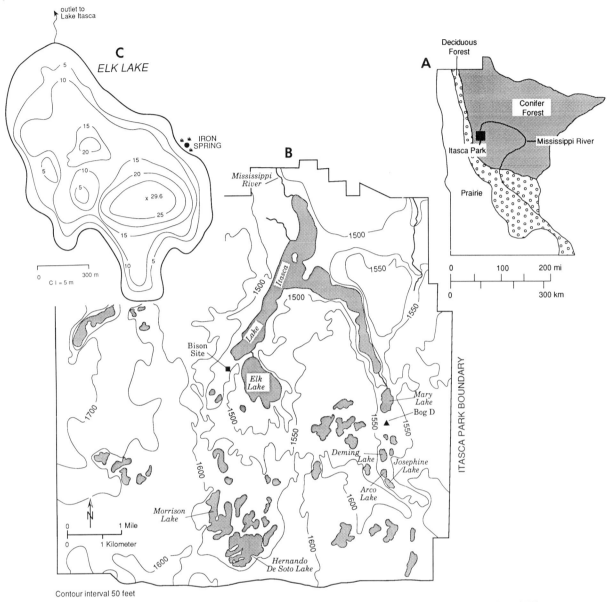

Figure 1. Maps of (A) Minnesota, (B) Itasca State Park, and (C) Elk Lake, showing the location of Elk Lake, general vegetation zones of Minnesota, topography and distribution of lakes in Itasca Park, bathymetry of Elk Lake, and location of the varved core (x) in the deepest part of the lake (29.6 m).

centrations, and by the large range of carbonate contents in surface sediments from both Elk Lake and Lake Itasca (Fig. 3). This large variability in carbonate content is due to local conditions of carbonate precipitation and preservation within individual lakes. The most important single factor causing this variability is the amount of precipitated carbonate that is subsequently dissolved in deeper, CO_2-charged bottom waters and within the sediments. The lower $CaCO_3$ content of sediments in the five lakes that have log cation values greater than 0.7 (Fig. 3) is due in part to the lower rate of $CaCO_3$ precipitation in these Ca-depleted prairie lakes, and in part to greater dilution by wind- and stream-borne clastic material.

Changes in relative proportions of dissolved calcium and magnesium can be used to monitor the precipitation of $CaCO_3$ in lakes across the steep precipitation-evaporation gradient between Wisconsin and North and South Dakota (Fig. 4). With increasing salinity (increasing log cations), the relative proportions of both calcium and magnesium increase at the expense of sodium and potassium. However, once carbonate begins to precipitate, at about 0.5 log total cations, the relative proportion of calcium decreases and the relative proportion of magnesium increases until magnesium replaces calcium as the dominant cation. The most drastic change in relative proportions of calcium and magnesium, caused by precipitation of large amounts of $CaCO_3$,

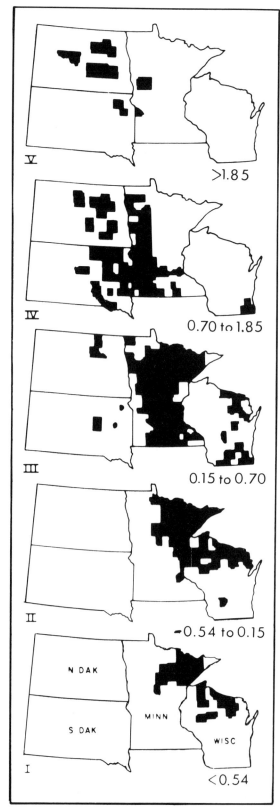

Figure 2. Maps of the geographic distribution (by county) of lakes in the north-central United States, grouped (I–V) in order of increasing log total cation concentration (in milliequivalents per liter, epm). Modified from Gorham and others (1983).

occurs at a log cation value of about 0.4 (2.5 epm) (Fig. 4). This point is what Eugster and Jones (1979) called the "calcite branch point" during closed-basin brine evolution. In a simplified model, lake waters in Minnesota can be viewed as the early stages of closed-basin brine evolution in which a beaker of calcium bicarbonate water from eastern Minnesota is transported across the steep precipitation-evaporation gradient of the state (e.g., Hall, 1972). As water in the beaker is evaporated, the calcite branch point is reached at a total cation concentration of about 0.4 epm, calcite is precipitated, and the water "evolves" from a calcium bicarbonate water to a calcium-magnesium bicarbonate water. This model corresponds to evolution path I of Eugster and Jones (1979), in which the bicarbonate concentration is greater than the calcium concentration (as in most Minnesota lakes), and the water becomes enriched in carbonate and depleted in calcium. This model is highly simplistic because Minnesota lakes do not represent a closed system, but it does describe the behavior of lake waters up to a log total cation value of about 0.5 epm (ionic strength of about 0.005). According to the Eugster and Jones model, the precipitation of calcite results in an increase in the Mg:Ca ratio in both the residual water and in the precipitating calcite. Notice that the log cation values of Elk Lake are just above the critical calcite branch point (0.42 to 0.56 log cations; Fig. 4). Therefore, carbonate precipitation in Elk Lake and the magnesium content of that precipitated carbonate should be sensitive indicators of salinity variations in the lake with time, in response to climatically induced changes in the delicate balance between precipitation and evaporation.

Mineralogy

In lakes with log cation values greater than 0.4, the absolute values of calcium and magnesium generally continue to increase with increasing salinity, but there is considerably more scatter in the data, particularly for calcium (Fig. 5). With the exception of about five lakes that are particularly enriched in sodium, the concentration of magnesium increases linearly with increasing total-cation concentration throughout the entire range, indicating that magnesium is conservative in that coprecipitation of magnesium with calcite has little effect on the concentration of magnesium in solution.

Most prairie lakes (lakes in Groups IV and V of Gorham and others, 1983) (Fig. 2), have Mg:Ca ratios greater than 1.0, and some are as high as 10. The ionic proportions in these prairie lakes are further complicated because concentrations of both sodium and sulfate increase considerably with increasing salinity until in the most-saline lakes (western Group IV and Group V of Gorham and others, 1983) (Fig. 2), sodium is the dominant cation and sulfate replaces bicarbonate as the dominant anion. It is because of the high Mg:Ca ratio in the prairie lakes that most sediments of these lakes contain high-Mg calcite and dolomite (Dean and Gorham, 1976a), although some of the dolomite in some of these lakes may be detrital. Empirical observations by Müller and others (1972) on the composition of sediments from

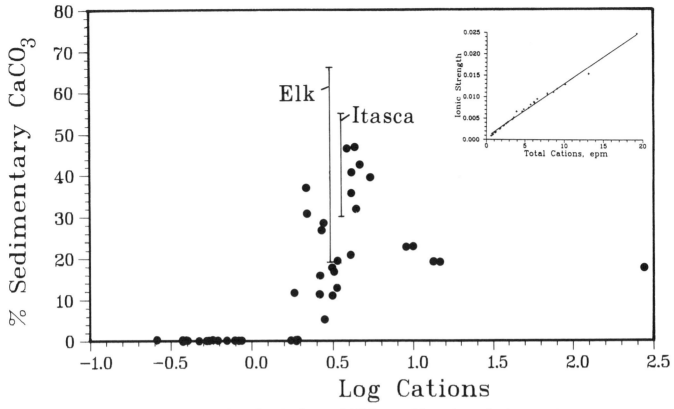

Figure 3. Percent CaCO$_3$ in profundal sediments of 46 Minnesota lakes vs. log total cation concentration (in milliequivalents per liter, epm) in surface waters from the same lakes. Ranges of CaCO$_3$ percentages in surface sediments from six localities in Elk Lake and six localities in Lake Itasca are also shown. Inset shows that ionic strength of these lake waters is linear with respect to total cations over the entire range of concentrations.

many lakes having considerable variation in salinity showed that primary high-Mg calcite precipitates when the Mg:Ca ratio of the water is between 2 and about 12; diagenetic dolomite forms in the sediments of those lakes in which high-Mg calcite is a primary carbonate mineral and the Mg:Ca ratio of the water is between about 7 and 12. Folk and Land (1975) concluded that dolomite can form in waters having a Mg:Ca ratio as low as 1.0 if the salinity is low. The importance of low salinity is to reduce the concentrations of other ions, especially sulfate, that interrupt precise ordering in the more complex dolomite crystal structure and favor the simpler structures of calcite and high-Mg calcite. Experimental studies by Fuchtbauer and Hardie (1980) showed that the Mg:Ca ratio of magnesian calcites is dependent on temperature as well as the Mg:Ca ratio of the solution, but is not dependent on total ionic strength of the solution. The effect of the Mg:Ca ratio in the water on the magnesium content of deposited CaCO$_3$ in Minnesota lakes is shown in Figure 6 for 17 high-carbonate lakes of Dean and Gorham (1976a). Although there is a considerable amount of scatter in the data, the mole percent Mg in CaCO$_3$ generally increases with increasing Mg:Ca in the water (Fig. 6), as predicted by the brine evolution model of Eugster and Jones (1979). Müller (1970) and Müller and Wagner (1978) also found that the Mg:Ca ratio of water in Lake Balaton, Hungary, determines the Mg content of high-Mg calcite precipitated by extensive growth of phytoplankton.

We used the program WATEQ (Truesdell and Jones, 1974; Plummer and others, 1976) to determine the degree of saturation of carbonate minerals in lakes across the steep precipitation-evaporation gradient from Wisconsin to North and South Dakota. As input we used the water chemistry data of Gorham and others (1982) for 28 lakes having a range in total cation concentration from 0.61 to 19.2 epm, and all having a bicarbonate concentration greater than calcium concentration. For all lakes we used a pH of 7.8 and a temperature of 20 °C, values typical of surface waters of central Minnesota lakes in early summer. WATEQ computes saturation indices (SI) for many minerals in which SI = log(IAP/K), where IAP is the ion activity product and K is the equilibrium constant of solubility (Truesdell and Jones, 1974). Negative SI values indicate that the calculated IAP is less than K and that the water is undersaturated with a particular mineral. Positive values of IAP indicate that the water is oversaturated. North-central United States lakes become saturated with calcite (SI = 0) at a log total cation concentration of about 0.5 (Fig. 7A); at that concentration there is a dramatic

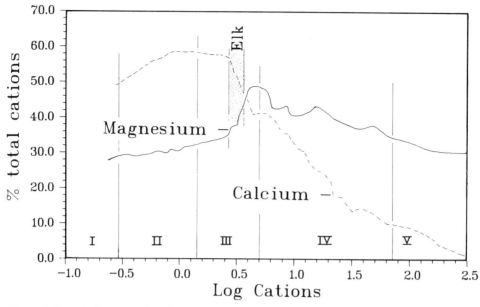

Figure 4. Smoothed curves, using a 21 point weighted moving average, showing relative proportions (as percent total cations) of dissolved calcium and magnesium vs. log total cation concentration (in milliequivalents per liter, epm) in surface waters of 219 lakes from Wisconsin, Minnesota, North Dakota, and South Dakota (Fig. 2; data of Gorham and others, 1982). Range of total cation concentration in surface waters of Elk Lake over a 10 month period is also shown (shaded area). Vertical lines and Roman numerals (I–V) denote the five salinity groups of Gorham and others (1982, 1983; Fig. 2).

decrease in calcium and increase in magnesium as percentages of total cations (Fig. 4). In other words, the lakes behave according to theory, at least with respect to calcite precipitation. Lake waters having log total cation concentrations of about 0.5 and 0.6 are theoretically saturated with dolomite and magnesite, respectively (Fig. 7). The greater variability of the saturation index for calcite at log total cation concentrations above 0.5 reflects the more complex chemistry of these more saline waters. Saturation indices for dolomite and magnesite remain almost linear with respect to log total cations to values of about 1.0 (10 epm; ionic strength of about 0.013; Fig. 7), further evidence for the conservative behavior of magnesium in all of these lakes.

According to Müller and others (1972), primary aragonite can form in lakes if the Mg:Ca ratio in the water is greater than about 12. Ratios of Mg:Ca this high are rare in temperate marl lakes, but occur in alkaline saline lakes such as those reported from Turkey and Afghanistan (Müller and others, 1972), Venezuela (Bradbury and others, 1981), and Nevada (Benson, 1984). Dean and Gorham (1976a) concluded that small amounts of aragonite they detected in surface sediments of some Minnesota lakes probably were derived from mollusc shells, because the Mg:Ca ratios in waters from these lakes generally are less than 2, and there was no relation between the presence of aragonite and salinity of the lake. WATEQ mineral saturation results show that Minnesota lake waters are always more saturated with calcite than with aragonite. Aragonite saturation is reached at a log total cation value of about 0.95 (9 epm; ionic strength of about 0.013), that is, at salinities greater than saturation with respect to either dolomite or magnesite.

Rhodochrosite ($MnCO_3$) is a minor but persistent carbonate mineral in most of the Holocene sediments in Elk Lake. As far as we know, the only other occurrence of rhodochrosite in lake sediments is in ferromanganese carbonate crusts in Green Bay, Lake Michigan (Callender and others, 1973). We were surprised, therefore, to find rhodochrosite in Elk Lake sediments until we obtained the results of water chemistry and sediment-trap studies (Nuhfer and others, Chapter 7). The water chemistry results for April to September, 1980, showed that in late April, after spring overturn, the concentration of dissolved Mn^{2+} was less than 3.0 parts per billion (ppb) throughout the lake (Fig. 8A). By late summer, the concentration of Mn^{2+} in the deepest waters of the lake was more than 1500 ppb (1.5 parts per million [ppm]; Fig. 8A), making Mn^{2+} a major cation. WATEQ mineral-saturation calculations showed that hypolimnetic waters in Elk Lake during

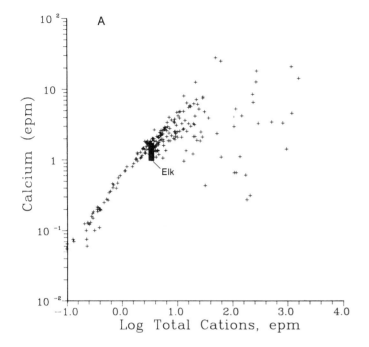

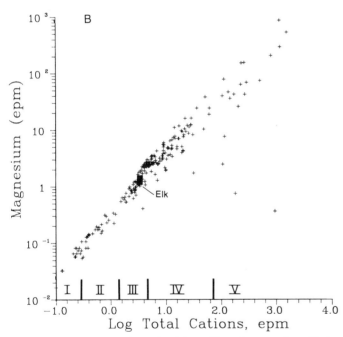

Figure 5. Log concentration of calcium (A) and magnesium (B) in milliequivalents per liter (epm) vs. log total cations for surface waters from 219 lakes in Wisconsin, Minnesota, North Dakota, and South Dakota (Fig. 2; data of Gorham and others, 1982). Ranges of concentrations in surface waters of Elk Lake over a 10 month period are shown by the black boxes. Vertical lines and Roman numerals (I–V) denote the five salinity groups of Gorham and others (1982, 1983) (Fig. 2).

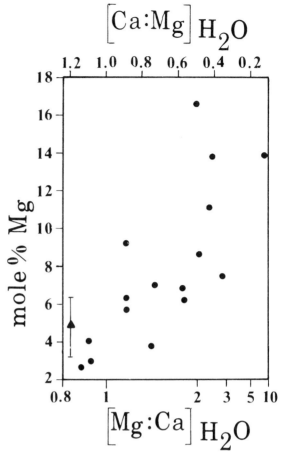

Figure 6. Mole% Mg in $CaCO_3$ in surface sediments from 16 high-carbonate Minnesota lakes vs. the Mg:Ca ratio in surface waters from the same lakes. The range (bar) and average (triangle) of mole% Mg in surface sediments from six localities in Elk Lake are also shown.

late summer were oversaturated with rhodochrosite at a Mn^{2+} concentration of about 1.0 ppm. After fall overturn, the Mn^{2+} concentration was less than 3.0 ppb throughout the lake (Fig. 8A). We anticipated, therefore, that rhodochrosite probably precipitated when manganese-rich bottom waters mixed with oxygenated, carbonate-rich surface waters. However, sediment-trap studies showed that manganese is a persistent and abundant component of the sediment rain throughout much of the year (Nuhfer and others, Chapter 7). Iron oxyhydroxides also precipitate in the same way throughout much of the year. The common appearance of a fairly thick seasonal lamina of ferric hydroxide gel in varves indicates that much of the precipitated iron oxyhydroxide is incorporated into the sediment. Ferric hydroxide colloids tend to be fairly stable, which may explain why they survive in reducing hypolimnion and sediment pore waters. Manganese oxyhy-

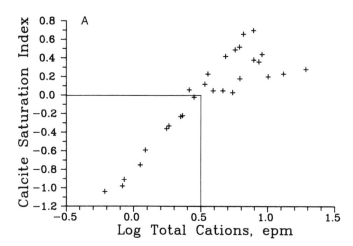
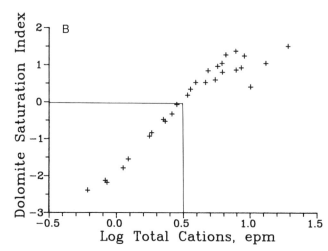
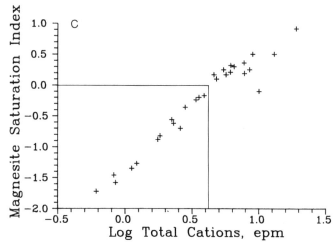

Figure 7. Saturation indices for calcite (A), dolomite (B), and magnesite (C) vs. log total cations in surface waters from 28 lakes from Wisconsin, Minnesota, North Dakota, and South Dakota (from data of Gorham and others, 1982).

droxides, on the other hand, are extremely unstable under reducing conditions, so it is unlikely that much of this material survives the trip to the bottom, particularly during the summer. In all probability the persistent presence of rhodochrosite in trap and bottom sediments represents conversion of manganese oxyhydroxides under anoxic conditions either in the hypolimnion or below the sediment-water interface.

Some of the manganese oxyhydroxides that precipitate may survive the trip to the bottom of the lake and become incorporated in the sediment. Unfortunately, hydrated manganese oxides are X-ray amorphous and cannot be detected by X-ray diffraction. Most, if not all, of the precipitated manganese oxyhydroxides that do reach the sediments are released from the sediments during summer and winter stagnation periods. For example, water samples collected from four different depths in Elk Lake on 14 February 1980 showed that water down to at least 16 m contained <3.0 ppb Mn^{2+}, but a sample from 26 m contained 780 ppb Mn^{2+}. The pattern of Mn^{2+} buildup in the hypolimnion during the summer of 1980 (Fig. 8A) shows a concentration gradient decreasing upward from the sediment-water interface, indicating that the manganese is coming from the sediments. Rhodochrosite is stable in the anoxic-sediment pore waters, so the only source of Mn^{2+} is manganese oxyhydroxide precipitated at overturn.

The distribution pattern of dissolved iron (Fig. 8B) is similar to that of dissolved manganese, increasing from less than 10 ppb after spring overturn to more than 500 ppb in the bottom waters by late July. Hypolimnetic waters of Elk Lake having an Fe^{2+} concentration of 500 ppb are slightly undersaturated with siderite ($FeCO_3$). Therefore, iron is not removed from the lake by carbonate precipitation, but is removed by precipitation of iron oxyhydroxides throughout much of the year, some of which is recrystallized as iron phosphate (Nuhfer and others, Chapter 7).

CARBONATE DEPOSITION IN ELK LAKE

Temperature and dissolved oxygen

The bottom of Elk Lake has two depressions, the deepest being about 30 m (Fig. 1). About 30% of the lake is below the thermocline during summer thermal stratification, which is significant because this bottom-water mass (hypolimnion) becomes anoxic and undersaturated with $CaCO_3$. The surface-water mass (epilimnion), however, becomes oversaturated with $CaCO_3$ during the summer, and is a favorable environment for the formation of $CaCO_3$.

Elk Lake usually freezes in November and ice cover persists until middle or late April. Water temperatures are less than 4°C at most depths during the winter, and the lake is isothermal at 4–5 °C after the ice melts (Fig. 9A). As surface-water temperatures increase to about 6°C in early May, the lake becomes thermally stratified into an epilimnion, metalimnion, and hypolimnion (Fig. 9A), and remains so until October or November when the water cools to 4–5°C at all depths.

The distribution of dissolved oxygen is related to thermal stratification (Fig. 9B). Oxygen concentrations in February are between 10 and 12 ppm above depths of 5 m, but decrease to about 1.0 ppm in the deepest water. Concentrations are high (10 to 11 ppm) at all depths after spring overturn when the lake is isothermal at 4–5°C. During summer, concentrations of dissolved oxygen usually are greater than 9 ppm in the epilimnion, and the hypolimnion is anoxic or nearly anoxic from late August until fall overturn in October or November (Fig. 9B).

The decrease in dissolved oxygen in the hypolimnion during May and June is particularly significant because it provides information about the fate of $CaCO_3$ sinking out of the epilimnion. The rate of oxygen decrease during this period is a measure of the rate of production of CO_2 from aerobic decomposition of organic

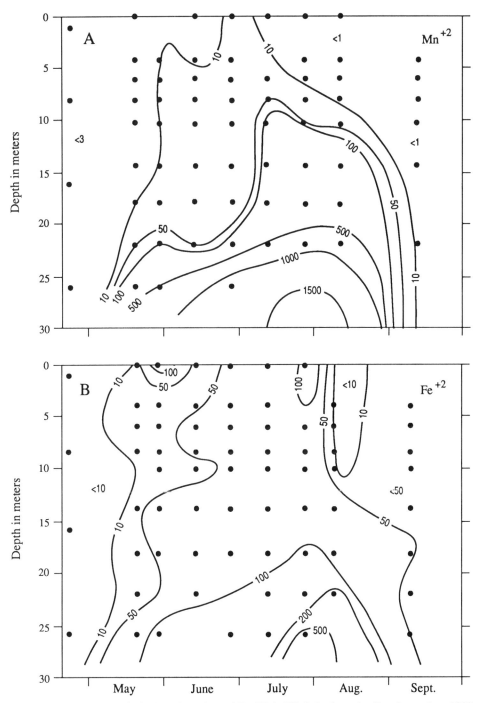

Figure 8. Distributions of dissolved Mn (A) and Fe (B) in Elk Lake from April to September, 1980. Values are in micrograms per liter (ppb).

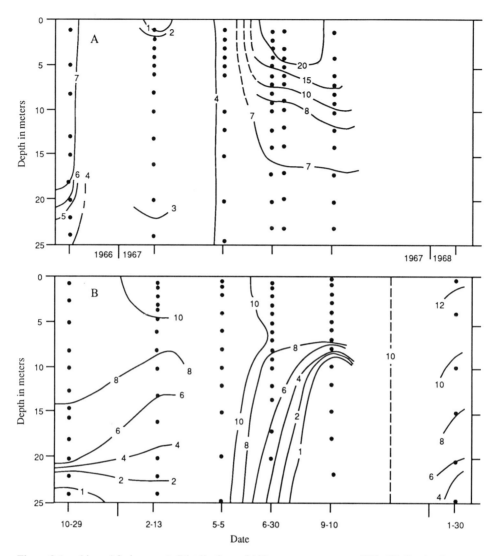

Figure 9 (on this and facing page). Distributions of (A) water temperature (°C); (B) dissolved oxygen (milligrams per liter); (C) pH; and (D) calcite saturation (ΔpH) in Elk Lake from October 1966 to January 1968. Negative values of ΔpH indicate oversaturation and positive values indicate undersaturation with respect to calcite.

matter during respiration. The production of CO_2 results in a lower pH, so that some of the carbonate minerals formed in the epilimnion will be dissolved in deeper water.

The quantity of oxygen in the water column below the epilimnion, at depths greater than 6 m, decreased 121 g · m^{-2} during the 56 days between 5 May and 30 June, 1967, for an average rate of decrease of 2.1 g O_2 · m^{-2} · day^{-1}. This rate of oxygen decrease is similar to the mean daily integral rate of oxygen production in the epilimnion by phytoplankton photosynthesis (2.6 g O_2 · m^{-2} · day^{-1} measured during 1966 and 1967; Megard and others, Chapter 3).

Carbonate saturation and primary productivity

Carbonate saturation was measured with a carbonate saturometer (Weyl, 1961) in both Elk Lake and Lake Itasca during 1966 and 1967 by placing crystalline calcite in contact with a glass pH electrode immersed in 20 ml of lake water. The degree of saturation is indicated by the change in pH (ΔpH) during two minutes after the addition of the calcite. If there are no inhibition or kinetic effects, the pH should not change if calcium carbonate in the water is in equilibrium with calcite. If the water is oversaturated, the pH decreases because the added calcite induces calcite precipitation and release of H^+. If the water is undersaturated, the pH increases because H^+ is consumed by dissolution of the added calcite.

As shown in Figure 10A, the degree of oversaturation or undersaturation can be determined simply by measuring the pH of the water, because there is a strong correlation ($r = 0.93$) between ΔpH and the initial pH of the sample (Fig. 10A). Calcite is in equilibrium at pH 7.73 in Elk Lake and Lake Itasca. The pH

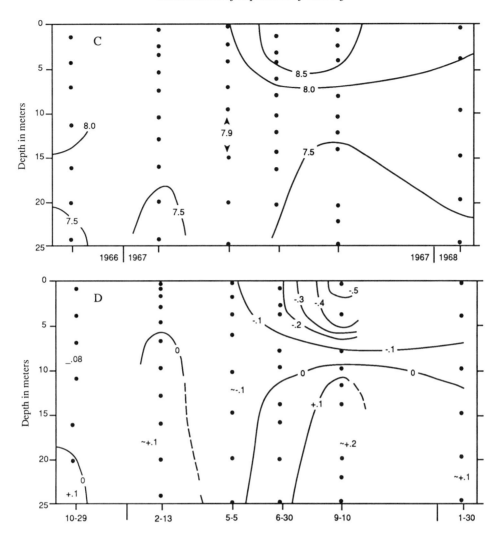

decreases after calcite addition if the pH is greater than 7.73, indicating that the water is oversaturated. The maximum decrease of about 0.5 pH units occurs at the highest pH (8.8). If the pH is less than 7.73, calcite addition causes the pH to increase, indicating that the water is undersaturated.

Computations of equilibrium saturation indices for calcite and dolomite provide independent theoretical confirmation that calcite is in equilibrium at pH 7.7 in Elk Lake. The computed equilibrium saturation indices for calcite and dolomite also increase with increasing pH at constant temperature (20°C; Fig. 10B). The indices were computed by WATEQ (Truesdell and Jones, 1974; Plummer and others, 1976) using the data for surface water of Elk Lake from Gorham and others (1982). The surface water of Elk Lake is theoretically saturated with both calcite and dolomite at a pH of about 7.7 (Fig. 10B), which, at least for calcite, is essentially the same as the equilibrium pH measured with the saturometer (Fig. 10A), indicating that any inhibition or kinetic effects are minor. Carbonate saturation is controlled by the concentrations of all individual species involved in carbonate equilibria (Ca^{2+}, CO_2, HCO_3^-, and H_2CO_3), but the CO_2 content of the water is the master variable that controls the pH and the proportions of the other species in solution (e.g., Plummer and Wigley, 1976; Stumm and Morgan, 1981; House, 1984).

The degree of $CaCO_3$ saturation at any depth depends on whether the lake is mixing or stratified, as illustrated by the depth profiles in Figures 11 and 12. The critical importance of stratification is shown by the data for 29 October 1966 (Fig. 11). At this time summer stratification had almost disappeared, and temperatures were near 7°C down to about 21 m, where there was still a weak thermocline. The pH was about 0.8 at all depths above the thermocline, and this water was oversaturated with calcite. However, the water below the thermocline contained little dissolved oxygen, had a pH of less than 7.7, and was undersaturated with calcite.

In contrast, during winter stratification (13 February 1967; Fig. 11) all depths below 5 m were oxygen deficient, had pH values of less than 7.7, and were undersaturated with calcite. Later, during spring circulation (5 May 1967), all depths were reaerated and oversaturated with calcite. The transition from cal-

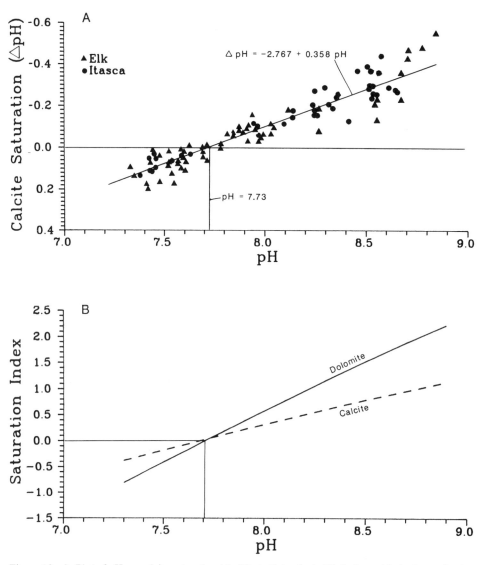

Figure 10. A: Plot of pH vs. calcite saturation (ΔpH) at all depths in Elk Lake and Lake Itasca for the period October 1966 to January 1968. B: Saturation indices for calcite and dolomite for Elk Lake surface water at 20 °C over a range of pH values.

cite undersaturation to oversaturation during spring circulation occurred while water temperatures were still only 4°C.

During summer stratification (e.g., 11 July and 10 September, 1967; 12 July 1980), the epilimnion (depths above 6 m) is oversaturated with both calcite and dolomite, but the hypolimnion (depths below 10 m) is undersaturated. The effects of lake stratification on pH and carbonate saturation (ΔpH) are best illustrated by the isopleth patterns of these two variables for the period from October, 1966, to January, 1968 (Fig. 9). As indicated in Figure 9, the epilimnion in Elk Lake is always oversaturated (i.e., ΔpH, is negative, pH is greater than 7.73). Deep water is undersaturated (ΔpH is positive, pH is less than 7.73) when concentrations of oxygen are low in winter and summer. The lowest values of pH, and, therefore, the greatest degree of undersaturation, occur in the anoxic hypolimnion during late summer (e.g., 10 September 1967, Fig. 11; 11 August 1980, Fig. 12). Oversaturation in the epilimnion increases progressively during summer; ΔpH increases from about −0.1 in May to about −0.5 in September. This is the period when photosynthetic activity is greatest in the epilimnion, and the phytoplankton are removing CO_2 most rapidly from the water.

The trends discussed above indicate that $CaCO_3$ saturation is

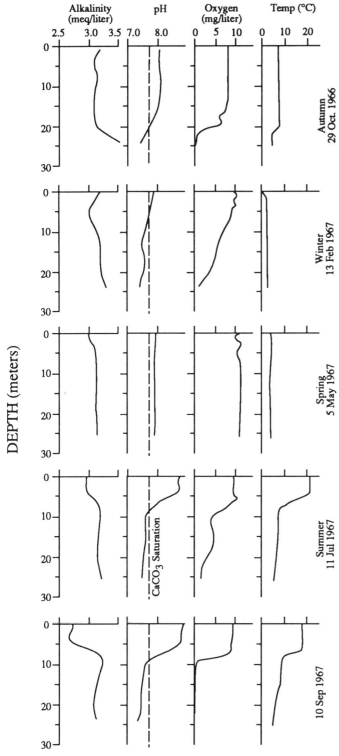

Figure 11. Depth profiles of alkalinity, pH, dissolved oxygen, and temperature in Elk Lake on five dates between 29 October 1966 and 10 September 1967.

controlled by CO_2 concentration, which, in turn, is directly related to the balance between plankton photosynthesis and respiration. Oversaturation occurs in the mixed layer in oxygenated water where the pH is high because CO_2 is removed by photosynthesis faster than it is added by respiration and exchange with the atmosphere. Undersaturation occurs in water where there is an accumulation of CO_2 and an oxygen deficit due to net respiration; the accumulation of CO_2 causes the pH to decrease below the equilibrium pH for carbonate saturation.

Stabel (1986) found that calcite precipitation in Lake Constance was associated with blooms of certain diatoms and other algae, indicating that calcite nucleation is catalyzed by phytoplankton. Stabel suggested that this nucleation might be a consequence of increased pH due to photosynthetic uptake of HCO_3^- and release of OH^-, and/or the result of algae acting as surface catalysts in the nucleation of calcite. The Elk Lake results indicate clearly that the dominant factor in calcite precipitation is the control of CO_2 by photosynthesis.

If photosynthetic activity controls carbonate precipitation, then there should be a quantitative relation between the rates of photosynthesis and calcium depletion. In theory, one mole of carbon consumed as CO_2 by photosynthesis should result in one mole of $CaCO_3$ formation in a solution containing only calcium bicarbonate. Figure 13 shows that there is indeed a relation between photosynthesis and calcium depletion in six lakes studied by Megard (1968), but the average ratio of calcium depletion to carbon consumption (slope of the line in Fig. 13) is 0.25; the range is about 0.1 to 0.4. We have no explanation as to why this average ratio is different from the theoretical ratio of 0.5, but the narrow range of values suggests that there is a simple fundamental stoichiometric relation that is distorted only slightly by the hydrologic balance of lakes or by macrophyte photosynthesis. For example, the high rate of calcium depletion in Lake Itasca (Fig. 13) is probably due to the fact that this lake has a larger proportion of the lake volume within the zone of rooted aquatic macrophytes, which precipitate carbonate on their leaves and stems. If the rates of calcium depletion in these lakes represents most of the $CaCO_3$ precipitation, then it is apparent that phytoplankton photosynthesis is sufficient to account for most of the $CaCO_3$ formed in summer. The role of macrophytes probably is minor in Elk Lake simply because their biomass and photosynthetic rate are relatively small in comparison to those of the phytoplankton.

MINERALOGY OF ELK LAKE CORE

Most of the carbonate that formed throughout the postglacial history of Elk Lake, particularly during the past 4000 years, is in the form of low-Mg calcite (Fig. 14). During the arid mid-Holocene prairie period (4000 to 8000 varve yr), however, dolomite composed a much larger proportion of the sediment. Some dolomite and calcite, particularly in the bottom part of the core (older than about 8000 varve yr), may be detrital, derived from Paleozoic limestone and dolomite rock flour in calcareous glacial drift. Most of the calcite, however, probably is a primary precipitate from the lake water just as it is today. Most of the dolomite probably is either a primary precipitate or an early diagenetic alteration of primary high-Mg calcite.

Dolomite and high-Mg calcite both are found today only in surface sediments of lakes in the western prairie regions of Minnesota where lake waters have a Mg:Ca ratio of at least 2.0

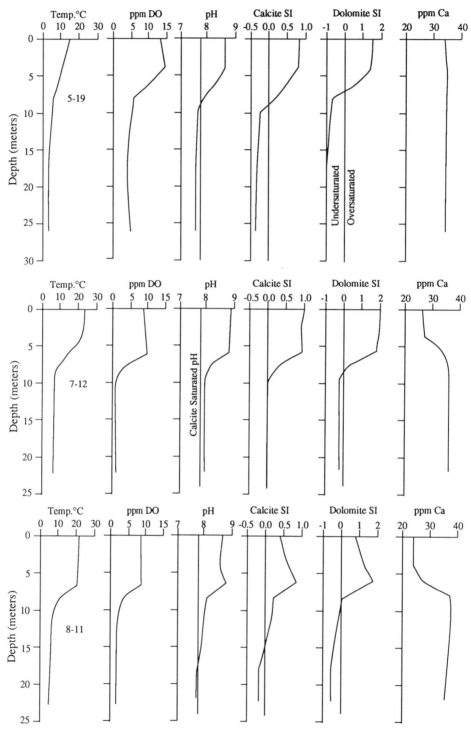

Figure 12. Depth profiles of temperature, dissolved oxygen, pH, calcite and dolomite saturation indices (SI), and calcium concentration in Elk Lake for 19 May, 12 July, and 11 August, 1980.

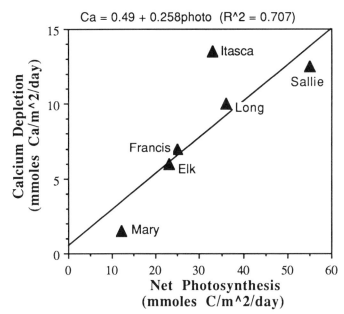

Figure 13. Rates of photosynthetic carbon assimilation and calcium depletion in six Minnesota Lakes during the summer of 1966 (modified from Megard, 1968).

(Dean and Gorham, 1976a; Gorham and others, 1983), but even in these sediments it is not clear whether the dolomite is primary or early diagenetic. High-Mg calcite forms by direct precipitation in Lake Manitoba, Canada, in response to increased pH brought about by intense phytoplankton photosynthesis during the ice-free season (Last, 1982; Last and Teller, 1983). Dolomite in the sediments of Lake Manitoba apparently is detrital. X-ray diffraction patterns (see Dean, Chapter 10) of some sediment samples from the Elk Lake core show peaks for both high-Mg calcite and dolomite, but most diffractograms that have peaks for dolomite do not have peaks for high-Mg calcite, which suggests either that high-Mg calcite did not form, or that it was completely converted to dolomite during diagenesis. During the prairie period Elk Lake probably had many of the characteristics of highly productive, saline lakes of the prairie regions of Minnesota today that are accumulating high-Mg calcite and dolomite in their sediments (Dean and Gorham, 1976a). Ostracod and diatom data (Forester and others, 1987) suggest that during much of the prairie period, Elk Lake had characteristics similar to those found today in Canadian prairie lakes.

The small amounts of aragonite that occur in Elk Lake sediments deposited during the prairie period between about 7200 and 6300 varve yr (Fig. 14) may be derived from mollusc shells, but the following observations suggest that this aragonite was a primary precipitate and therefore is additional evidence for much higher salinity during the prairie period. (1) The aragonite occurs in the interval of maximum salinity during the prairie period, as inferred from diatoms and ostracods (Forester and others, 1987; Bradbury and Dieterich-Rurup, Chapter 15). (2) The percentage of calcite is lower in this interval than in any other part of the core, but the fact that the percentage of $CaCO_3$ does not decrease suggests that at this time precipitation of $CaCO_3$ as aragonite replaced the precipitation of some $CaCO_3$ as calcite. (3) The aragonite occurs as needle-shaped crystals (Anderson, Chapter 5), which suggests that it formed as a chemical precipitate and is not mollusc debris. Laminae of aragonite needles in the sediments of saline meromictic Waldsea Lake in Saskatchewan, Canada, form by basin-wide precipitation (Last and Schweyen, 1985).

Minor amounts of rhodochrosite accumulated in Elk Lake throughout most of the Holocene just as it does today. Exceptions are between 200 and 1500 varve yr and during the period of maximum inferred salinity between 7200 and 6300 varve yr, when aragonite was precipitating. We assume that rhodochrosite precipitated as it does today by mixing of anoxic, Mn-rich bottom waters with carbonate-rich surface waters. Sediment-trap data (Nuhfer and others, Chapter 7) suggest that rhodochrosite forms throughout most of the periods of summer and winter stratification and not just at spring and fall overturn.

CONCLUSIONS

1. Regional studies on the precipitation of carbonate minerals in Minnesota lakes show that precipitation of low-Mg calcite does not occur until the lake waters have a calcium concentration of about 1.0 milliequivalent per liter (epm; equal to 20 parts per million, ppm), which is equivalent to a total cation concentration of 2.0 epm.

2. Lakes that have higher evaporation rates and higher water salinities in the western prairie regions of Minnesota have proportionally higher ratios of Mg:Ca. In these prairie lakes, magnesium usually is the dominant divalent cation, and Mg:Ca ratios range from 1.0 to 10. The sediments of these lakes typically contain high-Mg calcite and dolomite in addition to low-Mg calcite. The dolomite may be a primary precipitate or a diagenetic alteration of high-Mg calcite.

3. Studies of carbonate saturation in Elk Lake indicate that the epilimnion is almost always saturated or oversaturated with respect to calcite, and that the deep water is undersaturated during the summer and winter months when bottom-water oxygen concentrations are low. Oversaturation in the epilimnion increases progressively during the summer months, and reaches a maximum in September. In response to photosynthetic removal of CO_2, the pH of surface waters of Elk Lake increased from about 8.0 after spring overturn to almost 9.0 during late summer. That the rate of calcium depletion in the surface waters of Elk Lake by precipitation of calcite during the summer months is proportional to the rate of carbon fixation by photosynthesis indicates that carbonate precipitation is triggered by photosynthetic activity.

4. Low-Mg calcite has always been the dominant carbonate mineral in the Holocene sediments of Elk Lake, and rhodochrosite usually was a minor carbonate component. Today, and presumably in the past, precipitation of low-Mg calcite from

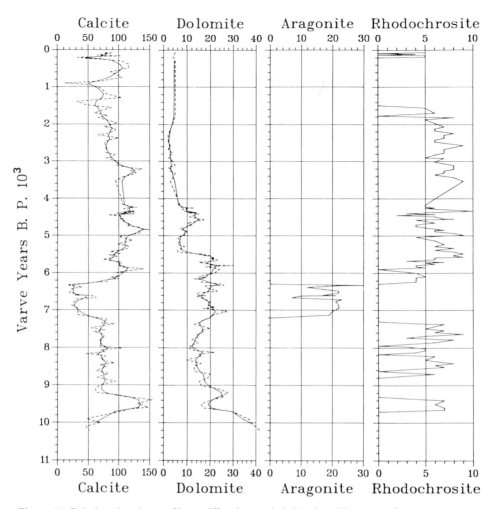

Figure 14. Relative abundances (X-ray diffraction peak height in millimeters) of calcite, dolomite, aragonite, and rhodochrosite in samples of varved sediment from the deep basin of Elk Lake.

oversaturated surface waters is triggered by phytoplankton photosynthesis during the summer months. Today, and presumably in the past, rhodochrosite forms at times of overturn when manganese-rich bottom waters mix with carbonate-rich surface waters, as well as throughout most of the periods of summer and winter stratification.

5. Dolomite is a common mineral in sediments deposited in Elk Lake during the early postglacial and the arid mid-Holocene prairie period. Some of this dolomite, particularly in the early postglacial sediments, may be derived from Paleozoic dolomite rock flour in calcareous glacial drift, but most of the dolomite probably formed in the lake, just as it is forming today in lakes of the semiarid prairie regions of western Minnesota. What is not clear, however, is whether the dolomite was a primary precipitate or formed by diagenetic alteration of high-Mg calcite.

6. Aragonite occurs as fine needle-like crystals only in sediments that are 6300 to 7200 years old. The occurrence of aragonite corresponds with other indicators of unusually high salinity, and suggests that the salinity of the lake and the Mg:Ca ratio at that particular time were higher than in any of the present prairie lakes of western Minnesota.

ACKNOWLEDGMENTS

We greatly appreciate the helpful suggestions made by L. V. Benson, C. J. Bowser, E. Callender, B. F. Jones, R. J. Spencer, and T. C. Winter. We thank Mary Bucherle and David Bosanko for making many of the field water chemistry measurements and collecting water samples for laboratory analyses.

REFERENCES CITED

Anderson, R. Y., Dean, W. E., Bradbury, J. P., and Love, D., 1985, Meromictic lakes and varved lake sediments in North America: U.S. Geological Survey Bulletin 1607, 19 p.

Benson, L. V., 1984, Hydrochemical data for the Truckee River drainage system, California and Nevada: U.S. Geological Survey Open-File Report 84-440, 33 p.

Bradbury, J. P., and eight others, 1981, Late Quaternary environmental history of

Lake Valencia, Venezuela: Science, v. 214, p. 1299–1305.

Callender, E., Bowser, C. J., and Rossman, R., 1973, Geochemistry of ferromanganese carbonate crusts from Green Bay, Lake Michigan [abs.]: Eos (Transactions, American Geophysical Union), v. 54, p. 340.

Dean, W. E., 1981, Carbonate minerals and organic matter in sediments of modern north temperate hard-water lakes: Society of Economic Paleontologists and Mineralogists Special Publication 31, p. 213–231.

Dean, W. E., and Fouch, T. D., 1983, Lacustrine environment, in Scholle, P. A., Bebout, D., and Moore, C., eds., Carbonate depositional environments: American Association of Petroleum Geologists Memoir 33, p. 97–130.

Dean, W. E., and Gorham, E., 1976a, Major chemical and mineral components of profound surface sediments in Minnesota lakes: Limnology and Oceanography, v. 21, p. 259–284.

——, 1976b, Classification of Minnesota lakes by Q-and R-mode factor analysis of sediment mineralogy and geochemistry, in Merriam, D. F., ed., Quantitative techniques for the analysis of sediments: Oxford, England, Pergamon Press, p. 61–71.

Dean, W. E., Bradbury, J. P., Anderson, R. Y., and Barnosky, C. W., 1984, Variability of Holocene climate change: Evidence from varved lake sediments: Science, v. 266, p. 1191–1194.

Eugster, H. P., and Jones, B. F., 1979, Behavior of major solutes during closed-basin brine evolution: American Journal of Science, v. 279, p. 609–631.

Folk, R. L., and Land, L. S., 1975, Mg/Ca ratio and salinity: Two controls over crystallization of dolomite: American Association of Petroleum Geologists Bulletin, v. 59, p. 60–68.

Forester, R. M., Delorme, L. D., and Bradbury, J. P., 1987, Mid-Holocene climate in northern Minnesota: Quaternary Research, v. 28, p. 263–273.

Fuchtbauer, H., and Hardie, L. A., 1980, Comparison of experimental and natural magnesian calcites: International Association of Sedimentologists, Abstracts for First European Meeting, Bochum, p. 167–169.

Gorham, E., Dean, W. E., and Sanger, J. E., 1982, The chemical composition of lakes in the north-central United States: U.S. Geological Survey Open-File Report 82-149, 65 p.

——, 1983, The chemical composition of lakes in the north-central United States: Limnology and Oceanography, v. 28, p. 287–301.

Hall, H. T., 1972, Environmental factors controlling the chemistry of Minnesota lakes: Geological Society of America Abstracts with Programs, v. 4, p. 524.

House, W. A., 1984, The kinetics of calcite precipitation and related processes: Ambleside, England, Freshwater Biological Association, Fifty-second Annual Report, p. 75–90.

Jones, B. F., and Bowser, C. J., 1978, The mineralogy and related chemistry of lake sediments, in Lerman, A., ed., Lakes—Chemistry, geology, and physics: New York, Springer-Verlag, p. 179–235.

Kelts, K., and Hsü, K. J., 1978, Freshwater carbonate sedimentation, in Lerman, A., ed., Lakes—Chemistry, geology, and physics: New York, Springer-Verlag, p. 295–323.

Last, W. M., 1982, Holocene carbonate sedimentation in Lake Manitoba, Canada: Sedimentology, v. 29, p. 691–704.

Last, W. M., and Schweyen, T. H., 1985, Late Holocene history of Waldsea Lake, Saskatchewan, Canada: Quaternary Research, v. 24, p. 219–234.

Last, W. M., and Teller, J. T., 1983, Holocene climate and hydrology of the Lake Manitoba basin, in Teller, J. T., and Clayton, L., eds., Glacial lake agassiz: Geological Association of Canada Special Paper 26, p. 333–353.

Megard, R. O., 1967, Limnology, primary productivity, and carbonate sedimentation of Minnesota lakes: University of Minnesota Limnological Research Center, Interim Report 1, 69 p.

——, 1968, Planktonic photosynthesis and the environment of carbonate deposition in lakes: University of Minnesota Limnological Research Center, Interim Report 2, 47 p.

Müller, G., 1970, High-magnesium calcite and protodolomite in Lake Balaton (Hungary) sediments: Nature, v. 226, p. 749–750.

Müller, G., and Wagner, F., 1978, Holocene carbonate evolution in Lake Balaton (Hungary): A response to climate and impact of man, in Matter, A., and Tucker, M. E., eds., Modern and ancient lake sediments: International Association of Sedimentologists Special Publication 2, p. 57–81.

Müller, G., Irion, G., and Forstner, U., 1972, Formation and diagenesis of inorganic Ca-Mg carbonates in the lacustrine environment: Naturwissenschaften, v. 59, p. 158–164.

Plummer, L. N., and Wigley, T.M.L., 1976, The dissolution of calcite in CO_2-saturated solutions at 25 °C and 1 atmosphere total pressure: Geochimica et Cosmochimica Acta, v. 40, p. 191–202.

Plummer, L. N., Jones, B. F., and Truesdell, A. H., 1976, WATEQF: A FORTRAN IV version of WATEQ, a computer program for calculating chemical equilibrium of natural waters: U.S. Geological Survey Water-Resources Investigations 76-13, 61 p.

Stabel, H. H., 1986, Calcite precipitation in Lake Constance: Chemical equilibrium, sedimentation, and nucleation by algae: Limnology and Oceanography, v. 31, p. 1081–1093.

Stumm, W., and Morgan, J. J., 1981, Aquatic chemistry: New York, Wiley Interscience, 585 p.

Truesdell, A. H., and Jones, B. F., 1974, WATEQ, a computer program for calculating chemical equilibria of natural waters: U.S. Geological Survey Journal of Research, v. 2, p. 233–248.

Weyl, P., 1961, The carbonate saturometer: Journal of Geology, v. 69, p. 32–44.

MANUSCRIPT ACCEPTED BY THE SOCIETY JULY 27, 1992

Geochemistry of surface sediments of Minnesota lakes

Walter E. Dean
U.S. Geological Survey, MS 939, Box 25046, Federal Center, Denver, Colorado 80225
Eville Gorham
Department of Ecology, Evolution, and Behavior, University of Minnesota, Minneapolis, Minnesota 55455
Dalway J. Swaine
CSIRO, North Ryde, New South Wales, 2113, Australia

ABSTRACT

Analyses of 36 trace, minor, and major elements were used to classify the sediments of 46 Minnesota lakes. Q-mode factor analyses grouped Minnesota lake sediments according to clastic-, carbonate-, organic-, and redox-related elements. Carbonate lakes occur in west-central Minnesota; their sediments have relatively high concentrations of $CaCO_3$, Ba, and Sr. Lakes with sediments containing more than 30% organic matter occur in east-central and northeastern Minnesota; these sediments have high concentrations of organic C, N, and H, and slightly elevated concentrations of Pb. Only three lakes have sediments included in the "redox" group, with relatively high concentrations of Fe, Mn, Mo, La, and Zn. High concentrations of redox-sensitive elements appear to be associated with oxidized iron and manganese minerals, but in Elk Lake, one of the three lakes in this group, a significant amount of iron and manganese is contained in iron phosphate, iron sulfide, and manganese carbonate. Clastic lake sediments are not diluted by large amounts of organic matter, carbonate minerals, or iron-manganese minerals, and are of two types: a western group derived largely from Cretaceous shales in the prairie regions, and a northeastern group derived from Precambrian crystalline rocks in the forested arrowhead region. Western clastic lake sediments have higher concentrations of Al, Na, K, B, Ba, V, Mg, and Sr. Most northeastern clastic lake sediments contain higher concentrations of Cu, Y, Be, and Ni, but several are chemically more similar to those of western prairie lakes.

INTRODUCTION

Dean and Gorham (1976a, 1976b) classified Minnesota lakes partly on water chemistry, and partly on the geochemistry and mineralogy of major components in surface sediments of 46 lakes. We have obtained more complete analyses of the same sediment samples at the U.S. Geological Survey (USGS) and the Commonwealth Scientific and Industrial Research Organization (CSIRO) of Australia. In this chapter we refine the geochemical classification of Minnesota lake sediments based on more extensive analyses of major, minor, and trace elements. This refined classification forms the basis for interpreting changes in geochemistry with depth (time) in the Elk Lake cores (Dean, Chapter 10).

We also consider relations between the geochemical classification of Minnesota lake sediments and the geochemical classification of Minnesota lake waters (Gorham and others, 1982, 1983). The water chemistry of Minnesota lakes is closely controlled by the steep climatic gradient from east to west across the state, and, to a lesser degree, by differences in surficial and bedrock geology. Lakes of northeastern Minnesota (group 1 and 2 lakes of Gorham and others, 1983; Fig. 1) contain dilute bicarbonate waters (conductivity <141 μmhos·cm^{-1} at 25°C) in Precambrian crystalline rocks (Fig. 2). This region is heavily forested and has a positive balance of precipitation minus evaporation. The most dilute northeastern lakes are close to rainwater in composition. Central Minnesota lakes (group 3)

Dean, W. E., Gorham, E., and Swaine, D. J., 1993, Geochemistry of surface sediments of Minnesota lakes, *in* Bradbury, J. P., and Dean, W. E., eds., Elk Lake, Minnesota: Evidence for Rapid Climate Change in the North-Central United States: Boulder, Colorado, Geological Society of America Special Paper 276.

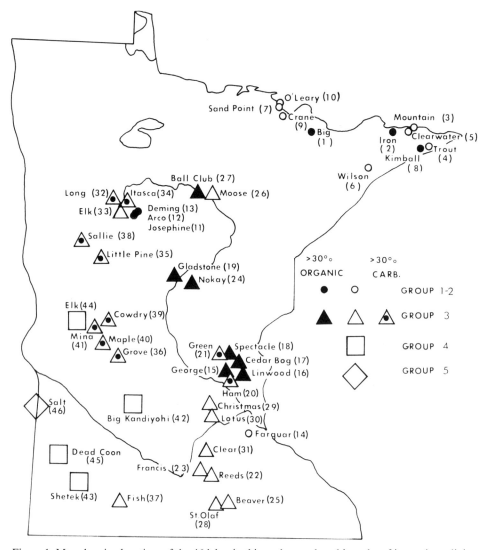

Figure 1. Map showing locations of the 46 lakes in this study, numbered in order of increasing salinity (Gorham and others, 1983). Symbols reflect water-salinity groups (groups 1–2, 3, 4, and 5) of Gorham and others (1983), modified to separate those lakes with profundal sediments rich in carbonate (>30% $CaCO_3$), those rich in organic matter (>30% organic matter), and those with intermediate contents of both carbonate and organic matter (Dean and Gorham, 1976a, 1976b).

contain calcium-magnesium bicarbonate waters of intermediate salinity (conductivity 141–500 $\mu mhos \cdot cm^{-1}$) in calcareous drift. All but the most dilute central Minnesota lakes precipitate $CaCO_3$ (Dean and Gorham, 1976a). Group 3 lakes occur in both forested and prairie regions with a slight positive or slight negative balance of precipitation minus evaporation. Western prairie lakes of Minnesota (group 4) have a distinctly negative balance of precipitation minus evaporation, and are the most saline (conductivity 501–7078 $\mu mhos \cdot cm^{-1}$). The surficial deposits of this region contain gypsum ($CaSO_4 \cdot 2H_2O$) derived from Cretaceous shale. As a result of the combination of net evaporation and the addition of sulfate from gypsum, sulfate replaces bicarbonate as the dominant anion in group 4 lakes. Because of loss of calcium by $CaCO_3$ precipitation, and gain of sodium and magnesium from Cretaceous shales, magnesium and sodium are the dominant cations in group 4 lakes. The group 5 lakes of Gorham and others (1983) lack outlets and have very high conductivities (>7078 $\mu mhos \cdot cm^{-1}$); they occur mainly in relatively dry North and South Dakota. They are strongly dominated by sulfate and sodium, and are represented in our study only by Salt Lake, on the Minnesota–South Dakota border (Fig. 1).

METHODS

An undisturbed sample of profundal sediment was collected from the deepest part of each lake with a Jenkin sampler. The top 10 cm of sediment from each lake was homogenized, oven dried

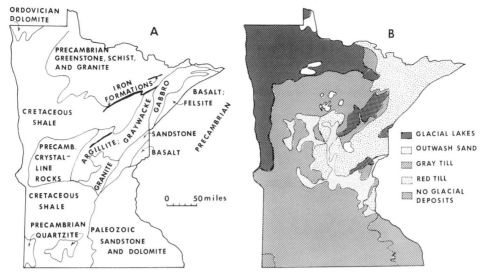

Figure 2. Maps of bedrock geology (A) and major surficial deposits (B) of Minnesota (modified from Ackroyd and others [1967] and Winter [1974]).

at 90°C, and ground to powder. Concentrations of total Ca, Mg, Fe, Mn, Zn, Cu, and Li were determined by atomic absorption spectrophotometry and Na and K by flame photometry. Solutions for these analyses were prepared by digesting a 500 mg aliquot of dried sample successively in HNO_3, HF, and HCl. The acid digest was evaporated to dryness and the residue taken up in 1.0N HCl. Total phosphorus was determined on a suitable dilution of the stock acid digestion for atomic absorption and flame photometry following the phosphomolybdate method of Mackereth (1963). Organic carbon (OC), total nitrogen (TN), and total hydrogen (TH) were measured chromatographically as CO_2, N_2, and H_2O, respectively, using a Hewlett-Packard CHN analyzer. Total sulfur (TS) was determined by igniting an aliquot of dried sample mixed with MnO_2 at 500°C and extracting sulfate with 0.001N NaOH. Sulfate concentration was determined by the ion-exchange method of Mackereth (1963). Estimates of the amounts of organic matter were made by loss on ignition at 550 °C (LOI-550°C), with excellent correlation between LOI at 550°C and OC (r = 0.97; Dean, 1974). The average LOI-550°C:OC ratio is 2.13. Estimates of amounts of $CaCO_3$ were made by LOI between 550 and 900°C with excellent agreement with estimates of $CaCO_3$ made from total Ca, acid-soluble Ca, and carbonate carbon (Dean, 1974). All the above analyses were performed at the University of Minnesota and will be referred to as UM analyses.

The major elements Si, Al, Fe, Ca, K, and Ti were determined by semiquantitative, wavelength-dispersive X-ray fluorescence spectrometry (XRF; Miesch, 1976a) at the USGS Denver. Seven major elements (Al, Fe, Mg, Ca, K, Ti, and Mn) and 16 trace elements (B, Ba, Be, Co, Cr, Cu, La, Mo, Nb, Ni, Pb, Sr, V, Y, Zn, and Zr) were determined by semiquantitative, optical emission spectroscopy (OES; Miesch, 1976a) at the USGS Denver. The precision and accuracy of the USGS XRF and OES methods were discussed in Miesch (1976a). An additional measure of analytical precision is provided in Table 1 as the mean and coefficient of variation (standard deviation as a percent of the mean) for a sample of deep-sea red clay that was analyzed seven times over a period of four years.

Semiquantitative estimates of Ca, Mn, B, Co, Cr, Cu, Ga, Mo, Ni, Pb, Sc, Sn, Sr, V, Y, Zn, and Zr were performed by optical emission spectroscopy at CSIRO, North Ryde, Australia. Arsenic was determined quantitatively on selected samples by atomic absorption graphite furnace at CSIRO. All values are reported on a dry-weight basis. Major elements are expressed either as percent of the element or the appropriate oxide; trace elements are expressed as parts per million (ppm).

RESULTS

Results for the major elements obtained from the three different laboratories (UM, USGS, and CSIRO) are given in the Appendix. For statistical analyses, we selected what we call "preferred results" for those elements for which we had more than one analysis. The philosophy behind the selection of preferred results was as follows. Because the USGS and CSIRO analyses are semiquantitative, we used the UM results when available (Fe, Mg, Ca, Na, K, Mn, OC, S, N, H, Cu, Li, and Zn); we used USGS XRF results for all of the other major elements (Si, Al, and Ti), and USGS emission spectrographic values for most of the other trace elements (B, Ba, Be, Co, Cr, La, Mo, Nb, Ni, Pb, Sr, V, Y, and Zr); we used CSIRO values only for those elements for which we did not have any analyses by another method (As, Ga, Sc, and Sn). Summary statistics for results obtained by the preferred analyses of sediments from all 46 lakes are shown in Table 2.

TABLE 1. COMPOSITION OF RED CLAY HATTERAS ABYSSAL PLAIN

Element	Mean	Coefficient of Variation*
	(wt.%)	(wt.%)
Si	25	0.0
Al	9.9	0.5
Fe	5.7	0.06
Mg	1.7	3.1
Ca	0.58	1.2
Na	1.5	3.5
K	3.3	1.3
Ti	0.50	2.4
P	0.07	6.3
Mn	0.11	0.7
	(ppm)	(ppm)
B	135	53
Ba	503	17
Be	110	20
Co	26	19
Cr	89	34
Cu	42	33
Ga	36	47
La	46	15
Mo	<4	
Ni	58	19
Pb	<20	
Sr	118	10
V	127	17
Y	27	15
Zn	120	26
Zr	137	11

*Mean and coefficient of variation ([standard deviation/mean] x 100) of seven analyses of a sample of red clay from Lamont-Doherty core V26-157.

TABLE 2. SUMMARY STATISTICS OF GEOCHEMICAL ANALYSES OF SURFACE SEDIMENTS IN 46 MINNESOTA LAKES

Element	Mean	Standard Deviation	N*
	(wt.%)	(wt.%)	
Si	18	4.5	46
Al	3.1	1.5	46
Fe	3.3	1.9	46
Mg	0.84	0.45	46
Ca	6.4	5.9	46
Na	0.47	0.26	46
K	0.85	0.35	46
Ti	0.18	0.095	46
P	0.15	0.13	46
Mn	0.25	0.59	46
Carb.-C	2.0	1.7	46
Org.-C	12	6.3	46
Total N	1.4	0.73	46
Total H	1.9	0.90	46
Total S	0.53	0.37	46
	(ppm)	(ppm)	
As	96	590	31
B	27	10	46
Ba	250	65	46
Be	1.2	0.65	46
Co	6.3	4.0	46
Cr	35	17	46
Cu	43	47	46
Ga	7.4	3.9	44
La	54	33	46
Mo	3.3	2.1	46
Nb	18	9.2	46
Ni	22	14	46
Pb	19.5	1.2	46
Sn	1.5	1.2	43
Sr	160	75	46
V	54	27	46
Y	14	8.0	46
Zn	150	50	46
Zr	130	66	46

*N is the total number of samples out of 46 that contained measurable concentrations greater than the lower limit of detection.

LAKE CLASSIFICATION

In Figure 1, the 46 Minnesota lakes are differentiated by symbol according to the water-chemistry classification of Gorham and others (1982, 1983). Dean and Gorham (1976a, 1976b) used Q-mode factor analysis to delineate three types of Minnesota lakes based on major chemical and mineral characteristics of their sediments. Factor 1 (carbonate) lakes contain more than 30% $CaCO_3$ (LOI-550–900 °C); factor 2 (organic) lakes contain more than 30% organic matter (LOI-550 °C); and factor 3 lakes all contain high proportions of clastic material with <30% $CaCO_3$ and <30% organic matter. The water-chemistry symbols in Figure 1 have been modified to reflect this three-factor sediment classification.

Our more extensive geochemical data on surface-sediment samples from the same 46 lakes were interpreted using Q-mode factor analysis to see if we could refine the factor-analysis classification of Dean and Gorham (1976a, 1976b), and determine the associations of additional major and trace components with the major components used in the earlier classification. We used the interactive version of the extended CABFAC computer program described by Klovan and Miesch (1976) that is applicable when row sums of the data matrix are constant. Descriptions and applications of interactive Q-mode programs are in Miesch (1976b,

TABLE 3. FACTOR VARIANCE DATA*

Element or Oxide	3-Factor Model	4-Factor Model	5-Factor Model
SiO_2	0.86	0.86	0.86
Al_2O_3	0.87	0.87	0.90
Fe_2O_3	0.17	0.80	0.89
MgO	0.56	0.58	0.59
CaO	0.95	0.95	0.97
Na_2O	0.57	0.58	0.58
K_2O	0.89	0.89	0.92
TiO_2	0.79	0.81	0.81
P_2O_5	0.03	0.04	0.09
MnO	0.01	0.30	0.32
OC	0.77	0.84	0.86
Total N	0.71	0.83	0.89
Total S	0.08	0.12	0.17
H_2O	0.63	0.62	0.64
B_2O_3	0.49	0.49	0.75
BaO	0.73	0.74	0.74
BeO	0.79	0.79	0.82
CoO	0.42	0.71	0.71
Cr_2O_3	0.83	0.90	0.90
CuO	0.08	0.17	0.50
Ga_2O_3	0.63	0.65	0.68
La_2O_3	0.07	0.75	0.75
Li_2O	0.36	0.49	0.53
MoO_3	0.07	0.66	0.65
Nb_2O_3	0.38	0.39	0.39
NiO	0.57	0.72	0.78
PbO	0.48	0.49	0.50
SnO_2	0.38	0.45	0.47
SrO	0.59	0.59	0.59
V_2O_5	0.87	0.90	0.90
Y_2O_3	0.45	0.47	0.71
ZnO	0.17	0.42	0.44
ZrO_2	0.65	0.66	0.66

*Proportions of variance in concentration of elements and element oxides explained using 3-, 4-, and 5-factor Q-mode models.

1976c). At that time we did not have the CSIRO data and used a combination of the UM and USGS data for 32 elements (columns) for each of the 46 samples (rows). All elements except OC, N, and S were converted to oxides and the sums of all oxides for each sample (row) were adjusted to 100% in order to satisfy the constant row-sum requirement of the program.

Factor variance data (Table 3) for a 46 lake, 33 element data matrix showed that three factors will explain at least 30% of the variance in concentrations of all elements except S, P, Mn, Cu, Mo, La, Fe, and Zn. Increasing the number of factors to four greatly increases the variance explained for all of these elements, except S and P, which were then eliminated from further analyses.

Four factor Q-mode model

We then tried a 4 factor model on the 46 lake, 33 element data matrix, first using the orthogonal varimax axes, and finally using oblique axes defined by end members. The end members may be theoretical samples, or they may be actual samples from the data matrix. For this study, the four reference axes in the Q-mode model represent four actual sediment samples. After a number of iterative runs through the main interactive Q-mode program (EQMAIN; Miesch, 1976b, 1976c) we chose sediments from Crane, Little Pine, Cedar Bog, and Sand Point lakes as end members for factor I, II, III, and IV axes, respectively. Measured concentrations of 37 chemical variables for these sediment samples are given in Table 4.

The factor loadings for each of the four end-member samples (Table 5) are essentially composite chemical variables expressed as mixing proportions of each of the four end members. In other words, the 30 observed chemical variables used as input to the Q-mode model have been reduced to four composite chemical variables. However, the loadings do not indicate which of the 30 measured variables had the most influence in selecting Crane Lake as the factor I end member, Little Pine Lake as the factor II end member, and so on. To determine which elements contributed to which end member, and to what degree, the factor loadings for each sample were treated as composite chemical variables, and correlation coefficients were computed between the loadings and concentrations of the 30 elements used as input variables. Results of the correlation analysis are given in Table 6.

To illustrate regional groups of lakes based on sediment chemistry, loadings for each of the four factors are plotted on a map of Minnesota in Figure 3. Results in Tables 5 and 6 and Figure 3 show that factor I is a "clastic" factor, predominant in sediments from ten southern lakes (lake numbers 14, 22, 23, 28, 29, 30, 31, 37, 42, and 46 in Fig. 1), and five northeastern lakes (lake numbers 4, 5, 6, 9, and 10 in Fig. 1). As a group, these sediments contain relatively high concentrations of the generally lithophilic elements K, Al, Si, Ti, Zr, V, Be, Cr, Na, B, Y, and Ni. Factor II alkes contain carbonate-rich sediments with relatively high concentrations of Ca, Sr, and Ba. Factor II predominates in sediments from 10 west-central lakes (numbers 20, 21, 32, 34, 35, 36, 38, 39, 40, and 41), and is negligible in northeastern lakes. Five scattered lakes (numbers 25, 26, 43, 44, and 45 in Fig. 1) have loadings that are >0.5 for both factors I and II; in all of them the loading for factor I is greater. Factor III is an "organic" factor that predominates in sediments from six east-central lakes (numbers 15, 1, 17, 19, 24, and 27 in Fig. 1) and three northeastern lakes (numbers 1, 2, and 8). Sediments of these lakes contain more than 30% organic matter (Fig. 1), and have high concentrations of OC, TN, TH, and Pb. Factor IV appears to be a "redox" factor that delineates lake sediments with relatively high concentrations of Mo, Fe, La, Mn, and Zn (Table 6). The factor IV end

TABLE 4. CHEMICAL COMPOSITION OF END-MEMBER LAKE SEDIMENTS*

Element	Crane	Trout	Salt	Dead Coon	Little Pine	Cedar Bog	Sand Point
	(wt.%)	(wt.%)	(wt.%)	(wt.%)	(wt.%)	(wt.%)	(wt.%)
SiO_2-X	51.0	55.4	47.3	49.7	23.8	22.9	41.5
Al_2O_3-X	11.6	5.0	8.9	8.1	1.2	1.8	9.1
Fe_2O_3-M	5.3	3.6	4.3	3.3	3.3	8.7	11.0
MgO-M	1.44	0.73	2.42	2.12	1.18	0.40	0.93
CaO-M	2.34	1.44	8.4	10.5	27.8	1.59	1.09
Na_2O-M	0.96	0.49	1.67	0.81	0.22	0.27	0.67
K_2O-M	1.64	0.7	1.81	1.53	0.41	0.49	1.3
TiO_2-X	0.43	0.29	0.42	0.47	0.14	0.11	0.36
MnO-M	0.18	0.07	0.17	0.14	0.24	0.09	5.08
CO_2-CC-M	1.83	1.83	7.69	8.4	20.5	1.46	1.83
OC-M	6.6	8.5	3.3	2.9	8.5	29.0	7.1
N-M	0.82	1.0	0.41	0.34	1.11	3.47	0.63
S-M	0.06	0.19	1.32	0.29	0.19	0.07	0.15
H_2O-M	10.9	12.5	8.0	4.6	10.5	35.3	15.5
LOI-550	16.9	21.0	11.3	6.4	17.8	60.0	17.9
	(ppm)	(ppm)	(ppm)	(ppm)	(ppm)	(ppm)	(ppm)
As-C	10	8	90	....	11	....	....
B-S	39	16	44	49	13	22	30
Ba-S	312	124	279	326	296	121	353
Be-S	2.1	3.2	1.7	1.4	0.4	0.5	1.8
Co-S	13	6	7	7	3	5	24
Cr-S	65	31	47	54	10	23	59
Cu-M	30	66	32	38	15	13	31
Ga-C	15	5	10	8	1	6	....
Ge-C	1	2	1	1	1	1	....
La-S	66	5	47	6	37	18	148
Li-M	5	2	6	3	1	2	7
Mo-S	2.9	1.7	1.2	3.4	2.5	5.5	7.7
Nb-S	30	28	15	48	26	12	24
Ni-S	39	22	25	36	3	7	38
Pb-S	19	27	1.5	4	1.4	30	19
Sc-C	10	10	10	7	7	7	....
Sn-C	1.0	2.0	1.0	1.0	0.6	3.0	....
Sr-S	149	83	477	261	186	55	120
V-S	99	46	92	97	17	14	94
Y-S	11	50	15	24	7	5	16
Zn-M	163	173	180	130	72	150	273
Zr-S	136	128	131	278	22	82	112

*Data from Appendix.

member is Sand Point (number 7), which is the only lake of the 46 included in our study that contains a clearly visible oxidized microzone (a reddish-brown layer with ferric hydroxide; Gorham, 1958, 1964; Gorham and Swaine, 1965) at the sediment surface. Other lakes with high loadings for factor IV are another northeastern lake, Mountain (number 3), and Elk Lake (number 33) in northwestern Minnesota.

Five factor Q-mode model

Intuitively we expected that there would be geochemical differences between the clastic components in sediments of lakes in northeastern Minnesota, where the bedrock is predominantly Precambrian crystalline rocks, and lakes in western Minnesota, where the bedrock is predominantly Cretaceous shale (Fig. 2A). Both clastic types were lumped together under factor I in the 4 factor Q-mode model. In an attempt to separate these two clastic types, we used a 5 factor model. The factor-variance data (Table 3) showed that adding a fifth factor explains a larger proportion of the variance in Y, B, Cu, and Li. These elements contributed little to the 4 factor model, and we anticipated that they might help to separate the northeastern and western Minnesota clastic types. After several iterative runs through the interactive Q-mode

TABLE 5. LOADINGS FOR 4-FACTOR END-MEMBER LAKES*

Lake	Factor I Crane	Factor II Little Pine	Factor III Cedar Bog	Factor IV Sand Point	Lake	Factor I Crane	Factor II Little Pine	Factor III Cedar Bog	Factor IV Sand Point
Big	0.46	-0.09	*0.75*	-0.13	Beaver	*0.64*	*0.57*	0.01	-0.21
Iron	0.19	-0.15	*0.72*	0.24	Moose	*0.71*	*0.63*	-0.07	-0.27
Mountain	0.21	-0.19	0.19	*0.79*	Ball Club	0.37	0.09	*0.63*	-0.09
Trout	*0.95*	-0.15	0.39	-0.19	St. Olaf	*0.9*	0.42	-0.01	-0.31
Clearwater	*0.77*	-0.16	-0.01	0.40	Christmas	*0.78*	0.20	0.15	-0.14
Wilson	*0.81*	-0.08	0.05	0.21	Lotus	*0.95*	0.46	-0.11	-0.29
Sand Point	0.00	0.00	0.00	*1.00*	Clear	*1.02*	0.22	0.04	-0.29
Kimball	0.43	-0.11	*0.93*	-0.25	Long	0.28	*0.85*	-0.09	-0.04
Crane	*1.00*	0.00	0.00	0.00	Elk	-0.56	0.38	0.30	*0.88*
O'Leary	*1.08*	0.07	0.07	-0.22	Itasca	-0.03	*0.80*	0.15	0.09
Josephine	*0.77*	-0.02	*0.62*	-0.37	Little Pine	0.00	*1.00*	0.00	0.00
Arco	*0.62*	0.02	*0.62*	-0.26	Grove	0.31	*0.73*	0.11	-0.15
Deming	*0.54*	-0.07	*0.85*	-0.31	Fish	*0.82*	0.44	0.01	-0.25
Farquar	*1.25*	-0.07	0.05	-0.24	Sallie	0.22	*0.90*	0.04	-0.17
George	0.46	-0.05	*0.67*	-0.08	Cowdry	-0.03	*0.71*	0.12	0.21
Linwood	0.18	0.21	*0.67*	-0.05	Maple	0.20	*0.93*	-0.02	-0.11
Cedar Bog	0.00	0.00	*1.00*	0.00	Mina	-0.07	*0.85*	0.04	0.18
Spectacle	*0.57*	-0.04	*0.79*	-0.32	Big Kandiyohi	*0.75*	0.45	-0.06	-0.14
Gladstone	0.15	-0.01	*0.76*	0.10	Shetek	*0.74*	*0.62*	0.00	-0.36
Ham	-0.07	*0.71*	0.31	0.06	Elk (Grant)	*0.90*	*0.60*	-0.09	0.40
Green	0.17	*0.58*	0.22	0.03	Dead Coon	*1.15*	*0.53*	-0.33	-0.35
Reeds	*0.92*	0.15	0.22	-0.29	Salt	*1.03*	0.48	-0.27	-0.25
Francis	*0.76*	0.18	0.28	0.22					
Nokay	0.08	0.19	*0.64*	0.10					

*Loadings >0.5 are in italics.

programs, we finally chose sediments from Salt, Little Pine, Cedar Bog, Sand Point, and Trout lakes as the end members for factors I, II, III, IV, and V, respectively. Chemical compositions of these sediments are given in Table 4. The factor loadings for each sediment were correlated against the geochemical input variables to determine which variables had the most influence on each factor (Table 7).

Factors II, III, and IV, with Little Pine, Cedar Bog, and Sand Point as end members, are essentially the same as factors II, III, and IV in the 4 factor model (Table 6). Factor I was strongest, with Salt Lake as the end member. Because Salt Lake is a very shallow and saline prairie pothole without an outlet, we tried a more typical prairie lake, Dead Coon Lake, as an end member with almost identical results. The composition of Dead Coon Lake sediment is included in Table 4 for comparison. The fifth and weakest factor has as its end member Trout Lake near Lake Superior (Fig. 1). It appeared, therefore, that by using chemical characteristics of the sediments from Salt (or Dead Coon) Lake, and those of sediments from Trout Lake, we might distinguish between the clastic sediments of northeastern lakes and those of the western prairie lakes. However, maps of factor loadings (Fig. 4) show that the sediments of about half of the northeastern lakes have more in common chemically with sediments from Salt Lake than they do with those from Trout Lake. The sediments of two lakes (Wilson and Clearwater) have chemical characteristics of both Salt and Trout lakes. The clastic affinities of sediments in three northeastern lakes (Big, Iron, and Kimball) are masked by their high organic content and, therefore, have high loadings for factor III (Fig. 3C). Also, the clastic affinity of sediment in Sand Point Lake is masked by its high iron and manganese contents.

Table 7 shows that the four elements (Y, B, Li, and Cu) whose proportion of variance explained increased markedly by going from a 4- to 5-factor model contributed significantly in determining loadings for factor I (B and Li) and factor V (Y and Cu). Al, Mg, Na, K, Ba, and Zr all contribute significantly to Factor I but not to Factor V (Table 7). The chemical characteristics of the sediments of factor V lakes, particularly those of Trout and Wilson (and probably Kimball if we could strip away the organic influence), are probably derived from sills of Duluth Gabbro that crop out around Lake Superior (Fig. 2A), and are heavily mineralized, particularly with Cu and Ni (Sims, 1968); hence the relatively high concentrations of these two elements in factor V lake sediments. The chemical characteristics of clastic materials being deposited in lakes near the Canadian border (such as Mountain, Clearwater, O'Leary, Sand Point, and Crane) are derived from acidic Precambrian crystalline rocks, and are more similar in composition to clastic material derived from Cretaceous shale.

TABLE 6. CORRELATION COEFFICIENTS BETWEEN FACTOR LOADINGS AND OXIDE CONCENTRATIONS (IV-FACTOR MODEL)*

Oxide	Factor I Crane	Factor II Little Pine	Factor III Cedar Bog	Factor IV Sand Point
SiO_2	0.82	-0.58	-0.21	-0.19
Al_2O_3	0.82	-0.47	-0.32	-0.20
Fe_2O_3	-0.38	-0.44	0.24	0.77
MgO	0.43	0.44	-0.68	-0.34
CaO	-0.46	0.96	-0.46	0.01
Na_2O	0.66	-0.35	-0.33	-0.12
K_2O	0.86	-0.31	-0.48	-0.27
TiO_2	0.80	-0.46	-0.33	-0.19
MnO_2	-0.19	-0.06	-0.18	0.54
CO_2-carb	-0.42	0.97	-0.49	-0.02
OC	-0.35	-0.44	0.95	-0.06
Total N	-0.33	-0.37	0.88	0.15
H_2O	-0.35	-0.54	0.96	0.05
B_2O_3	0.64	-0.20	-0.35	-0.24
BaO	0.33	0.51	-0.81	-0.14
BeO	0.75	-0.60	-0.18	-0.12
CoO	0.33	0.55	-0.15	0.37
Cr_2O_3	0.70	-0.61	-0.23	0.02
CuO	0.36	0.51	-0.03	0.14
La_2O_3	-0.19	-0.38	-0.07	0.77
Li_2O	0.35	-0.49	-0.11	0.22
MoO_3	-0.57	-0.03	0.00	0.81
Nb_2O_5	0.37	0.23	-0.60	-0.12
NiO	0.52	-0.58	-0.19	0.19
PbO	-0.01	-0.61	0.67	-0.02
Sr	0.22	0.62	-0.74	-0.21
V_2O_5	0.77	-0.43	-0.43	-0.08
Y_2O_3	0.54	-0.53	-0.10	0.00
ZnO	-0.07	-0.47	0.14	0.51
ZrO_2	0.80	-0.37	-0.29	-0.33

*Correlation coefficients >0.5 are in italics.

DISCUSSION

The results of the Q-mode factor analyses have enabled us to compare and contrast the compositions of lake sediments from different climatic and geologic regions of Minnesota, and to define compositional end members. We were somewhat surprised that there was not a greater difference in clastic composition between sediments in lakes in igneous terranes in northeastern Minnesota and those of the western prairie lakes. The main differences that emerge from the factor analyses between the western-type clastic material and northeastern-type clastic material are higher concentratins of Al, Na, K, B, Ba, Li, Mg, and Sr in western-type clastic material, and higher concentrations of Pb, Y, Be, Cu, and, to a lesser extent, Ni in northeastern-type clastic material. The higher concentrations of Mg, Ba, and Sr in the prairie lake sediments may be due to the presence of carbonate minerals, but high concentrations of these elements in sediment of Crane Lake (as well as those of other lakes near the Canadian border) suggest that carbonate minerals do not explain all of the difference. The higher concentrations of Pb, Y, and Be in the northeastern lake sediments are probably associated with potash feldspars in Precambrian granite (Fig. 2A). The higher concentrations of Cu and Ni in the northeastern lakes most likely are derived from Cu and Ni mineralization. However, concentrations of copper are just as high in sediments of some prairie lakes (Fig. 5; Appendix). Accumulations of Cu in the surface sediments from the Great Lakes was shown experimentally by Cline and Upchurch (1973) to be due mainly to bacterial release of Cu from organic complexes. However, we found no organic Cu association in our samples of Minnesota lake sediments. A possible explanation for high concentrations of Cu in sediments of some prairie lakes is application of $CuSO_4$ for algal control in the highly productive waters of these lakes, although those sediments that have high concentrations of Cu do not have high concentrations of total sulfur (Appendix I). We obtained information on which lakes had been approved for $CuSO_4$ application, but could not find out whether $CuSO_4$ had actually been applied.

The redox association of elements (Fe, Mn, La, Mo, and Zn; factor IV of the 4 factor model, Fig. 3D) only appears in three lakes, Sand Point, Mountain, and Elk (Clearwater County). The sediment of the factor IV end-member lake, Sand Point, contains a distinct reddish-brown oxidized microzone at the surface, and high concentrations of iron and, particularly, manganese (Appendix; Fig. 6). An oxidized microzone is not readily distinguishable in the sediment of Mountain Lake, which is poorer in iron than that of Sand Point Lake (Fig. 6), but it may form seasonally or during certain years, as it does in the sediments of some lakes in the English Lake District (Gorham, 1958; Mackereth, 1966). Gorham (1964) and Gorham and Swaine (1965) showed that oxidized surface layers in sediments of the English lakes typically are greatly enriched in Mn, Mo, and Fe relative to reduced sediments a few centimeters below the oxidized zone.

Identifying the mineral residences of iron and manganese in fine-grained sediments usually is difficult because many of the minerals cannot be detected by XRD or microscopic techniques (Jones and Bowser, 1978). The inferred or identified mineral phases in sediments from Elk Lake (Clearwater County) include X-ray amorphous iron and manganese oxyhydroxides, pyrite observed in scanning electron micrographs, one or more iron phosphate minerals, and manganese carbonate (rhodochrosite) (Dean, Chapter 10). The ultimate source of iron and manganese in Elk Lake is from influx of iron- and manganese-rich ground waters. However, the direct source of iron and manganese in the sediments of Elk Lake is from reduction of iron and manganese minerals that survived the trip through a reducing hypolimnion during the previous few years (Dean, Chapter 10; Nuhfer and others, Chapter 7). This reduced iron and manganese builds up in the hypolimnion during summer stratification. Precipitation of iron and manganese minerals occurs at fall overturn when iron- and manganese-rich dysaerobic hypolimnetic waters mix with oxygen- and bicarbonate-rich epilimnetic waters. Precipitation

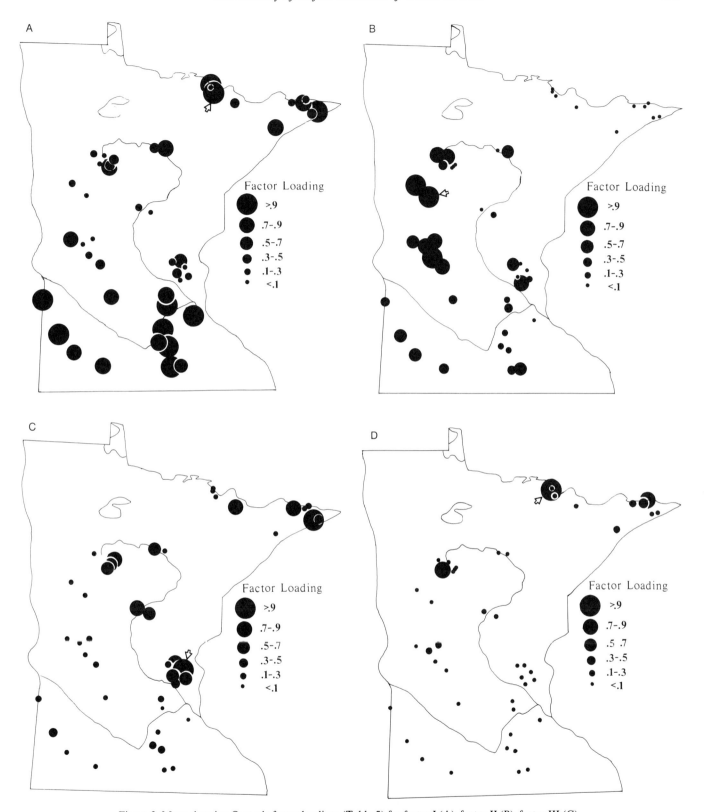

Figure 3. Maps showing Q-mode factor loadings (Table 5) for factor I (A), factor II (B), factor III (C), and factor IV (D) resulting from a 4 factor Q-mode model of the geochemistry of surface sediments from 46 Minnesota lakes. End-member lakes for each factor are indicated by arrows.

TABLE 7. CORRELATION COEFFICIENTS BETWEEN FACTOR
LOADINGS AND OXIDE CONCENTRATIONS
(V-FACTOR MODEL)*

Oxide	Factor I Salt	Factor II Little Pine	Factor III Cedar Bog	Factor IV Sand Point	Factor V Trout
SiO_2	0.71	-0.73	-0.23	0.07	0.70
Al_2O_3	0.90	-0.72	-0.22	-0.07	0.42
Fe_2O_3	0.04	-0.44	-0.06	0.78	0.16
MgO	0.51	0.22	-0.42	-0.25	-0.20
CaO	-0.37	0.96	-0.33	-0.16	-0.60
Na_2O	0.75	-0.56	-0.27	0.10	0.37
K_2O	0.96	-0.59	-0.32	0.00	0.33
TiO_2	0.77	-0.64	-0.30	0.07	0.58
MnO_2	0.24	-0.18	-0.32	0.60	-0.01
CO_2-carb	-0.34	0.96	-0.35	-0.18	-0.59
OC	-0.56	-0.20	0.89	-0.21	-0.13
Total N	-0.55	-0.14	0.93	-0.30	-0.23
H_2O	-0.49	-0.31	0.85	-0.08	-0.06
B_2O_3	0.81	-0.46	-0.14	-0.04	0.05
BaO	0.57	0.22	-0.59	-0.02	-0.21
BeO	0.62	-0.71	-0.25	0.13	0.76
CaO	0.59	-0.70	-0.34	0.57	0.52
Cr_2O_3	0.82	0.82	-0.27	0.29	0.58
CuO	0.24	-0.50	-0.24	0.28	0.74
La_2O_3	0.21	-0.43	-0.40	0.84	0.38
Li_2O	0.66	-0.67	-0.15	0.40	0.25
MoO_3	-0.18	0.00	-0.30	0.75	0.05
Nb_2O_3	0.39	0.08	-0.52	0.00	0.17
NiO	0.59	-0.70	-0.35	0.40	0.70
PbO	-0.15	-0.48	0.58	-0.02	0.18
SrO	0.38	0.40	-0.50	-0.10	-0.31
V_2O_3	0.86	-0.66	-0.41	0.19	0.54
Y_2O_3	0.30	-0.52	-0.31	0.18	0.88
ZnO	0.13	-0.47	-0.12	0.56	0.38
ZrO_2	0.68	-0.53	-0.20	-0.11	0.48

*Correlation coefficients >0.5 are in italics.

also occurs in the epilimnion during the winter months, and in the metalimnion during the summer by mixing of epilimnetic and hypolimnetic waters (Nuhfer and others, Chapter 7). Therefore, the unusually high concentrations of iron and manganese in the sediments of Elk Lake would be even higher if all of the iron and manganese minerals that precipitated from the water column remained in the sediments.

Although the redox association of Fe, Mn, and Mo suggests that these elements are closely related, we found that, in fact, Fe and Mn are not particularly well correlated ($r = 0.37$, $n = 46$), and that Mo is more highly correlated with Fe than with Mn. This association may be due to predominantly sulfide residences for Fe and Mo, and a predominantly oxyhydroxide residence for Mn. The correlation between Fe and Mo is illustrated in Figure 7. We have found a similar situation in sediments of lakes in the English Lake District (Dean and others, 1988), and the English lake data are included in Figure 7 for comparison. Although more variability and scatter characterize the data for Minnesota lake sediments, most of the Minnesota sediments fall along the same regression line as the English sediments.

In redox-sensitive deep-sea sediments Mo is usually closely associated with Mn, and both elements are usually not associated with iron (Gardner and others, 1982; Dean and Gardner, 1989). The difference in the behavior of Fe in fresh-water and marine sediments is probably related to the low availability of sulfate in fresh-water systems. In marine sediments, once solid manganese oxides and hydroxides are reduced at some depth below the sediment-water interface, the resulting Mn(II) diffuses out of the sediment. However, if ferric hydroxide is reduced, the resulting Fe(II) usually combines with sulfide produced by bacterial sulfate reduction to form various solid phases of ferrous sulfide (such as mackinawite, greigite, pyrrhotite, and pyrite; Berner, 1971, 1984; Sweeney and Kaplan, 1973). In most marine sediments, therefore, most of the sulfur is present in sulfide minerals, and Fe and S are usually highly correlated. In most fresh-water lake sediments, however, interstitial waters have low sulfate concentrations so that reduced Fe(II) escapes from the sediment into the water column, unless there is an abundance of some other anionic complex such as phosphate or carbonate to form solid ferrous phosphate (vivianite) and ferrous carbonate (siderite). Sulfate reduction in deep-sea sediments is more likely to be limited by organic carbon than by sulfate. In fresh-water sediments, however, sulfate is more likely to be limiting than either organic carbon or iron. As a result, the S:C and S:Fe ratios in anoxic fresh-water sediments are much lower than in anoxic marine sediments (e.g., see Berner and Raiswell, 1984).

Mothersill (1976) reported a close association among Fe, Mn, and Co in sediments from Lake Victoria. Cline and Chambers (1977) concluded that an association between Mn and Co, Cd, Cu, and Zn in surface sediments from northeastern Lake Michigan was redox related. Mackereth (1966) suggested that coprecipitation of Co, Ni, Cu, and Zn with manganese and iron may be the dominant mechanism of transferring these metals from water to sediments in the English lakes. However, these elements were also inversely related to grain size, so that the association may be more of a clay-organic-trace element relation as described by Shimp and others (1971) for sediments from southern Lake Michigan, by Glenn and Van Atta (1973) for Columbia River sediments, by Oliver (1973) for Ottawa and Rideau River sediments, by Baker-Blocker and others (1975) for sediments in Grand Traverse Bay, Lake Michigan, and by Fitchko and Hutchinson (1975) for river sediments entering the Great Lakes. In other words, organic matter is usually concentrated in fine-grained sediments, and both materials may be enriched in trace elements, particularly the transition elements, by adsorption on clay minerals and/or organic complexing.

The sediment of Elk Lake (number 33 in Fig. 1) is enriched in both iron (mainly as iron hydroxide, iron sulfide, and iron phosphate) and manganese (as both hydrated manganese oxide and manganese carbonate, rhodochrosite; Dean, Chapter 10; Nuhfer and others, Chapter 7). These components have distinct seasonal cycles as evidenced by varve and sediment-trap studies

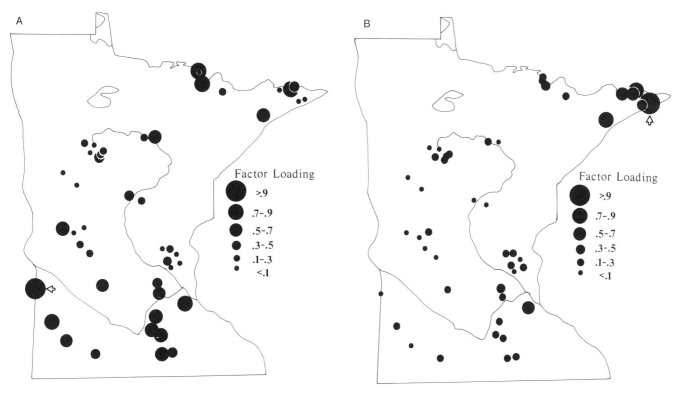

Figure 4. Maps showing Q-mode factor loadings for factor I (southwestern clastic sediments; A) and factor V (northeastern clastic sediments; B) resulting from a 5 factor Q-mode model of the geochemistry of surface sediments from 46 Minnesota lakes. End-member lakes for each factor are indicated by arrows.

(Anderson, Chapter 5; Nuhfer and others, Chapter 7), and water chemistry (Dean and Megard, Chapter 18). Other lakes, chiefly in the forested regions of Minnesota, are enriched in iron or manganese, but usually not both (Fig. 6).

CONCLUSIONS

1. Q-mode factor analyses of the elemental compositions of sediments from 46 Minnesota lakes delineate five groups of sediment types. Sediments in lakes of west-central Minnesota that contain high concentrations of $CaCO_3$ also contain high concentrations of Ba and Sr. Organic-rich sediments occur in lakes in east-central and northeastern Minnesota; they contain high concentrations of organic C, N, and H, and relatively high concentrations of Pb. There are two clastic sediment types. Those of the western prairie lakes, derived mostly from Cretaceous shales, are enriched in Al, Na, K, B, Ba, Li, Mg, and Sr. Clastic sediments in most lakes in the arrowhead region of northeastern Minnesota are enriched in Cu, Pb, Y, and Be, probably associated with potash feldspars in clastic materials derived from Precambrian granite. Sediments from some other northeastern Minnesota lakes, how-

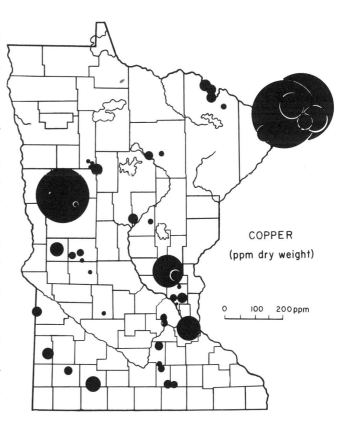

Figure 5. Map showing concentrations of Cu in surface sediments from 46 Minnesota lakes.

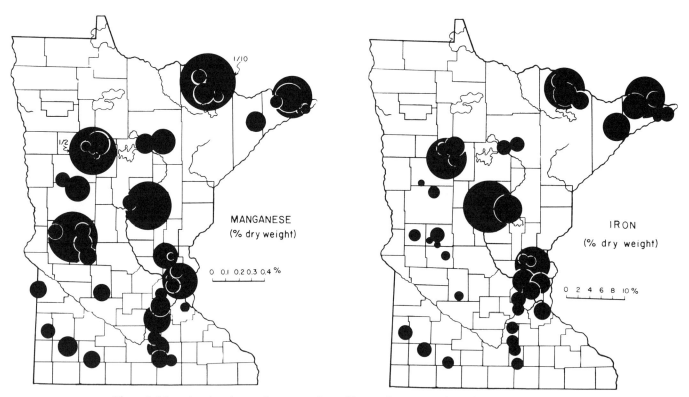

Figure 6. Maps showing elemental concentrations of iron and manganese in surface sediments from 46 Minnesota lakes.

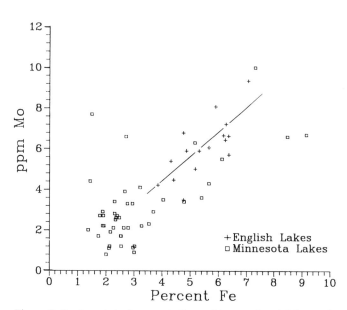

Figure 7. Scatter plot of concentrations of iron and molybdenum in surface sediments from 46 Minnesota lakes and 15 English lakes.

ever, have chemical characteristics that are similar to those of the western prairie lakes. A fifth minor group of lakes was distinguished on the basis of high concentrations of Mo, Fe, La, Mn, and Zn, reflecting unusual redox conditions in the sediments of these lakes.

2. High concentrations of Cu in sediments of extreme northeastern Minnesota lakes probably derive from Cu-mineralized gabbro sills that crop out in that area. The sediments of several lakes of western and southern Minnesota, however, contain equally high concentrations of Cu, which in these highly productive prairie lakes may be due to application of $CuSO_4$ for algal control.

3. Concentrations of iron and/or manganese, and several trace elements, particularly Mo, Zn, and La, are elevated in several lakes in central and northeastern Minnesota. In general, the concentrations of Fe and Mn are not particularly well correlated, and Mo, Zn, and La are more closely associated with Fe than with Mn.

ACKNOWLEDGMENTS

We greatly appreciate the helpful suggestions by C. J. Bowser, D. Engstrom, R. C. Severson, M. L. Tuttle, and T. C. Winter on earlier versions of this paper. We thank P. H. Briggs, J. E. Taggart, R. G. Havens and J. S. Wahlberg for providing the X-ray fluorescence and emission spectrographic analyses.

APPENDIX. CHEMICAL ANALYSES OF SEDIMENT SAMPLES FROM MINNESOTA LAKES*

Lake	Si-x (wt.%)	Al-x (wt.%)	Al-s (wt.%)	Fe-x (wt.%)	Fe-s (wt.%)	Fe-m (wt.%)	Mg-s (wt.%)	Mg-m (wt.%)	Ca-x (wt.%)	Ca-s (wt.%)	Ca-m (wt.%)
Big	18.0	4.00	4.03	3.00	1.23	3.0	0.56	0.29	0.70	0.69	0.93
Iron	17.9	2.21	2.05	4.78	2.98	4.8	0.15	0.44	0.96	1.46	0.84
Mountain	20.4	3.32	3.26	7.32	4.94	7.8	0.45	0.74	0.79	1.06	0.81
Trout	25.9	2.63	2.52	2.52	2.08	2.6	0.23	0.44	0.74	0.83	1.03
Clearwater	24.9	5.32	3.68	5.66	3.58	6.9	0.74	0.88	1.38	2.05	0.84
Wilson	24.3	4.22	4.42	5.16	3.65	5.2	0.74	0.82	2.38	3.62	0.79
Sand Point	19.4	4.79	3.37	7.71	4.53	7.5	0.11	0.56	1.02	1.18	0.78
Kimball	17.6	2.23	4.64	1.88	1.37	1.8	0.28	0.53	1.03	1.41	0.88
Crane	23.8	6.12	6.66	2.69	2.71	4.0	0.51	0.87	1.22	1.61	0.96
O'Leary	23.5	5.85	4.37	2.96	2.09	2.9	0.64	0.93	1.70	2.64	1.21
Josephine	18.6	4.10	3.65	2.00	1.50	2.1	0.34	0.51	1.20	1.53	0.87
Arco	18.7	3.64	3.67	2.40	1.61	2.5	0.33	0.51	1.22	1.64	0.91
Deming	18.0	2.70	2.74	2.10	1.51	2.1	0.20	0.43	0.70	1.23	0.63
Faquar	25.3	5.12	5.73	3.27	2.60	3.2	0.46	0.62	1.03	1.14	0.80
George	17.1	3.19	3.31	4.04	2.66	4.1	0.38	0.52	1.56	2.96	1.15
Linwood	17.1	1.32	3.62	3.53	2.44	3.5	0.10	0.32	5.73	6.80	5.80
Cedar Bog	10.7	0.97	2.74	6.12	5.32	6.0	0.11	0.24	1.28	1.57	1.14
Spectacle	19.2	3.07	2.60	2.30	1.51	1.9	0.23	0.44	1.59	2.29	1.11
Gladstone	14.0	2.04	2.07	9.15	6.56	9.4	0.23	0.45	2.00	2.49	1.61
Ham	8.9	1.12	3.17	2.94	2.11	3.2	0.87	0.48	17.0	13.5	16.4
Green	15.4	1.31	4.94	1.90	1.40	1.9	0.97	0.48	13.3	12.1	12.7
Reeds	21.1	4.45	3.76	2.36	1.88	2.3	0.63	0.98	4.34	5.40	3.44
Francis	20.7	3.93	3.25	2.53	1.85	2.4	0.92	1.20	4.21	6.17	3.51
Nokay	18.0	3.30	3.32	5.40	3.86	5.4	0.09	0.46	8.00	8.62	7.6
Beaver	18.5	3.24	7.00	1.92	1.40	2.1	0.55	1.13	11.7	12.7	10.3
Moose	16.7	3.55	3.10	2.79	2.17	2.8	1.10	1.45	12.8	12.6	11.1
Ball Club	21.7	1.82	5.23	2.76	1.83	2.8	0.21	0.55	2.28	2.89	1.94
St. Olaf	21.3	4.37	3.48	2.12	1.75	2.1	0.78	0.98	8.49	11.9	6.80
Christmas	23.2	3.75	3.09	2.67	1.89	2.7	0.68	0.84	4.42	7.78	3.61
Lotus	21.7	3.93	7.62	2.15	1.50	2.3	0.77	1.00	7.85	13.2	6.90
Clear	22.9	5.08	3.55	2.54	1.82	2.3	0.81	1.09	5.80	7.19	4.80
Long	14.5	2.44	5.28	2.47	1.51	2.3	0.60	1.08	17.3	17.9	15.2
Elk	14.9	1.09	2.86	8.47	4.90	8.8	0.87	0.47	8.93	10.2	8.80
Itasca	8.1	0.80	2.33	3.21	2.29	3.2	0.13	0.62	18.5	12.8	19.4
Little Pine	11.1	0.65	3.49	2.33	1.61	2.4	0.24	0.71	18.6	15.5	19.9
Grove	12.4	2.11	2.37	1.78	1.47	1.8	0.56	0.93	15.2	11.2	14.1
Fish	20.9	3.83	2.50	2.26	1.61	2.4	0.68	1.11	8.03	8.9	7.1
Sallie	10.5	0.95	4.17	1.37	1.14	1.4	0.68	1.23	18.6	14.9	16.5
Cowdry	14.3	1.21	4.24	2.71	1.74	2.6	0.43	0.94	12.7	12.7	12.0
Maple	13.4	2.03	4.86	1.40	1.01	1.4	0.67	1.16	18.5	13.9	15.9
Mina	13.7	1.20	4.88	1.50	1.24	1.5	0.20	0.90	15.9	10.0	15.3
Shetek	19.5	4.13	7.92	2.63	1.79	2.8	0.54	1.14	10.6	11.5	9.3
Big Kandiyohi	19.3	3.38	5.51	1.73	1.24	1.8	1.03	1.66	10.1	10.9	9.0
Elk-Grant Co.	21.0	3.57	6.95	1.88	1.49	2.1	1.84	2.79	7.70	12.2	6.8
Dead Coon	23.2	4.26	4.69	2.30	1.87	2.5	1.17	1.28	8.69	10.5	7.5
Salt	22.1	4.71	3.77	3.02	1.97	3.1	0.95	1.46	7.46	12.3	6.0

APPENDIX. CHEMICAL ANALYSES OF SEDIMENT SAMPLES FROM MINNESOTA LAKES* (continued)

Lake	Na-m (wt.%)	K-x (wt.%)	K-s (wt.%)	K-m (wt.%)	Ti-x (wt.%)	Ti-s (wt.%)	Ti-c (wt.%)	Mn-s (wt.%)	Mn-m (wt.%)	S-m (wt.%)	P-m (wt.%)
Big	0.38	0.70	1.11	0.66	0.10	0.12	0.6	0.009	0.07	0.27	0.10
Iron	0.32	0.47	0.84	0.40	0.14	0.18	0.8	0.067	0.08	0.37	0.13
Mountain	0.63	1.11	1.81	1.08	0.21	0.30	0.5	0.21	0.28	0.28	0.15
Trout	0.36	0.72	1.26	0.58	0.17	0.23	0.5	0.024	0.06	0.29	0.14
Clearwater	0.88	1.38	2.38	1.26	0.38	0.37	0.6	0.093	0.22	0.24	0.10
Wilson	1.33	1.21	1.97	1.05	0.46	0.48	0.8	0.045	0.14	0.24	0.10
Sand Point	0.50	1.31	1.95	1.08	0.22	0.23	....	0.76	3.94	0.21	0.26
Kimball	0.35	0.56	0.93	0.51	0.17	0.20	0.6	0.018	0.03	0.61	0.20
Crane	0.71	1.57	2.29	1.36	0.25	0.28	0.5	0.096	0.14	0.18	0.13
O'Leary	0.82	1.61	2.63	1.41	0.26	0.26	0.8	0.068	0.09	0.21	0.10
Josephine	0.62	1.13	1.74	0.97	0.22	0.24	1.5	0.022	0.04	0.34	0.10
Arco	0.57	1.07	1.58	0.88	0.21	0.21	0.6	0.023	0.03	0.44	0.14
Deming	0.40	0.90	1.37	0.77	0.20	0.16	0.6	0.024	0.03	0.51	0.18
Faquar	0.73	1.91	2.60	1.57	0.43	0.44	1.5	0.042	0.06	0.24	0.09
George	0.60	1.09	1.76	0.91	0.21	0.24	0.8	0.030	0.04	0.59	0.11
Linwood	0.25	0.53	0.86	0.48	0.079	0.075	0.4	0.16	0.25	0.59	0.25
Cedar Bog	0.20	0.48	0.86	0.41	0.065	0.083	0.4	0.070	0.07	0.44	0.23
Spectacle	0.49	0.92	1.43	0.80	0.20	0.19	0.6	0.029	0.05	0.67	0.25
Gladstone	0.37	0.82	1.17	0.64	0.13	0.16	0.6	0.090	0.10	0.38	0.23
Ham	0.25	0.46	0.71	0.36	0.069	0.053	0.1	0.066	0.09	0.56	0.11
Green	0.27	0.53	0.85	0.42	0.090	0.087	0.1	0.24	0.20	0.46	0.14
Reeds	0.41	1.22	1.83	1.10	0.23	0.29	0.6	0.13	0.17	0.33	0.11
Francis	0.47	1.24	2.01	1.06	0.25	0.24	0.6	0.089	0.09	0.34	0.13
Nokay	0.25	0.90	0.94	0.54	0.070	0.073	0.25	0.19	0.32	0.42	0.16
Beaver	0.38	1.08	1.68	0.93	0.23	0.16	0.3	0.051	0.09	0.43	0.09
Moose	0.69	1.17	1.78	1.04	0.18	0.23	0.4	0.15	0.19	0.19	0.08
Ball Club	0.38	0.70	1.15	0.60	0.10	0.096	0.25	0.099	0.15	0.65	0.15
St. Olaf	0.44	1.22	2.09	1.12	0.23	0.24	0.4	0.11	0.13	0.32	0.10
Christmas	0.51	1.24	2.02	1.11	0.22	0.25	0.3	0.051	0.09	0.87	0.07
Lotus	0.50	1.25	1.96	1.11	0.23	0.31	0.4	0.11	0.15	0.38	0.08
Clear	0.45	1.39	2.04	1.24	0.27	0.27	0.6	0.14	0.21	0.26	0.12
Long	0.45	0.76	1.35	0.65	0.12	0.089	0.15	0.089	0.07	0.38	0.21
Elk	0.20	0.41	0.75	0.34	0.040	0.062	0.20	0.26	0.43	0.52	0.38
Itasca	0.13	0.36	0.73	0.28	0.032	0.025	0.03	0.081	0.15	0.21	0.18
Little Pine	0.16	0.45	0.76	0.34	0.080	0.037	0.10	0.13	0.19	0.27	0.11
Grove	0.27	0.67	1.35	0.65	0.12	0.10	....	0.085	0.12	0.52	0.10
Fish	0.44	1.22	2.00	1.11	0.23	0.23	0.60	0.10	0.13	0.99	0.94
Sallie	0.24	0.55	1.10	0.50	0.089	0.071	0.20	0.059	0.11	0.40	0.08
Cowdry	0.24	0.57	1.04	0.50	0.090	0.075	0.25	0.087	0.12	1.51	0.10
Maple	0.27	0.68	1.17	0.60	0.13	0.078	0.15	0.085	0.13	0.90	0.07
Mina	0.22	0.59	1.06	0.51	0.080	0.070	0.20	0.24	0.35	1.09	0.07
Shetek	0.40	1.28	1.89	1.12	0.26	0.31	0.40	0.069	1.4	1.48	0.09
Big Kandiyohi	0.83	1.11	1.72	1.08	0.17	0.17	0.40	0.067	0.12	1.14	0.08
Elk-Grant Co.	0.60	1.20	2.07	1.16	0.19	0.26	0.30	0.081	0.11	0.60	0.08
Dead Coon	0.60	1.46	2.29	1.27	0.28	0.35	0.40	0.079	0.11	0.42	0.08
Salt	1.24	1.73	2.58	1.50	0.25	0.22	0.40	0.061	0.13	1.82	0.08

APPENDIX. CHEMICAL ANALYSES OF SEDIMENT SAMPLES FROM MINNESOTA LAKES* (continued)

Lake m	N-m (wt.%)	H-m (wt.%)	TC-m (wt.%)	CC-m (wt.%)	OC-m (wt.%)	LOI-m (wt.%)	CaCO3 (wt.%)
Big	1.78	2.75	18.09	0.4	17.6	38.1	3.1
Iron	1.79	3.55	25.3	0.4	24.9	40.2	3.7
Mountain	0.86	1.69	10.0	0.4	9.6	19.2	3.3
Trout	1.00	1.40	9.0	0.5	8.5	21.0	4.5
Clearwater	0.50	1.03	6.11	0.3	5.8	15.8	2.8
Wilson	0.60	1.70	8.0	0.3	7.7	17.4	2.4
Sand Point	0.63	1.73	7.6	0.5	7.1	17.9	4.2
Kimball	2.85	3.58	23.9	0.4	23.5	46.9	3.6
Crane	0.82	1.22	7.1	0.5	6.6	16.9	4.0
O'Leary	0.97	1.38	9.0	0.3	8.7	19.2	2.2
Josephine	2.09	2.89	20.9	0.3	20.6	40.2	2.5
Arco	2.36	2.85	21.6	0.4	21.3	41.6	3.6
Deming	2.74	3.31	25.0	0.5	24.6	47.9	3.8
Faquar	0.92	1.05	7.0	0.3	6.7	16.3	2.6
George	2.16	3.47	17.6	0.4	17.3	39.5	3.1
Linwood	1.97	3.24	17.1	1.3	15.9	35.2	10.4
Cedar Bog	3.47	3.95	28.9	0.4	28.6	60.0	4.2
Spectacle	2.66	3.11	21.4	0.4	21.1	43.0	3.5
Gladstone	2.61	3.12	21.1	0.3	20.9	43.4	3.8
Ham	1.39	1.78	17.1	4.5	12.7	27.2	37.3
Green	1.26	1.36	14.2	3.7	10.6	20.9	30.9
Reeds	1.45	1.83	13.7	1.2	12.5	24.3	10.3
Francis	1.60	1.87	14.0	1.3	12.8	26.2	11.1
Nokay	2.48	2.76	21.4	1.9	19.5	38.4	15.7
Beaver	0.99	1.27	12.1	3.2	9.0	16.6	26.6
Moose	0.90	1.19	11.8	3.4	8.4	16.8	28.3
Ball Club	1.85	2.50	17.8	0.7	17.1	30.6	5.6
St. Olaf	0.96	1.22	10.4	2.2	8.2	16.1	18.0
Christmas	0.99	1.25	9.5	1.2	8.3	18.8	10.3
Lotus	0.87	0.58	9.1	2.1	7.0	16.4	17.1
Clear	0.97	1.48	10.3	1.5	8.8	16.5	12.1
Long	0.64	0.82	10.8	4.8	6.0	12.5	40.3
Elk	0.99	1.92	13.1	2.3	10.8	23.8	20.1
Itasca	1.60	1.79	18.9	5.4	13.4	23.8	45.2
Little Pine	1.11	1.17	14.1	5.6	8.5	17.8	46.3
Grove	1.25	1.43	15.5	4.3	11.2	21.9	35.6
Fish	0.92	1.23	10.0	2.5	7.5	16.0	21.0
Sallie	1.20	1.51	15.0	5.6	9.5	18.8	46.4
Cowdry	1.35	1.59	14.6	3.8	10.8	21.5	31.3
Maple	1.13	1.22	14.7	5.1	9.6	18.4	42.6
Mina	1.04	1.49	13.5	4.7	8.8	17.1	39.4
Shetek	0.72	1.13	10.3	2.7	7.6	13.3	22.5
Big Kandiyohi	0.91	1.26	12.1	3.1	8.9	15.4	22.5
Elk-Grant Co.	0.81	1.35	10.1	2.3	7.8	15.6	19.0
Dead Coon	0.34	0.51	5.2	2.3	2.9	6.4	19.1
Salt	0.41	0.89	5.4	2.1	3.3	11.3	17.1

APPENDIX. CHEMICAL ANALYSES OF SEDIMENT SAMPLES FROM MINNESOTA LAKES* (continued)

Lake	As-c (ppm)	B-s (ppm)	B-c (ppm)	Ba-s (ppm)	Be-s (ppm)	Co-s (ppm)	Co-c (ppm)	Cr-s (ppm)	CCr-c (ppm)	Cu-S (ppm)	Cu-m (ppm)
Big	....	15	20	142	1.12	6	20	27	60	63	20
Iron	10	13	10	133	1.01	12	30	42	100	308	224
Mountain	....	32	15	215	1.90	11	20	63	150	330	222
Trout	8	16	15	124	3.18	6	10	31	60	113	66
Clearwater	....	30	20	235	1.89	12	15	71	100	184	88
Wilson	....	18	10	214	2.36	14	15	60	60	234	46
Sand Point	....	30	....	353	1.82	24	....	59	....	36	31
Kimball	....	13	20	105	1.00	6	20	23	60	114	53
Crane	10	39	30	311	2.14	13	25	65	100	226	30
O'Leary	10	29	30	284	1.48	10	15	61	60	92	36
Josephine	....	24	40	244	1.42	6	15	44	80	76	37
Arco	....	22	15	245	1.58	5	<8	39	60	58	39
Deming	....	23	30	168	1.09	4	20	33	80	134	40
Faquar	....	37	10	308	2.10	10	15	61	80	203	84
George	75	32	10	236	1.55	8	15	40	40	56	24
Linwood	....	18	15	200	0.51	3	<8	14	60	93	30
Cedar Bog	....	22	25	121	0.54	5	15	23	25	51	13
Spectacle	4,000	25	30	220	1.28	5	15	41	60	71	48
Gladstone	....	45	25	161	1.01	4	6	42	40	10	20
Ham	14	12	15	232	0.39	2	6	9	15	9	9
Green	....	14	<10	225	0.51	4	6	16	<10	157	113
Reeds	....	46	60	298	2.06	7	20	44	60	159	25
Francis	....	36	60	292	1.48	7	20	40	80	52	19
Nokay	....	27	10	166	0.70	4	<8	21	25	27	18
Beaver	....	25	30	328	1.33	4	<8	33	40	87	22
Moose	7	36	40	290	1.32	6	20	42	60	56	19
Ball Club	....	23	30	174	0.84	4	8	16	20	178	26
St. Olaf	....	32	40	331	1.49	6	6	44	40	110	28
Christmas	....	33	20	309	1.57	8	6	40	40	153	25
Lotus	....	31	20	332	1.70	6	6	45	40	115	20
Clear	7	39	40	331	2.06	6	8	50	60	153	28
Long	....	15	<10	264	0.49	3	8	15	25	11	14
Elk	120	20	10	203	0.34	2	<8	17	10	118	18
Itasca	....	15	<10	275	0.36	2	<8	11	<10	10	7
Little Pine	11	13	<10	296	0.40	3	<8	10	15	9	15
Grove	....	21	....	282	0.82	4	....	23	....	34	16
Fish	....	32	20	291	1.35	5	<8	40	40	180	51
Sallie	45	19	25	310	0.46	3	8	14	20	163	139
Cowdry	....	19	25	272	0.59	3	<8	13	20	15	24
Maple	....	16	10	300	0.58	3	8	15	20	13	15
Mina	....	16	10	271	0.46	3	<8	15	25	13	23
Shetek	15	40	15	261	1.86	6	20	45	40	122	31
Big Kandiyohi	9	31	15	253	0.67	5	<8	26	40	12	16
Elk-Grant Co.	....	42	20	295	1.28	6	<8	38	40	120	46
Dead Coon	....	49	20	326	1.37	7	10	54	40	98	38
Salt	90	44	200	279	1.68	7	20	47	40	88	32

APPENDIX. CHEMICAL ANALYSES OF SEDIMENT SAMPLES FROM MINNESOTA LAKES* (continued)

Lake	Ga-c (ppm)	La-s (ppm)	Li-m (ppm)	Mo-s (ppm)	Mo-c (ppm)	Nb-s (ppm)	Ni-s (ppm)	Ni-c (ppm)	Pb-s (ppm)	Pb-c (ppm)
Big	5	34	2	0.9	3	7	16	40	21	60
Iron	6	99	2	3.4	3	14	45	90	19	25
Mountain	10	156	3	10.0	8	18	71	80	29	40
Trout	5	5	2	1.7	1	28	22	30	27	40
Clearwater	10	124	4	4.3	3	16	53	80	24	15
Wilson	8	116	1	6.3	2	20	32	40	10	15
Sand Point	....	148	7	7.7	....	24	38	....	19	....
Kimball	10	30	2	2.2	3	3	22	60	39	60
Crane	15	66	5	2.9	1	30	39	40	18	20
O'Leary	15	58	4	1.1	<1	20	37	50	10	20
Josephine	15	39	3	0.8	3	16	27	60	34	30
Arco	10	40	2	2.7	3	15	23	40	26	40
Deming	15	36	2	1.1	4	3	22	60	38	40
Faquar	15	67	3	2.2	1	23	39	40	25	40
George	10	85	3	3.5	2	19	22	30	85	100
Linwood	6	62	1	2.3	1	15	7	15	37	60
Cedar Bog	6	18	2	5.5	10	12	7	15	30	80
Spectacle	8	42	2	2.8	6	14	20	30	86	200
Gladstone	6	20	2	6.7	8	13	12	25	48	80
Ham	1	43	1	3.3	1	17	6	10	28	40
Green	3	38	1	2.7	<1	7	9	6	6	10
Reeds	10	46	3	2.6	3	20	32	40	12	25
Francis	10	50	2	1.7	<1	10	25	30	16	40
Nokay	4	80	3	3.6	3	6	9	15	18	40
Beaver	6	36	1	2.2	<1	28	19	20	5	30
Moose	8	48	1	2.1	<1	34	21	30	8	25
Ball Club	6	46	1	3.3	3	8	12	15	19	60
St. Olaf	10	46	1	1.2	<1	24	25	20	8	25
Christmas	8	50	2	3.9	4	25	24	20	48	60
Lotus	8	42	2	1.9	<1	34	21	20	11	25
Clear	10	52	3	1.2	<1	15	28	40	3	15
Long	5	34	1	2.6	<1	17	8	10	5	20
Elk	4	119	1	6.6	8	5	6	40	12	30
Itasca	1	53	1	4.1	<1	21	5	<3	4	15
Little Pine	1	37	1	2.5	<1	26	9	<3	2	10
Grove	....	40	1	2.7	....	11	13	....	6	....
Fish	8	48	2	2.1	3	22	20	20	6	15
Sallie	4	34	1	2	<1	9	8	15	1	15
Cowdry	6	43	1	6.6	15	36	14	15	11	30
Maple	6	30	1	4.4	<1	18	11	10	3	20
Mina	5	29	1	7.7	20	13	10	15	3	15
Shetek	7	47	2	2.1	3	22	26	40	5	40
Big Kandiyohi	8	30	1	1.7	3	14	15	20	4	20
Elk-Grant Co.	10	42	1	2.9	<1	21	25	20	8	20
Dead Coon	8	6	3	3.4	<1	48	36	20	4	15
Salt	10	47	6	1.2	3	15	25	30	2	20

APPENDIX. CHEMICAL ANALYSES OF SEDIMENT SAMPLES FROM MINNESOTA LAKES* (continued)

Lake	Sc-c (ppm)	Sn-c (ppm)	Sr-s (ppm)	V-s (ppm)	V-c (ppm)	Y-s (ppm)	Y-c (ppm)	Zn-m (ppm)	Zn-s (ppm)	Zr-s (ppm)	Zr-c (ppm)
Big	10	4	68	31	100	11	20	151	91	26	100
Iron	10	1	105	57	100	16	15	162	147	97	200
Mountain	15	3	118	87	80	32	20	281	254	166	150
Trout	10	2	83	46	60	50	40	173	127	128	150
Clearwater	10	1	173	90	80	25	20	251	216	164	100
Wilson	10	1	183	96	100	25	15	126	127	264	200
Sand Point	....	....	120	94	....	16	....	273	161	112	....
Kimball	10	3	74	40	80	18	20	164	131	91	150
Crane	10	1	149	99	150	11	10	163	155	126	200
O'Leary	10	0.8	196	71	100	15	10	139	66	149	150
Josephine	10	4	125	60	150	15	10	167	140	161	300
Arco	10	2	121	54	150	14	20	206	146	169	150
Deming	10	2	106	48	150	11	20	181	124	92	200
Faquar	10	1	135	85	100	21	15	136	113	277	300
George	10	3	124	55	60	14	15	225	180	155	400
Linwood	10	0.8	90	23	25	8	10	93	81	98	150
Cedar Bog	10	3	55	14	20	5	<10	150	86	82	150
Spectacle	10	6	103	49	100	13	10	160	153	150	200
Gladstone	10	2	78	34	60	9	10	147	126	131	250
Ham	10	0.8	151	18	15	5	10	141	84	28	80
Green	10	4	158	28	15	7	7	169	68	137	60
Reeds	10	0.8	127	90	100	15	15	134	97	214	150
Francis	10	1	138	65	100	14	15	247	106	170	150
Nokay	10	1	88	22	20	7	10	118	80	84	60
Beaver	10	0.8	167	62	40	12	10	183	78	158	60
Moose	10	0.8	237	61	60	12	10	98	97	163	80
Ball Club	10	0.8	87	31	40	9	<10	96	66	87	100
St. Olaf	10	2	185	81	60	15	10	116	115	159	100
Christmas	10	1	153	72	60	15	10	148	133	175	100
Lotus	10	0.8	176	72	60	14	10	112	99	190	80
Clear	10	1	157	93	100	17	10	141	100	193	80
Long	10	0.8	164	26	25	9	10	148	95	126	60
Elk	10	0.8	113	16	20	4	<10	202	66	27	60
Itasca	10	0.8	157	17	10	6	<10	125	74	30	60
Little Pine	10	0.6	185	17	15	7	10	72	70	22	60
Grove	....	....	233	35	....	9	....	108	109	97	....
Fish	10	0.8	194	69	100	14	15	100	87	165	100
Sallie	10	0.6	294	23	20	7	<10	124	77	110	60
Cowdry	10	0.8	187	30	40	7	10	248	95	29	60
Maple	10	0.8	283	27	25	8	10	78	71	30	60
Mina	10	0.8	227	29	40	6	10	160	71	26	60
Shetek	10	2	221	76	150	15	15	127	92	193	60
Big Kandiyohi	10	0.8	190	47	60	9	10	105	66	112	60
Elk-Grant Co.	10	1	242	58	60	13	10	106	99	198	80
Dead Coon	10	1	261	97	80	24	15	130	99	278	80
Salt	10	1	477	92	150	15	10	180	90	131	60

*If there are results of concentrations of an element by more than one method or laboratory, the preferred method is in italics in the column heading. x = USGS, X-ray fluorescence; s = USGS optical emission spectroscopy; m = University of Minnesota, various methods; c = CSIRO, optical emission spectroscopy; TC = total carbon; CC = carbonate carbon; OC = organic carbon; LOI = loss on ignition at 550°C; ellipses indicate no analysis.

REFERENCES CITED

Ackroyd, E. A., Walton, W. C., and Hills, D. L., 1967, Ground-water contribution to stream flow and its relation to basin characteristics in Minnesota: Minnesota Geological Survey Report of Investigations, 6, 36 p.

Baker-Blocker, A., Callender, E., and Josephson, P. D., 1975, Trace-element and organic carbon content of surface sediment from Grand Traverse Bay, Lake Michigan: Geological Society of America Bulletin, v. 86, p. 1358–1362.

Berner, R. A., 1971, Principles of chemical sedimentology: New York, McGraw-Hill, 240 p.

Berner, R. A., 1984, Sedimentary pyrite formation—An update: Geochimica et Cosmochimica Acta, v. 48, p. 605–615.

Berner, R. A., and Raiswell, R., 1984, C/S method for distinguishing freshwater from marine sedimentary rocks: Geology, v. 12, p. 365–368.

Cline, J. T., and Chambers, R. L., 1977, Spatial and temporal distribution heavy metals in lake sediments near Sleeping Bear Point, Michigan: Journal of Sedimentary Petrology, v. 47, p. 716–727.

Cline, J. T., and Upchurch, S. B., 1973, Mode of heavy metal migration in the upper strata of lake sediment: Proceedings of the 16th Conference on Great Lakes Research, 1973: Toronto, International Association for Great Lakes Research, p. 349–356.

Dean, W. E., 1974, Determination of carbonate and organic matter in calcareous sediments and sedimentary rocks by loss on ignition: Comparison with other methods: Journal of Sedimentary Petrology, v. 44, p. 242–248.

Dean, W. E., and Gardner, J. V., 1989, Changes in redox conditions in deep-sea sediments of the subarctic North Pacific Ocean: Possible evidence for the presence of North Pacific deep water: Paleoceanography, v. 4, p. 639–653.

Dean, W. E., and Gorham, E., 1976a, Major chemical and mineral components of profundal surface sediments in Minnesota lakes: Limnology and Oceanography, v. 21, p. 259–284.

Dean, W. E., and Gorham, E., 1976b, Classification of Minnesota lakes by Q- and R-mode factor analysis of sediment mineralogy and geochemistry, in Merriam, D. F., ed., Quantitative techniques for the analysis of sediments: Oxford, Pergamon Press, p. 61–71.

Dean, W. E., Moore, W. S., and Nealson, K., 1980, Manganese cycles and the origin of manganese nodules, Oneida Lake, New York: Chemical Geology, v. 34, p. 53–64.

Dean, W. E., Gorham, E., and Swaine, D. J., 1988, Geochemistry of surface sediments of lakes in the English Lake District, in Round, F. E., ed., Algae and the aquatic environment: Bristol, England, Biopress, Ltd., p. 244–271.

Fitchko, J., and Hutchinson, T. C., 1975, A comparative study of heavy metal concentrations in river mouth sediments around the Great Lakes: Journal of Great Lakes Research, v. 1, p. 46–78.

Gardner, J. V., Dean, W. E., Klise, D. K., and Baldauf, J. G., 1982, A climate-related oxidizing event in deep-sea sediments from the Bering Sea: Quaternary Research, v. 18, p. 91–107.

Glenn, J. L., and Van Atta, R. O., 1973, Relations among radionuclide content and physical, chemical and mineral characteristics of Columbia River sediments: U.S. Geological Survey Professional Paper 433-M, 52 p.

Gorham, E., 1958, Observations on the formation and breakdown of the oxidized microzone at the mud surface in lakes: Limnology and Oceanography, v. 3, p. 291–298.

Gorham, E., 1964, Molybdenum, iron, and manganese in lake muds: Verhandlung International Vereinigung für Theoretische und Angewandte Limnologie, v. 15, p. 330–333.

Gorham, E., and Swaine, D. J., 1965, The influence of oxidizing and reducing conditions upon the distribution of some elements in lake sediments: Limnology and Oceanography, v. 10, p. 268–279.

Gorham, E., Dean, W. E., and Sanger, J. E., 1982, The chemical composition of lakes in the north-central United States: U.S. Geological Survey Open-File Report 82-149, 65 p.

Gorham, E., Dean, W. E., and Sanger, J. E., 1983, The chemical composition of lakes in the north-central United States: Limnology and Oceanography, v. 28, p. 287–301.

Jones, B. F., and Bowser, C. J., 1978, The mineralogy and related chemistry of lake sediments, in Lerman, A., ed., Lakes—Chemistry, geology, physics: New York, Springer-Verlag, p. 179–235.

Klovan, J. E., and Miesch, A. T., 1976, Extended CABFAC and QMODEL computer programs for Q-mode factor analysis of compositional data: Computers and Geosciences, v. 1, p. 161–178.

Mackereth, F.J.H., 1963, Some methods of water analysis for limnologists: Ambleside, England, Freshwater Biological Association Scientific Publication 21, 70 p.

Mackereth, F.J.H., 1966, Some chemical observations on postglacial lake sediments: Royal Society of London Philosophical Transactions, ser. B, v. 250, p. 165–213.

Miesch, A. T., 1976a, Geochemical survey of Missouri—Methods of sampling, laboratory analysis, and statistical reduction of data: U.S. Geological Survey Professional Paper 954-A, p. A1–A39.

Miesch, A. T., 1976b, Interactive computer programs for petrologic modeling with extended Q-mode factor analysis: Computers and Geosciences, v. 2, p. 439–492.

Miesch, A. T., 1976c, Q-mode factor analysis of geochemical and petrologic data matrices with constant row-sums: U.S. Geological Survey Professional Paper 574-G, p. G1–G47.

Mothersill, J. S., 1976, The mineralogy and geochemistry of the sediments of northwestern Lake Victoria: Sedimentology, v. 23, p. 553–565.

Oliver, B. G., 1973, Heavy metal levels of Ottawa and Rideau river sediments: Environmental Science and Technology, v. 7, p. 135–137.

Schultz, L. G., Tourtelot, H. A., Gill, J. R., and Boerngen, J. G., 1980, Composition and properties of the Pierre Shale and equivalent rocks, northern Great Plains region: U.S. Geological Survey Professional Paper 1064-B, 114 p.

Shimp, N. F., Schleicher, J. A., Ruch, R. R., Heck, D. B., and Leland, H. V., 1971, Trace element and organic carbon accumulation in the most recent sediments of southern Lake Michigan: Illinois State Geological Survey Environmental Geology Notes, no. 41, 25 p.

Sims, P. K., 1968, Copper and nickel developments in Minnesota: Mining Congress Journal, v. 58, p. 29–34.

Sweeney, R. E., and Kaplan, I. R., 1973, Diagenetic sulfate reduction in marine sediments: Marine Chemistry, v. 9, p. 165–174.

Winter, T. C., 1974, The natural quality of ground water in Minnesota: Minnesota Department of Natural Resources, Division of Waters, Soils, and Minerals Bulletin 26, 25 p.

MANUSCRIPT ACCEPTED BY THE SOCIETY JULY 27, 1992

Geological Society of America
Special Paper 276
1993

Physical properties, mineralogy, and geochemistry of Holocene varved sediments from Elk Lake, Minnesota

Walter E. Dean
U.S. Geological Survey, MS 939, Box 25046, Federal Center, Denver, Colorado 80225

ABSTRACT

Elk Lake in northwestern Minnesota is situated close to a climatically sensitive ecotone, the forest-prairie border, that migrated back and forth over the drainage basin of the lake during the Holocene. The entire postglacial (Holocene) sediment record in the deepest part of Elk Lake is composed of annual layers (varves) that record the seasonal pulses of many sediment components, and, most important, provide high-resolution (seasonal) time calibration of rates and timing of environmental change. These varved sediments contain many allochthonous and autochthonous components that are sensitive to changing environmental conditions in the drainage basin and the lake.

The mineral components of Elk Lake sediments consist mainly of authigenic calcium, magnesium, and manganese carbonate minerals (low-Mg calcite, high-Mg calcite, dolomite, and rhodochrosite), opaline silica (from diatoms), X-ray amorphous iron and manganese oxyhydroxides, and an iron phosphate mineral tentatively identified as rockbridgeite [$(Fe, Mn)Fe_4(PO_4)_3(OH)_5$], plus minor contributions from fine-grained detrital quartz, feldspar, illite, and kaolinite. The most notable characteristic of the sediments in Elk Lake is that most of the components were formed in the lake.

Q-mode factor analysis of sediment geochemistry reduced 23 observed compositional variables, expressed as percent or parts per million of elements, to three composite variables (factor loadings) whose "concentrations" are expressed on a scale of –1.0 to 1.0. Factor 1 expresses the composition of the inorganic clastic fraction based on concentrations of Mg, Na, Al, Cr, V, Y, Sc, Ni, Sr, Co, and Cu. Factor 2 expresses the similarities in variations of Fe, Mn, P, organic carbon, and Mo. Factor 3 loadings are a synthesis of concentrations of Mn, S, Ca, La, and Ba.

Geochemical characteristics define three distinct chemical stages in the development of Elk lake: (1) a carbonate-, manganese-, iron-, sulfur-rich early-lake stage that lasted from 10,400 to 8200 varve yr; (2) a clastic- and diatom-rich mid-Holocene prairie-lake stage that lasted from 8200 to 4000 varve yr; and (3) a final iron-, manganese-, phosphate-, and organic-rich modern-lake stage that developed over the past 4000 yr. The chemical characteristics of the sediments deposited during these three lake phases can be represented by the average compositions of three groups of samples: (1) the average composition of sediments deposited over the past 2000 yr, representing the modern-lake stage; (2) the composition of a 50 varve sample centered on 5700 varve yr that represents the maximum clastic influx into the lake during the prairie period; and (3) the average composition of sediments deposited over the initial 2000 yr of the lake's existence (10,400 to 8400 varve yr).

Dean, W. E., 1993, Physical properties, mineralogy, and geochemistry of Holocene varved sediments from Elk Lake, Minnesota, *in* Bradbury, J. P., and Dean, W. E., eds., Elk Lake, Minnesota: Evidence for Rapid Climate Change in the North-Central United States: Boulder, Colorado, Geological Society of America Special Paper 276.

The amplitudes of climatic oscillations, as reflected by changes in concentration of many climatically sensitive elements, were greatest during the prairie period, pronounced cycles having periodicities of several hundred years. The extremes of these oscillations occurred within several centuries or less, which suggests that significant changes in climate may occur abruptly and rapidly. Most notably, the end of the mid-Holocene prairie period, marked by a sudden decrease in the influx of detrital clastic material, occurred within a few decades.

INTRODUCTION

Lake sediments are an integration of components from two distinct environments—the external terrestrial environment of the drainage basin and beyond (allochthonous components) and the internal lake environment (autochthonous components). The allochthonous components are principally an integration of soils within the drainage basin plus any wind-borne material from outside the drainage basin and distinctive biological components such as needles, seeds, and pollen grains. The autochthonous components are the organic constituents produced by the lake such as organic matter produced by plankton and biogenic silica produced by diatoms, and any authigenic minerals (e.g., calcite) precipitated from the lake water. These components are then modified by processes within the lake and within the postdepositional (diagenetic) environment.

To the extent that the allochthonous components of lake sediments are an integration of soils in the drainage basin, and to the extent that chemical characteristics of soils reflect the chemical characteristics of geologic materials from which they were derived, then allochthonous components of lake sediments may be examined as an integration of the geochemistry of bedrock and surficial geologic materials, modified by decomposition and disintegration in the weathering and transport environments. To the extent that the chemical and biological characteristics of the autochthonous components of lake sediments reflect the chemical and biological characteristics of the lake and the drainage basin, the sediments may be viewed as recorders of short-term (seasonal) and long-term (years to thousands of years) variations in conditions within the lake and its environment.

In an attempt to relate the chemical and mineralogical characteristics of sediments from Minnesota lakes to bedrock geology, surficial geology, climate, chemical, and biological characteristics of the lake, and early diagenetic changes, Dean and Gorham (1976a, 1976b) and Dean and others (Chapter 9) analyzed the major, minor, and trace components in surface profundal sediments from 46 lakes throughout the state. The drainage basins of these lakes are situated in extremely different climatic regimes ranging from boreal coniferous forest in the northeastern part of Minnesota to semiarid prairie in the western and southern parts of the state. This extreme climatic gradient is reflected in extreme differences in hydrologic regime and water chemistry of these lakes (Gorham and others, 1983). These regional studies of the geochemistry and mineralogy of surface sediments and waters of lakes across Minnesota provide the baseline information for interpreting Holocene climatic fluctuations as recorded in the chemical and mineralogical characteristics of sediment cores from Elk Lake and other lakes in Minnesota. An intermediate connection between seasonal fluxes of allochthonous and autochthonous components and the actual accumulation of these components in the anoxic bottom waters of the deep basin of Elk Lake is provided by sediment-trap studies (Nuhfer and others, Chapter 7).

Several characteristics of Elk Lake and its sediments together have provided a rare, if not unique, high-resolution paleoclimatic record of the Holocene. First, Elk Lake is situated close to a climatically sensitive ecotone (the forest-prairie border) that migrated back and forth over the drainage basin of Elk Lake during the Holocene (Fig. 1). Minnesota is located in a region of climatic transition between latitudinal climatic gradients characteristic of the Great Lakes region and longitudinal climatic gradients of the Great Plains and Rocky Mountains. This transition is dramatically illustrated by the change in vegetation zones in Minnesota from north-south in south-central Minnesota to east-west in western Minnesota (Fig. 1). The present prairie-forest border also roughly defines the zero precipitation minus evaporation line, with net evaporation to the west and net precipitation to the east (Bright, 1968). Northwestern Minnesota is at a "climatic triple junction" between Pacific air masses, Arctic air masses, and Atlantic-Gulf air masses (Anderson and others, Chapter 1).

The sediments of Elk Lake contain many allochthonous and autochthonous components that are sensitive to changing environmental conditions in the drainage basin and lake. Most of these components are formed in the lake (see Nuhfer and others, Chapter 7). The entire postglacial (Holocene) sediment record in the deepest part of Elk Lake is composed of annual layers (varves) that record the seasonal pulses of many of the sediment components, and, most important, provide a high-resolution (seasonal) time calibration of rates and timing of environmental change over the past 10,000 yr (Anderson, Chapter 5).

MATERIALS AND METHODS

Samples of sediment from the 22.4-m-long 1978 deep-basin (29.5 m water depth; Fig. 1) core was sampled at 50 year intervals for analyses of geochemistry, mineralogy, and diatoms. Each channel or strip sample was about 1 cm on a side and 50 varves (average of about 10 cm) long. Similar samples were collected from part of an oriented core collected in 1980 for paleomagnetic studies (Sprowl and Bannerjee, Chapter 11) in order to cover a coring gap (see Anderson and others, Chapter 4). An aliquot of each 50 year sample was mixed with distilled water into a homogeneous slurry, and slides were made for diatom analyses (Brad-

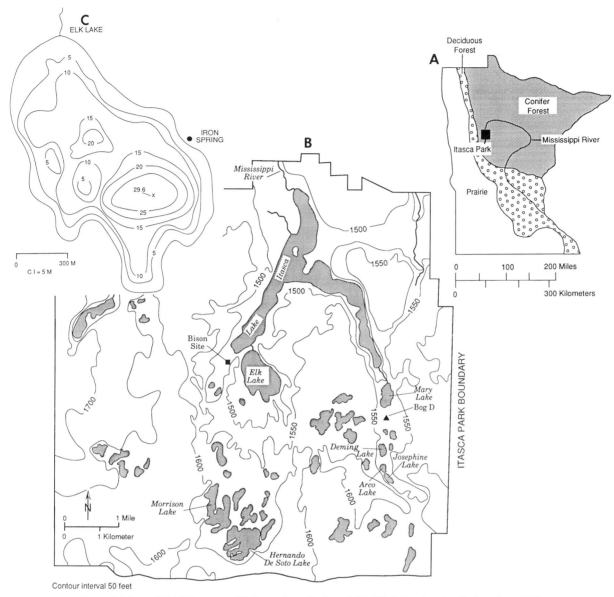

Figure 1. Maps of (A) Minnesota, (B) Itasca State Park, and (C) Elk Lake, showing the location of Elk Lake, general vegetation zones of Minnesota, topography and distribution of lakes in Itasca Park, bathymetry of Elk Lake, and location of the varved core (x) in the deepest part of the lake (29.6 m).

bury and Dieterich-Rurup, Chapter 15). The remainder of each 50 year sample was then dried and ground for mineral and geochemical analyses.

A 1.0 cm^3 sample for measurement of water content and bulk density was collected every 50 varves using a constant-volume sampler that is rectangular in cross section. The sampler was centered on the 50th varve, and sampled from one to five varves depending upon varve thickness. Each 1.0 cm^3 sample was placed in a preweighed container, weighted, dried, and reweighed to determine water content and bulk density. A 0.5 cm^3 sample for pollen analysis was collected adjacent to the physical properties samples using the same constant-volume sampler (Whitlock and others, Chapter 17).

Every other 50 year geochemistry-mineralogy sample collected was actually analyzed, resulting in the analysis of one 50 year composite sample every 100 years. Bulk mineralogy was determined by X-ray diffraction (XRD) using Ni-filtered, Cu-Kα radiation. Relative abundances of minerals are reported as XRD peak heights in millimeters.

Examples of X-ray diffractograms with peaks identified for minerals present are shown in Figure 2. Oriented glass slides were prepared of clay-size (<2 μm) fractions of samples at 1 m inter-

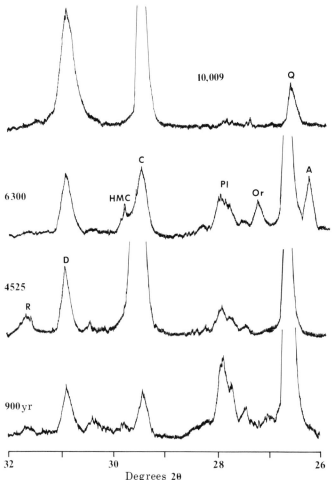

Figure 2. Representative X-ray diffraction patterns of untreated samples of sediment from 900, 4525, 6300, and 10,009 varve yr in the Elk Lake core. Identified peaks are A, aragonite, Q, quartz, Or, orthoclase, Pl, plagioclase, C, calcite, HMC, high-Mg calcite, D, dolomite, and R, rhodochrosite.

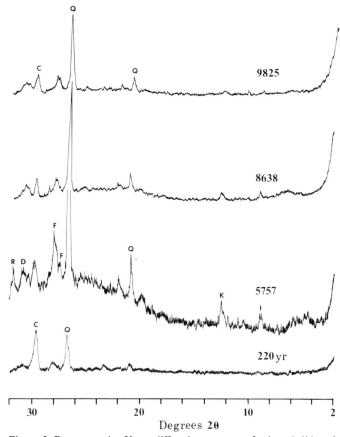

Figure 3. Representative X-ray diffraction patterns of oriented slides of the clay-size (<2 μm) fraction of sediment from 220, 5757, 8638, and 9825 varve years from the Elk Lake core. Identified peaks are I, illite, K, kaolinite, Q, quartz, F, feldspar, C, calcite, D, dolomite, R, rhodochrosite.

vals in an attempt to identify clay minerals present. The clay minerals illite and kaolinite were detected in a few samples in the interval between about 5500 and 7000 varve yr, but diffractograms of most samples contained only peaks for calcite, quartz, feldspar, and/or dolomite, and perhaps an amorphous "hump" between about 20° and 30° 2θ that is probably due to amorphous silica from diatom debris. Examples of XRD diffractograms of clay-size material are shown in Figure 3.

Of 136 sediment samples selected for inorganic geochemical analyses, 20 were chosen at random for duplicate analyses. The basic sampling scheme consisted of one 50 year composite sample every 100 years; some additional samples were taken to ensure complete overlap with the 1978 and 1980 cores (see Anderson and others, Chapter 4, for discussion of the coring gap in the 1978 core). The samples were air dried and ignited at 525 °C and at 950 °C. Weight losses during each ignition were recorded. The weight loss on ignition at 525 °C is proportional to loss of CO_2 from ignition of organic matter plus H_2O lost from clays and opaline silica from diatoms (Fig. 4; Dean, 1974). Weight loss on ignition between 525 and 950 °C is due mainly to loss of CO_2 from carbonate minerals (mostly calcite and dolomite).

The ashed sample was ground in a ceramic mill to pass a 100 mesh (149 μm) sieve. Concentrations of the major elements Si, Al, Fe, Ca, K, P, Ti, and Mn were analyzed by wavelength-dispersive X-ray fluorescence spectrometry (XRF). Fluorescence from high concentrations of manganese in all samples predating about 8600 varve yr prevented detection of any of the other elements in these samples. Concentrations of Mg and Na were determined by atomic absorption spectrophotometry. Concentrations of 15 trace elements (B, Ba, Co, Cr, Cu, La, Mn, Nb, Ni, Pb, Sr, V, Y, Zn, and Zr) and four major elements (Al, Fe, Mn, and Ti) were analyzed by optical emission spectroscopy.

Total sulfur was determined with a LECO sulfur analyzer on an aliquot of the dried, unashed sample. Total carbon was determined with a LECO carbon analyzer on another aliquot of the unashed sample. Organic carbon (OC) also was determined with a LECO carbon analyzer after an aliquot of the unashed

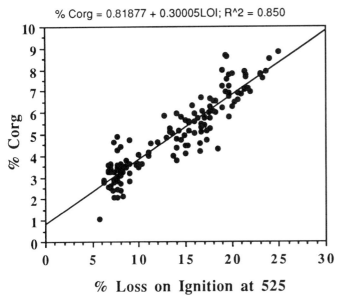

Figure 4. Plot of loss on ignition at 525 °C vs. organic carbon in sediments from Elk Lake.

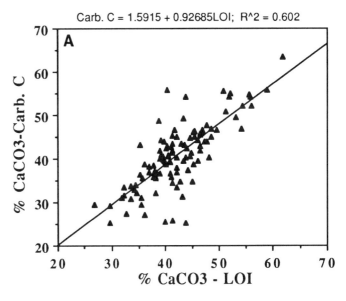

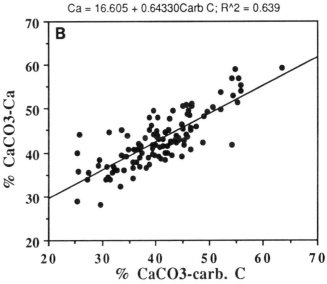

Figure 5. Plots of (A) $CaCO_3$ calculated from carbonate-carbon vs. $CaCO_3$ calculated from loss on ignition (LOI) between 525 °C and 950 °C (Dean, 1974); (B) $CaCO_3$ calculated from carbonate-carbon and $CaCO_3$ calculated from total Ca.

sample was acidified to remove carbonate minerals. Carbonate carbon was determined by difference between total carbon and OC. The relations between OC and organic matter determined by loss on ignition at 525 °C is shown in Figure 4; in general, the organic-matter content is about three times the OC content. Comparison of percent $CaCO_3$ calculated from loss on ignition between 525 °C and 950 °C (Dean, 1974), carbonate-carbon, and total Ca concentration is shown in Figure 5. Details of all analytical methods, including discussions of precision and accuracy, were described in Baedecker (1987).

RESULTS

Physical properties

Results of water-content (percent water) and dry bulk-density (g/cm^3) measurements (Fig. 6) show general effects of compaction with increasing depth; water content decreases from 80% at the top of the section to <60% at the base of the section, and bulk density increases from 0.2 to 0.3 g/cm^3 at the top of the section to 0.6 g/cm^3 at the base. Both curves, however, also show considerable high-frequency cyclic variations (decades to hundreds of years) that are mirror images. The higher water content, lower density parts of the cycles probably represent periods of greater diatom abundance (see Bradbury and Dieterich-Rurup, Chapter 15), and the lower water content, higher density parts of the cycles may represent periods of greater clastic-mineral influx and/or lower diatom productivity. For example, the large cycle of high to low water content, low to high bulk density between 5200 and 5800 varve yr (Fig. 6) corresponds to a change from low to high clastic influx as interpreted from varve thickness and geochemistry. Abundances of diatoms are subject to extreme short-term variations (Bradbury and Dieterich-Rurup, Chapter 15), so that the highest frequency fluctuations in physical properties probably are related to content of highly porous diatom frustules.

Mineralogy

Plots of relative intensity of XRD peak heights for the minerals quartz, orthoclase feldspar (27° feldspar), plagioclase feldspar (28° feldspar), calcite, dolomite, rhodochrosite, and arago-

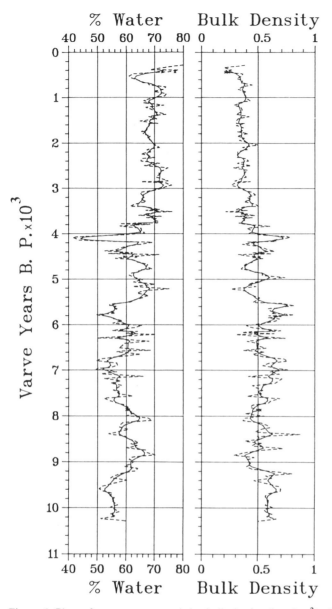

Figure 6. Plots of percent water and dry bulk density (in g/cm^3) of sediment vs. varve years in the Elk Lake core. The raw data are connected by a dashed line; the solid line through the raw data is a smoothed curve computed using 11-sample weighted moving average.

nite (Fig. 7) show that the clastic mineral fraction is dominated by quartz. Orthoclase feldspar is detectable by XRD in a few samples of sediment over the past 4000 yr, but is most abundant in samples from between 4000 and 9000 varve yr. Plagioclase feldspar, mostly albite, is detectable only in samples from between about 5700 and 8400 varve yr.

A detailed discussion of the carbonate minerals in Elk Lake is presented elsewhere (Dean and Megard, Chapter 8) and is reviewed only briefly here. The dominant carbonate mineral throughout the core is low-Mg calcite, which also is, by far, the most common carbonate mineral in hard-water lakes (Jones and Bowser, 1978; Kelts and Hsü, 1978; Dean, 1981; Dean and Fouch, 1983). The second most abundant carbonate mineral in the Elk Lake core is dolomite, which is most abundant during the mid-Holocene prairie period (Fig. 7). Most of the calcite in the Holocene sediments of Elk Lake probably formed as a primary precipitate from the lake water just as it does today. Most of the dolomite is probably either a primary precipitate or, more likely, an early diagenetic alteration of primary high-Mg calcite. Regardless of the timing of dolomite formation, the presence of dolomite in greatest abundance during the prairie period indicates that the waters of Elk Lake had characteristics that were similar to those of present prairie lakes in western Minnesota; i.e., a relatively high salinity and an Mg:Ca ratio of at least 2.0 (Dean and Gorham, 1976a; Gorham and others, 1983; Dean and Megard, Chapter 8).

Aragonite occurs as fine needle-shaped crystals in the sediments of Elk Lake only between 7200 and 6300 varve yr (Fig. 7; Anderson, Chapter 5). A comparison of the profiles of calcite and aragonite in Figure 7 shows clearly that aragonite formed at the expense of calcite. If the empirical observations of Müller and others (1972) are correct, the presence of primary aragonite implies a high salinity and an Mg:Ca ratio >12. Such conditions are rare in temperate lakes and are usually found only in alkaline saline lakes. It would appear, therefore, that the maxima of salinity and Mg:Ca ratio occurred in Elk Lake during the period between 7200 and 6300 varve yr.

Rhodochrosite ($MnCO_3$) is a rare carbonate mineral in either lacustrine (Jones and Bowser, 1978) or marine environments. Therefore, its presence in the Elk Lake core, even in low abundances, between 9700 and 7300 varve yr and between 6300 and 1500 varve yr was unexpected. As far as I know, the only other reported occurrence of rhodochrosite in lake sediments is in ferromanganese carbonate crusts in Green Bay, Lake Michigan (Callender and others, 1973). Theoretical mineral equilibrium calculations based on epilimnetic and hypolimnetic water chemistry of Elk Lake (Dean and Megard, Chapter 8) indicate that hypolimnetic waters of Elk Lake are saturated with rhodochrosite at a manganese concentration of about 1.0 ppm, which is achieved in the hypolimnion during late summer. I predicted, therefore, that rhodochrosite should precipitate at fall overturn, when manganese-rich bottom waters mix with bicarbonate-rich surface waters. However, observations from sediment-trap investigations (Nuhfer and others, Chapter 7) show that rhodochrosite is abundant in traps, particularly deep traps, throughout the summer. It may be that rhodochrosite forms in anoxic hypolimnetic or sediment pore waters by alteration of manganese oxide that apparently forms at the base of the metalimnion when dissolved Mn^{2+} in the hypolimnion mixes with oxygenated, epilimnetic waters (see discussion below and in Nuhfer and others, Chapter 7).

Geochemistry

Results of analyses for major, minor, and trace elements in sediment samples from the deep-basin varved core are plotted versus varve years in Figure 8. Some elements, particularly the

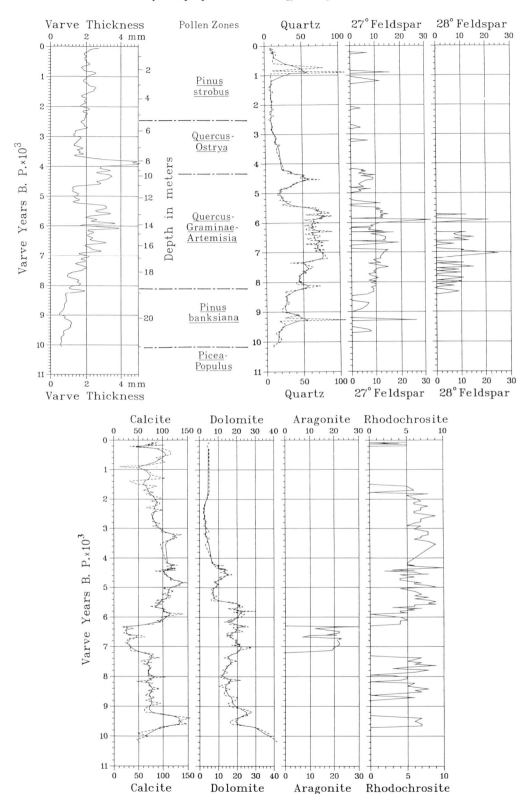

Figure 7. Plots of XRD relative peak heights (in millimeters) for quartz, orthoclase feldspar (27° feldspar), plagioclase feldspar (28° feldspar), calcite, dolomite, rhodochrosite, and aragonite vs. time in varve years for sediment samples from the Elk Lake core. A plot of varve thickness vs. varve years, and the major pollen zones are shown for reference. The raw data for quartz, calcite, and dolomite are connected by dashed lines; the solid lines through the raw data are smoothed curves computed using 11-sample weighted moving averages.

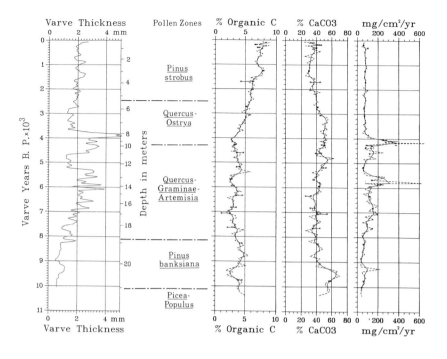
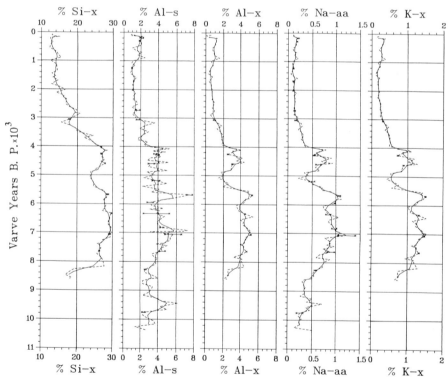

Figure 8 (on this and following three pages). Plots of percent organic carbon, percent CaCO$_3$, sediment mass accumulation rate (MAR in mg dry sediment/cm^2/yr), percentages of major elements, and parts per million (ppm) of trace elements vs. varve years for samples from the varved Elk Lake core. A plot of varve thickness vs. varve years and the major pollen zones are shown for reference. The suffixes x, aa, and s following an element indicate that the element concentrations were determined by X-ray fluorescence, atomic absorption, or optical emission spectrophotometry, respectively. All concentrations, except those of organic carbon, CaCO$_3$, are Sr are expressed on a carbonate-free basis. The raw data are connected by dashed lines; the solid lines through the raw data are smoothed curves computed using 11-sample weighted moving averges. Duplicate analyses are indicated by two points plotted as x at the same depth and connected by a horizontal line.

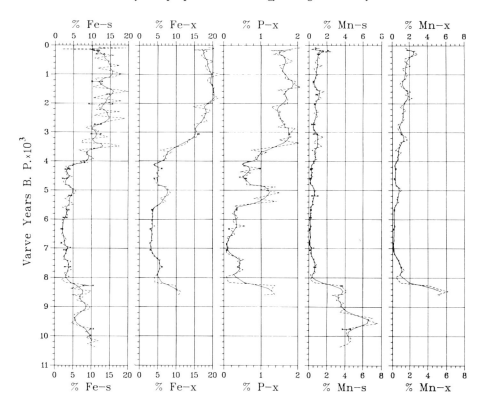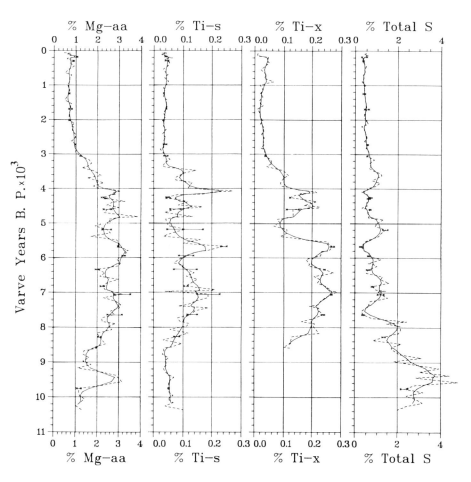

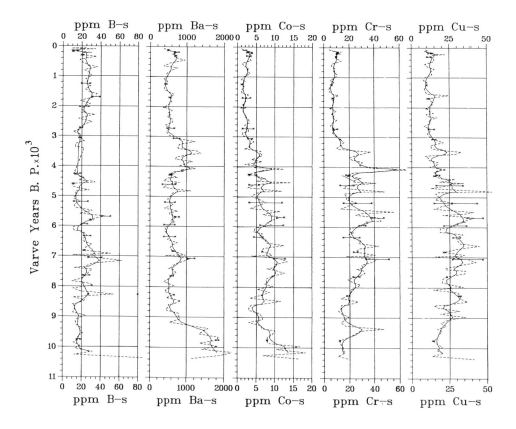

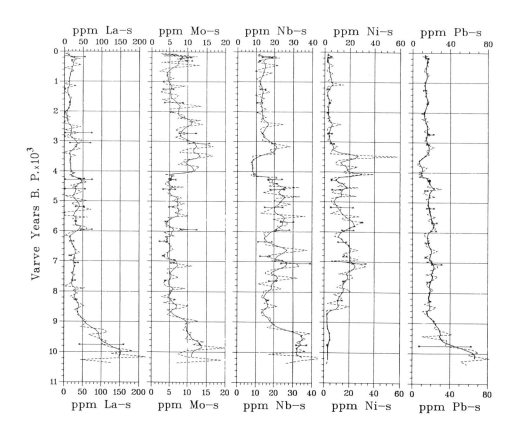

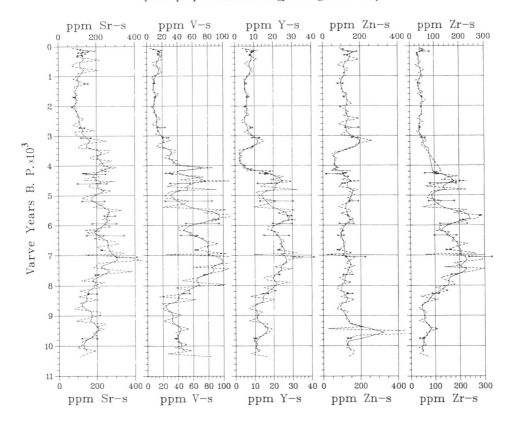

alkaline earth elements such as Ba, Mg, and Sr, may substitute for Ca in the carbonate fraction, but most elements reside mainly in the clastic or organic fractions, which are diluted by the carbonate fraction. Consequently, it is necessary to consider the variations of these noncarbonate elements on a carbonate-free basis. Therefore, the element concentration profiles shown in Figure 8 are plotted on a carbonate-free basis, except for concentrations of OC, $CaCO_3$, and Sr, which are plotted on a bulk-sediment, dry-weight basis. Concentrations of Mg and Ba were plotted on a carbonate-free basis in Figure 8 because, as discussed below, they mainly reside in phases other than carbonate minerals. Figure 9 shows smoothed profiles of variations in concentration with depth for five selected elements on both bulk-sediment and carbonate-free bases. The carbonate-free concentrations are, of course, all higher, but because the amount of carbonate in the sediment does not vary too much from bottom to top of the section (Fig. 8), the general patterns of change with time are not significantly different between the two profiles. Consequently, relative changes in concentration with time based on the carbonate-free profiles (Fig. 8) would not be any different if they were based on bulk-sediment concentration profiles.

The bulk-sediment mass accumulation rate (MAR, in mg dry sediment/cm^2/yr; Fig. 8) is the dry bulk density (Fig. 6; in g/cm^3) normalized to time (bulk density divided by sedimentation time in yr/cm to give g/cm^2/yr). The bulk-sediment MAR can be used to compute MARs for individual components simply by multiplying the bulk-sediment MAR by the fraction of the component present in the sediment. Examples of the results of MAR computations for OC, Al, Mn, Ni, and Fe are shown in Figure 10. These five components were chosen for illustration because they represent different patterns of variation in concentration with time (Fig. 8). Components with similar variations in concentration would have similar patterns of variation in MAR. For example, variations in MAR of sodium would be similar to those of aluminum (Fig. 10) because the variations in concentrations of these two elements are similar (Fig. 8); variations in MAR of phosphorus would be similar to those of iron, and so on.

All of the component MARs (Fig. 10) are influenced by the very pronounced peaks in bulk-sediment MAR centered on 4200 and 5700 varve yr (Fig. 8). These peaks are enhanced for MARs of those elements that also have high concentrations in sediment from those two parts of the section (e.g., Al, Na, K, Ti, Cr, V, Y, and Zr). Some elements also had high MARs between the present and 3200 varve yr (e.g., OC, Fe, Mn, and P), and 8000 and 10,000+ varve yr (e.g., Mn, S, Ba, and Pb), because the sediment deposited at those times had high concentrations of these elements, even though the bulk-sediment MARs were low (Fig. 8).

DISCUSSION

Element associations and geochemical zonation by Q-mode factor analysis

The concentration profiles in Figure 8 show that certain groups of elements have similar patterns of variations in concentration with time, and these different groups of elements have

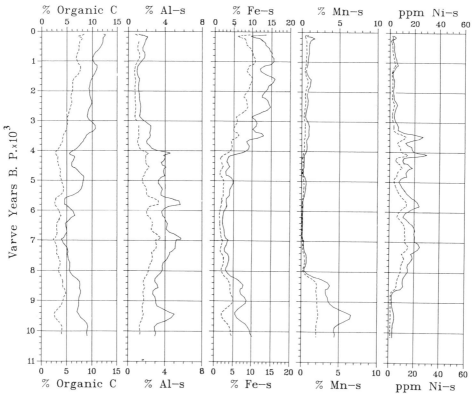

Figure 9. Smoothed curves of percentages of organic carbon, Al, Fe, and Mn, and parts per million (ppm) Ni vs. varve years on a bulk-sediment basis (dashed curve) and a carbonate-free basis (solid curve) for the Elk Lake core; s = determined by optical emission spectrophotometry.

concentration minima and maxima in different parts of the section. Q-mode factor analyses were run on the bulk-sediment concentration data for 23 variables from the deep-basin varved core in order to determine which variables were most closely related to each other, and if distinctive zones within the cores could be objectively delineated based on variations in element concentrations. The computer program used for the analyses was a modified version of CABFAC described by Klovan and Miesch (1976). Prior to running the analyses, concentrations of each element were transformed to proportions of the range of concentrations of that element so that all data were expressed on a scale of 0.0 to 1.0. After trying several different sets of reference axes in multidimensional space, I ultimately chose three orthogonal reference axes that maximize the variance of the transformed data in each of the three dimensions (varimax solution; Klovan and Miesch, 1976). The three-factor varimax solution explained 73% of the total variance in the 23 transformed variables. Each factor expresses some compositional attribute of the sediments based on a synthesis of several measured compositional variables (elements). In order to determine which elements had the largest influence on creating the composite compositional variables (factor loadings), the factor loadings were treated as chemical variables, and correlation coefficients were computed between the loadings and the 23 observed chemical variables (elements). Results of this correlation analysis are given in Table 1. Variations in factor loadings for the varved core are plotted versus varve years in Figure 11.

Table 1 shows that factor 1 expresses the composition of the inorganic clastic fraction. The elements that had the greatest influence on determining factor 1, in order of decreasing importance (decreasing correlation coefficient), are Mg, Na, Al, Cr, V, Y, Sc, Ni, Sr, Co, and Cu. Consequently, the variations in factor 1 loadings with time (Fig. 11) are similar to the variations in concentration with time of any of the observed compositional variables (elements) that are synthesized by factor 1 (e.g., compare the profile of factor 1 loadings in Fig. 11 with that of percent sodium in Fig. 8).

Factor 2 expresses the similarities in variations of iron and OC, and these two elements had the greatest influence on factor 2 loadings. Thus, the profile of factor 2 loadings (Fig. 11) is similar to those of percent iron and OC (Fig. 8). Factor 2 loadings also contain a minor contribution from molybdenum. Factor 3 loadings are a synthesis of concentrations of Mn, S, Ca, La, and Ba in decreasing order of importance (decreasing correlation coefficient, Table 1), and the profile of factor 3 loadings (Fig. 11) are similar to the profiles of concentrations of these five elements (Fig. 8).

The profiles of factor loadings (Fig. 11) do not really pro-

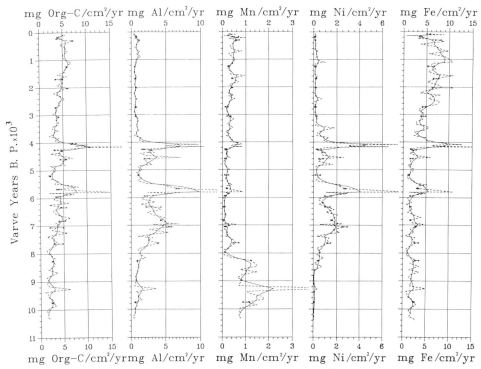

Figure 10. Plots of mass accumulation rates (MAR, in mg element/cm²/yr) of organic carbon, Al, Mn, Ni, and Fe vs. varve years in the varved Elk Lake core. The raw data are connected by dashed lines; solid lines through the raw data are smoothed curves computed using 11-sample weighted moving averages.

TABLE 1. CORRELATION COEFFICIENTS BETWEEN Q-MODE FACTOR LOADINGS AND CONCENTRATIONS OF ELEMENTS*

Variable	Factor 1	Factor 2	Factor 3
Organic C	-0.78	0.90	-0.17
CaCO$_3$	0.14	-0.37	0.63
Total S	-0.03	-0.29	0.68
Mg	0.96	-0.90	-0.24
Na	0.91	-0.84	-0.33
Al	0.82	-0.79	-0.20
Fe	-0.83	0.88	-0.11
Ti	0.72	-0.65	-0.41
Mn	-0.44	0.09	0.84
Ba	-0.08	-0.18	0.48
Co	0.64	-0.72	0.01
Cr	0.81	-0.78	-0.26
Cu	0.63	-0.58	-0.21
La	0.01	-0.28	0.59
Mo	-0.41	0.35	0.36
Nb	0.27	-0.32	0.09
Ni	0.71	-0.62	-0.39
Pb	-0.02	-0.17	-0.35
Sc	0.75	-0.66	-0.39
Sr	0.70	-0.69	-0.10
V	0.81	-0.77	-0.26
Y	0.78	-0.68	-0.31
Zn	-0.14	0.18	0.00

*Correlation coefficients >0.5 are in italics.

vide any additional information that cannot be gained from the profiles of element concentrations (Fig. 8). However, because the factor loadings are composite variables that synthesize the attributes of more than one element, the boundaries between chemical zones tend to be sharper. For example, the profiles of element concentration (Fig. 8) indicate that there is a marked change in the chemical composition of the sediment in Elk Lake that occurred between about 8400 varve yr and 7500 varve yr, but plots for most elements show a gradual change in concentration (e.g., the profile of percent sodium in Fig. 8). The profiles of factor loadings (Fig. 11), however, define this change more closely at about 8200 varve yr. Thus, the factor loadings indicate that the main chemical changes in Elk Lake, probably associated with the onset of dry conditions of the mid-Holocene prairie period, began 8200 yr ago. Similarly, the profiles of factor loadings, as well as those of most element concentrations, indicate that changes in chemical conditions associated with the return to moisture conditions at the end the prairie period occurred fairly abruptly at about 4000 varve yr.

Notice in Figure 11 that the minimum in concentrations of most clastic variables centered at about 5000 varve yr (e.g., profile for Na; Fig. 8) is not very prominent in the profile of factor 1 ("clastic factor") loadings. Part of the reason for this is that most of the major element data used in this factor analysis were spectrographic data, which generally are more noisy because of greater analytical imprecision. For example, compare the amount

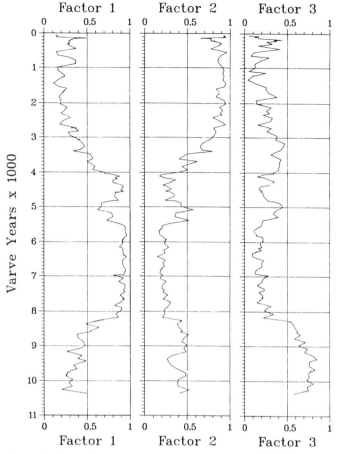

Figure 11. Plots of factor loadings vs. varve years for the results of a 3 factor Q-mode analysis of 23 geochemical variables, excluding analyses by X-ray fluorescence, in samples from the varved core.

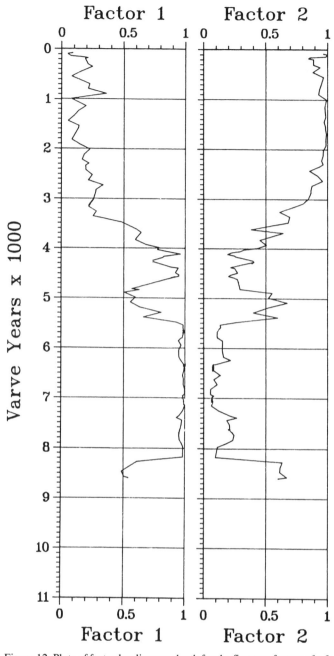

Figure 12. Plots of factor loadings vs. depth for the first two factors of a 3 factor Q-mode analysis of 27 geochemical variables, including analyses by X-ray fluorescence, in samples from the upper part of the section in the varved core (last 8500 varves).

of variation and, particularly, the amount of difference between duplicate analyses for aluminum by XRF and emission spectroscopy in Figure 8. From this data set, and from numerous other data sets, I have found that XRF and atomic absorption results are more precise (less variable) than emission spectrographic results for the same elements. However, the XRF data do not go to the bottom of the core because of interference due to Mn fluorescence in samples from the lower part of the section. In order to better define relations among variables for the upper part of the section (last 8500 varves), I incorporated the major element results from XRF and atomic absorption, and the trace element data from spectrographic analyses, and ran a Q-mode factor analysis of the combined data set. This analysis also has the advantage of incorporating the results for Si, K, and P. Four factors explained 82% of the total variance in the transformed data. Factors 1 and 2 alone explained 68% of this variance. Loadings for these two factors are plotted versus varve years in Figure 12. Factor 1 is a clastic factor that synthesizes the following elements in decreasing order of importance (decreasing correlation coefficient between element concentration and factor 1 loading): Mg, Al, Ti, Na, K, Si, V, Y, Co, Cr, Sc, Cu, Ni, and Sr. Factor 2 loadings synthesize variations in P, Fe, and OC, with minor contributions from Mn and Mo. Factors 3 and 4 were made up mainly of variations in calcium and manganese, respectively. The plots of factor loadings in Figure 12 clearly show lower clastic loadings (factor 1) and higher Fe-P-C loadings (factor 2) centered on about 5000 varve yr.

In addition to providing an objective means of zoning the

cores, the Q-mode factor analyses provide an objective analysis of the associations among geochemical parameters. I was surprised to find that magnesium is the best indicator of the clastic fraction as judged by having the highest correlation with factor 1 loadings in both Q-mode models (Table 1). I expected Al, Na, Ti, and K to contribute heavily to a synthesis of clastic variables, but I expected more magnesium to be in the carbonate fraction, either in dolomite or substituting for calcium in calcite. For example, from their study of the geochemistry and mineralogy of surface sediments of Minnesota lakes, Dean and Gorham (1976a, 1976b) concluded that magnesium was present in the carbonate fraction and in the clastic fraction, mainly in chlorite. In Elk Lake, however, it appears that most of the magnesium resides in the clastic fraction. Similarly, strontium appears to be associated with the clastic fraction (highest correlations with factor 1 loadings; Table 1), although I expected this alkaline earth cation to be present mainly substituting for calcium in carbonate minerals. Both barium and manganese appear to have their main affinities with carbonate minerals, as indicated by correlation coefficients of 0.84 and 0.48, respectively, with factor 3 loadings (Table 1).

Most of the trace transition elements (V, Co, Cr, Cu, and Ni) contribute to the chemical attributes summarized by factor 1 loadings, which suggests that these elements reside mainly in the clastic fraction. With the exception of Mo, which has a weak association with Fe-organic factor 2, there is no indication that organic particles or iron and manganese oxyhydroxides are important carriers of trace metals, as has been found in Lake Zurich (Sigg, 1985; Sigg and others, 1987) and in polluted lakes in eastern Ontario (Tessier and others, 1985).

Iron and manganese geochemistry

One of the most striking features of the geochemical record in Elk Lake sediment is the abundance of iron and manganese, both in the early and late phases of Holocene lake development. Identifying the mineral residences of iron and manganese in fine-grained sediments usually is difficult because many of the minerals cannot be detected by XRD or microscopic techniques. The inferred or identified mineral phases in Elk Lake sediments include X-ray amorphous iron and manganese oxyhydroxides, pyrite observed in scanning electron micrographs, an iron phosphate mineral (tentatively identified as rockbridgeite [(Fe, Mn)Fe$_4$(PO$_4$)$_3$(OH)$_5$] based on weak XRD peaks), and manganese carbonate (rhodochrosite). However, all of these minerals are probably related to the precipitation of the oxyhydroxides, and the intense redox cycling of iron and manganese within the epilimnion, hypolimnion, and sediments. In addition, iron appears to be associated closely with organic matter, probably as organic complexes, as indicated by its grouping in factor 2 of the Q-mode factor analysis (Table 1). Engstrom and Wright (1984) found that concentrations of authigenic iron and manganese in the sediments of two Labrador lakes increased markedly, coincident with a rapid transition from tundra to conifer forest in the drainage basins of the lakes. They interpreted this change as increased mobilization of organometallic complexes through humus accumulation in coniferous vegetation. Such an explanation might also explain the early and late Holocene accumulations of iron in Elk Lake. Much of the iron and manganese that is involved in formation of iron and manganese minerals in Elk Lake is recycled by reduction in the hypolimnion and sediment pore waters, but there is a large net burial of iron that is currently on the order of 5–10 mg Fe/cm^2/yr (Fig. 10). The manganese mass accumulation rate is presently only about 0.5 mg Mn/cm^2/yr, but was about ten times higher during the early Holocene (Fig. 10). High accumulation rates of iron and manganese, similar to the highest values calculated for Elk Lake, were reported for recent eutrophic lake sediments in Harvey's Lake, Vermont by Engstrom and others (1985). Harvey's Lake is seasonally anoxic, and Engstrom and others speculated that seasonal anoxia may be critical in preventing net loss of iron and manganese from profundal sediments during stratification. Some iron and manganese will be released from profundal sediments during summer stratification, but much of this will be returned to the sediments by precipitation at fall overturn. Sediment-trap studies at Elk Lake (Nuhfer and others, Chapter 7) show that much of sedimentation of iron- and manganese-rich particles occurs at fall overturn, but they accumulate in sediment traps throughout most of the year. In lakes like Elk Lake and Harvey's Lake, seasonal anoxia serves as an iron and manganese pump; iron and manganese released by reduction during summer stratification is returned by precipitation of oxyhydroxide, phosphate, and carbonate minerals. In Elk Lake it would appear that the efficiency of the pump has been increasing over the past 4000 yr, producing higher percentages and accumulation rates of both elements (Fig. 10).

The ultimate source of iron and manganese probably is from ground water entering the lake through porous littoral sediments. There is no direct eivdence for this, but there are several iron-rich springs around the lake, a prominent one of which appears about 10 m above lake level on the southeast side of the lake (Fig. 1; see chemical data in Megard and others, Chapter 3). High concentrations of iron and manganese were also found by LaBaugh and others (1981) in ground-water monitoring wells near Williams Lake on an outwash plain just south of the Itasca moraine south and east of Elk Lake. The concentration of iron in lake water from Williams Lake is only about 1% of the average concentration in ground water; the concentration of manganese in lake water is 17% of the average manganese concentration in ground water. Williams Lake has no surface inflow or outflow, and ground-water inflow and outflow is between 10% to 20% of the total water budget. LaBaugh and others (1981) concluded that oxidizing conditions in Williams Lake caused iron and manganese to precipitate, although no measurements of this were made. The much higher concentrations of manganese in lower Holocene sediments in Elk Lake suggest that most of the manganese, which is much more easily mobilized than iron by reduction in soils and sediments (Mackereth, 1966), was leached from glacial till in the drainage basin of the lake early in the lake's history.

Relations among iron, sulfur, organic carbon, and phosphorus are illustrated graphically by cross plots in Figure 13. The lack

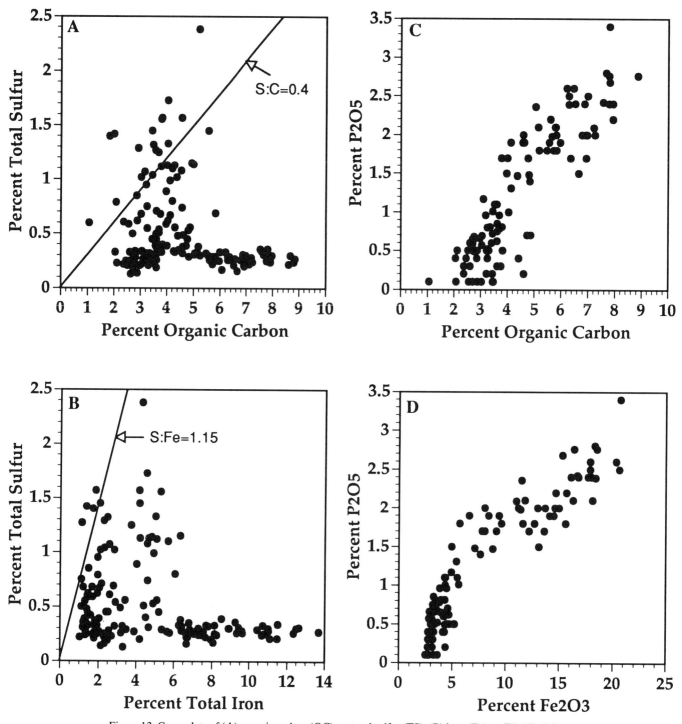

Figure 13. Cross plots of (A) organic carbon (OC) vs. total sulfur (TS); (B) iron (Fe) vs. TS; (C) OC vs. phosphorus (P); and (D) Fe vs. P. The diagonal line in (A) represents an S:OC ratio of 0.4, the average for normal reducing marine sediments (Berner and Raiswell, 1984). The diagonal line in (B) represents a S:Fe ratio of 1.15, that of stoichiometric pyrite (FeS_2).

of correlation between OC and S (Fig. 13A) and Fe and S (Fig. 13B) is typical in fresh-water environments where sulfate is severely limited and, therefore, there is little sulfate reduction and only limited formation of iron sulfides (Berner and Raiswell, 1984). A strong correlation between OC and S (with an S:C ratio of about 0.4; line in Fig. 13A) occurs in nearshore marine environments dominated by terrigenous clastic sediments where iron is not limiting, sulfate is not limiting, and the main factor limiting sulfate reduction and iron sulfide formation is amount of reactive organic matter (e.g., Berner, 1984; Raiswell and Berner, 1986. In such environments, iron is present in excess so there is a poor correlation between total Fe and total S. A good correlation between total Fe and total S usually occurs in environments where iron is limiting, such as in clastic-poor, carbonate-rich sedimentary regimes, with abundant reactive organic matter and sulfate. In such environments, there is abundant reactive sulfide by sulfate reduction, all of the iron reacts to form iron sulfides, and there is net loss of sulfide across the sediment-water interface (Dean and Arthur, 1989).

Another method of examining associations between elements and changes in association with time is with the moving correlation coefficient technique described by Dean and Anderson (1974); examples are shown in Figure 14 using a 9 sample moving correlation window. By this technique, a correlation coefficient between concentrations of two elements is computed for samples 1 through 9, 2 through 10, 3 through 11 and so on, and the results are plotted versus time.

As in most fresh-water environments, the sulfate concentration in Elk Lake is low (3.3 ppm), so that sulfate reduction, generation of hydrogen sulfide, and formation of pyrite are all relatively minor processes, the sulfur content of the sediments is very low over a large range of organic carbon concentrations (Berner and Raiswell, 1984), and there is little correlation between organic carbon and sulfur, or between iron and sulfur (Figs. 13, A and B, and 14, A and B). Pyrite formation is distinctly sulfur limited, and there is considerable excess iron (Fig. 13B). The only sediments that have S:OC ratios even close to the value of 0.4 typical of normal marine sediments (Fig. 13A; Berner and Raiswell, 1984) are those deposited during the first 2000 yr of the Holocene. This suggests that sulfate was not so limiting during the early phase of Elk Lake's development. The source of this sulfate was probably gypsum in glacial drift derived from the Cretaceous Pierre Shale, which underlies much of northwestern Minnesota (Ackroyd and others, 1967; Schultz and others, 1980). Diatom preparations from the lower part of the varved core contain abundant specimens of Cretaceous marine diatoms (Bradbury and Dieterich-Rurup, Chapter 15), indicating that Cretaceous shale is an important component of glacial drift in the Itasca area. Cretaceous gypsum is also the main reason for the high sulfate concentrations in modern lakes of western Min-

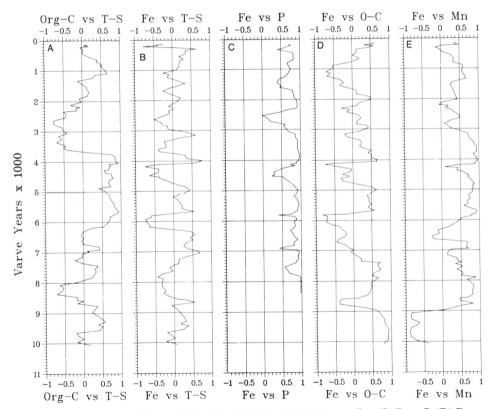

Figure 14. Moving correlation coefficients for (A) OC vs. S; (B) Fe vs. sulfur; (C) Fe vs. P; (D) Fe vs. O-C; and (E) Fe vs. Mn.

nesota and North and South Dakota (Gorham and others, 1983). The early sediments in Elk Lake contain unusually high concentrations of sulfur (Fig. 8), but there is surprisingly little correlation between either sulfur and iron or sulfur and organic carbon (Fig. 14). The highest degree of correlation between sulfur and organic carbon is during the last stages of the prairie period, between 6100 and 3800 varve yr (Fig. 14A). It is not clear, therefore, what the mineral residence time is of the high concentration of sulfur that accumulated during the early-lake phase.

There is a good correlation between organic carbon and phosphorous in Elk Lake sediments (Fig. 13C). This is the relation one would expect in a lake where phosphorus probably is the main nutrient limiting primary productivity. The excellent, persistent correlation between iron and phosphorus (Figs. 13D and 14C) emphasizes the importance of iron phosphate mineralization on the cycling of phosphorus in Elk Lake and, hence, on primary productivity. The burial of large amounts of phosphorus each year (currently about 1.2 mg/cm^2/yr), especially over the past 3000 yr (Fig. 8), undoubtedly has retarded the eutrophication of Elk Lake and is responsible for its present oligotrophic to mesotrophic state and the decline of the phosphorus-dependent diatom *Stephanodiscus minutulus* during the past 4000 yr (Bradbury and Dietrich-Rurup, Chapter 15). Because of the close association of cycles of iron, phosphorus, and organic carbon, as indicated by the similar patterns of concentration versus time (Fig. 8), association in factor analyses (Table 1), and the high degree of correlation between carbon and phosphorus, and iron and phosphorus, there should be a good correlation between iron and organic carbon. As seen in Figure 14D, this is true for the past few hundred years in Elk Lake and has been true throughout most of the Holocene, but not always. The correlation between iron and organic carbon shows distinct asymmetrical cycles with periodicities of about 1000 yr (Fig. 14D). These cycles suggest that the control of iron on organic productivity (and hence organic carbon production) deteriorated over periods of about 1000 yr, and then was quickly reestablished within a few hundred years. Periods when there was a negative correlation between iron and carbon sometime corresponds to periods when the correlation between iron and phosphorus tended to break down, but certainly this is not always true (cf. Fig. 14, C and D).

Concentration gradients of dissolved iron and manganese in the epilimnion and hypolimnion of Elk Lake through the summer (Fig. 15) show that concentrations of both begin to build just above the sediment-water interface soon after stratification, indicating that the dissolved iron and manganese diffuse out of the sediments. The most likely source of iron and manganese in the sediments is from reduction of iron and manganese minerals that survived the trip through a reducing hypolimnion during the previous few years. By the end of summer stratification, the concentration of dissolved manganese in the hypolimnion is as high as 1500 ppb closest to the sediment-water interface, having increased from <3 ppb after spring overturn (Fig. 15B). I anticipated that most iron and manganese oxyhydroxide precipitation

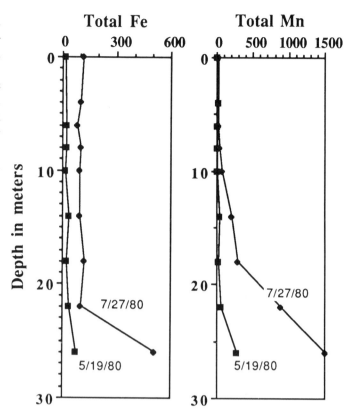

Figure 15. Plots of concentrations (in parts per billion) of total dissolved iron and manganese in Elk Lake for two dates (5/19/80 and 7/27/80).

should occur in late summer as the thermocline deepens, or at fall overturn. Sediment-trap studies show that there is a fall ferric hydroxide pulse, but iron and manganese also are persistent and abundant components of the sediment rain throughout much of the year (Nuhfer and others, Chapter 7).

The common appearance of a fairly thick seasonal lamina of ferric hydroxide gel in varves (Anderson, Chapter 5) indicates that much of the precipitated iron hydroxide is incorporated into the sediment. Ferric hydroxide colloids tend to be fairly stable (e.g., Tipping and others, 1981; Davison and Seed, 1983; Stauffer and Armstrong, 1986), which may explain why they survive in reducing hypolimnion and sediment pore waters. Manganese oxyhydroxides, however, are extremely unstable under reducing conditions, so it is unlikely that much of this material survives the trip to the bottom, particularly during the summer. Observations of thin sections of varves (Anderson, Chapter 5) indicate that some manganese oxide is preserved, particularly in varves from the lower Holocene, and this may represent manganese oxyhydroxides that formed during the winter when the hypolimnion is not anoxic. Unfortunately, iron and manganese oxyhydroxides are X-ray amorphous, so it is difficult to document their presence, let alone what the mineral phases are. The persistent presence of rhodochrosite in trap and bottom sediments most likely represents conversion of manganese oxyhydroxides under anoxic conditions either in the hypolimnion or below the sediment-water interface.

Like iron and manganese, phosphate also builds up in the hypolimnion during stratification periods (Megard and others, Chapter 3), but unlike iron and manganese the concentration of phosphate (and other nutrients such as silica and nitrate) is high throughout the hypolimniom and not just above the sediments. Iron phosphate as one or several poorly crystallized minerals, probably lipscombite [$Fe^{2+}Fe_2^{3+}(PO_4)_2(OH)$] or rockbridgeite [$(Fe, Mn)Fe_4(PO_4)_3(OH)_5$], based on weak XRD peaks, is a common and persistent component in sediment traps. It is particularly abundant in deep traps where it composed more than 40% of the crystalline fraction during the summer, fall, and winter (Nuhfer and others, Chapter 7). Ferrous phosphate may form by reaction of ferric hydroxide in a reducing, phosphate-rich hypolimnion (or sediment pore waters; e.g., Hearn and others, 1983), or by coprecipitation with ferric hydroxide when Fe^{2+} is oxidized within the metalimnion, as apparently occurs in culturally eutrophic Shagawa Lake, Minnesota (Stauffer and Armstrong, 1986). Nriagu and Dell (1974) suggested that ferric phosphate (as the minerals tinticite and/or cocoxenite) should form under oxidizing conditions by the adsorption of phosphate onto ferric hydroxide. It is not known whether ferric phosphate is a necessary precursor to mixed ferric-ferrous phosphate minerals such as lipscombite and rockbridgeite, or whether these minerals in Elk Lake have been "caught in the act" of transforming into a ferrous phosphate phase such as vivianite [$Fe_3(PO_4)_2 \cdot 8H_2O$], which is the most common and most stable iron phosphate mineral found in recent sediments (see reviews by Nriagu, 1972; Hearn and others, 1983). Vivianite has not been identified in XRD patterns of Elk Lake sediment, but is common as a blue coating on fish debris.

The stoichiometry of the iron phosphate in Shagawa Lake was calculated by Stauffer and Armstrong (1986) as $Fe_3(OH)_3PO_4$. Buffle and others (1989) determined the characteristics of a colloidal iron-calcium-phosphate species that formed at the seasonal oxic-anoxic boundary in a eutrophic lake in Switzerland. This material had a size range of 0.4 to 0.04 mm and a composition of Fe(II)/Fe(III)/PO_4/Ca(II) in molar proportions of 2.5/2.5/1.25/1.0, respectively, and an unknown OH content. Another possibility for the iron-phosphorus-organic association is the adsorption of phosphorus by organic-ferric hydroxide colloids (Koenings and Hooper, 1976; Francko and Heath, 1979, 1982; Sholkovitz and Copland, 1981; Tipping and others, 1981; Hearn and others, 1983). However, these iron-organic colloids are most likely to form in humic-rich distrophic waters such as bogs.

To try to sum up the inferred distribution of iron in Elk Lake sediments, the average sediment deposited in Elk Lake over the past 2000 yr contains 11.5% Fe (or 16% computed as Fe_2O_3), 1.0% P (or 2.3% computed as P_2O_5 and 3.1% computed as PO_4), and 0.3% S. If all of the 1% P is incorporated in rockbridgeite, this would account for 2.7% Fe (5.4% as Fe_2O_3). If all of the 0.3% S is incorporated in pyrite, this would account for an additional 0.37% Fe_2O_3. Therefore 16% Fe_2O_3 total, minus 0.4 as FeS_2, minus 5.4% Fe_2O_3 as rockbridgeite, equals a remaining 10.2% Fe_2O_3 that is assumed to be present as ferric hydroxide or ferric oxyhydroxides. This is in line with the observation of a relatively thick seasonal lamina of what appears to be a ferric hydroxide gel in thin sections from the upper part of the varved section in Elk Lake (Anderson, Chapter 5). Other possible residences for iron include iron carbonate (siderite) in quantities below detection by XRD, or coprecipitated with Ca in calcite, as suggested by Nriagu (1967) for excess iron in sediments in Lake Mendota.

Weathering indices

Detrital clastic minerals are minor components in Elk Lake sediments, even those deposited during the prairie period. However, these minerals contain sensitive indicators of weathering intensity (rates of decomposition, or available moisture) in the drainage basin and beyond. Plagioclase feldspar is particularly sensitive to chemical weathering, and is quickly decomposed when there is sufficient available moisture. Orthoclase is less sensitive to chemical weathering but also is readily decomposed under conditions of high available moisture. The presence of a relatively high abundance of plagioclase in the sediments of Elk Lake between 4000 and 9000 varve yr (Fig. 7), therefore, would argue for much drier conditions and less decomposition of feldspars in rock-weathering products in the drainage basin. This interval corresponds to the *Quercus*-Gramineae-*Artemesia* pollen zone (Fig. 7; Whitlock and others, Chapter 17), indicating that the drainage basin of Elk Lake was largely covered by drier prairie vegetation. During this period, prairie vegetation expanded eastward to cover much of the forested regions of Minnesota (McAndrews, 1966; Stark, 1976; Wright, 1976, and Chapter 2; Whitlock and others, Chapter 17). The mid-Holocene is widely recognized as a time of unusually warm and dry climate in Europe and North America (Wright, 1976, and Chapter 2). The prairie period in Minnesota was distinctly drier than at present, but actually may have been colder rather than warmer (see Forester and others, 1987; Dean and Stuiver, Chapter 12; Bradbury and others, Chapter 20). The presence of plagioclase in the sediments between 5700 and 8400 varve yr suggests that this period was the driest in the lake's postglacial history.

The concentration of sodium may be taken as an index of lack of decomposition of plagioclase feldspar and hence lack of available moisture for decomposition in soils of the drainage basin (Mackereth, 1966; Engstrom and Wright, 1984; Dean and others, 1984). With abundant available moisture in the drainage basin, plagioclase feldspar weathers rapidly and sodium remains in solution, which, of course, is why the sea is sodium rich. Potassium in orthoclase feldspars, however, is more conservative and forms K-rich clays on decomposition. The concentration of sodium in lake sediments should provide an index to the degree of chemical weathering in the drainage basin. The concentration of sodium increases considerably during the prairie period, as do the concentrations of many other elements associated with the clastic fraction, especially K, Ti, Al, Mg, Co, Cr, Cu, Ni, V, Y, and Zr.

Geochemical history of Elk Lake

In the introductory chapter to this volume (Anderson and others, Chapter 1), the developmental history of Elk Lake is subdivided into three phases. A cold, postglacial lake period, a dry prairie-period lake, and a cooler, wetter modern-period lake that concludes the past 4000 yr of history in Elk Lake and the surrounding region. Plots of element concentration (Fig. 8) and Q-mode factor loadings (Figs. 11 and 12) show that these three phases correspond to distinct chemical stages in the development of Elk Lake based on chemical composition of the sediments. These three stages can be represented in a ternary plot showing the relative proportions of Al, Fe, and Mn (Fig. 16). This ternary plot also serves to emphasize the abrupt transitions between lake stages.

In order to compare the bulk composition of Elk Lake sediments from the three different geochemical stages recorded in the varved core, I made the following assumptions regarding the major elements. (1) All carbonate is present as $CaCO_3$. Three is minor dolomite and rhodochrosite present but these would have little effect on the assumption that calcite is the dominant carbonate mineral. (2) The contributions of Si, Al, Fe, Mg, Mn, Na, K, and P can be represented for computational purposes by their appropriate oxides (SiO_2, Al_2O_3, Fe_2O_3, MgO, Na_2O, K_2O, and PO_4). Most of the Al, Mg, Na, and K are probably present in aluminosilica minerals; Si also is in aluminosilicate minerals, but a large fraction of the Si is present as biogenic opaline silica from diatom debris. An estimate of the amount of detrital SiO_2 that is present in any given sample can be obtained by multiplying the Al_2O_3 content by four, the SiO_2: Al_2O_3 ratio in average shale. Some magnesium is present in carbonate minerals, but the Q-mode factor analyses showed that magnesium had a strong association with the clastic fraction. Some of the iron is present in the reduced Fe(II) state (see discussion below) but this has little effect on the assumption that total iron can be represented as Fe_2O_3. As discussed below, it appears that most of the phosphorus is present as phosphate (PO_4). The organic fraction of lake sediments usually can be represented by loss on ignition (LOI) at 525 °C (Fig. 4; Dean, 1974). In Elk Lake sediments, however, LOI at 525 °C also contains a significant contribution from water, mostly from biogenic silica and iron and manganese oxyhydroxides.

Based on the above assumptions, I computed the average composition of Elk Lake sediments for the past 2000 yr to represent the modern lake, and the average composition for the period between 10,360 and 8470 varve yr to represent the early Holocene lake. I chose the composition of a sample from 5700 varve yr to represent the maximum of clastic influx as judged by maxima on the down-core plots of Al, Na, and K (Fig. 8). Results of these computations are given in Table 2.

Perhaps the most striking observation from Table 2 is that during the earliest and latest stages of Elk Lake, more than 90% of the sediment components were produced in the lake in the form of $CaCO_3$, organic matter, biogenic SiO_2, and authigenic iron and manganese minerals. Even during the prairie period, most of the sediment was still produced in the lake; detrital clastic material, represented as a *maximum* by sample 5700 varve yr (Table 2) was <30%.

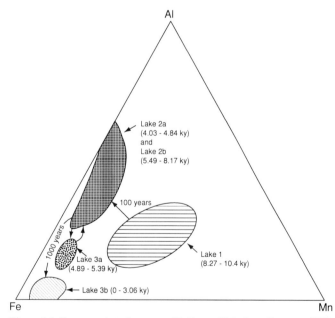

Figure 16. Ternary plot of percent Al, Fe, and Mn in sediments of Elk Lake that define the three geochemical stages in the development of Elk Lake.

TABLE 2. MAJOR COMPONENTS IN ELK LAKE SEDIMENTS FROM A SINGLE SAMPLE FROM ABOUT 5700 VARVE YEARS, AND AVERAGED FOR THE LAST 2000 YEARS AND FOR THE PERIOD BETWEEN 10,360 AND 8470 VARVE YEARS*

Component	Last 2000 yr (%)	5700 yr (%)	10,360 to 8470 yr (%)
$CaCO_3$	40	38	70
Org. matter + H_2O(OC)	21 (7.5)	7.1 (2.7)	17 (4.0)
SiO_2	18	39	....
Al_2O_3	0.8	6.4	2.9
Fe_2O_3	16	3.1	5.7
PO_4	3.1	0.7	....
MnO	1.4	0.2	2.6
MgO	0.8	3.2	1.3
Total S	0.3	0.2	1.3
K_2O	0.2	1.2	....
Na_2O	0.1	0.9	0.2
Total	101.7	100.0	

*Values in parentheses following values of organic matter +H_2O are percent organic carbon. Ellipses indicate insufficient data.

The manganese- and iron-rich early-lake stage lasted from 10,400 to 8200 varve yr (Fig. 16). During this stage, the lake was about 20 m deeper than at present if one simply subtracts the amount of sediment deposited in the lake over the past 10,000 yr. This greater depth, together with a greater influx of dissolved materials from unweathered glacial drift, suggests that the lake may have been meromictic. As discussed earlier, the high concentration of sulfur (as pyrite) that is so characteristic of the sediments of this early stage of the lake was probably the result of dissolution of gypsum in unweathered Cretaceous shale in the glacial drift.

The steep sides of the lake produced turbidites that occasionally interrupt the varved sequence. Although discrete, identifiable turbidites were eliminated from sampling, reworked shallow-water material in the varves undoubtedly was included in sampling. Turbidites are most abundant in the interval between 10,000 and 9000 varve yr, and the distinct peaks in $CaCO_3$ (calcite), Mn (rhodochrosite), and detrital clastic elements (Al, Mg, Na, Cr) (Figs. 7 and 8) probably represent an influx of reworked shallow-water material.

The clastic- and diatom-rich mid-Holocene lake stage with high concentrations of silica and many clastic-related elements, represented by Al in Figure 16, lasted from 8200 to 4000 varve yr. The transition between the first two lake stages occurred over a period of only about 100 yr. The removal of the protective tree cover in the drainage basin during the prairie period resulted in the influx of detrital clastic material. The high silica content of the prairie-period sediments (Fig. 8) reflects this high detrital influx as well as an increase in total diatom flux (Bradbury and Dietrich-Rurup, Chapter 15). Rapid cyclic variations with periods of 100 to 300 yr in abundances of the diatoms *Stephanodiscus minutulus* and *Fragilaria crotonensis* are interpreted as representing fluctuations in wind stress and turbulence in the lake (Bradbury and Dietrich-Rurup, Chapter 15). Concentrations of the clastic-related major, minor, and trace elements show distinct cyclic fluctuations with periodicities of several hundreds of years (e.g., see profiles for Na and Cu in Fig. 8) that reflect variation in influx of clastic material, much of which probably was wind borne. The cycles in diatom abundance and clastic influx are most certainly related, with wind intensity as the common denominator. The best example of a correlation between clastic sediment flux and diatom productivity is between abundance of *Aulacoseira ambigua* and concentrations of clastic elements (e.g., Na). *A. ambigua* blooms during summer and early fall in response to extreme wind-driven turbulence (Bradbury and Dietrich-Rurup, Chapter 15). It only occurs in Elk Lake between 6400 and 3800 varve yr, but during this interval there is an exact correspondence between its abundance and concentration of any of the clastic variables.

The change from prairie-lake conditions to modern-lake conditions began abruptly about 5500 varve yr, lasted about 500 yr, then, just as abruptly, reverted back to prairie-lake conditions about 4900 varve yr (Figs. 8 and 16). Prairie-lake conditions lasted another 800 yr and then abruptly switched to modern geochemical characteristics. The varve-thickness data show that this final transition to modern-lake conditions occurred over a few tens of years (Anderson and others, Chapter 4).

The general climatic model that has been developed for the mid-Holocene in the midwestern United States is one of gradual increases in warmth and dryness from about 8.5 ka to maxima about 7.0 ka, and then gradual decreases to more or less present conditions about 4.0 ka (Wright, 1976). In contrast, Bryson and others (1970) strongly suggested that Holocene climate was punctuated by rapid transitions between numerous dry and moist periods, each lasting hundreds of years. The geochemical and other data from Elk Lake certainly lend support to the punctuated climate model of Bryson and others (1970).

The final iron-rich modern lake stage began at about 4000 varve yr with an abrupt decrease in clastic influx but required 1000 yr or so to develop its full characteristics; i.e., high concentrations of iron, manganese, phosphorus, and organic carbon. The increases in concentrations of these elements were due in part to the decrease in dilution by clastic material, but are also due to real increases in burial or mass accumulation rates (Fig. 10). The rapid burial of large amounts of iron and phosphorus must have had a marked effect on phytoplankton populations in the lake. Burial of phosphorus undoubtedly is responsible for the change in dominance from *S. minutulus* to *F. crotonensis*, *Asterionella formosa* and *Synedra,* an assemblage indicative of low phosphorus conditions (Bradbury and Dietrich-Rurup, Chapter 15). Although the diatom data suggest that Elk Lake became more oligotrophic following the prairie period, the increase in the organic carbon burial rate suggests that productivity must have increased. The plant pigment data (Sanger and Hay, Chapter 13) indicate that cyanopigments became abundant, indicating that greater primary productivity must have shifted to the cyanobacteria (blue-green algae), which can better compete for available phosphorus than the diatoms (Bradbury and Dietrich-Rurup, Chapter 15).

CONCLUSIONS

Three distinct chemical stages in the development of Elk Lake can be defined based on chemical composition of the sediments. A manganese-, carbonate-, iron-, and sulfur-rich early-lake stage lasted from 10,400 to 8200 varve yr. A clastic- and diatom-rich mid-Holocene prairie-lake stage with high concentrations of silica and many clastic-related elements lasted from 8200 to 4000 varve yr. A final iron-rich modern-lake stage, that also is enriched in sedimentary manganese, phosphate and organic matter, developed over the past 4000 yr. During the earliest and latest stages of Elk Lake, more than 90% of the sediment components were produced in the lake in the form of $CaCO_3$, organic matter, biogenic SiO_2, and authigenic iron and manganese minerals. Even during the prairie period, most of the sediment was still produced in the lake. The geochemical data show that the changes between lake stages, and even within stages, were very abrupt, usually occurring within decades or, at most, within a few hundred years.

One of the most striking features of the geochemical record in Elk Lake sediment is the abundance of iron and manganese, both in the early and late phases of Holocene lake development. The inferred or identified mineral phases in Elk Lake sediments include X-ray amorphous iron and manganese oxyhydroxides, pyrite, manganese carbonate (rhodochrosite), and an iron phosphate mineral (tentatively identified as rockbridgeite by XRD). The persistent presence of rhodochrosite in trap and bottom sediments likely represents conversion of manganese oxyhydroxides under anoxic conditions, either in the hypolimnion or below the sediment-water interface. The burial of large amounts of phosphorus each year as iron phosphate (currently about 1.2 mg/cm^2/yr for phosphorus and 5 mg/cm^2/yr for iron), especially over the past 3000 yr, undoubtedly has retarded the eutrophication of Elk Lake and is responsible for its present oligotrophic to mesotrophic state.

REFERENCES CITED

Ackroyd, E. A., Walton, W. C., and Hills, D. L., 1967, Ground-water contribution to stream flow and its relation to basin characteristics in Minnesota: Minnesota Geological Survey Report of Investigations 6, 36 p.

Baedecker, P. A., ed., 1987, Methods for geochemical analysis: U.S. Geological Survey Bulletin 1770, 129 p.

Berner, R. A., 1984, Sedimentary pyrite formation: An update: Geochimica et Cosmochimica Acta, v. 48, p. 605–615.

Berner, R. A., and Raiswell, R., 1984, C/S method for distinguishing fresh-water from marine sedimentary rocks: Geology, v. 12, p. 365–368.

Bright, R. C., 1968, Surface water chemistry of some Minnesota lakes, with preliminary notes on diatoms: University of Minnesota Limnological Research Center, Interim Report 3, 58 p.

Bryson, R. A., Baerreis, D. A., and Wendland, W. M., 1970, The character of late-glacial and post-glacial climatic changes, in Dart, W., Jr., and Jones, J. K., Jr., eds., Pleistocene and recent environments of the central Great Plains: Lawrence, Kansas, University of Kansas Press, p. 53–74.

Buffle, J., De Vitre, R. R., Perret, D., and Leppard, G. G., 1989, Physico-chemical characteristics of a colloidal iron phosphate species formed at the oxic-anoxic interface of a eutrophic lake: Geochimica et Cosmochimica Acta, v. 53, p. 399–408.

Callender, E., Bowser, C. J., and Rossman, R., 1973, Geochemistry of ferromanganese carbonate crusts from Green Bay, Lake Michigan [abs.]: Eos (Transactions, American Geophysical Union), v. 54, p. 340.

Davison, W., and Seed, G., 1983, The kinetics of oxidation of ferrous iron in synthetic and natural waters: Geochimica et Cosmochimica Acta, v. 47, p. 67–79.

Dean, W. E., 1974, Determination of carbonate and organic matter in calcareous sediments and sedimentary rocks by loss on ignition: Comparison with other methods: Journal of Sedimentary Petrology, v. 44, p. 242–248.

—— , 1981, Carbonate minerals and organic matter in sediments of modern north temperate hard-water lakes, in Ethridge, F., and Flores, F., eds., Recent and ancient nonmarine depositional environments: Models for exploration: Society of Economic Paleontologists and Mineralogists Special Publication 31, p. 213–231.

Dean, W. E., and Anderson, R. Y., 1974, Application of some correlation coefficient techniques to time-series analysis: Mathematical Geology, v. 6, p. 363–372.

Dean, W. E., and Arthur, M. A., 1986, Iron-sulfur-carbon relationships in organic-carbon-rich sequences I: Cretaceous Western Interior Seaway: American Journal of Science, v. 289, p. 708–743.

Dean, W. E., and Fouch, T. D., 1983, Lacustrine environments, in Scholle, P. A., Bebout, D., and Moore, C., eds., Carbonate depositional environments: Tulsa, Oklahoma, American Association of Petroleum Geologists Memoir 33, p. 97–130.

Dean, W. E., and Gorham, E., 1976a, Major chemical and mineral components of profundal surface sediments in Minnesota lakes: Limnology and Oceanography, v. 21, p. 259–284.

—— , 1976b, Classification of Minnesota lakes by Q-and R-mode factor analysis of sediment mineralogy and geochemistry, in Merriam, D. F., ed., Quantitative techniques for the analysis of sediments: Oxford, England, Pergamon Press, p. 61–71.

Dean, W. E., Bradbury, J. P., Anderson, R. Y., and Barnosky, C. W., 1984, Variability of Holocene climate change: Evidence from varved lake sediments: Science, v. 266, p. 1191–1194.

Engstrom, D. R., and Wright, H. E., Jr., 1984, Chemical stratigraphy of lake sediments as a record of environmental change, in Howarth, E. Y., and Lund, J.W.G., eds., Lake sediments and environmental history: Studies in palaeolimnology and palaeoecology in honour of Winifred Tutin: Leicester, Leicester University Press, p. 11–67.

Engstrom, D. R., Swain, E. B., and Kingston, J. C., 1985, A paleolimnological record of human disturbance from Harvey's Lake, Vermont: Geochemistry, pigments, and diatoms: Freshwater Biology, v. 15, p. 261–288.

Forester, R. M., DeLorme, L. D., and Bradbury, J. P., 1987, Mid-Holocene climate in northern Minnesota: Quaternary Research, v. 28, p. 263–273.

Francko, D. A., and Heath, R. T., 1979, Functionally distinct classes of complex phosphorus compounds in lake water: Limnology and Oceanography, v. 24, p. 463–473.

—— , 1982, UV-sensitive complex phosphorus: Association with dissolved humic material and iron in a bog lake: Limnology and Oceanography, v. 27, p. 564–569.

Gorham, E., Dean, W. E., and Sanger, J. E., 1983, The chemical composition of lakes in the north-central United States: Limnology and Oceanography, v. 28, p. 287–301.

Hearn, P. P., Parkhurst, D. L., and Callender, E., 1983, Authigenic vivianite in Potomac River sediments: Control by ferric oxy-hydroxides: Journal of Sedimentary Petrology, v. 53, p. 165–177.

Jones, B. F., and Bowser, C. J., 1978, The mineralogy and related chemistry of lake sediments, in Lerman, A., ed., Lakes—Chemistry, geology, and physics: New York, Springer-Verlag, p. 179–235.

Kelts, K., and Hsü, K. J., 1978, Freshwater carbonate sedimentation, in Lerman, A., ed., Lakes—Chemistry, geology, and physics: New York, Springer-Verlag, p. 295–323.

Klovan, J. E., and Miesch, A. T., 1976, Extended CABFAC and QMODEL computer programs for Q-mode factor analysis of compositional data: Computers and Geosciences, v. 1, p. 161–178.

Koenings, J. P., and Hooper, F. F., 1976, The influence of colloidal organic matter on iron and iron-phosphorus cycling in an acid bog lake: Limnology and Oceanography, v. 21, p. 684–696.

LaBaugh, J. W., Groschen, G. E., and Winter, T. C., 1981, Limnological and geochemical survey of Williams Lake, Hubbard County, Minnesota: U.S. Geological Survey Water-Resources Investigations 81-41, 38 p.

Mackereth, F.J.H., 1966, Some chemical observations on post-glacial lake sediments: Royal Society of London Philosophical Transactions, ser. B, v. 250, p. 165–213.

McAndrews, J. H., 1966, Postglacial history of prairie, savanna, and forest in northeastern Minnesota: Torrey Botanical Club Memoirs, v. 22, p. 1–72.

Müller, G., Irion, G., and Forstner, U., 1972, Formation and diagenesis of inorganic Ca-Mg carbonates in the lacustrine environment: Naturwissenschaften, v. 59, p. 158–164.

Nriagu, J. O., 1967, The distribution of iron in lake sediments: Wisconsin Academy of Science, Arts, and Letters Proceedings: v. 56, p. 153–163.

—— , 1972, Stability of vivianite and ion-pair formation in the system $Fe_3(PO_4)_2$-H_3PO_4-H_2O: Geochimica et Cosmochimica Acta, v. 36, p. 459–470.

Nriagu, J. O., and Dell, C. I., 1974, Diagenetic formation of iron phosphates in recent lake sediments: American Mineralogist, v. 59, p. 934–946.

Raiswell, R., and Berner, R. A., 1986, Pyrite and organic matter in Phanerozoic normal marine shales: Geochimica et Cosmochimica Acta, v. 50, p. 1967–1976.

Schultz, L. G., Tourtelot, H. A., Gill, J. R., and Boerngen, J. G., 1980, Composition of the Pierre Shale and equivalent rocks, northern Great Plains region: U.S. Geological Survey Professional Paper 1064-B, 114 p.

Sholkovitz, E. R., and Copland, D., 1981, The coagulation, solubility, and adsorption properties of Fe, Mn, Cu, Ni, Cd, Co, and humic acids in river water: Geochimica et Cosmochimica Acta, v. 45, p. 181–189.

Sigg, L., 1985, Metal transfer mechanisms in lakes: The role of settling particles, *in* Stumm, W., ed., Chemical processes in lakes: New York, Wiley, p. 283–310.

Sigg, L., Sturm, M., and Kistler, D., 1987, Vertical transport of heavy metals by settling particles in Lake Zurich: Limnology and Oceanography, v. 32, p. 112–130.

Stark, D., 1976, Paleolimnology of Elk Lake, Itasca State Park, northwestern Minnesota: Archiv für Hydrobiologie, Supplement 50 (Monographische Beitrang), p. 208–274.

Stauffer, R. E., and Armstrong, D. E., 1986, Cycling of iron, manganese, silica, phosphorus, calcium and potassium in two stratified basins of Shagawa Lake, Minnesota: Geochimica et Cosmochimica Acta, v. 50, p. 215–229.

Tessier, A., Rapin, F., and Carignan, R., 1985, Trace metals in oxic lake sediments: Possible adsorption onto iron oxyhydroxides: Geochimica et Cosmochimica Acta, v. 49, p. 183–194.

Tipping, E., Woof, C., and Cooke, D., 1981, Iron oxide from a seasonally anoxic lake: Geochimica et Cosmochimica Acta, v. 45, p. 1411–1419.

Wright, H. E., Jr., 1976, The dynamic nature of Holocene vegetation, a problem in paleoclimatology, biogeography, and stratigraphic nomenclature: Quaternary Research, v. 6, p. 581–596.

MANUSCRIPT ACCEPTED BY THE SOCIETY JULY 27, 1992

Geologic implications of the Elk Lake paleomagnetic record

Donald R. Sprowl*
Department of Geology, 120 Lindley Hall, University of Kansas, Lawrence, Kansas 66045-2124
Subir K. Banerjee
Department of Geology and Geophysics, University of Minnesota, Minneapolis, Minnesota 55455

ABSTRACT

Comparison of the bulk magnetic properties of the Elk Lake sediments with varve thickness and sediment density suggests that most of the magnetic carrier in the sediments is allochthonous to the lake. An exception may be the fine-grained magnetite in the upper three meters of section. Correlation of the Elk Lake directional record with the records from Lake St. Croix and Kylen Lake demonstrates the accuracy of the Elk Lake varve chronology.

INTRODUCTION

The seasonal laminations in the Elk Lake sediments provide continuous time control for sedimentologic studies and indicate a low-energy depositional environment, free from bioturbation. This made Elk Lake a prime site for paleomagnetic studies because it was presumed that the low-energy depositional environment would permit the magnetic recording mechanism to operate at maximum efficiency. For reasons that are not well understood, Elk Lake appears to have recorded magnetic field variations less well than other Minnesota lakes (Sprowl and Banerjee, 1989). Even so, the Elk Lake record has contributed to the description and dating of magnetic field secular variation in central North America. In addition, the magnetic record permits external confirmation of the Elk Lake chronology.

There are two classes of magnetic information in the Elk Lake sediment: vector information regarding the secular variation of the local magnetic field and bulk magnetic properties that are controlled by sedimentologic and biologic factors. These are dealt with individually in the following sections. Measurements reported here were performed primarily on material from the 1983 Elk Lake cores, and the depth and age references are to the 1983 depth scale and the 1983 varve chronology, respectively (Sprowl and Banerjee, 1989; Sprowl, Chapter 6).

VARIATION IN BULK MAGNETIC PROPERTIES WITH DEPTH

Two bulk magnetic measurements that are useful for demonstrating depth variations in sediment magnetic properties are magnetic susceptibility (χ) and anhysteretic remanent magnetization (ARM).

The measure of the magnetization induced in the sediment by an applied magnetic field is χ, which is a function of the mineralogy, grain size, and concentration of magnetic minerals in the sediment. Curie temperature experiments on magnetic extracts and remanence acquisition experiments on bulk samples (Sprowl and Banerjee, 1989) indicate that magnetite is the dominant magnetic mineral in Elk Lake. In magnetite-rich sediments such as those of Elk Lake, χ is determined mostly by total concentration of magnetite. χ is a weak function of magnetite grain size, larger grains being somewhat more susceptible (King and others, 1982). Saturation magnetization experiments confirm that χ is a good proxy of total magnetite concentration throughout the Elk Lake section (Sprowl and Banerjee, 1989, Fig. 4, Table 3).

ARM is the remanent magnetization imparted to the sample by exposure to an unbalanced, alternating magnetic field. In contrast to χ, ARM is strongly dependent on grain size, the smallest grains (<0.1 μm) acquiring up to 50 times as much ARM per volume as larger grains (>25 μm) (King and others, 1982). Thus, with χ and ARM, we can monitor both magnetite concentration in general and changes in magnetite grain size through the sedimentary section.

---

*Present address: Department of Chemistry, Louisiana College, Pineville, LA 71359.

Sprowl, D. R., and Banerjee, S. K., 1993, Geologic implications of the Elk Lake paleomagnetic record, *in* Bradbury, J. P., and Dean, W. E., eds., Elk Lake, Minnesota: Evidence for Rapid Climate Change in the North-Central United States: Boulder, Colorado, Geological Society of America Special Paper 276.

Figure 1 shows side-by-side plots of χ and ARM, along with sediment density, as a function of varve age for Elk Lake core 83b (Sprowl, Chapter 6). Figure 2 plots the ratio of the ARM and χ curves. Figure 3 plots varve thickness and χ. The following sections discuss the implications of these curves.

1. The most dramatic feature of the magnetic curves is the eight-fold increase in ARM at the 2000 varve years level (see Sprowl, Chapter 6, Fig. 2), an increase that sends it well off scale in the presentation of Figure 1. In contrast, the rise in χ (and total magnetite concentration) between 3000 varve years and 1000 varve years is less than three fold. The ratio of ARM to χ, plotted in Figure 2, clearly indicates that the additional magnetite present in the sediment above 2000 varve years is significantly finer grained than the material throughout the rest of the section (King and others, 1982). The origin of the fine-grained magnetite in the upper section is uncertain. It is doubtful that the decrease in clastic influx noted at 3500 varve years would be accompanied by a significant increase in allochthonous magnetite influx. Two explanations seem possible. First, the fine-grained magnetite in the upper section may be biogenic (Blakemore, 1975, and many others), suggesting that we have a record of the advent of mag-

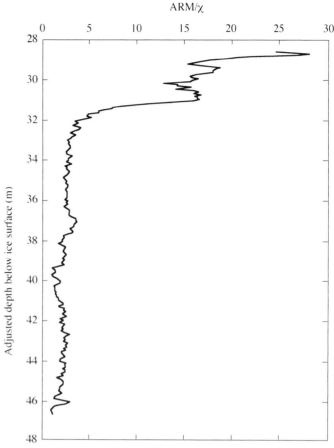

Figure 2. The ratio of ARM to χ, versus depth, from core 83b.

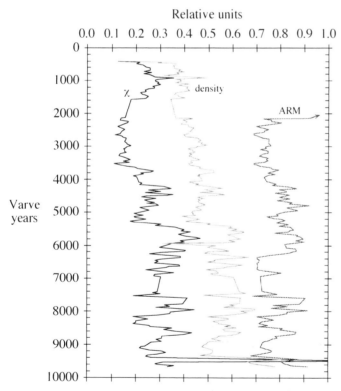

Figure 1. Magnetic susceptibility (χ), sediment density, and ARM from core 83b are plotted to demonstrate correspondence of high-frequency signals. Horizontal axis is relative for each curve as follows. Susceptibility runs full scale (0 to 1) from 0 to 20×10^{-6} in cgs units. Density runs full scale from 0.85 to 1.85 g/cm^3. ARM runs full scale from -3.6×10^{-4} to 2.4×10^{-4} emu. ARM values were acquired using an alternating field sweep from 1000 to 0 Oe with a bias field of 0.25 Oe. Above 2000 varve years, ARM values average about 8.0×10^{-4} emu, about 1.8 units on the horizontal scale.

netotactic activity in Elk Lake. Second, originally deposited fine-grained magnetite may have been dissolved from the lower part of the section (Karlin and Levi, 1983).

2. Below the 7000 varve year level, ARM and χ vary together, indicating an unchanging magnetite grain size. Between 7000 and 2000 varve years, ARM and χ show similarities but also significant differences. The reductions in χ at 5400 varve years and 3600 varve years, which correlate with reduction in varve thickness (Fig. 3) at the advent of the intra-prairie period oscillation and at the modern period, respectively, are much less pronounced in the ARM curve. The two curves together indicate a decrease in total magnetite concentration and average magnetite grain size as sedimentation rate decreases. This can be explained without difficulty by decreased clastic influx.

3. There is a striking correspondence between the high-frequency signals in χ and density in Figure 1 and a reasonable correspondence between the high-frequency signals in χ and varve thickness in Figure 3. Clearly, the short-term variations in χ and density are driven by the same mechanism. High-frequency variations in water content could produce the observed relation between χ and density, but should produce an inverse relation with varve thickness. Alternatively, assuming a constant percentage of magnetite in the allochthonous clastic component of the

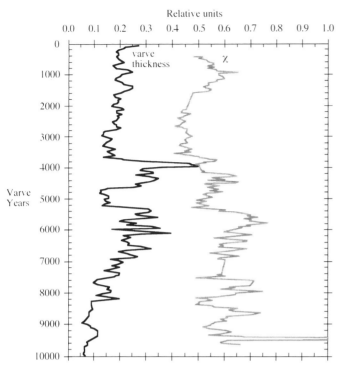

Figure 3. Plots of varve thickness and magnetic susceptibility (χ) from core 83b. Horizontal axis is relative for each curve as follows. Susceptibility runs full scale (0 to 1) from -6×10^{-6} to 14×10^{-6} in cgs units. Varve thickness runs full scale from 0 to 1 cm/yr.

sediment, variations in percent clastics should give rise to covariation of susceptibility, density, and varve thickness. There is no a priori reason to expect χ and varve thickness to vary together, because χ is a volume-independent bulk property. The observed correlation seems to require the primary magnetite source to be allochthonous to the lake. Thus, an increase in clastic sedimentation that increases varve thickness also increases the concentration of magnetite in the bulk sediment.

EXTERNAL CONFIRMATION OF THE ELK LAKE CHRONOLOGY

The accuracy of the Elk Lake varve chronology can be evaluated against remote radiocarbon chronologies by correlating the Elk lake directional magnetic record with those from other Minnesota lakes. The measurement and analysis of the Elk Lake paleomagnetic record are detailed elsewhere (Sprowl and Banerjee). Parallel, oriented cores were collected in 1982 and 1983; the two 1983 cores covered the entire sedimentary section (Sprowl, Chapter 6). Pairs of oriented samples were collected every 2 cm for measurement of magnetic properties and remanent directions. The directional record is represented by two angles that the remanent magnetic vector makes with a reference direction. Inclination is the angle measured down from horizontal to the remanent vector. Declination is the angle measured east from due north to the remanent vector.

Because secular variation of the magnetic field occurs on regional rather than local scales, the inclination and declination curves are useful for time correlation between sites. Within several hundred kilometers of Elk Lake, maxima and minima in the directional curves will occur at the same time at all sites. If all sites record faithfully the field variations, then the directional curves can be used for time correlation (King and others, 1983).

Figure 4 plots the Elk Lake directional record along with the records from Lake St. Croix and Kylen Lake from eastern Minnesota (Lund and Banerjee, 1985). The Elk Lake curves are plotted versus varve years; the St. Croix and Kylen curves are plotted versus times scales derived from radiocarbon dates. Table 1 lists the radiocarbon dates used. Through the upper 4000 years, correlation of features between the Elk and St. Croix records is obvious and time offsets are less than 300 years. At 4500 years, correlation is ambiguous. Between 6000 and 8000 years, the three records correlate well with time offsets of less than 500 years. Below 8000 years, both the Elk and St. Croix records are suspect, because of coring difficulties at Elk Lake and because of poor time control at Lake St. Croix. At 8000 years, the declination curves suggest that the varve age is perhaps 300 years too old, but the inclination curves do not correlate well at this offset. The data do not permit us to evaluate the Elk Lake bottom age of 10,500 years, but clearly indicate the validity of the varve counting process and suggest an overall error in the varve counts of less than 500 years. Because the varve counts are continuous and are not affected by errors associated with radiocarbon dating, the varve ages are preferred for the maxima and minima in the magnetic directional curves.

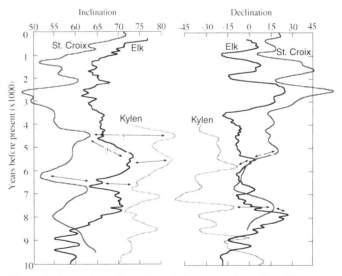

Figure 4. Comparison of secular variation curves from three Minnesota lakes. The bold curves are from Elk Lake. The medium gray lines are from Lake St. Croix. The light gray lines are from Kylen Lake. Time scale is with reference to the varve chronology for the Elk Lake curves and to radiocarbon dates for the St. Croix and Kylen curves. Units of inclination and declination are degrees. Four of the curves are shifted horizontally for visual clarity as follows: St. Croix inclination, 4° left; Kylen inclination, 9° right; St. Croix declination, 5° right; Kylen declination, 6° left.

TABLE 1. RADIOCARBON DATES FROM KYLEN AND ST. CROIX LAKES

Depth below ice surface (cm)	Corrected radiocarbon age
Lake St. Croix, core 75B	
1,187 to 1,197	100
1,310 to 1,320	416
1,480 to 1,490	802
1,696 to 1,708	1,096
2,005 to 2,015	1,944
2,305 to 2,315	2,928
2,503 to 2,613	4,683
2,791 to 2,801	6,964
2,993 to 3,003	9,634
Kylen Lake, core 76A	
390 to 400	4,260
435 to 440	4,880
515 to 520	6,350
595 to 600	7,190
666 to 671	7,790
756 to 766	8,740
786 to 791	8,940
846 to 851	9,840
906 to 911	10,470

REFERENCES CITED

Blakemore, R. P., 1975, Magnetotactic bacteria: Sicence, v. 190, p. 377–379.

Karlin, R., and Levi, S., 1983, Diagenesis of magnetic minerals in recent hemipelagic sediments: Nature, v. 303, p. 327–330.

King, J. W., and Banerjee, S. K., Marvin, J., and Ozdimir, O., 1982, A comparison of different magnetic methods for determining the relative grain size of magnetite in natural materials: Some results from lake sediments: Earth and Planetary Science Letters, v. 59, p. 404–419.

King, J. W., Banerjee, S. K., Marvin, J., and Lund, S., 1983, Use of small amplitude paleomagnetic fluctuations for correlation and dating of continental climate changes: Palaeogeography, Palaeoclimatology, Palaeoecology, v. 42, p. 167–183.

Lund, S., and Banerjee, S. K., 1985, Late Quaternary paleomagnetic field variation from two Minnesota lakes: Journal of Geophysical Research, v. 90, p. 803–825.

Sprowl, D. R., and Banerjee, S. K., 1989, The Holocene paleosecular variation record from Elk Lake, Minnesota: Journal of Geophysical Research, v. 94, p. 9369–9388.

Manuscript Accepted by the Society July 27, 1992

Stable carbon and oxygen isotope studies of the sediments of Elk Lake, Minnesota

Walter E. Dean
U.S. Geological Survey, MS 939, Box 25046, Federal Center, Denver, Colorado 80225
Minze Stuiver
Quaternary Research Center, University of Washington, Seattle, Washington 98105

ABSTRACT

Variations in the ratios of $^{18}O:^{16}O$ and $^{13}C:^{12}C$ in calcite throughout the Holocene in Elk Lake, Minnesota, are recorded in three varve-calibrated carbonate cores. Marl in a varved deep-basin (29.6 m) core consists mainly of calcite precipitated from surface waters during the summer and probably provides the least complicated isotope record. Marl in a sublittoral (10 m) core consists of calcite contributed from several inorganic and organic sources and probably is the most complicated of the three isotope records. Calcite from shells of the ostracod *Candona ohioensis* in the sublittoral core provides a record of shallow-water conditions in Elk Lake for the period between 10,500 and 5500 varve yr. Variations in the $^{13}C:^{12}C$ ratio of organic carbon deposited in Elk Lake during the Holocene are recorded in organic matter in the deep-basin core.

All three oxygen isotope records show that, in general, the $^{18}O:^{16}O$ ratio in carbonate was enriched in ^{18}O by several parts per mil during the mid-Holocene relative to the past few thousand years. This pattern of oxygen isotope variation is similar to that observed for carbonate materials from other lakes in the northeastern and north-central United States. Oxygen isotope records from these other lakes also show that the $^{18}O:^{16}O$ ratio during the early Holocene was lower than during the mid-Holocene, and this pattern has been interpreted as representing a response to a generally warmer and drier climate during the mid-Holocene beginning about 8000 varve yr (the so-called hypsithermal). Ostracod and diatom assemblages from Elk Lake cores show, however, that the lake was colder and more saline than at present until at least 6700 varve yr, with conditions similar to those that exist today in cold prairie lakes of Canada. It may be more appropriate, therefore, to refer to the mid-Holocene in northwestern Minnesota as the "prairie period" rather than the hypsithermal, indicating that the climate was drier, but with no connotation regarding temperature. The oxygen isotope data from the three Elk Lake records for this period are somewhat equivocal. Values of $\delta^{18}O$ in the marl from the sublittoral core and shells of *Candona* increase from 10,000 to about 6800 varve yr. However, values of $\delta^{18}O$ in the marl that accumulated in the deepest part of the lake over the same interval (10,000–6800 varve yr) are more or less constant and enriched in ^{18}O; this probably reflects the cold, saline prairie-lake conditions predicted from the ostracod and diatom assemblage data. All three oxygen isotope records show decreases in $^{18}O:^{16}O$ ratios after about 6800 varve yr in response to an increase in temperature and decrease in salinity of the lake.

The $^{13}C:^{12}C$ ratios in carbonates from all three Elk Lake records show a distinct pattern; the ratio increased gradually from 10,000 to 8000 varve yr going into the

Dean, W. E., and Stuiver, M., 1993, Stable carbon and oxygen isotope studies of the sediments of Elk Lake, Minnesota, *in* Bradbury, J. P., and Dean, W. E., eds., Elk Lake, Minnesota: Evidence for Rapid Climate Change in the North-Central United States: Boulder, Colorado, Geological Society of America Special Paper 276.

mid-Holocene prairie period and then decreased gradually coming out of the prairie period between about 5500 and 2500 varve yr. These changes in the $^{13}C:^{12}C$ ratio could have been related to temperature through its effect on solubility of carbon dioxide; however, this interpretation is not supported by the oxygen isotope data. Another possibility is that changes in the $^{13}C:^{12}C$ ratio are related to organic productivity that removes ^{13}C-depleted organic carbon and results in ^{13}C-enriched surface waters. This interpretation implies that organic productivity was higher in Elk Lake during the mid-Holocene prairie period.

Support for the high-productivity, ^{13}C-enriched surface-water model for the mid-Holocene prairie period in Elk Lake is provided by changes in the $^{13}C:^{12}C$ ratio of organic carbon in the deep-basin core. These changes parallel almost exactly those in the $^{13}C:^{12}C$ ratio of carbonate carbon, but are about $2^0/_{00}$ larger (about $6^0/_{00}$ as opposed to about $4^0/_{00}$ for carbonate carbon). The difference of about $2^0/_{00}$ may represent ^{13}C depletion due to CO_2 limitation. The percentage of organic carbon in the sediment did not increase during the prairie period because it was diluted by an increased flux of detrital clastic material. The ultimate burial rate of organic carbon increased considerably, however, indicating that organic productivity was higher and/or the degree of preservation increased. Diatom assemblages and plant-pigment concentrations indicate that productivity was higher during the prairie period. Pyrolysis hydrogen and oxygen indices show that the ^{13}C-enriched organic matter that accumulated during the prairie period was hydrogen rich and oxygen poor relative to organic matter that accumulated before and after. These two indices demonstrate that the organic matter that accumulated during the prairie period was much better preserved.

INTRODUCTION

Stable isotopes of oxygen and carbon are powerful indicators of environmental conditions in both lacustrine and marine environments. The ratio $^{18}O:^{16}O$ in marine carbonate material, particularly in the shells of foraminifers, can be a precise paleothermometer. In lakes, the oxygen isotope composition of carbonate material generally is not as useful as a paleothermometer because of the small reservoir of a lake, the scarcity of pelagic calcareous organisms, and the possibility of extreme variations in salinity and precipitation through time that could also affect the oxygen isotope composition (Stuiver, 1970). If used with caution, however, oxygen isotope stratigraphy in lacustrine carbonate sequences can provide an indication of temperature change, mostly through the effect of temperature on evaporation.

The $^{13}C:^{12}C$ ratio in marine and lacustrine carbonate material has been used to monitor changes in carbon reservoirs in response to preferential fixing of ^{12}C during photosynthesis and release of isotopically light (^{13}C depleted) CO_2 during respiration and decay. It also has been used as a monitor of the CO_2 exchange between the lake and the atmosphere: as such, it has been used both as an indicator of paleoproductivity and a tracer of water masses. The $^{13}C:^{12}C$ ratio of organic carbon has been used as an indicator of source of organic matter and, to a lesser extent, as a measure of paleoproductivity.

The purpose of this study was to examine the oxygen and carbon isotope stratigraphies of the sediments of Elk Lake and interpret these stratigraphies in terms of changes in temperature, salinity, organic productivity, and sources of organic matter in Elk Lake throughout the Holocene. To aid in this interpretation, we draw on other indicators of temperature (ostracods; Forester and others, 1987), salinity and lake level (ostracods and diatoms; Bradbury and Dieterich-Rurup, Chapter 15), productivity (diatom flux, pyrolysis hydrogen index, organic-carbon content and flux, and plant pigments; Sanger and Hay, Chapter 13), and organic-matter source (pyrolysis hydrogen index and plant pigments).

METHODS

Samples were collected from the varved deep-basin core (29.6 m water depth; Fig. 1; Anderson and others, Chapter 4) at 50 varve intervals for inorganic geochemistry and diatom analyses, although only every other sample was analyzed for geochemistry (Dean, Chapter 10). We selected 23 of the geochemistry samples for isotopic and Rock-Eval pyrolysis analyses. Samples were collected from the sublittoral core (10 m water depth; Fig. 1) at 10 cm intervals for inorganic geochemistry (Dean, Chapter 10). We selected 20 of the geochemistry samples for isotopic analyses. Aliquots of dried and ground geochemistry samples from each of these two cores were used for stable isotope analysis and Rock-Eval pyrolysis (Tables 1 and 2). Age values in varve years (varve yr) in Tables 1 and 2, and used for plotting in illustrations, are the age of the midpoint of the sample interval relative to the varve time (T) scale (T_0 = A.D. 1927; Anderson and others, Chapter 4).

Stable carbon isotope ratios were determined using standard techniques (Pratt and Threlkeld, 1984). Powdered samples for carbon isotope determinations of organic carbon were oven dried at 40°C and reacted with an excess of 0.5N HCl for 24 hr to

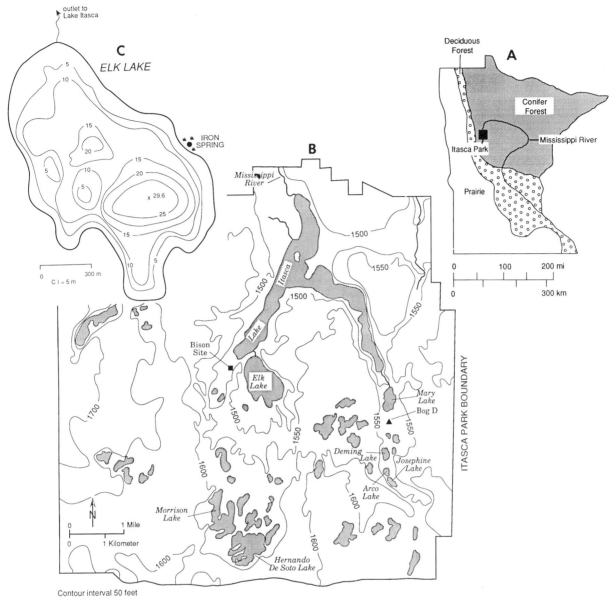

Figure 1. Maps of (A) Minnesota, (B) Itasca State Park, and (C) Elk Lake showing general vegetation zones of Minnesota, the location of Elk Lake, bathymetry of Elk Lake, and location of the varved core (x in C) in the deepest part of the lake.

dissolve carbonate minerals. The residue was centrifuged, decanted, washed three times, dried under flowing nitrogen at 50°C, and combusted at 1000°C under oxygen pressure in an induction furnace. The resulting CO_2 was dehydrated and purified in a high-vacuum gas-transfer system. Samples of bulk sediment for determination of carbon and oxygen isotope ratios in carbonate were reacted with 100% phosphoric acid, and the evolved CO_2 was dehydrated and purified in a high-vacuum gas-transfer system. All isotope ratios were determined using a Finnigan MAT 251, 6 in, 90°-sector, isotope ratio mass spectrometer. Results are reported in the standard per mil (‰) δ notation relative to the University of Chicago Pee Dee belemnite (PDB) marine-carbonate standard:

$$\delta ‰ = [R_{sample}/R_{PDB}) - 1] \times 10^3,$$

where R is the ratio (^{13}C:^{12}C) or (^{18}O:^{16}O).

The Rock-Eval pyrolysis method provides a rapid determination of hydrogen contentm and degree of preservation of organic matter in sediments and sedimentary rocks (Espitalié and others, 1977; Peters, 1986). Programmed heating of a sample in a helium atmosphere results in generation of hydrocarbons and

TABLE 1. ORGANIC CARBON, ISOTOPIC, AND PYROLYSIS RESULTS FROM ELK LAKE DEEP-BASIN VARVED CORE*

Varve	% Corg	δ ^{13}Corg	δ ^{18}O	δ ^{13}C	H-Index	O-Index
203	6.97	-33.7	-7.64	-2.83	188	228
398	6.71	-34.6	-7.18	-3.10	277	268
943	5.71	-34.5	-6.80	-2.15	218	358
1,194	5.95	-34.0	-7.00	-2.51	201	304
1,644	5.66	-34.5	-7.06	-2.55	174	384
1,800	5.66	-34.5	-7.21	-2.20	193	321
2,147	4.64	-32.9	-7.83	-2.71	170	287
2,776	4.30	-33.8	-8.21	-2.64	178	331
3,315	3.72	-33.1	-7.91	-1.94	235	253
3,537	3.83	-33.2	-7.54	-0.91	139	335
4,305	2.56	-30.4	-7.83	0.30	149	195
4,390	2.06	-29.9	....	....	223	176
4,648	2.81	-30.2	-7.07	0.48	140	143
5,338	2.82	-29.8	-6.48	1.35	257	150
5,776	2.22	-29.4	-7.26	0.66	249	137
5,893	2.40	-29.4	....	....	298	134
6,437	2.03	-29.2	-6.46	0.93	241	161
6,871	2.47	-28.9	-5.84	1.35	268	136
7,207	2.40	-28.0	-6.33	0.34	219	116
7,687	2.22	-29.6	-6.37	-0.05	204	185
8,422	2.78	-31.8	-6.32	-3.22	162	221
9,125	3.45	-34.0	-6.00	-0.88	180	290
9.909	2.84	-35.0	-6.08	-6.23	173	449

*Varve year is the midpoint of 50 year sample. Values of δ ^{13}Corg, and values of δ ^{13}C and δ ^{18}O of bulk-sediment carbonate, are all expressed in the per mil difference, ‰, relative to the PDB marine carbonate standard. Rock-Eval pyrolysis hydrogen index (H-Index, in milligrams of hydrocarbons per gram of organic carbon) and oxygen index (O-Index, in milligrams of carbon dioxide per gram of organic carbon) in samples from the deep basin varved core from Elk Lake.

TABLE 2. ISOTOPIC RESULTS FROM ELK LAKE SUBLITTORAL CORE*

Depth (cm)	δ ^{13}C	δ ^{18}O	Varve Year
9	-2.59	-7.87	130
59	-2.62	-8.68	900
110	-2.43	-8.14	1,700
159	-2.54	-8.54	2,500
210	-0.68	-7.65	3,500
239	0.13	-8.40	4,050
259	-0.16	-8.30	4,150
309	-0.03	-9.06	4,550
359	0.51	-7.89	5,100
379	0.99	-6.82	5,460
410	-0.02	-7.55	5,715
460	0.54	-7.54	6,100
509	1.14	-6.50	6,680
559	0.23	-6.40	7,110
609	-0.37	-7.32	8,200
659	-0.67	-6.57	9,010
709	-3.28	-8.33	9,490
759	-1.63	-7.38	10,100
769	-1.85	-7.79	10,190
789	-0.85	-6.14	10,400

*Values of δ ^{13}C and δ ^{18}O of bulk-sediment carbonate are all expressed in the per mill difference, ‰, relative to the PDB marine carbonate standard.

CO_2. The instrument was calibrated by analysis of reference rock samples, a synthetic standard (n-$C_{20}H_{42}$), and CO_2 gas. The generated hydrocarbons are measured with a flame-ionization detector, and the CO_2 yield is monitored using a thermal conductivity detector. Free or adsorbed hydrocarbons (HC) are generated by heating the sample in flowing helium at a relatively low temperature (250°C) for 5 min, and the yield is recorded as the area under the first peak on a pyrogram (S_1 in mgHC/g sample). The S_1 peak is roughly proportional to the content of organic compounds that can be extracted from the sediment using organic solvents. The area under the second peak on a pyrogram is proportional to hydrocarbons generated by thermal breakdown of kerogen and, to a small degree, by cracking of resins and asphaltenes as the sample is heated from 250 to 550°C at 25°C per minute (S_2 in mgHC/g sample). CO_2 is also generated by degradation of organic matter, and is retained during the heating interval from 250 to 390°C and is analyzed as the integrated area under the third peak on a pyrogram (S_3 in mgCO$_2$/sample). Our data are expressed in terms of a hydrogen index (HI), which is the S_1 peak normalized to the organic carbon (C_{org}) content of the sample (mgHC/gC_{org}), and an oxygen index (OI), which is the S_3 peak normalized to the C_{org} content of the sample (mgCO$_2$/g C_{org}) (Tissot and Welte, 1984). The Rock-Eval pyrolysis HI and OI correlate well with atomic H/C and O/C ratios measured in the same samples by other methods (Espitalié and others, 1977; Tissot and Welte, 1984).

RESULTS AND DISCUSSION

Results of carbon and oxygen isotope analyses of bulk-sediment carbonate and carbon isotope analyses of organic carbon from sediment samples from the varved deep-basin core are listed in Table 1 and plotted versus varve yr in Figure 2 along with values of pyrolysis hydrogen and oxygen indices, percent organic carbon, organic carbon mass-accumulation rate (MAR), and total diatom flux. Results of carbon and oxygen isotope analyses of bulk-sediment carbonate (Table 2) and shells of the ostracod *Candona ohioensis* (Table 3) from the sublittoral ostracod core (Forester and others, 1987) are shown in Figure 3. An estimate of average salinity in Elk Lake derived from ostracod species-composition data (R. Forester, 1989, written commun.) is also plotted in Figure 3. The estimated time scale in Figure 3 was derived by correlation of geochemical data between the varved deep-basin core and the sublittoral core (Forester and others, 1987). To aid the reader in visualizing this correlation, percent sodium, one of the main elements used for the geochemical correlation, is plotted in both Figures 2 and 3 (see Dean and others, 1984).

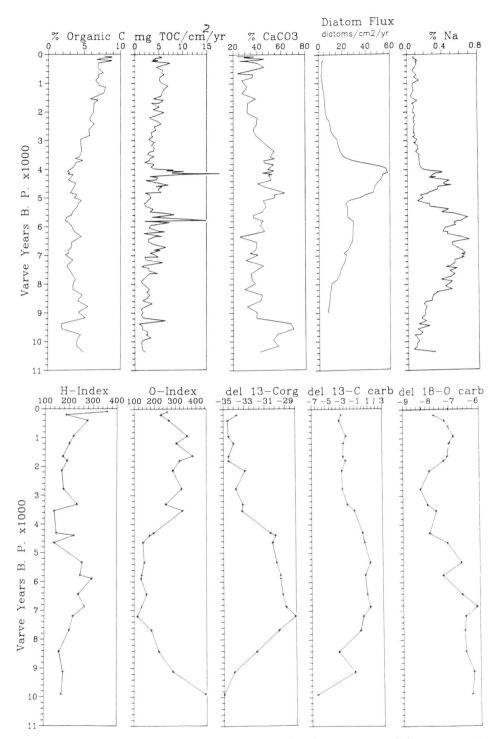

Figure 2. Distributions of percent organic carbon, organic-carbon mass-accumulation rate, percent $CaCO_3$, total diatom flux, percent sodium, pyrolysis hydrogen and oxygen indices, values of $\delta^{13}C$ for organic carbon and carbonate carbon, and values of $\delta^{18}O$ for carbonate vs. time (varve yr) in the varved deep-basin core. Percentages of $CaCO_3$, organic carbon, and sodium are from Dean (Chapter 10); diatom-flux data are from Bradbury and Dieterich-Rurup (Chapter 15); pyrolysis results and carbon and oxygen isotope data are from Table 1.

TABLE 3. ISOTOPIC RESULTS FROM *CANDONA OHIOENSIS* IN ELK LAKE SUBLITTORAL CORE*

Estimated Varve Years	$\delta^{18}O$	$\delta^{13}C$
5,750	-5.90	-4.20
5,800	-5.81	-2.92
5,900	-4.75	-3.40
6,300	-3.26	-2.74
6,450	-4.02	-2.80
6,600	-4.32	-2.20
6,800	-3.94	-4.60
6,900	-3.89	-2.47
7,000	-2.63	-3.10
7,200	-4.06	-2.54
7,650	-4.44	-2.44
7,750	-4.84	-3.47
8,700	-4.00	-3.83
10,100	-7.65	-7.83
10,250	-7.70	-5.16
10,350	-6.95	-4.66

*Values of $\delta^{13}C$ and $\delta^{18}O$ of bulk-sediment carbonate are all expressed in the per mil difference, ‰, relative to the PDB marine carbonate standard.

Oxygen isotopes

By measuring values of $\delta^{18}O$ in bulk-sediment $CaCO_3$ (marl) in both the varved deep-basin core and the sublittoral core and in calcite from ostracod shells, we had hoped to differentiate variations in oxygen isotopic composition due to temperature from those due to salinity (evaporation-precipitation balance). Marl in the deep-basin core consists mainly of $CaCO_3$ precipitated in the water column from July through September (Dean and Megard, Chapter 8; Nuhfer and others, Chapter 7). Marl in the sublittoral core probably consists of a mixture of $CaCO_3$ precipitated in the water column and on the leaves of rooted aquatic vegetation, and biogenic $CaCO_3$ derived from plants (charophytes) and animals (molluscs and ostracods). The $\delta^{18}O$ record in marl from the sublittoral core should be more difficult to interpret than that from the varved deep-basin core because of these mixed signals. Therefore, the isotopic composition of $CaCO_3$ from the deep-basin core and from ostracod shells from the sublittoral core should provide the cleanest signals of conditions in the lake during the summer months.

The species composition of the ostracod population provides information on the salinity and ionic composition of lake water (Forester and others, 1987). The average salinity of Elk Lake, as estimated from ostracod species composition in the sublittoral core (Fig. 3), together with other salinity indicators in the deep-basin core, can be used to help interpret the oxygen isotope data.

In general, variations in $\delta^{18}O$ are greatest in the sublittoral core. If all of the variation in $\delta^{18}O$ was the result of equilibrium precipitation of $CaCO_3$ at different temperatures, then a total variation of about 2‰ (e.g., from -6 between 10,000 and 7000 varve yr to -8‰ between 4000 and 2000 varve yr in the deep-basin core; Fig. 2) would imply a change in average summer temperature of about 8°C assuming a decrease of 0.24‰ for an increase of 1°C (McCrea, 1950; Stuiver, 1970) and a constant salinity. It would also imply that the early to middle Holocene in northwestern Minnesota, between about 4000 and 8000 varve yr, was colder than today.

The mid-Holocene time interval is widely recognized in Europe and northeastern United States, mainly from phytogeographical evidence, as a period that was several degrees Celsius warmer than at present, and it has been referred to as the altithermal, hypsithermal, or climatic optimum by various authors. However, the evidence for climatic warming during the mid-Holocene in the upper midwest of the United States is equivocal. Evidence obtained from modeling of pollen data (e.g., Webb and Bryson, 1972; Bartlein and others, 1984, 1986; Winkler and others, 1986) suggests that during the mid-Holocene the midwestern United States was considerably drier and as much as 2°C warmer than at present. In Minnesota, prairie vegetation expanded eastward probably by more than 100 km into forested regions between 8.5 and 4.0 ka (Wright, 1976). Ostracod and diatom data from Elk Lake (Forester and others, 1987) show that northwestern Minnesota was drier during the mid-Holocene than at present, but the ostracod populations indicate that it was cooler than at present, at least until about 6700 varve yr, with conditions similar to those found today in Canadian prairie lakes. Therefore, we prefer to use "prairie period" (Wright, 1976) to refer to the mid-Holocene in Minnesota and North and South Dakota, with the implication of conditions that were drier than at present but with no implication as to temperature.

The effect of hydrologic balance on the oxygen isotope composition of surface waters of lakes is a complex balance of the interplay of ground-water influx, amount and timing of precipitation (January snow versus August rain), and rate of evaporation. The net effect of these variables for lakes in the Great Lakes region is illustrated by the data of Stuiver (1970; Fig. 4), which show that the surface waters of lakes in July at lat 42°N are 9‰ heavier than those at lat 47°N. The considerable scatter in the data probably reflects local effects, but it is clear from the empirical relation that the overall net effect would be that, on average, the surface waters of a lake at lat 42°N (and, therefore any $CaCO_3$ formed from these waters) would be more enriched in ^{18}O (less negative values of $\delta^{18}O$) by about 9‰ relative to those of a lake at lat 47°N. Stuiver (1970) concluded that this difference is due mainly to differences in rate of evaporation. In other words, as you go north in the Great Lakes region, the surface waters of lakes in July evaporate less and the oxygen in those lakes gets isotopically lighter. Stuiver's data suggest that this evaporation effect would decrease the $^{18}O:^{16}O$ ratio 1.3‰ per degree Celsius from south to north. The ratio would aso *increase* 0.25‰ per degree Celsius going from south to north due to the direct temperature effect on equilibrium isotopic fractionation during calcite precipitation, or a net change of 1.05‰ per degree Celsius.

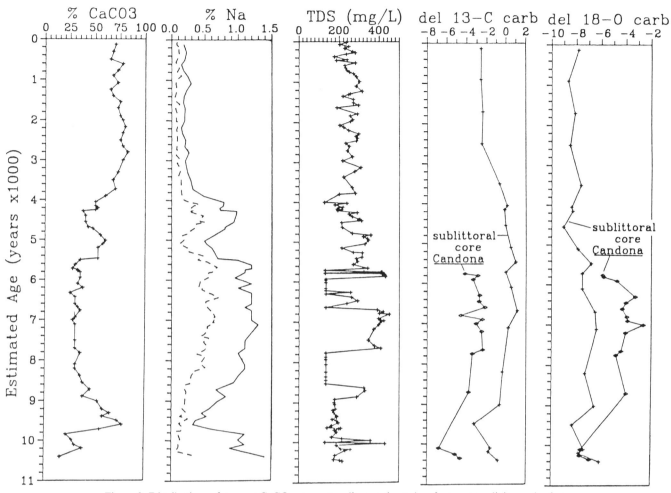

Figure 3. Distributions of percent $CaCO_3$, percent sodium, estimated surface-water salinity, and values of $\delta^{13}C$ and $\delta^{18}O$ for carbonate from marl and the ostracod *Candona ohioensis* vs. time (estimated varve yr) in the sublittoral (10 m) core from Elk Lake. Percentages of $CaCO_3$ and sodium are from Dean (Chapter 10). Estimates of surface-water salinity are based on ostracod population studies (R. M. Forester, 1989, written commun.). Carbon and oxygen isotope data for marl are from Table 2, and those for *Candona ohioensis* are from Table 3. Estimated time in varve yr is based on correlations of geochemical profiles, mostly using concentration of sodium, between the deep-basin core (dashed line) and the littoral core (Dean and others, 1984; Forester and others, 1987).

In the prairie regions of Minnesota and North and South Dakota the climatic gradients are very different than in the Great Lakes region. The salinity gradient of surface waters of lakes is more longitudinal than latitudinal in response to more arid air masses in the lee of the Rocky Mountains. This change in climate gradients can be seen in the bend in vegetation zones in Minnesota (Fig. 1) from east-west in southeastern Minnesota to north-south in western Minnesota. As a result, the salinity of surface waters of modern lakes increases from east to west across Minnesota and into North and South Dakota (Gorham and others, 1983). The zero net precipitation minus evaporation line passes just west of Elk Lake and corresponds roughly to the forest-prairie border (Fig. 1). Today, Elk Lake is in the group II salinity classification of Gorham and others (1983; specific conductivities of 142 to 501 μmhos, equivalent to salinities of about 85 to 300 mg/l), typical of lakes throughout central Minnesota. Ostracod assemblage data (Forester and others, 1987) suggest that the maximum salinity of Elk Lake was attained between 7700 and 6700 varve yr (Fig. 3). The average salinity during that period was about 400 mg/l, which would place Elk Lake in the group IV salinity classification of Gorham and others, typical of lakes in the prairie regions of western Minnesota and eastern North and South Dakota.

The very general characteristics of the oxygen isotope records for $CaCO_3$ in the Elk Lake cores (Figs. 2 and 3) for the past 8000 yr are similar to those presented by Stuiver (1970) for marl and/or molluscs from four lakes in Maine, New York, Indiana, and South Dakota, in that values of $\delta^{18}O$ are $1^0/_{00}-2^0/_{00}$ higher

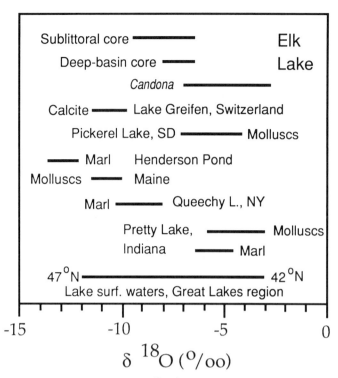

Figure 4. Ranges of values of $\delta^{18}O$ for dissolved inorganic carbon in surface waters of lakes in the Great Lakes region, marl and molluscs from Pretty Lake, Indiana, and Henderson Pond, Maine, marl from Queechy Lake, New York, molluscs from Pickerel Lake, South Dakota, sedimentary calcite from Lake Greifen, Switzerland, shells of *Candona ohioensis* and marl from the sublittoral (10 m) core, and marl from the deep-basin core of Elk Lake. Data for lake waters in the Great Lakes region, and for carbonate materials from Pretty Lake, Queechy Lake, Henderson Pond, and Pickerel Lake, are from Stuiver (1970); data for calcite from Lake Greifen are from McKenzie (1982); and data for Elk Lake are from Tables 1, 2, and 3. All values are relative to the marine carbonate standard (PDB), except those for water which are relative to standard mean ocean water (SMOW).

(^{18}O enriched) in $CaCO_3$ formed during the mid-Holocene than during the past several thousand years (Fig. 4). However, the absolute values of $\delta^{18}O$, the magnitude of the positive excursion during the mid-Holocene, and when the excursion began and ended are different for each lake, as well as for the three oxygen isotope records from Elk Lake. The extreme variations in $\delta^{18}O$ between and within lakes (Fig. 4) suggest that the main cause of the variations was not just the direct effect of temperature on the $^{18}O:^{16}O$ ratio of $CaCO_3$. Variations are greatest in the oxygen isotope records from Pickerel Lake, South Dakota (Fig. 4), and Elk Lake, which are in regions of net evaporation (Pickerel) or a slight positive water balance (Elk). Presumably both lakes had much greater negative water balances (net evaporation) during the prairie period. At least for Elk and Pickerel lakes, the effect of evaporation on change in oxygen isotope composition of lake water, perhaps as modified by the source of precipitation, is the most likely cause of within-lake variations in the oxygen isotope compositions of $CaCO_3$ in sediments deposited over the past 8000 yr. Values of $\delta^{18}O$ for molluscs from Pickerel Lake are constant at $-4^0/_{00}$ between 8000 and 4000 yr, then decrease to about $-6^0/_{00}$ at about 4,300 yr, and remain more or less constant at about $-6^0/_{00}$ for the past 4000 yr. Because Pickerel Lake is within the east-west evaporation gradient of the prairie regions of the upper midwest, it would have responded more rapidly and drastically to variations in hydrologic balance with time; the changes in ^{18}O content of molluscs may reflect these hydrologic variations.

Stuiver (1970) did not report any values of $\delta^{18}O$ in marl from Pickerel Lake, but values of $\delta^{18}O$ from marl and molluscs from two other lakes (Pretty Lake, Indiana, and Henderson Lake, Maine) show that, in general, the mollusc and marl data have the same patterns of change but that the mollsucs are about $2^0/_{00}$ more enriched in ^{18}O (Fig. 4). If marl $\delta^{18}O$ values in Pickerel Lake are about $2^0/_{00}$ lighter than mollusc values, then the Pickerel Lake $\delta^{18}O$ record would be similar in general characteristics to the $\delta^{18}O$ records for marl in both the deep-basin and sublittoral cores for the past 8000 yr, decreasing from about $-6^0/_{00}$ in prairie-period sediments to about $-8^0/_{00}$ at about 4000 varve yr (Figs. 2, 3, and 5A). Using the salinity estimates from ostracod assemblage data, a decrease of $2^0/_{00}$ in $\delta^{18}O$ corresponds to a decrease in total dissolved solids of about 200 mg/l (from about 400 to about 200 mg/l) (Fig. 3).

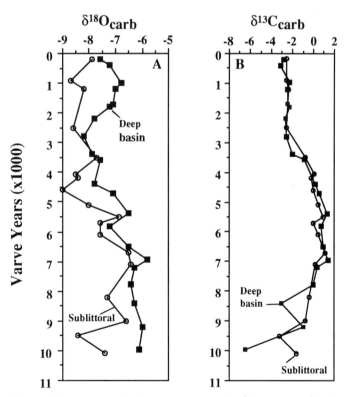

Figure 5. Comparative distributions of values of $\delta^{18}O$ (A) and $\delta^{13}C$ (B) for marl vs. time (in varve yr) for the deep-basin and sublittoral cores of Elk Lake.

Although the two Elk Lake marl records (sublittoral and deep basin) are similar in general characteristics, they are different in detail. The values of $\delta^{18}O$ are remarkably constant at about $-6.4‰$ in the deep-basin core between 10,000 and 6800 varve yr (Figs. 2 and 5A). Values of $\delta^{18}O$ in four samples from a 1 m section of unlaminated sediment below the varved sequence have a narrow range, from $-7.2‰$ to $-6.0‰$. The ages of these four samples are not known, but the results show that the constant values of $\delta^{18}O$ between 10,000 and 6800 varve yr extends back probably at least beyond 11,000 yr.

Part of the difference between the $\delta^{18}O$ records in the deep-basin and sublittoral cores may be due to the more complex mineralogy of carbonate in the deep-basin core, particularly during the first 5000 years of the lake's Holocene history (Dean, Chapter 10; Dean and Megard, Chapter 8). Some rhodochrosite ($MnCO_3$) is present throughout the deep-basin core, but its concentration is very minor and probably had little effect on isotopic composition of the carbonate. Low-magnesium calcite has always been the dominant carbonate mineral in the sediments of Elk Lake, but between 7200 and 6300 varve yr, about half of the $CaCO_3$ present in the deep-basin core was present as the polymorph aragonite (Dean, Chapter 10). A single value of $\delta^{18}O$ in this interval is slightly heavier (Fig. 2), but, because this time also corresponds to the time of highest salinity in the lake's history (Forester and others, 1987) (Fig. 3), the increase is probably due to salinity rather than difference in mineralogy. Dolomite makes up a minor but significant part of the carbonate fraction in samples of varved sediment between 10,000 and 5400 varve yr (Dean, Chapter 10). Because dolomite usually is enriched in ^{18}O relative to coexisting calcite, particularly in hypersaline environments (e.g., Anderson and Arthur, 1983; Talbot and Kelts, 1986, 1990; Talbot, 1990), the generally ^{18}O enriched values of $\delta^{18}O$ in the deep-basin core between 10,000 and 6800 varve yr may be due, at least in part, to the presence of dolomite. Between 9500 varve yr and the bottom of the varved section, the abundance of dolomite increases, but this increase is not matched by an increase in $\delta^{18}O$. Although dolomite usually is enriched in ^{13}C as well as ^{18}O (Anderson and Arthur, 1983), there is no enrichment in ^{13}C at the bottom of the deep-basin core and there is no difference in values of $\delta^{13}C$ in marl between the deep-basin and sublittoral cores. We conclude, therefore, that isotopic differences due to carbonate mineralogy are minor and that changes in isotopic composition within and between cores are due to other factors.

Values of $\delta^{18}O$ in the sublittoral core decrease from about $-7.4‰$ to $-8.5‰$ from 10,200 to 9500 varve yr, and then increase to values that average about $-6.4‰$ during the prairie period, with variations of about $1‰$ (Figs. 3 and 5A). The amount of variation in the sublittoral core is not surprising considering the difference sources of $CaCO_3$.

The mollusc $\delta^{18}O$ record from Pickerel Lake and, to a lesser degree, the marl $\delta^{18}O$ record from the sublittoral core in Elk Lake, indicate that the ^{18}O content of carbonate increased from about 10,000 to 8000 varve yr. Values of $\delta^{18}O$ in the ostracod *Candona ohioensis* from the sublittoral core also show an enrichment in ^{18}O from the early Holocene to the prairie period (Fig. 3), but there are not enough specimens of *Candona* present to clearly document this change from 10,500 to 8800 varve yr. The oxygen isotopic composition of *Candona* is always heavier than that of contemporary marl, generally by about $3‰$, similar to the difference between the oxygen isotopic composition of molluscs and marl in Pretty and Henderson lakes (Stuiver, 1970), although the difference between molluscs and marl in sediments from these two lakes is only about $2‰$.

The difference between heavier values of $\delta^{18}O$ in *Candona* relative to those of marl in the sublittoral core may simply reflect a difference in time of year of precipitation of the $CaCO_3$. The $CaCO_3$ in marl is precipitated mostly during the late summer (Dean and Megard, Chapter 8; Nuhfer and others, Chapter 7), whereas *Candona* secretes its shell during the spring and early summer (R. M. Forester, 1989, personal commun.) and, therefore, would record a generally lower temperature. The lack of a difference between $\delta^{18}O$ of marl and that of *Candona* in the sublittoral core prior to 10,000 varve yr may indicate that *Candona* only grew in late summer, at the same time that marl was precipitated.

The biggest fundamental difference among the $\delta^{18}O$ records for marl and ostracods in the sublittoral core and for marl from the deep-basin core is that the latter do *not* show an increase in $\delta^{18}O$ between at least 10,000 and 8000 varve yr (Fig. 2). If we apply the standard interpretation based on transfer functions developed for eastern to midwestern United States pollen data to the prairie regions of Minnesota and eastern North and South Dakota, then the mid-Holocene was warmer and drier beginning about 8000 varve yr (Bartlein and Whitlock, Chapter 18). If the ostracod and diatom assemblage data (Forester and others, 1987) are correct, however, then the lake was *colder* and more saline than at present until at least 6700 varve yr. Oxygen and hydrogen isotope data from cellulose in fossil wood from an esker in eastern Ontario (Edwards and Fritz, 1986) indicate that cold, dry conditions persisted there at least until 7.4 ka, when the climate became somewhat warmer but remained dry. The main hypsithermal period, with distinctly warmer and *wetter* conditions, did not begin in eastern Ontario until about 6.0 ka. It is likely, therefore, that there are conflicts between the pollen data and the isotope data of Edwards and Fritz (1986) for eastern North America and between the pollen data and ostracod data for northwestern Minnesota. Because the ostracods monitor conditions in the water and because the pollen interpretation is extended from the midwest where climatic gradients are different than in Minnesota and North and South Dakota, we tend to believe the ostracod data; that is, Elk Lake stayed cold until about 6700 varve yr. The ostracod data are supported by the oxygen isotope data from marl in the deep-basin core, which show that values of $\delta^{18}O$ vary little until about 6800 varve yr (Fig. 5A). We conclude that there could not have been very large variations in temperature or salinity, or these variations would be recorded in the oxygen isotope record. We further conclude that between the early Holocene and about 6800 varve yr, the oxygen isotope composition of carbonate in Elk Lake was influenced by both colder temperatures and higher salinity, both of which would tend to produce enrichment in ^{18}O in carbonate precipitated from the water.

It may be that the discrepancies between warmer and colder interpretations for the prairie period obtained from transfer functions of pollen data based on midwestern models versus those obtained from ostracod data are due to the difference between the latitudinal water-balance gradient in the Great Lakes region and the longitudinal gradient in Minnesota and North and South Dakota. This difference in water balance, together with the interplay of three different air masses (Pacific, Arctic, and Atlantic-Gulf) (Bartlein and others, 1984; Anderson and others, Chapter 1), tends to make the region of northwestern Minnesota and eastern North and South Dakota a climatic "triple junction." The extreme variability in temperature and water balance is obvious even in modern weather conditions in this region (see discussion of weather data from Fargo, North Dakota, by Megard and others, Chapter 3). Changes in the relative interplay of these air masses with time undoubtedly had an effect on the timing and isotopic composition of precipitation that reached the Elk Lake drainage basin.

As discussed earlier, it is possible, but unlikely, that the ^{18}O-enriched marl values in the deep-basin core between 10,000 and 5400 varve yr may be due to the presence of a larger amount of dolomite. All of the variables—temperature, salinity, and carbonate mineralogy—apparently acted to different degrees in different carbonate materials to produce the observed differences between the three oxygen isotope records in Elk Lake. Unfortunately, the isotopic record for *Candona* is incomplete, and interpretation of the deep-basin marl may be complicated by variations in carbonate mineralogy. It is probably significant that the values of $\delta^{18}O$ in marl and *Candona* from the sublittoral core converge in samples older than 10,000 varve yr (Fig. 3). At that time, the $CaCO_3$ in the two records was apparently recording the same environmental conditions, probably at the same time of the year.

The critical date appears to be about 6700 varve yr. After this date the ostracod species-composition data indicate that Elk Lake became warmer and less saline (Forester and others, 1987). In response to these conditions, carbonate in all three $\delta^{18}O$ records became depleted in ^{18}O by about $2^0/_{00}$. Values of $\delta^{18}O$ in marl from the sublittoral core reached a minimum of $-9^0/_{00}$ at about 4.5 varve yr, and then varied between $-9^0/_{00}$ and $-8^0/_{00}$ from 4.5 varve yr to the present. Values of $\delta^{18}O$ in marl from the deep-basin core reached a minimum of $-8.2^0/_{00}$ at about 2800 varve yr, then increased steadily to about $-6.8^0/_{00}$ at 900 varve yr, and then declined again from 900 varve yr to the present (Fig. 5A). If the increase in $\delta^{18}O$ between 2800 and 900 varve yr is due entirely to temperature, it would indicate a decrease of a little less than 6°C. The pollen data (Whitlock and others, Chapter 17) indicate that *Pinus strobus* began replacing the mixed deciduous forest in the Elk Lake region about 3000 varve yr, probably in response to cooler and moister climates (Jacobsen, 1979). Changes in diatom assemblages at this time also suggest cooler conditions (Bradbury and Dieterich-Rurup, Chapter 15). The combination of pollen, diatom, and oxygen isotope data, all of which indicate cooler conditions after 2.8 varve yr, may be recording the neoglacial period of expansion of arctic and alpine glaciers that culminated in the "little ice age" between A.D. 1450 and 1850 (Denton and Karlén, 1973).

In summary, the oxygen isotope compositions of marl in the deep-basin and sublittoral cores and of the ostracod *Candona ohioensis* probably were influenced by both colder temperature and higher salinity in Elk Lake, relative to the present, between 10,000 and 6800 varve yr. Values of $\delta^{18}O$ in marl from the two cores are generally similar, but values in *Candona* are about $3^0/_{00}$ more enriched in ^{18}O. All three oxygen isotope records show a decrease in values of $\delta^{18}O$ after about 6800 varve yr in response to an increase in temperature and decrease in salinity. Values of $\delta^{18}O$ in marl from the deep-basin core increase steadily between 2800 and 900 varve yr, possibly in response to a decrease in temperature coincident with the neoglacial period.

Carbon isotopes

Carbonate carbon. The isotopic composition of carbon in carbonate minerals precipitated from lake water is mainly a function of the isotopic composition of dissolved inorganic carbon (DIC), which, in turn, is mainly controlled by temperature and partial pressure of dissolved CO_2 (pCO_2). The effect of temperature is not a direct isotope effect but mainly the effect of temperature on pCO_2—the higher the temperature, the lower the pCO_2. During the summer growing season in a lake, the temperature effect on pCO_2 generally is enhanced by photosynthetic removal of CO_2 and by evasion to the atmosphere of dissolved CO_2 that is isotopically lighter than dissolved HCO_3^- (Mook and others, 1974; Herczeg and Fairbanks, 1987). The net effect is that DIC is enriched in ^{13}C in the surface waters of a lake. Conversely, DIC in the bottom waters of a lake is depleted in ^{13}C due to respiration of ^{13}C-depleted CO_2. For example, McKenzie (1982) found that values of $\delta^{13}C$ of DIC in surface waters of Lake Greifen, Switzerland, increased from $-11^0/_{00}$ in winter to $-7^0/_{00}$ in summer, but the DIC in bottom waters decreased from $-11^0/_{00}$ to $-13^0/_{00}$ during the same period (Fig. 6). Values for sediment calcite were even more enriched in ^{13}C (by $-7.1^0/_{00}$ during the past 200 yr). Lee and others (1987) reported that values of $\delta^{13}C$ in DIC in the surface waters of Lake Greifen increased by $1.8^0/_{00}$ in one day in June during a sediment-trap experiment. Quay and others (1986) measured an increase in $\delta^{13}C$ of DIC in the epilimnion in Lake Washington, from $-8.7^0/_{00}$ in January to $-4.5^0/_{00}$ in July, when the pCO_2 of the epilimnion was almost undetectable. This was accompanied by depletion of ^{13}C in the hypolimnion that reached a minimum of $-11^0/_{00}$ in October. Herczeg and Fairbanks (1987) measured an increase of about $7^0/_{00}$ in $\delta^{13}C$ of DIC between April and June in a soft-water lake in New York.

The oxygen isotope records for marl from the deep-basin and sublittoral cores show several distinct differences (Fig. 5A). In contrast, the $\delta^{13}C$ records for marl in these two cores are

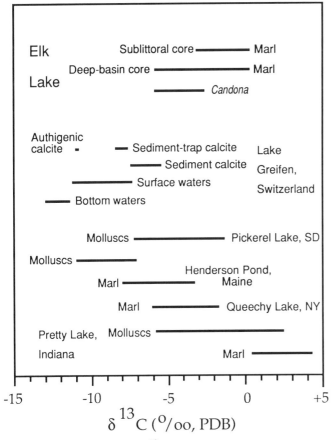

Figure 6. Ranges of values of $\delta^{13}C$ for marl and molluscs from Pretty Lake, Indiana, and Henderson Pond, Maine; marl from Queechy Lake, New York; molluscs from Pickerel Lake, South Dakota; surface and bottom waters, sediment calcite, sediment-trap calcite, and authigenic calcite from Lake Greifen, Switzerland; and shells of *Candona ohioensis* and marl from the sublittoral core, and marl from the deep-basin core of Elk Lake. Data for carbonate materials from Pretty Lake, Queechy Lake, Henderson Pond, and Pickerel Lake are from Stuiver (1970); data for calcite and waters from Lake Greifen are from McKenzie (1982) and Lee and others (1987); and data for Elk Lake are from Tables 1, 2, and 3. All values are relative to the marine carbonate standard (PDB), except those for water which are relative to standard mean ocean water (SMOW).

almost identical (Fig. 5B), except for a few differences at the base of the core. The limited $\delta^{13}C$ data for the ostracod *Candona ohioensis* also show the same trends from early to middle Holocene, but values of $\delta^{13}C$ are consistently about $3‰$ more depleted in ^{13}C (Fig. 3). We suggested that the ^{18}O-enriched characteristics of *Candona ohioensis* relative to marl in the sublittoral core may be due to secretion of shell material over a longer period of the year. This would also explain the ^{13}C-depleted characteristics of ostracod carbonate, which would incorporate isotopically lighter spring and early summer DIC, rather than just the heavier late summer DIC that controls the isotopic composition of most of the marl.

Using the deep-basin core as the standard, values of $\delta^{13}C$ increased by about $8‰$ between 10,000 and 6800 varve yr during the time when the oxygen isotope composition remained essentially unchanged and enriched in ^{18}O in response to cold, saline lake conditions. Carbonate deposited during most of the prairie period remained enriched in ^{13}C until about 5500 varve yr, when values of $\delta^{13}C$ began to steadily decrease; the same rate of decrease is in marl from both cores. The decrease in values of $\delta^{13}C$ at 5500 varve yr corresponds to marked changes in a number of parameters in Elk Lake that indicate the beginning of the shift from warm-dry conditions between 6800 and 5500 varve yr to cool-wet conditions of the past 3000 yr (Dean and others, 1984; Bradbury and others, Chapter 20). It is interesting that neither the oxygen nor carbon isotope ratios record the minor ca. 600 year return to prairie-lake conditions between about 4700 and 4100 varve yr (e.g., see profile for Na in Fig. 2). Values of $\delta^{13}C$ have remained very constant at $-2.5‰$ to $-3‰$ from 3000 varve yr to the present, a decrease of about $4‰$ from maximum values during the prairie period.

The carbon isotope records in carbonate material from Elk Lake (Figs. 2, 3, and 5B) all show a distinct, unequivocal pattern; values of $\delta^{13}C$ increased gradually going into the prairie period, then decreased gradually coming out of the prairie period. Values of $\delta^{13}C$ in molluscs from Pickerel Lake, South Dakota, also increased from $-7‰$ to $-1‰$ going into the prairie period, but remained enriched throughout the middle to late Holocene (Stuiver, 1970). The gradual changes in values of $\delta^{13}C$ for marl in both the sublittoral and deep-basin cores and the fact that both cores responded so similarly suggest that the carbon isotope system in Elk Lake was well buffered against the high-frequency climatic cycles that characterize many other proxy variables in the Elk Lake cores (see Dean and others, 1984; discussion by Bradbury and others, Chapter 20).

These gradual changes in $\delta^{13}C$ may be related to the general long-term temperature changes from a cold postglacial (10,000 to 6800 varve yr), to a warm mid-Holocene (6800 to 5500 varve yr), and finally to cooler conditions during the past 3000 varve yr. These changes in carbon isotopic composition as a function of temperature would not be so much a direct isotope effect, but rather an effect of temperature on the concentration of carbon dioxide in the water (higher pCO_2 in colder water), as has been argued by Stuiver (1975) and Håkansson (1985). Higher pCO_2 would cause depletion of both ^{18}O and ^{13}C of carbonate material formed in the lake, and there should be a positive correlation between values of $\delta^{18}O$ and $\delta^{13}C$. Talbot (1990) and Talbot and Kelts (1990), however, argued that primary carbonate minerals precipitated from hydrologically open lakes should show little or no linear relation between oxygen and carbon isotopic variations, because variations in oxygen isotopic composition are due mainly to changes in water balance (i.e., the precipitation:evaporation ratio), whereas carbon isotopic variations are due mainly to outgassing of CO_2 from the lake surface. The observed values of $\delta^{18}O$ and $\delta^{13}C$ in marl from both the varved and sublittoral cores are what we would expect in modern primary carbonate

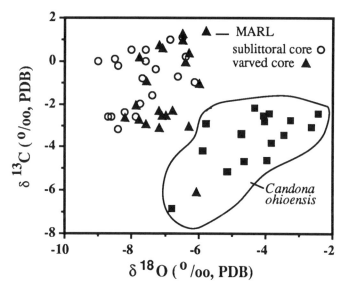

Figure 7. Plot of values of $\delta^{18}O$ vs. $\delta^{13}C$ for marl from the sublittoral and deep-basin cores and for samples of *Candona ohioensis* from the sublittoral core.

minerals precipitated from surface waters of open lakes (Talbot and Kelts, 1990) with no covariant trend (Fig. 7). In contrast, there is a good covariant trend between values of $\delta^{18}O$ and $\delta^{13}C$ for samples of *Candona* ($r = 0.64$, $n = 15$) (Fig. 7). Talbot (1990) suggested that covariance between values of $\delta^{18}O$ and $\delta^{13}C$ is characteristic of primary carbonate material that formed in hydrologically closed lakes; however, Elk Lake is clearly hydrologically open, and the carbon isotopic composition of the calcite in *Candona* probably is responding mainly to temperature through the effect on pCO_2.

The gradual changes in $\delta^{13}C$ of carbonate in *Candona* and in marl from both cores also might be related to organic productivity through removal of ^{13}C-depleted organic carbon. According to this scenario, photosynthesis preferentially removes ^{13}C-depleted carbon from the surface waters and fixes this isotopically light carbon in phytoplankton biomass. If all of the carbon fixed by photosynthesis in the surface waters was released by respiration and decay in bottom waters, there would be no net change in the isotopic composition of the carbon reservoir of the lake. However, some fraction of this ^{13}C-depleted organic carbon is buried in the sediments, and, as a first approximation, we assume that the rate of organic-carbon burial is proportional to the rate of surface-water organic productivity. During periods of high productivity, the surface-water carbon reservoir should become highly enriched in ^{13}C, and this change in isotopic composition should be reflected in the isotopic composition of calcite formed in the surface waters.

This model was used by Scholle and Arthur (1980) to explain the correlation between periods of rapid carbon burial throughout the Phanerozoic (so-called 'oceanic anoxic events') and positive excursions in the secular curve of ^{13}C in marine carbonate material. McKenzie (1982) suggested that this model might apply to lakes and pointed out that values of $\delta^{13}C$ in carbonate precipitated in Lake Greifen, Switzerland, increased by about $2^0/_{00}$ with an increase in productivity that accompanied cultural eutrophication. To explore this possibility further, we need to examine the organic side of the carbon cycle in Elk Lake and any indications for changes in productivity.

Organic carbon. The two main factors that control the isotopic composition of organic carbon in lakes are temperature, mostly through the control of temperature on pCO_2, and organic-matter source (e.g., Stuiver, 1975; Håkansson, 1985). Fractionation of carbon isotopes due to pCO_2 can be considerable if the range in pCO_2 is large (see review by Buchardt and Fritz, 1980). If the pCO_2 of surface water is much greater than the pCO_2 of the atmosphere, organic carbon produced from this CO_2 can be depleted in ^{13}C by as much as $20^0/_{00}$ (e.g., Rau, 1978; Rau and others, 1989). However, if the pCO_2 of surface water is less than that of the atmosphere, as might occur during an intense algal bloom, then the organic carbon fixed may be only depleted by a few per mil (e.g., Deuser and others, 1968; Calder and Parker, 1973; Herczeg and Fairbanks, 1987). For example, Herczeg and Fairbanks (1987) found that the difference between $\delta^{13}C$ of particulate organic carbon (POC) and that of dissolved CO_2 in a soft-water lake in New York was $-19^0/_{00}$ in June, but that during an intense blue-green algal bloom in July, the POC was actually more enriched in ^{13}C than the dissolved CO_2, such that the difference was $+0.4^0/_{00}$. The lake studied by Herczeg and Fairbanks (1987) is a soft-water lake wherein the main source of CO_2 is from the atmosphere (e.g., Emerson and others, 1973), but in a hard-water lake like Elk Lake the partial pressure of dissolved CO_2 is much greater than that of atmospheric CO_2 because of the large bicarbonate reservoir, and there is a net loss of CO_2 to the atmosphere throughout the year (e.g., Otsuki and Wetzel, 1974; Stuiver, 1975). Consequently, fractionation between dissolved CO_2 and organic carbon is more constant. For two hard-water lakes in New England, Oana and Deevey (1960) found that the fractionation between plankton and surface-water bicarbonate was $27.3^0/_{00}$ and $28.8^0/_{00}$.

Comparison of the plots of $\delta^{13}C$ versus time for both organic carbon and carbonate carbon in the same samples from the deep-basin core (Fig. 2) shows that the two records correlate very well (correlation coefficient, $r = 0.83$, $n = 21$). If we assume that values of $\delta^{13}C$ for marl are representative of values of $\delta^{13}C$ in dissolved inorganic carbon (DIC) in late summer in Elk Lake, then the difference between $\delta^{13}C$ values of organic carbon and those of carbonate carbon ($\Delta^{13}C$) gives an estimate of the change in fractionation between organic carbon and the inorganic reservoir. Changes in $\Delta^{13}C$ with time (Fig. 8A) show that this fractionation increased gradually by about $3^0/_{00}$ (from $-29^0/_{00}$ to $-32^0/_{00}$) between 8500 varve yr and the top of the deep-basin core. Alternatively, if we assume that the organic-inorganic carbon fractionation remained about constant, we can calculate a value of $\delta^{13}C_{org}$ from $\delta^{13}C_{carb}$ (Fig. 8B). The difference between the measured and calculated values of $\delta^{13}C_{org}$ (shaded area in

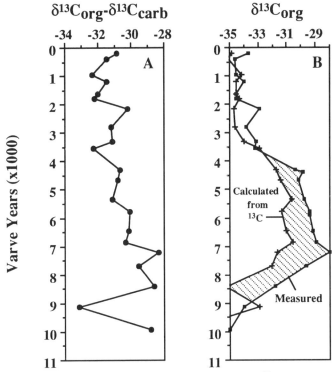

Figure 8. A: Plot of the difference between values of $\delta^{13}C$ for organic carbon and those for carbonate carbon ($\Delta^{13}C$) vs. varve yr in the deep-basin core. B: Plot of measured values of $\delta^{13}C_{org}$ and values of $\delta^{13}C_{org}$ calculated from $\delta^{13}C_{carb}$, assuming a constant fractionation of $32^o/_{oo}$ between organic and inorganic carbon. The difference between measured and calculated values of $\delta^{13}C_{org}$ (shaded area) represents ^{13}C enrichment due to CO_2 limitation.

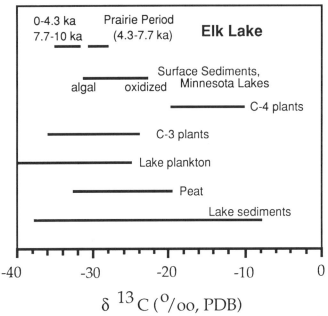

Figure 9. Ranges of values of $\delta^{13}C$ for organic carbon from lake sediments, peat, lake plankton, C-3 plants, C-4 plants, surface sediments of Minnesota lakes (ranging from algal rich to oxidized), and the deep-basin core of Elk Lake. Data for lake sediments, peat, lake plankton, C-3 plants, and C-4 plants were compiled by Bender (1971) and Deines (1980); data for surface sediments of Minnesota lakes are from Table 4; and data for Elk Lake are from Table 1. All values are relative to the marine carbonate standard (PDB).

Fig. 8B) may be taken as a measure of enrichment of ^{13}C due to CO_2 limitation. In other words, of the $6^o/_{oo}$ positive excursion in $\delta^{13}C_{org}$ during the prairie period, about $4^o/_{oo}$ is due to the reservoir DIC ^{13}C enrichment and about $2^o/_{oo}$ is due to ^{13}C enrichment resulting from drawdown of dissolved CO_2 and a greater utilization of HCO_3^- by some phytoplankton.

The sources of organic matter to lake sediments include planktonic algal debris, rooted aquatic macrophytes, and terrestrial vegetation washed in from the drainage basin. Although there are lakes in which the sedimentary organic matter is mainly allochthonous, the organic matter in the sediments of most temperate lakes is autochthonous and most of that is from phytoplankton production. The range in values of $\delta^{13}C$ of organic carbon in lake sediments is quite large (Fig. 9), but the range of values for most temperate lake sediments is much narrower (e.g., Stuiver, 1975). The available data for $\delta^{13}C$ values in the organic component of lake sediments suggest that there is a latitude effect and that organic matter in lakes above lat 30°N have lighter values of $\delta^{13}C$ (Stuiver, 1975; Buchardt and Fritz, 1980). The observed effect of latitude on isotopic composition of organic matter in lakes is empirical, but may reflect the combined effects of temperature and rate of supply of CO_2 (Håkansson, 1985).

The range of values of $\delta^{13}C_{org}$ in surface sediments of many but not all of the lakes for which Gorham and Sanger (1975) measured concentrations of plant pigments is shown in Figure 9 along with ranges of $\delta^{13}C$ values for peat, lake plankton, C-3 plants (Calvin-Benson photosynthetic pathway), and C-4 plants (Hatch-Slack photosynthetic pathway). Gorham and Sanger (1975) concluded from their plant pigment studies that most organic matter in the sediments of Minnesota lakes was autochthonous. As an independent measure of terrestrial versus algal origin of organic matter in Minnesota lake sediments, we use the pyrolysis hydrogen index (HI). A plot of HI versus total pigments (data of Gorham and Sanger, 1975) for surface sediments of 18 Minnesota lakes (Fig. 10) shows that, in general, those lakes having the highest total-pigment concentrations (most algal rich as interpreted by Gorham and Sanger [1975]) also contain the most hydrogen-rich organic matter. In addition, those Minnesota surface sediments that contain the most hydrogen-rich (algal) organic matter generally have the most ^{13}C-depleted organic matter (Fig. 11A); i.e., the organic matter is isotopically more like lake plankton and less like peat (Fig. 9). The Minnesota lakes that have surface sediments with the most ^{13}C-depleted organic carbon (lightest values of $\delta^{13}C_{org}$; Table 4; Figs. 9 and 11A), the most hydrogen-rich organic matter, and the highest concentrations of pigments (Table 4; Fig. 10) are Lakes Itasca, Arco, and Josephine, which are close to Elk Lake in Itasca State Park; two

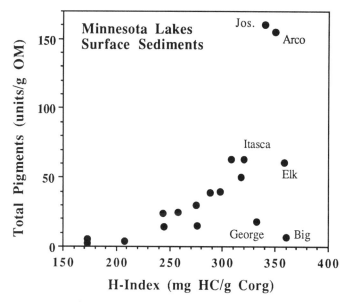

Figure 10. Plot of pyrolysis hydrogen index (HI) vs. total pigments (in spectrophotometric units per gram of organic matter [OM] as measured by loss on ignition at 550 °C; Gorham and Sanger [1975]) in surface sediments of 18 Minnesota lakes (see Table 4).

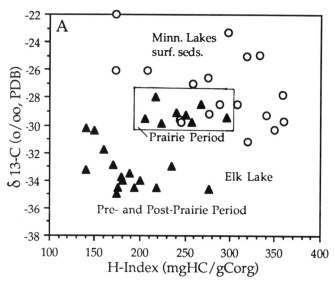

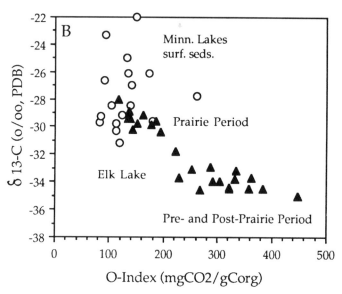

Figure 11. Plots of pyrolysis hydrogen index (A) and oxygen index (B) vs. $\delta^{13}C$ for organic matter in surface sediments from 18 Minnesota lakes (Table 4) (open circles) and from the deep-basin core from Elk Lake (Table 1) (solid triangles).

of these lakes (Arco and Josephine) are meromictic and their sediments consist almost entirely of algal debris.

From the above examples and discussion, it would appear that lake sediments containing the most ^{13}C-depleted organic carbon and the most hydrogen-rich organic matter would be characteristic of lakes having high algal productivity. A plot of pyrolysis HI versus $\delta^{13}C$ can be a useful method of illustrating a source trend of organic matter in sediments. We interpret the trend for surface sediments in Minnesota lakes (Fig. 11A) as representing a range of sources of organic matter, from more terrestrial components (hydrogen poor and ^{13}C enriched) to more algal components (hydrogen rich and ^{13}C depleted). The sediments in Elk Lake deposited before and after the prairie period show a similar trend (Fig. 11A), although the sediments in Elk Lake are more ^{13}C depleted than even the most algal-rich sediments of other lakes. More important, however, is that the sediments of Elk Lake deposited during the prairie period are among the most hydrogen rich, but plot well above the trend and clearly are anomalous (samples within box in Fig. 11A).

Several lines of evidence suggest that Elk Lake was more productive during the prairie period, and had characteristics similar to lakes of the prairie regions of Minnesota today. The plant pigment data of Sanger and Hay (Chapter 13) suggest that, although Elk Lake was not particularly eutrophic during the prairie period, it was more mesotrophic than it is today. There is, however, an abrupt increase in average concentration of pigments between 9200 and 8200 varve yr that suggests an increase in productivity. Considerable variation in pigment characteristics during most of the prairie period lead Sanger and Hay (Chapter 13) to conclude that there were fluctuating algal blooms in response to considerable environmental stress.

Most of the primary productivity in Elk Lake today is from diatoms, and presumably this was the case in the past (Bradbury and Dieterich-Rurup, Chapter 15). The total flux of diatoms was substantially higher during the prairie period (Fig. 2), and the diatom population was dominated by *Stephanodiscus minutulus,* which requires abundant phosphorus (Bradbury and Dieterich-Rurup, Chapter 15). There is an excellent correlation between the concentration of phosphorus and organic carbon in the varved

TABLE 4. ORGANIC CARBON, PYROLYSIS, PIGMENT, AND
ISOTOPIC RESULTS OF SURFACE SEDIMENTS
FROM MINNESOTA LAKES*

Lake	% Corg	H-Index	O-Index	Pigments	$\delta^{13}C_{org}$
Big	16.4	359	82	7.0	-29.7
Iron	18.5	245	113	13.8	-29.8
Mountain	7.2	173	173	2.3	-26.1
Clearwater	5.5	207	134	4.3	-26.1
Wilson	7.1	276	123	14.9	-29.1
Kimball	19.7	275	91	30.0	-26.6
Specitcle	17.7	298	93	40.1	-23.4
Salt	2.7	173	149	5.0	-22.0
Green	9.3	288	103	39.1	-28.6
Itasca	8.7	320	118	63.1	-31.3
Crane	11.9	244	178	23.9	-29.6
Josephine	22.2	340	84	160.0	-29.4
Mina	6.4	308	139	63.0	-28.6
Arco	15.8	350	112	155.0	-30.3
Linwood	15.3	258	140	25.5	-27.0
Lotus	7.0	318	131	50.1	-25.1
George	17.3	333	151	17.7	-32.2
Elk	10.2	358	260	60.8	-27.8

*Rock-Eval pyrolysis hydrogen index (H-Index) is in milligrams of hydrocarbons per gram of organic carbon, and oxygen index (O-Index) is in milligrams of carbon dioxide per gram of organic carbon. Concentration of total pigments is in units per gram organic matter. Values of $\delta^{13}C_{org}$ are expressed as the per mil difference, ‰, relative to the PDB marine carbonate standard.

sediments of the deep-basin core (Dean, Chapter 10). There is also an excellent correlation between phosphorus and iron, indicating that most of the phosphorus is incorporated in an iron phosphate mineral, tentatively identified as rockbridgeite (Nuhfer and others, Chapter 7; Dean, Chapter 10). During the past 3.0 ka, a large amount of phosphorus has been locked up in the sediments as iron phosphate, and apparently this has retarded eutrophication of Elk Lake (see discussion by Bradbury and others, Chapter 20). During the prairie period, however, there was more phosphorus available for diatom production and increased wind-driven turbulence to keep the diatoms within the euphotic zone (Bradbury and Dieterich-Rurup, Chapter 15).

One commonly used measure of change in organic productivity in a lake is the organic-carbon content of the sediment; however, because changes in organic-carbon content also can be caused by dilution with other sediment components, a better measure of the flux of organic carbon is the mass accumulation rate (MAR) in milligrams or grams of organic carbon per unit of bottom area per unit of time. The percentage of organic carbon did not increase in sediments of Elk Lake during the prairie period (Fig. 2), presumably because of dilution with the increased flux of detrital clastic material (Dean, Chapter 10), but overall the organic-carbon MAR (expressed in mg C/cm^2/yr; Dean, Chapter 10) more than doubled (Fig. 2), and there were several pulses when the flux of organic carbon was more than five times the average flux prior to the prairie period. The organic-carbon MAR decreased markedly after 4000 varve yr and then began a progressive increase reaching levels during the past 2000 years that were as high as during the latter part of the prairie period (Fig. 2).

Several other geochemical parameters indicate that organic productivity in Elk Lake was higher during the prairie period. The pyrolysis hydrogen index (HI) of organic matter that was deposited in Elk Lake during the prairie period is generally higher, particularly between 5500 and 7000 varve yr, and the oxygen index (OI) is distinctly lower (Fig. 2). The lower values of HI between 4500 and 3500 varve yr may be due to a decrease in preservation of organic matter during a time of high wind stress, possibly lower lake level, and influx of detrital clastic material (see Bradbury and others, Chapter 20), perhaps under conditions similar to but not as extreme as those described by Talbot and Livingstone (1989) for several lakes in east Africa. The very marked increase in the OI during the prairie period suggests either that there was a change in source of organic matter during the prairie period or that there was a change in the conditions of oxidation of organic matter. Because the prairie-period sediments plot above the source trend on a plot of HI versus $\delta^{13}C$ (Fig. 11A), a change in the source of organic matter is not likely. Whatever caused a decrease in the degree of oxidation of the organic matter (decrease in OI) also resulted in a marked increase in the ^{13}C content of that organic matter because there is a distinct inverse relation between OI and values of $\delta^{13}C$ of organic carbon in Elk Lake sediments (Fig. 11B). We suggest that this correlation is the result of increased productivity and the burial of well-preserved, hydrogen-rich, oxygen-poor, ^{13}C-enriched organic matter.

Studies by McKenzie and colleagues at the Swiss Federal Institute of Technology (McKenzie, 1982, 1985; Lee and others, 1987; Hollander and others, 1988) clearly demonstrate the relation between changes in phytoplankton productivity (eutrophication) in Lake Greifen, northeastern Switzerland, and the carbon isotopic composition of organic carbon and precipitated calcite in sediments of the lake. With increased eutrophication of Lake Greifen there was an increase in sedimentary organic-carbon content, an increase in $\delta^{13}C$ values in both calcite and organic carbon, an increase in pyrolysis hydrogen index, and a decrease in oxygen index. McKenzie and her colleagues recognized that this relation has important implications for interpreting the carbon isotopic composition of ancient marine carbonates and suggested that lakes are excellent natural beakers for studying geochemical processes (McKenzie, 1985).

As mentioned earlier, periods of ^{13}C-enriched marine carbonate throughout the Phanerozoic have been correlated with periods of rapid burial of organic carbon (Scholle and Arthur, 1980; Arthur and others, 1985). During periods of unusually high marine organic productivity, extremely rapid burial of ^{13}C-depleted organic carbon also can affect the isotopic composition of *organic* carbon fixed by photosynthesis on a global scale, as has been demonstrated for the so-called oceanic anoxic event (OAE) at the Late Cretaceous Cenomanian-Turonian boundary (Pratt and Threlkeld, 1984; Arthur and others, 1988). If such an organic-carbon–burial event can alter the isotopic composition of the surface-water carbon reservoir of the world ocean, surely high productivity in a lake should alter the isotopic composition of the much smaller surface-water reservoir of a lake.

CONCLUSIONS

Variations in oxygen-isotope composition of marl from the profundal and sublittoral zones of Elk Lake and of ostracods from the sublittoral zone share some general characteristics, particularly the ^{18}O-depleted character of carbonate during the mid-Holocene, but differ in details. Variations in temperature, salinity, and carbonate mineralogy apparently acted to different degrees in different carbonate materials to produce the observed differences between the three oxygen isotope records. The isotope record from the varved deep-basin core shows that values of δ^{18}O were remarkably constant at $-6.0‰$ to $-6.5‰$ between 10,000 and 6800 varve yr, then decreased to a minimum of $-8‰$ at 4500 varve yr. These constant, ^{18}O-enriched values of δ^{18}O during the early to middle Holocene indicate that Elk Lake was colder or more saline than today; we conclude that it was both. There could not have been very large changes in temperature or salinity during the interval from 10,000 to 6800 varve yr, or these changes would be recorded in the oxygen isotope record. The oxygen isotope results support the conclusions of ostracod and diatom population studies that Elk Lake was colder and more saline than today until about 6800 varve yr, after which it became warmer and less saline.

An abrupt decrease in values of δ^{18}O in marl from both the deep-basin and sublittoral cores between 5500 and 4500 varve yr shows that there was a fundamental change in climate at this time, from warmer-drier to cooler-wetter, probably in response to a major shift in air masses. A distinct increase in ^{18}O between 2800 and 900 varve yr probably is a response to a decrease in temperature coincident with the neoglacial period.

The climatic record from Elk Lake does not entirely fit the hypsithermal model of the Holocene interpreted from pollen records of the Great Lakes region, which suggests that the climate became warmer and drier between 10.0 and about 8.0 ka and then became cooler and wetter again about 4.0 ka. The prairie regions of western Minnesota and North and South Dakota expanded greatly during the mid-Holocene, but ostracod assemblage data and oxygen isotope data suggest that this expanded prairie was cold until about 6800 varve yr, with conditions more like those found today in prairie regions of western Canada. The difference in climatic response between the northeastern Great Plains and the Great Lakes region is that the Great Lakes region has a north-south precipitation-evaporation gradient, whereas the northeastern Great Plains has an east-west gradient. In addition, the northeastern Great Plains is under the influence of three different air masses that produce tremendous climatic variability.

Carbon isotope records from carbonate-carbon in marl from the profundal and sublittoral cores are virtually identical and show a gradual increase in values of δ^{13}C of about 8‰ between 10,000 and 6800 varve yr, then a gradual decrease of about 4‰ from about 5500 to 3000 varve yr. The decrease in δ^{13}C at 5500 varve yr corresponds to marked changes in a number of parameters in Elk Lake that signal the beginning of the shift from warm-dry conditions that existed at least between 6800 and 5500 varve yr and the cool-wet conditions that characterize the last 3000 varve yr. The gradual changes into and out of the prairie period in values of δ^{13}C for marl in both the sublittoral and deep-basin cores and the fact that both cores responded so similarly suggest that the carbon isotope system in Elk Lake was well buffered against the high-frequency climatic cycles that characterize many other proxy variables in the Elk Lake cores. Lack of a change in oxygen isotope composition between 10,000 and 6800 varve yr argues against the carbon isotope change being due to variations in one or more environmental conditions such as temperature. We conclude that the ^{13}C-enriched surface-water inorganic-carbon reservoir during the mid-Holocene was caused by increased burial of ^{13}C-depleted organic carbon.

The parallel records for δ^{13}C in organic carbon and in carbonate carbon in the profundal core indicate that the total-carbon reservoir, not just the inorganic-carbon reservoir, was enriched in ^{13}C during the mid-Holocene prairie period in Elk Lake. The positive carbon isotope excursion in the prairie period is larger for organic carbon than for carbonate carbon (6‰ in contrast to 4‰ for carbonate carbon), and the 2‰ difference may represent ^{13}C enrichment due to drawdown of dissolved CO_2 and a greater utilization of HCO_3^- by some phytoplankton. The excess organic-carbon burial required to produce such a large increase in values of δ^{13}C in both inorganic and organic carbon must have required a large increase in primary productivity. There is not an increase in the percentage of organic carbon in the profundal sediments during the prairie period, presumably due to dilution with the increased flux of detrital clastic material at the same time. The flux of organic carbon (in mg/cm^2/yr), however, more than doubled during the prairie period, and there were several intervals of very large increases in organic-carbon flux. Diatom and plant-pigment data also indicate that primary productivity was considerably higher during the prairie period. Pyrolysis hydrogen and oxygen indices lend additional support for increased productivity and the burial of well-preserved, hydrogen-rich, oxygen-poor, ^{13}C-enriched algal organic matter during the prairie period.

ACKNOWLEDGMENTS

We thank Chuck Threlkeld and April Vuletich ((USGS Denver) for providing the carbon and oxygen isotope analyses of marl and organic carbon from the two Elk Lake cores, and Ted Daws (USGS Denver) for providing the Rock-Eval pyrolysis analyses. Rick Forester (USGS Denver) kindly provided the samples of *Candona ohioensis* for isotope analyses as well as unpublished data on lake salinity and temperature interpreted from ostracod populations, based on a total dissolved solids curve for ostracod populations kindly provided by L. Denis Delorme. We are very grateful for reviews by Mike Arthur, Rick Forester, and Jim White.

REFERENCES CITED

Anderson, T. F., and Arthur, M. A., 1983, Stable isotopes of oxygen and carbon and their application to sedimentologic and paleoenvironmental problems, *in* Stable isotopes in sedimentary geology: Society of Economic Paleontologists and Mineralogists, Short Course Notes No. 10, p. 1.1–1.151.

Arthur, M. A., Dean, W. E., and Schlanger, S. O., 1985, Variations in the global carbon cycle during the Cretaceous related to climate, volcanism, and changes in atmospheric CO_2, *in* Sundquist, E. T., and Broecker, W. S., eds., The carbon cycle and atmospheric CO_2—Natural variations Archean to present: Washington, D.C., American Geophysical Union Geophysical Monograph 32, p. 504–529.

Arthur, M. A., Dean, W. E., and Pratt, L. M., 1988, Geochemical and climatic effects of increased marine organic carbon burial at the Cenomanian/Turonian boundary: Nature, v. 335, p. 714–717.

Bartlein, P. J., Webb, T., III, and Fleri, E., 1984, Holocene climatic change in the northern midwest—Pollen-derived estimates: Quaternary Research, v. 22, p. 361–374.

Bartlein, P. J., Prentice, I. C., and Webb, T., III, 1986, Climatic response surfaces from pollen data for some eastern North American taxa: Quaternary Research, v. 13, p. 35–57.

Bender, M. M., 1971, Variations in the $^{13}C/^{12}C$ ratios of plants in relation to the pathway of photosynthetic carbon dioxide fixation: Phytochemistry, v. 10, p. 1239–1343.

Buchardt, B., and Fritz, P., 1980, Environmental isotopes as environmental and climatological indicators, *in* Fritz, P., and Fontes, J. C., eds., Handbook of environmental isotope geochemistry, Volume 1, The terrestrial environment: Amsterdam, Elsevier, p. 473–504.

Calder, J. A., and Parker, P. L., 1973, Geochemical implications of induced changes in ^{13}C fractionation by blue-green algae: Geochimica et Cosmochimica Acta, v. 37, p. 133–140.

Dean, W., Bradbury, J. P., Anderson, R. Y., and Barnosky, C. W., 1984, The variability of Holocene climate change—Evidence from varved lake sediments: Science, v. 226, p. 1191–1194.

Deines, P., 1980, The isotopic composition of reduced organic carbon, *in* Fritz, P., and Fontes, J. C., eds., Handbook of environmental isotope geochemistry, Volume 1, The Terrestrial Environment: Amsterdam, Elsevier, p. 329–406.

Denton, G. H., and Karlén, W., 1973, Holocene climatic variations—Their pattern and possible cause: Quaternary Research, v. 3, p. 155–205.

Deuser, W. G., Degens, E. T., and Guillard, R.R.L., 1968, Carbon isotope relationships between plankton and seawater: Geochimica et Cosmochimica Acta, v. 32, p. 657–660.

Edwards, T.W.D., and Fritz, P., 1986, Assessing meteoric water composition and relative humidity from ^{18}O and ^2H in wood cellulose—Paleoclimatic implications for southern Ontario, Canada: Applied Geochemistry, v. 1, p. 715–723.

Emerson, S. E., Broecker, W. S., and Schindler, D. W., 1973, Gas exchange rates in a small Canadian shield lake as determined by the radon method: Fisheries Research Board of Canada Journal, v. 30, p. 1475–1493.

Espitalié, J., Laporte, J. L., Madec, M., Marquis, F., Leplat, P., Paulet, J., and Bouteleu, A., 1977, Methode rapide de characterisation des roches mértes de leur potential petrolier en de leur degré d'evolution: Revue de l'Institut Francais de Petrole, v. 32, p. 23–42.

Forester, R. M., DeLorme, L. D., and Bradbury, J. P., 1987, Mid-Holocene climate in northern Minnesota: Quaternary Research, v. 28, p. 263–273.

Gorham, E., and Sanger, J. E., 1975, Fossil pigments in Minnesota lake sediments and their bearing upon the balance between terrestrial and aquatic inputs to sedimentary organic matter: Verhandlungen International Vereinigun für Limnologie, v. 19, p. 267–273.

Gorham, E., Dean, W. E., and Sanger, J. E., 1983, The chemical composition of lakes in the north-central United States: Limnology and Oceanography, v. 28, p. 287–301.

Håkansson, S., 1985, A review of various factors influencing the stable carbon isotope ratio of organic lake sediments by the change from glacial to postglacial environmental conditions: Quaternary Science Reviews, v. 4, p. 135–146.

Herczeg, A. L., and Fairbanks, R. G., 1987, Anomalous carbon isotope fractionation between atmospheric CO_2 and dissolved inorganic carbon induced by intense photosynthesis: Geochimica et Cosmochimica Acta, v. 51, p. 895–899.

Hollander, D. J., McKenzie, J. A., and Vandenbroucke, M., 1988, Secular variation in the geochemistry of organic carbon-rich sediments of a eutrophic lake model to the Cenomanian/Turonian event (OAE): Geological Society of America Abstracts with Programs, v. 20, no. 6, p. A220.

Jacobsen, G. L., 1979, The paleoecology of white pine (*Pinus strobus*) in Minnesota: Journal of Ecology, v. 67, p. 697–726.

Lee, C., McKenzie, J. A., and Sturm, M., 1987, Carbon isotope fractionation and changes in the flux and composition of particulate matter resulting from biological activity during a sediment trap experiment in Lake Greifen, Switzerland: Limnology and Oceanography, v. 32, p. 83–96.

McCrea, J. M., 1950, The isotopic chemistry of carbonates and a paleotemperature scale: Journal of Chemical Physics, v. 18, p. 849–857.

McKenzie, J. A., 1982, Carbon-13 cycle in Lake Greifen—A model for restricted ocean basins, *in* Schlanger, S. O., and Cita, M. B., eds., Nature and origin of Cretaceous carbon-rich facies: New York, Academic Press, p. 197–207.

McKenzie, J. A., 1985, Carbon isotopes and productivity in the lacustrine and marine environment, *in* Stumm, W., ed., Chemical processes in lakes: New York, Wiley-Interscience, p. 99–118.

Mook, W. G., Bommerson, J. C., and Staverman, W. H., 1974, Carbon isotope fractionation between dissolved bicarbonate and gaseous carbon dioxide: Earth and Planetary Science Letters, v. 22, p. 169–176.

Oana, S., and Deevey, E. S., 1960, Carbon 13 in lake waters and its possible bearing in paleolimnology: American Journal of Science, v. 258, p. 153–268.

Otsuki, A., and Wetzel, R. G., 1974, Calcium and total alkalinity budgets and calcium carbonate precipitation of a small hardwater lake: Archive für Hydrobiologie, v. 73, p. 14–30.

Peters, K. E., 1986, Guidelines for evaluating petroleum source rock using programmed pyrolysis: American Association of Petroleum Geologists Bulletin, v. 70, p. 318–329.

Pratt, L. M., and Threlkeld, C. N., 1984, Stratigraphic significance of $^{13}C/^{12}C$ ratios in mid-Cretaceous rock of the western interior, U.S.A., *in* Stott, D. F., and Glass, D. J., eds., The Mesozoic of middle North America: Canadian Society of Petroleum Geology Memoir 9, p. 305–312.

Quay, P. D., Emerson, S. R., Quay, B. M., and Devol, A. H., 1986, The carbon cycle for Lake Washington—A stable isotope study: Limnology and Oceanography, v. 31, p. 596–611.

Rau, G. H., 1978, Carbon-13 depletion in a subalpine lake—Carbon flow implications: Science, v. 201, p. 901–902.

Rau, G. H., Takahashi, T., and DesMarais, D. J., 1989, Latitude variation in plankton $\delta^{13}C$—Implications for CO_2 and productivity in past oceans: Nature, v. 341, p. 516–518.

Scholle, P. A., and Arthur, M. A., 1980, Carbon isotopic fluctuations in pelagic limestones; potential stratigraphic and petroleum exploration tool: American Association of Petroleum Geologists Bulletin, v. 64, p. 67–87.

Stuiver, M., 1970, Oxygen and carbon isotope ratios of fresh-water carbonates as climatic indicators: Journal of Geophysical Research, v. 75, p. 5247–5257.

—— , 1975, Climate versus changes in ^{13}C content of the organic component of lake sediments during the late Quaternary: Quaternary Research, v. 5, p. 251–262.

Talbot, M. R., 1990, A review of the palaeohydrological interpretation of carbon and oxygen isotope ratios in primary lacustrine carbonates: Chemical Geology (Isotope Geoscience Section), v. 80, p. 261–279.

Talbot, M. R., and Kelts, K., 1986, Primary and diagenetic carbonates in the anoxic sediments of Lake Bosumtwi, Ghana: Geology, v. 14, p. 912–916.

—— , 1990, Paleolimnological signatures from carbon and oxygen isotopic ratios in carbonates from organic carbon-rich lacustrine sediments, *in* Katz, B. J., ed., Lacustrine basin exploration—Case studies and modern analogs: American Association of Petroleum Geologists Memoir 50, p. 99–112.

Talbot, M. R., and Livingstone, D. A., 1989, Hydrogen index and carbon isotopes of lacustrine organic matter as lake level indicators: Palaeogeography, Palaeoclimatology, Palaeoecology, v. 70, p. 121–137.

Tissot, B. P., and Welte, D. H., 1984, Petroleum formation and occurrence (second edition): Berlin, Springer Verlag, 538 p.

Webb, T., III, and Bryson, R. A., 1972, Late- and post-glacial climatic change in the northern Midwest USA—Quantitative estimates derived from fossil spectra by multivariate statistical analysis: Quaternary Research, v. 2, p. 70115.

Winkler, M. G., Swain, A. M., and Kutzbach, J. E., 1986, The middle Holocene dry period in the northern midwestern United States—Lake levels and pollen stratigraphy: Quaternary Research, v. 25, p. 235–250.

Wright, H. E., Jr., 1976, The dynamic nature of Holocene vegetation: Quaternary Research, v. 6, p. 581–596.

MANUSCRIPT ACCEPTED BY THE SOCIETY JULY 27, 1992

Printed in U.S.A.

Fossil pigments in Holocene varved sediments in Elk Lake, Minnesota

Jon E. Sanger and Ruth J. Hay
Department of Botany-Microbiology, Ohio Wesleyan University, Delaware, Ohio 43015

ABSTRACT

Fossil pigments were examined in a 22 m core of varved sediment from the deep basin of Elk Lake, Minnesota. The lake appears to have evolved gradually from oligotrophic mesotrophic conditions in the earliest period (ca. 10,000+ years ago), to mesotrophic eutrophic conditions at present. Variations in productivity, species diversity, and relative importance of individual plant groups are related to changing climate and water level. The cyanobacteria gradually became more important in the planktonic flora; the least biomass and greatest variation occurred during the first 1000 years of the lake's postglacial history, and again during the mid-Holocene prairie interval. The ratio of chlorophyll derivatives to carotenoids indicates that there were no periods of large-scale influx of allochthonous detritus, nor were there major slumps of littoral detritus into the deep profundal zone. In addition, there is no evidence to indicate that excessive drying of the lake occurred during the mid-Holocene prairie interval.

INTRODUCTION

Sedimentary chlorophyll derivatives and carotenoids were examined in a 22 m core of varved sediment from the deep basin of Elk Lake, Itasca Park, Minnesota. This long sequence of highly organic, laminated sediment was deposited in a bottom-water environment that was anoxic or near-anoxic throughout most of the postglacial history of the lake, an environment that was highly conductive to the preservation of pigment molecules.

Much of the organic detritus that contributes to lake sediments contains a collection of pigments from organisms living within the lake and the surrounding drainage basin. Pigments are contained in leaves from terrestrial plants and in soil humus layers that are washed in during periods of runoff (Gorham and Sanger, 1975). An abundance of pigments is transported to the sediments in fragments of aquatic macrophytes, phytoplankton detritus, and the corpses of zooplankton that have ingested the primary producers. Decomposition of the pigment molecules occurs in the water column by photo-oxidation and bacterial and fungal attack as the fragments slowly sink to the bottom (Daley, 1973; Gray, 1974). Sediment-trap studies suggest that as much as 90% of the organic matter synthesized at the surface of a lake undergoes some decomposition prior to incorporation into surface sediment (Kleerekoper, 1953; Tutin, 1955; Ohle, 1962). Decomposition continues at the mud surface, particularly if the bottom waters are oxygenated, until the organic material is covered by sediment and protected from further destruction by continuous anoxia, darkness, and constant cold conditions (Vallentyne, 1962; Pennington and others, 1973). The chlorophyll molecules are converted to a collection of derivatives including pheophytins, pheophorbides, and chlorophyllides in an assortment of isomers and allomers (Daley and others, 1977).

Undecomposed chlorophyll is found occasionally in sediments (Daley, 1973; Daley and Brown, 1973; Daley and others, 1977; Engstrom and others, 1985), but electron microscopy of sedimentary materials (Sanger, unpublished) indicates that its occurrence depends on the presence of undecomposed chloroplastids and membrane systems where pigment molecules reside. Undecomposed chlorophyll is therefore most likely to accompany undecomposed leaf fragments of eukaryotes. The multitude of chlorophyll derivatives that can occur in response to varying conditions of decomposition makes a study of the individual chlorophyll compounds too complex to be of much paleoecological significance (Sanger, 1988). Except for some isomerization and esterification, carotenoids appear to be preserved largely unchanged from their original structure in living tissue, and many

Sanger, J. E., and Hay, R. J., 1993, Fossil pigments in Holocene varved sediments in Elk Lake, Minnesota, *in* Bradbury, J. P., and Dean, W. E., eds., Elk Lake, Minnesota: Evidence for Rapid Climate Change in the North-Central United States: Boulder, Colorado, Geological Society of America Special Paper 276.

are highly specific to certain taxa (Brown, 1968; Sanger, 1971a, 1971b; Zullig, 1981; Swain, 1985).

The value of sedimentary plant pigments as paleoecological indicators has been well documented in a variety of aquatic habitats over diverse geographic regions. They can serve as indices of present and past trophic conditions in lakes that are deep enough to stratify (Vallentyne, 1956, 1957, 1960; Gorham, 1960, 1961; Brown, 1969; Sanger and Gorham, 1970; Gorham and others, 1974; Murray and Douglas, 1976; Santelman, 1981; Zullig, 1981; Guilizzoni and others, 1983; Engstrom and others, 1985). Pigments are also useful for revealing successional changes in postglacial ecology and limnology, especially changes in aquatic postglacial productivity and phytoplankton diversity. They can provide information on the balance between allochthonous and autochthonous organic contributions to the sediments, postglacial water-level changes, past periods of meromixis, and the invasion of peatlands into the drainage basin (Fogg and Belcher, 1961; Belcher and Fogg, 1964; Czeczuga, 1965; Czeczuga and Czerpak, 1968; Wetzel, 1970; Gorham and Sanger, 1972; Sanger and Gorham, 1972; Whitehead and others, 1973; Handa, 1975; Brown and others, 1977; Sanger and Crowl, 1979; Culver and others, 1981).

METHODS

Elk Lake core segments, 50 varves long, were stored frozen until processing according to methods described by Sanger and Gorham (1970, 1972). Total pigments were extracted with 90% acetone, and 200 ml of extract were divided into three aliquots for separation and measurement of various pigments. First, the light absorption of a 5 ml aliquot of the acetone extract was measured spectrophotometrically in a 1 cm cell at 660 nm to determine concentrations of total chlorophyll derivatives, and at 750 nm to determine concentrations of bacteriochlorophyll. A 50 ml second aliquot was saponified for two hours with 20% weight/volume (w/v) methanolic potassium hydroxide to allow the removal of chlorophyll derivatives. Total carotenoids were extracted with 30°–60° petroleum ether, and epiphasic and hypophasic components were partitioned in a separatory funnel using 30°–60° petroleum ether and 90% methanol as the two phases. Carotenoid concentrations were determined by measuring the light absorbance at a peak near 450 nm. For comparison with previous studies, pigments were calculated as absorbance units per gram organic matter; one unit is equivalent to an absorbance of 1.0 in a 10 cm cell when that unit is dissolved in 100 ml of solvent.

A third aliquot of 100 ml of the acetone extract was transferred to diethyl ether in a separatory funnel and the solution was evaporated to 1.0 ml in order to concentrate the pigments. Pigments were chromatographed and identified spectrophotometrically in two dimensions on 20 × 20 cm, thin-layer plates coated with silica gel 7G according to techniques and spot locations on chromatograms described by Sanger and Gorham (1970, 1972). Lutein, β-carotene, and the entire collection of prokaryote pigments, mainly myxoxanthophyll, canthaxanthin, and echinenone, were scraped from the plate, the pigments were dissolved from the silica gel with 100% acetone, and the relative pigment concentrations were measured spectrophotometrically by light adsorption at their respective peaks near 450 nm. The ubiquitous lutein and β-carotene served as a base against which the relative importance of the prokaryote pigments could be ascertained. The pigment spots were extracted immediately on removal of the plate from the second-dimension tank to avoid loss from fading, and a facsimile of the chromatogram was drawn on paper with colored pencils. Close-interval sampling was done whenever there were substantial changes in the sedimentary pigments over a short time interval.

INTERPRETATION OF PIGMENT DATA

The general interpretation of pigment data was discussed by Sanger and Gorham (1970, 1972), Sanger and Crowl (1979), Santelman (1981), Swain (1985), and Sanger (1988). It has been demonstrated that total sedimentary chlorophyll derivatives and carotenoids undergo little diagenesis once they are deposited and buried beneath the oxidized layer at the sediment-water interface (Gray, 1974). Concentrations and diversity of total sedimentary pigments from contemporary sediment as determined by thin-layer chromatography commonly remain similar to those from early postglacial and late-glacial sediments in cores from lakes. A greater production rate of autochthonous organic matter increases the concentration of pigments in the sedimenting organic matter. A greater productivity rate also tends to produce more oxygen-deficient conditions in the bottom waters, conducive to pigment preservation. Sediments in Elk Lake accumulated throughout the Holocene in bottom waters that were always close to being anoxic. Consequently, the presence of high concentrations of pigments in these sediments reflects a high pigment input as well as an environment that was more conducive to pigment preservation than typical highly productive dimictic lakes (also see Gorham and others, 1974; Tarapchak, 1973).

Accurate interpretation of the paleoecological record, based on pigment data, depends to a large extent on the sources of organic matter that contributed to lake sediments. An examination of the ratio of chlorophyll derivatives to total carotenoids (CD/TC) provides important clues to allochthonous versus autochthonous sources of organic matter. Allochthonous plant detritus, although poor in total pigments, tends to be relatively richer in chlorophyll derivatives than carotenoids because carotenoids are more susceptible to total oxidation than the pheo-chlorophyll derivatives (Gorham and Sanger, 1964, 1967, 1975; Sanger and Gorham, 1970, 1972). A CD/TC ratio >0.6 in the sediment indicates aerobic decomposition at the sediment-water interface, or a predominantly allochthonous source of organic matter. CD/TC ratios >2.0, and less commonly as high as 7.0, have been found in peaty sediments. Low ratios (<0.3) are common in reducing lake sediments and reflect relatively better preservation of carotenoids where the organic matter has had little chance to be oxidized.

When considering the diversity of sedimentary carotenoids, much can be learned about the variety of organisms present in a lake by examining pigments specific to those organisms. There are at least four carotenes and 14 xanthophylls common in mixed blooms of plankton in Minnesota lakes, all of which are deposited in the sediments (Sanger and Gorham, 1970). Carotenoids that are uniquely characteristic of the prokaryotes are particularly useful because the cyanobacteria (blue-green algae) tend to dominate the plankton in very productive lakes. The cyanophycean carotenoids myxoxanthophyll, canthaxanthin, and echinenone were found by Santelman (1981) in the sediments of a collection of Minnesota lakes ranging in trophic status from oligotrophic to eutrophic. Myxoxanthophyll and oscillaxanthin in recent sediments derived from increasing cyanobacteria populations have provided much evidence for cultural eutrophication in a variety of lakes (Zullig, 1961, 1981, 1982; Griffiths and others, 1969; Griffiths and Edmondson, 1975; Gorham and Sanger, 1976, and unpublished; Griffiths, 1978; Guilizzoni and others, 1981, 1982, 1983; Engstrom and others, 1985).

Pigment diversity, expressed as total number of spots from two-dimensional thin-layer chromatograms (Sanger and Gorham, 1970), reflects the biotic diversity of the organic contribution, and, more important, the way anaerobic, dark, cold sedimentary environments preserve organic molecules. Long periods of anaerobic conditions protect and preserve organic molecules whereas aerobic conditions promote rapid and sometimes preferential destruction of pigments. Because carotenoids are more labile than chlorophyll derivatives (Goodwin, 1980), they are more sensitive indicators of redox conditions. The presence of relatively high concentrations of cyanobacterial carotenoids that fall into protective anaerobic bottom waters of Elk Lake should provide clear evidence for true variations in cyanophycean populations through time, rather than variations in the degree of preservation that occurs in lakes with frequently changing redox conditions at the sediment-water interface. Traces of the carotenoids canthaxanthin and echinenone (myxoxanthin) occur in the eye spot of the Euglenophyta (Goodwin, 1980), and some canthaxanthin has been isolated from fungi, other photosynthetic bacteria, and crustaceans, but it is unlikely that any of these sources would make significant sedimentary contributions. Compared to higher plants, the cyanobacteria have higher cellular concentrations of carotenoids, and are therefore the overwhelming contributors of carotenoids.

The ratio of carotene plus lutein to the prokaryote carotenoids (myxoxanthophyll, canthaxanthin, and echinenone) (C+L/P) should be a measure of the importance of the cyanobacteria relative to all other plants that contribute sedimentary pigments. Lutein and β-carotene are the most ubiquitous carotenoids in the plant kingdom. The prokaryote carotenoids (P) considered above are restricted mainly to the cyanobacteria and other common aquatic prokaryotes.

Substantial concentrations of bacterial carotenoids have been found in sediments from Sunfish Lake, Ontario (McIntosh, 1983), and from some meromictic lakes such as Green Lake, Fayetteville, New York. There is evidence that abundant bacterial populations, in addition to the cyanobacteria, below the chemocline in Fayetteville Green Lake may contribute as much as 20% of the total sedimentary pigment complex (Sanger, unpublished data). In Elk Lake, however, the prokaryote pigments (P) apparently are largely derived from cyanobacteria because concentrations of bacteriochlorophyll from spectrophotometric readings at the 750 nm peak remain below 1.0 units per gram organic matter throughout the core.

Data presented in Table 1 provide a comparison of pigments in sediments from the Elk Lake core with those from eight oligotrophic and 21 eutrophic holomictic lakes, and from nine meromictic lakes with oligotrophic to eutrophic surface waters. Average pigment concentrations are distinctly higher in the meromictic lakes. Gorham and Sanger (1972) found that pigment concentrations and diversity in meromictic Stewart's Dark Lake, Wisconsin, increased greatly from the shallowest to the deepest sediments. Their data demonstrated that sedimentary pigments

TABLE 1. SOME CHARACTERISTIC VALUES FOR SEDIMENTARY PIGMENTS IN MINNESOTA LAKE SEDIMENTS

Lake Type	Number of lakes sampled	Igneous loss 550 °C	Chlorophyll derivatives	Total carotenoids	Total pigments	Spot number	Chlorophyll deriv. total carotenoids
			——————Units/gram organic matter——————				
Oligotrophic (holomictic)	8	21.3	4.6	7.3	11.9	30	0.70
Eutrophic (holomictic)	21	23.5	10.2	30.9	41.1	38	0.35
Eutrophic-oligotrophic (meromictic)	9	45.8	24.6	73.6	98.2	35	0.61
Elk Lake (mean for entire core)		10.6	7.6	25.7	33.3	33	0.31

increase two to three times in sediments in the monimolimnion compared to those in the deeper parts of the mixolimnion. Concentrations of monimolimnetic chlorophyll derivatives are about three times higher, and those of carotenoids are two times higher, than the maximum concentrations in the hypolimnetic sediments of eutrophic holomictic Minnesota lakes.

RESULTS AND DISCUSSION

In the context of the above background on the general interpretation of sedimentary pigments, we will discuss the data from the Elk Lake core beginning with the inception of varved sediment about 10,400 ka; times refer to varve years before present (varve yr). The earliest varves in Elk Lake are thinner than at any point in the core (Fig. 1), indicating a slow sedimentation rate. They contain relatively high concentrations of $CaCO_3$ (Fig. 1) due to fresh calcareous tills being eroded. The vegetation was dominated initially by spruce and aspen that were rapidly replaced by pine as the glacial climate began to moderate.

The concentration of organic matter in sediments of Elk Lake, as measured by loss on ignition at 550 °C (LOI-550, Fig. 1), is distinctly lower than in surface sediments from most of the Minnesota Lakes (Table 1), and considerably lower than in sediments from the nine meromictic lakes. The mean content of organic matter from the entire Elk Lake core is 10.6%, which is even lower than the organic contents of surface sediments in southwestern Minnesota Lakes (low-carbonate, low-organic group 2 lakes of Dean and Gorham, 1976; mean LOI at 550 °C 18.6%). Unlike Elk Lake, these lakes commonly have intensive oxidation of organic matter by resuspension of sediment, but they also receive considerable wind-blown material from prairie areas. It is not likely that any pigmented organic matter would accompany these silty wind-deposited materials containing glass-derived opal phytoliths. The greatest period of wind erosion in the Elk Lake area is likely to have occurred during the dry prairie period, from 4000 to 8000 varve yr (as indicated by the thicker varves in Fig. 1; Dean, Chapter 10).

The concentrations of total chlorophyll derivatives and carotenoids are presented in Figure 2 and Table 1. The mean value of chlorophyll derivatives for the core is 7.6 units (absorbance units per gram organic matter). This value is higher than the average for oligotrophic holomictic Minnesota lakes (4.6 units), but well

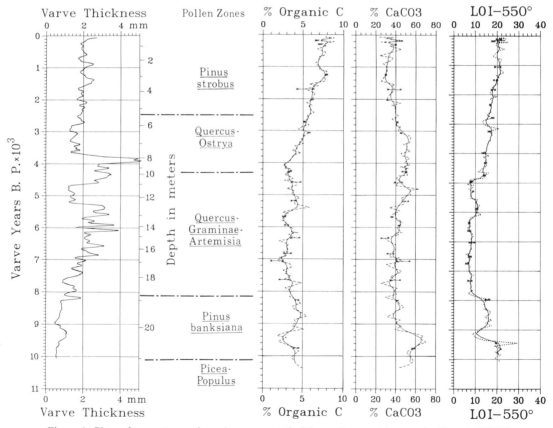

Figure 1. Plots of percent organic carbon, percent $CaCO_3$, and percent loss on ignition at 550 °C (LOI-550) vs. varve years before present (B.P.) for the Elk Lake core. A plot of varve thickness vs. varve years and the major pollen zones are shown for reference. The raw data are connected by dashed lines; the solid lines through the raw data are smoothed curves computed using 11 sample weighted moving averages.

below the averages of eutrophic holomictic lakes (10.2 units) and meromictic lakes (24.6 units). The mean carotenoid concentration for the entire Elk Lake core is 25.7 units, which is higher than the highest value for the eight oligotrophic lakes (14.4 units). The carotenoid concentrations in Elk Lake are very similar to those in eutrophic lakes with partly oxidized bottom waters, but lower than the average for meromictic lakes with totally anoxic bottom waters (73.6 units). Considering the protection afforded by the anoxic to near-anoxic bottom waters of Elk Lake, the pigment concentrations in the Elk Lake core suggest that throughout its history Elk Lake remained largely oligotrophic to mesotrophic, and probably never developed intense phytoplankton blooms. Primary productivity studies (Megard, 1967) show that Elk Lake is still mesotrophic.

An important consideration is that the relatively deep water of Elk Lake (currently about 30 m in the deep basin) could result in greater oxidation of organic compounds during sedimentation. In addition, the anoxic deep basin of Elk Lake represents a relatively small portion of the lake bottom; sedimentary organic detritus may reside on aerobic sediment surfaces for some time before slowly sinking into the deepest part of the basin. Together these considerations may contribute to greater decomposition of pigments, giving the impression of lesser productivity as suggested by pigment concentrations. This may explain partly why Elk Lake has much lower total sedimentary pigments when compared to shallower meromictic lakes that undergo continuous anoxia.

Elk Lake, like many hard-water lakes with sedimentary $CaCO_3$ contents in excess of 20% (Fig. 1), has undergone suppressed productivity caused by complex nutrient removal interactions. Adsorption to, and coprecipitation with carbonates tends to sequester nutrients, especially phosphorus, iron, manganese, and a variety of micronutrients (Wetzel, 1972; Otsuki and Wetzel, 1974; Wetzel and Otsuki, 1974). These interactions undoubtedly contributed to the sustained oligotrophy-mesotrophy of the lake and are important when considering the data on variations in planktonic blooms.

In the earliest part of the varved sediment section in the Elk Lake core, pigment concentrations decrease significantly to their lowest values at about 10,000 varve yr (Fig. 2). The rapid decline in productivity, as evidenced by the pigments, was probably due largely to loss of phosphorus and other nutrients by coprecipitation with carbonate minerals (Otsuki and Wetzel, 1972; White and Wetzel, 1975; Manny and others, 1978). Productivity was probably also limited by low temperatures and high turbidity.

The lowest values of total chlorophyll derivatives and total carotenoids are 4 and 10 units, respectively, at about 10,000 varve yr (Fig. 2). These concentrations are typical of values for sediments from oligotrophic lakes, and are well below the mean values for eutrophic lakes (Table 1). This suggests that the trophic level during the early history of the lake was oligotrophic to mesotrophic, or that oxic conditions prevailed in the water column and at the mud surface for much of the time, causing oxidation of the pigments. During the same time period the CD/TC ratio was nearly twice as high as in most of the period from 9000 varve yr to the present (Fig. 3), suggesting a rapid and significant influx of terrestrial detritus before forests and soil-humus layers developed, or that aerobic decomposition selectively destroyed the carotenoids in sedimenting autochthonous organic detritus. The maximum CD/TC ratio of 0.49 overlaps with the range for oligotrophic lakes, again suggesting mesotrophy.

Pigment diversity (number of spots on a two-dimensional thin-layer chromatogram; Fig. 3) in the lower part of the core is about 32, and does not change much throughout the entire Holocene section. This constancy of pigment diversity indicates that, although plankton populations may have changed, the diversity of the total flora in the lake remained much the same.

The importance of cyanobacteria can be assessed from the C+L/P ratio (Fig. 3). A change to higher values reflects a relative increase in abundance of eukaryotic organisms, whereas a change to lower values reflects increased importance of prokaryotes (cyanobacteria). In the lower part of the core, the C+L/P ratio varies considerably from 7.9 to 99, and the average is several times higher than in the remainder of the core. This variability suggests that there was considerable variation in cyanophycean populations. It is significant, however, that prokaryote pigments appeared periodically in abundance throughout the early history of the lake, but populations of cyanophycean organisms apparently went through blooms and crashes. There were four periods

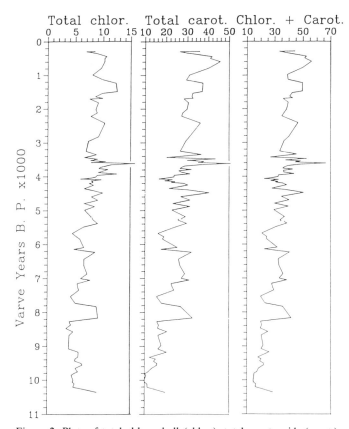

Figure 2. Plots of total chlorophyll (chlor.), total carotenoids (carot.), and total pigments (as absorbance units per gram organic matter) vs. varve years before present (B.P.) for the Elk Lake core.

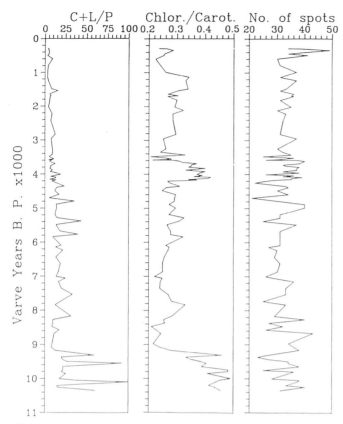

Figure 3. Plots of the ratios of carotene + lutein to total prokaryote pigments (C+L/P), chlorophyll derivatives to total carotenoids (chlor./carot.), and number of spots on two-dimensional thin-layer chromatograms before present (B.P.) for the Elk Lake core.

when numbers of cyanophytes apparently were relatively sparse: at the base of the core at 10,360, at 10,110, at 9560, and at 9325 varve yr. These four periods may represent times of climatic deterioration during overall climatic amelioration. The sharpness and slight variation of the peaks, together with slightly higher CD/TC ratios (Fig. 3), may signify greater runoff during these periods with turbid, high-water conditions that reduced total productivity and brought in greater concentrations of allochthonous organic material. The evidence for climatic variability suggested by the pigment data may possibly reflect small ice advances in areas to the north that were followed by melting and rapid runoff. After 10,000 varve yr, each successive peak in C+L/P and in CD/TC is less intense, which suggests that each successive period of climatic deterioration was less in magnitude.

Following the erratic period of the early postglacial from about 10,400 to 9000 varve yr, the concentration of organic matter (LOI at 550 °C) in the sediments declines and reaches a minimum between 7000 and 6000 varve yr (Fig. 1), during the dry prairie period when forests in the drainage basin of Elk Lake were replaced by grasses, sagebrush, and oak savanna (Gramineae, *Artemisia,* and *Quercus* pollen zone; Whitlock and others, Chapter 17). At about 10,000 varve yr, concentrations of pigments begin a slow, gradual increase that continues, with considerable variation, especially between 5000 and 3000 varve yr to the top of the core. Between about 9200 and 8200 varve yr, concentrations of chlorophyll derivatives are lower than at any other time: the average, 4.2 units (Fig. 2), is about the expected amount for sediments in an oligotrophic lake (Table 1). Concentrations of carotenoids are at their lowest about 10,000 varve yr. A significant change in the nature of the sedimentary organic matter occurs at about 9300 varve yr and is clearly evident in the CD/TC ratio (Fig. 3), which by 9000 varve yr has a value that is about half that at 10,000 varve yr. Concurrent with this change in CD/TC ratio is a sharp drop of about 50% in the C+L/P ratio which indicates a large, more stable population of prokaryotes. These data, taken together, suggest a sharp decrease in the supply of allochthonous organic matter, and an increase in in situ organic productivity, possibly dominated by cyanophyceans. The C+L/P ratio and concentration of total pigments during the period between 9300 and 8200 varve yr suggest that there was considerable stability of primary production that is common to mesotrophic lakes. An average total pigment concentration of 21.8 units (Fig. 2) is about twice that of sediments of oligotrophic lakes (Table 1), and about half that of sediments of eutrophic lakes. Because the highly anoxic sediments of Elk Lake preserve organic molecules better than the more oxic sediments of oligotrophic lakes, the pigment values are perhaps typical of the more productive of the oligotrophic lakes. In a core from Browns Lake, Ohio, pigment concentrations during the time when the lake was meromictic were about double those of the holomictic conditions that followed, a clear illustration of the degree to which anoxic sediment preserves pigments (Sanger and Crowl, 1979).

After the apparent stability of all parameters between 9200 and 8200 varve yr, the average concentrations of pigments more than doubled between 8200 and 7800 varve yr (Fig. 2). This abrupt increase in sedimentary pigments suggests that there was a rapid increase in productivity, which is common in the early postglacial history of lakes when vegetation and soil humus layers become established. Evidence from other chapters in this volume (e.g., Anderson, Chapter 5; Bradbury and Dieterich-Rurup, Chapter 15) indicates that erosion of the drainage basin and turbidity of the lake water decreased and allowed the phytoplankton to utilize the better light conditions and available nutrients. Pigment concentrations decreased again between 7800 and 7500 varve yr, as did the CD/TC ratio and concentration of organic matter (Fig. 1). These decreases suggest that primary productivity decreased, probably in response to a decreasing supply of available nutrients from erosion of the drainage basin.

The interval between 7000 and 6000 varve yr appears to have been another period of relative stability in primary productivity. After about 6300 varve yr, however, concentrations of both chlorophyll derivatives and carotenoids began to decrease, reaching low values at about 5600 varve yr. The geochemical and diatom data indicate that these low pigment concentrations correspond to a period of high rate of erosion in the drainage basin and high turbulence in the lake (Bradbury and Dieterich-Rurup,

Chapter 15), resulting in thick varves (Fig. 1). The organic matter that was deposited at that time did not, however, include abundant allochthonous material from humus layers that would be low in total pigments but would have more chlorophyll derivatives than carotenoids. This may represent a period of extreme drought when water levels were low and wave action slowly eroded shallow-water sediments into the profundal regions of the lake. The pigments in the entire section from 6000 to 4000 varve yr are characterized by considerable variation in the C+L/P ratio (Fig. 3), which suggests that there were fluctuating algal blooms in response to considerable environmental stress from climatic variations. Perhaps the lake shifted back and forth from phytoplankton domination to conditions favoring higher standing crops of aquatic macrophytes.

The concentration of organic matter (Fig. 1) begins to increase dramatically at about 4000 varve yr, and continues to increase to the top of the core. Erratic fluctuations in pigment concentrations begin at about 4500 varve yr and continue to about 3200 varve yr. Within this same interval, spot number is more erratic than at any other period in the history of the lake, and the CD/TC ratio shows a sharp increase, remains stable for several hundred years, and then decreases abruptly (Fig. 3). Taken together, these data suggest that there were extremes in climatic variability that may have caused wide variations in lake levels due to evaporation, the influx of eroded allochthonous organic matter from the drainage basin, fluctuating plankton blooms, and perhaps continued episodes of erosion of littoral detritus into the profundal region. It is clear that the cyanobacteria were an important part of the biota throughout these years. As a group, they are noted for their rapid turnover rates, erratic blooms, and population crashes in response to environmental stress.

Between 3200 and 600 yr B.P., there is an increase in concentrations of pigments to values that are more than twice those in the early postglacial sediments, and are about the same as those from the average of 21 eutrophic lakes shown in Table 1 (41 units). There is little variation in the C+L/P ratio, however, which suggests that there was little fluctuation in cyanophycean populations. At about 600 yr B.P., there is an abrupt decrease in total pigments (Fig. 2), and a concomitant increase in CD/TC ratio and spot number (Fig. 3). This may reflect a return to a more moist climate that increased water levels, bringing in a somewhat greater proportion of allochthonous materials.

SUMMARY

The sedimentary pigment data indicate that the trophic level of Elk Lake has gradually increased from oligotrophic mesotrophic to mesotrophic-eutrophic during the Holocene. Variations in productivity occurred throughout the Holocene in response to changing climate and water levels. With the exception of the early postglacial part of the section, the CD/TC ratio remains below 4.0, which indicates that the lake has never received large quantities of terrestrial organic detritus. In addition, there is no evidence for intensive slumping of the littoral sediments as occurred in Stewart's Dark Lake, Wisconsin (Gorham and Sanger, 1972), or desiccation from excessive low water levels as occurred in Kirchner Marsh in southeastern Minnesota (Sanger and Gorham, 1972) due to drought conditions during the mid-Holocene. In addition, there is no evidence from the sedimentary pigment data that the anoxic conditions that permitted the accumulation of finely laminated sediments were ever interrupted.

Considerable instability of plankton populations as evidenced by the C+L/P ratio apparently occurred between 10,400 and 9000 varve yr, and from 6000 to 4000 varve yr (Fig. 3), indicating that there was considerable environmental stress from climatic variations during late glacial and early postglacial and again during the latter part of the prairie period. The C+L/P ratio also shows that the cyanobacteria, which contribute largely to the concentration of prokaryote pigments, always have been an important component of the biota. Fluctuations in the concentrations of prokaryote pigments were greatest during periods of most rapid change in climate. Apart from this, the plant ecology of Elk Lake probably has not changed substantially since the lake formed more than 10,000 years ago.

REFERENCES CITED

Belcher, J. H., and Fogg, G., 1964, Chlorophyll derivatives and carotenoids in the sediments of two English lakes, *in* Miyake, Y., and Koyama, T., eds., Recent researches in the fields of hydrosphere, atmosphere, and nuclear geochemistry: Tokyo, Maruzen Co., p. 39–48.

Brown, S. R., 1968, Bacterial carotenoids from freshwater sediments: Limnology and Oceanography, v. 13, p. 233–241.

——, 1969, Paleolimnological evidence from fossil pigments: Verhandlungen International Vereinigung für Limnologie, v. 17, p. 95–103.

Brown, S. R., Daley, D. J., and McNeeley, R. N., 1977, Composition and stratigraphy of the fossil phorbin derivatives of Little Round Lake, Ontario: Limnology and Oceanography, v. 22, p. 336–348.

Culver, D. A., Vaga, R. M., Munch, C., and Harris, S. M., 1981, Paleoecology of Hall Lake, Washington—A history of meromixis and disturbance: Ecology, v. 62, p. 848–863.

Czeczuga, B., 1965, Quantitative changes in sedimentary chlorophyll in bed sediment of the Mikolajaki Lake during the postglacial period: Schweizerische Zeitschrift Hydrologie, v. 27, p. 88–89.

Czeczuga, B., and Czerpak, R., 1968, Investigations on vegetable pigments in postglacial bed sediments of lakes: Hydrologie, v. 30, p. 217–231.

Daley, R. J., 1973, Experimental characterization of lacustrine chlorophyll diagenesis. II. Bacterial, viral, and herbivore grazing effects: Archiv für Hydrobiologie, v. 72, p. 409–439.

Daley, R. J., and Brown, S. R., 1973, Experimental characterization of lacustrine chlorophyll diagenesis. I. Physiological and environmental effects: Archiv für Hydrobiologie, v. 72, p. 277–204.

Daley, R. J., Brown, S. R., and McNeeley, R. N., 1977, Chromatographic and SCEP measurements of fossil phorbins and the postglacial history of Little Round Lake, Ontario: Limnology and Oceanography, v. 22, p. 349–360.

Dean, W. E., and Gorham, E., 1976, Major chemical and mineral components of profundal surface sediments in Minnesota lakes: Limnology and Oceanography, v. 21, p. 259–284.

Engstrom, D. R., Swain, E. B., and Kingston, J. C., 1985, A palaeolimnological record of human disturbance from Harvey's Lake, Vermont: Geochemistry, pigments, and diatoms: Freshwater Biology, v. 15, p. 261–288.

Fogg, G. E., and Belcher, J. H., 1961, Pigments from the bottom deposits of an English Lake: New Phytology, v. 60,, p. 129–142.

Goodwin, T. W., 1980, The biochemistry of the carotenoids: New York, Chap-

man and Hall, 491 p.
Gorham, E., 1960, Chlorophyll derivatives in surface muds from the English lakes: Limnology and Oceanography, v. 5, p. 29–33.
——, 1961, Chlorophyll derivatives, sulfur, and carbon in sediment cores from two English lakes: Canadian Journal of Botany, v. 39, p. 333–338.
Gorham, E., and Sanger, J. E., 1964, Chlorophyll derivatives in woodland, swamp, and pond soils of Cedar Creek Natural History Area, Minnesota, U.S.A., in Miyake, Y., and Koyama, T., eds., Recent researches in the field of hydrosphere, atmosphere, and nuclear geochemistry: Tokyo, Maurzen Co., p. 1–12.
——, 1967, Plant pigments in woodland soils: Ecology, v. 48, p. 306–308.
——, 1972, Fossil pigments in the surface sediments of a meromictic lake: Limnology and Oceanography, v. 17, p. 618–622.
——, 1975, Fossil pigments in Minnesota lake sediments and their bearing upon the balance between terrestrial and aquatic inputs to sedimentary organic matter: Verhandlungen International Vereinigung für Limnologie, v. 19, p. 2267–2273.
——, 1976, Fossilized pigments as stratigraphic indicators of cultural eutrophication in Shagawa Lake, northeastern Minnesota: Geological Society of America Bulletin, v. 87, p. 1638–1640.
Gorham, E., Lund, J.W.G., Sanger, J. E., and Dean, W. E., 1974, Some relationships between algal standing crop, water chemistry, and sediment chemistry in the English lakes: Limnology and Oceanography, v. 19, p. 601–617.
Griffiths, M., 1978, Specific blue-green algal carotenoids in sediments of Esthwaite Water: Limnology and Oceanography, v. 23, p. 777–784.
Griffiths, M., and Edmondson, W. T., 1975, Burial of oscillaxanthin in the sediment of Lake Washington: Limnology and Oceanography, v. 20, p. 945–952.
Griffiths, M., Perrot, P. S., and Edmondson, W. T., 1969, Oscillaxanthin in the sediment of Lake Washington: Limnology and Oceanography, v. 14, p. 317–326.
Gray, C.B.J., 1974, Distributions of chlorophyll derivatives in sediments of the Great Lakes [M.S. thesis]: Kingston, Ontario, Queens University, 93 p.
Guilizzoni, P., Bonomi, G., Galanti, G., Ruggiu, D., and Saraceni, C., 1981, Relazione tra L'evoluzione trofica del Lago di Mergozzo ed il contenuto in pigmenti vegetali, sostanza organica, carbonio e azoto dei suoi sedimenti: Memorie dell'Instituto Italiano di Idrobiologia, v. 39, p. 119–145.
——, 1982, Basic trophic status and recent development of some Italian lakes as revealed by plant pigments and other chemical components in sediment cores: Memorie dell'Instituto Italiano di Idrobiologia, v. 40, p. 79–98.
——, 1983, Relationship between sedimentary pigments and primary production—Evidence from core analyses of twelve Italian lakes: Hydrobiologia, v. 103, p. 103–106.
Handa, N., 1975, Organogeochemical studies of a 200-meter core sample from Lake Biwa. III. The determination of chlorophyll derivatives and carotenoids: Japan Academy of Sciences Proceedings, v. 51, p. 442–446.
Kleerekoper, H., 1953, The mineralization of plankton: Fisheries Research Board of Canada Journal, v. 10, p. 283–291.
Manny, B. A., Wetzel, R. G., and Bailey, R. E., 1978, Paleolimnological sedimentation of organic carbon, nitrogen, phosphorus, fossil pigments, pollen, and diatoms in a hypereutrophic hardwater lake—A case history of eutrophication: Polska Archwm Hydrobiologie, v. 25, p. 243–267.
McIntosh, H. J., 1983, A paleolimnological investigation of the bacterial carotenoids of Sunfish Lake [M.S. thesis]: Kingston, Ontario, Queens University, 129 p.
Megard, R. O., 1967, Limnology, primary productivity, and carbonate sedimentation of Minnesota lakes: University of Minnesota, Limnological Research Center Interim Report 1, 69 p.
Murray, D. A., and Douglas, D. J., 1976, Some sedimentary pigment determinations in a 1.0 metre core from L. Ennell—A eutrophic lake in the Irish Midlands, in Horie, S., ed., Paleolimnology of Lake Biwa and the Japanese Pleistocene, Volume 4 (Paleoecology Symposium, 2nd, Copenhagen): Societe Internationale Vereinigung für Limnologie, Special Publication, p. 703–714.

Ohle, W., 1962, Der Stoffhaushalt der Seen als Grundlage einer allgemeinen Stoffwechseldynamik der Gewasser: Kieler Meerforsch, v. 18, p. 107–120.
Otsuki, A., and Wetzel, R. G., 1972, Coprecipitation of phosphate with carbonates in a marl lake: Limnology and Oceanography, v. 17, p. 762–767.
——, 1974, Calcium and total alkalinity budgets and calcium carbonate precipitation of a small hardwater lake: Archive für Hydrobiologie, v. 73, p. 14–30.
Pennington, W., Cambray, R. S., and Fisher, E. M., 1973, Observations on lake sediments using fallout ^{137}Cs as a tracer: Nature, v. 242, p. 324–326.
Sanger, J. E., 1971a, Identification and quantitative measurement of plant pigments in soil humus layers: Ecology, v. 52, p. 959–963.
——, 1971b, Quantitative investigations of leaf pigments from their inception of buds through autumn coloration to decomposition in falling leaves: Ecology, v. 52, p. 1075–1089.
——, 1988, Fossil pigments in paleoecology and paleolimnology: Palaeogeography, Palaeoclimatology, Palaeoecology, v. 62, p. 343–359.
Sanger, J. E., and Crowl, G. H., 1979, Fossil pigments as a guide to paleolimnology of Browns Lake, Ohio: Quaternary Research, v. 11, p. 342–352.
Sanger, J. E., and Gorham, E., 1970, The diversity of pigments in lake sediments and its ecological significance: Limnology and Oceanography, v. 15, p. 59–69.
——, 1972, Stratigraphy of fossil pigments as a guide to the postglacial history of Kirchner Marsh, Minnesota: Limnology and Oceanography, v. 17, p. 840–954.
Santelman, E. P., 1981, Fossilized plant pigments in sediments from four Minnesota lakes [M.S. thesis]: Minneapolis, University of Minnesota, 78 p.
Swain, E. B., 1985, Measurement and interpretation of sedimentary pigments: Freshwater Biology, v. 15, p. 53–75.
Tarapchak, S. J., 1973, Studies of phytoplankton distribution and indicators of trophic state in Minnesota lakes [Ph.D. thesis]: Minneapolis, University of Minnesota, p. 21–47.
Tutin, W., 1955, Preliminary observations on a year's cycle of sedimentation in Windermere, England: Memorie dell'Instituto Italiano di Idrobiologia, Supplemente 8, p. 447–484.
Vallentyne, J. R., 1956, Epiphasic and hypophasic carotenoids in postglacial lake sediments: Limnology and Oceanography, v. 1, p. 252–262.
——, 1957, Carotenoids in a 20,000-year old sediment from Searles Lake, California: Archives of Biochemistry and Biophysics, v. 70, p. 29–34.
——, 1960, Fossil pigments, in Allen, M. B., ed., Comparative biochemistry of photoreactive systems: New York, Academic Press, p. 83–105.
——, 1962, Solubility and decomposition of organic matter in nature: Archiv für Hydrobiologie, v. 58, p. 423–434.
Wetzel, R. G., 1970, Recent and postglacial production rates of a marl lake: Limnology and Oceanography, v. 15, p. 491–503.
——, 1972, The role of carbon in hard-water marl lakes, in Likens, G. E., ed., Nutrients and eutrophication: American Society of Limnology and Oceanography Special Publication 1, p. 84–97.
Wetzel, R. G., and Otsuki, A., 1974, Allochthonous organic carbon of a marl lake: Archiv für Hydrobiologie, v. 73, p. 31–56.
White, W. S., and Wetzel, R. G., 1975, Nitrogen, phosphorus, particulate and colloidal carbon content of sedimenting seston of a hardwater lake: Verhandlungen International Vereinigung für Limnologie, v. 19, p. 330–339.
Whitehead, D. R., Rochester, H., Jr., Rissing, S. W., Douglass, C. B., and Sheehan, M. S., 1973, Late glacial and postglacial productivity changes in a New England pond: Science, v. 181, p. 744–746.
Zullig, H., 1961, Die Bestimmung von Myxoxanthophyll in Bohrprofilen zum nachweis vergangener Blaualgenentfatung: International Vereinigung für Theoretische und Angewandte Limnologie Verhandlungen, v. 14, p. 263–270.
——, 1981, On the use of carotenoid stratigraphy in lake sediments for detecting past developments of phytoplankton: Limnology and Oceanography, v. 26, p. 970–976.
——, 1982, Die Entwicklung von St. Moritz zum Kurort im Spiegel der Sedimente des St. Moritzersee: Wasser, Energie, Luft, v. 74, p. 177–183.

MANUSCRIPT ACCEPTED BY THE SOCIETY JULY 27, 1992

Printed in U.S.A.

Surface sample analogues of Elk Lake fossil diatom assemblages

Richard B. Brugam
Department of Biological Sciences, Southern Illinois University, Edwardsville, Illinois 62026

ABSTRACT

Abundance maps of diatom percentages from 174 Minnesota lakes sediment surface samples show that many diatom species have centers of abundance in lake types from particular regions of the state. Small *Stephanodiscus* species characterize lakes in the southwestern prairies and in urbanized areas where trophic status is high. *Aulacoseira granulata* and *Stephanodiscus niagarae* are most abundant in the shallow, eutrophic lakes of southwestern Minnesota. These geographic associations result from environmental optima for species. Although correlations between particular species and environmental factors show high variance, clear relations can be demonstrated. In particular, the DECORANA program for ordination analysis shows that many species have clearly defined optima either in low- or high-alkalinity lakes. The relations discovered from the surface sample data set can be used to understand the fossil assemblages from Elk Lake, Minnesota. The dominance of small *Stephanodiscus* species in Elk Lake suggests that the lake has been somewhat eutrophic for most of its history. The appearance of *Aulacoseira ambigua* and *Aulacoseira granulata* during the prairie period at Elk Lake implies that the lake was shallower or more turbulent at that time. DECORANA ordination shows that the fossil diatom assemblages of Elk Lake have not changed much since the lake was formed. Thus, environmental changes at Elk Lake were probably very subtle.

INTRODUCTION

Diatoms are important indicators of past lake environments. Because they are well preserved as fossils, they are present in numbers that are suitable for statistical analysis, and their ecology is moderately well known. Investigators have found that sediment surface sample analogues are useful in understanding diatom ecology and in reconstructing past lake environments (Brugam, 1983, 1988; Dixit, 1986; Gasse, 1986; Charles and Norton, 1986; Huttunen and Meriläinen, 1986). Surface sample analogues are spatially and temporally integrated fossil diatom assemblages taken from the sediment-water interface of lakes whose environmental conditions are well known. Presumably, these surface samples represent the modern diatom communities in the lake. The living communities are, in turn, controlled by the currently prevailing environmental conditions. Thus, the fossils from the surface samples should be representative of modern environmental conditions. When a broad range of surface sample analogues is assembled, past lake environments can be reconstructed with considerable precision.

A growing surface sample analogue data set exists for Minnesota (Brugam, 1981; Brugam and Patterson, 1983; Brugam and others, 1988). The data set now contains 174 samples representing most of the different kinds of lakes in the state. This data set is useful in interpreting past environmental conditions at Elk Lake.

The Minnesota diatom data set reflects variations in lake water chemistry across the state. A gradient in water chemistry exists across the state; low alkalinity, somewhat acid lakes dominate in the northeast (Fig. 1), and high alkalinity, alkaline lakes dominate in the southwest (Gorham and others, 1983; Dean and others, Chapter 8). This variation correlates with the climatic gradient from high rainfall and low evapotranspiration in the

---

Brugam, R. B., 1993, Surface sample analogues of Elk Lake fossil diatom assemblages, *in* Bradbury, J. P., and Dean, W. E., eds., Elk Lake, Minnesota: Evidence for Rapid Climate Change in the North-Central United States: Boulder, Colorado, Geological Society of America Special Paper 276.

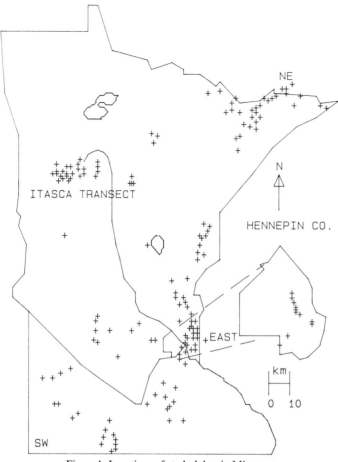

Figure 1. Locations of study lakes in Minnesota.

northeastern mixed conifer-hardwood forests to low rainfall and high evapotranspiration in the prairie southwest. The northeast is also underlain by granitic rocks with little $CaCO_3$, whereas the southwest has glacial tills high in carbonates.

Along the Wisconsin border, on the eastern edge of the state (EAST in Fig. 1), glacial tills of different origin produce surface waters of widely varying chemistry. Low-carbonate Superior Lobe till was deposited by glaciers coming from the northeast and contains kettle lakes low in alkalinity. In contrast, Grantsburg till came from a sublobe of the Des Moines glacier, which flowed from the west producing lakes containing highly alkaline water (Brugam, 1981). Lakes along the Wisconsin border are variable; low-carbonate lakes exist within a few kilometers of high-carbonate lakes (Brugam, 1981). Some of the lakes are located near suburban housing developments or farms that increase their trophic status.

Lakes near Minneapolis (population 371,000, Hennepin County, Fig. 1) and St. Paul (population 266,000) and their sub-urbs are heavily influenced by human activities: many of the lakes have eutrophied for this reason (Brugam and Speziale, 1983).

The Itasca transect (Fig. 1), a line extending directly west from Itasca State Park for 40 km in northwestern Minnesota, represents a location where the effects of watershed vegetation on lakes can be studied. Although the entire transect is underlain by carbonate-rich glacial tills, which cause the lakes to be alkaline and rich in nutrients, it touches three vegetation formations. The eastern end is located in a mixed conifer-hardwood forest. The central part of the transect is located in a deciduous forest, and the western end of the transect is prairie. The western lakes are shallow and eutrophic, but the eastern lakes are deep, and have variable trophic statuses. Elk Lake is located in the eastern end of the Itasca transect. Paleoecological studies of the transect were begun by McAndrews (1966) and continued by Birks (1973) and Synerholm (1979).

Surface sample assemblages from modern lakes in all regions of Minnesota are used to interpret the past environments of

Elk Lake. The data set represents a broad range of modern environments, including some that are analogous to past environments at Elk Lake.

METHODS

Sampling of surface sediments

Diatom samples were collected over a long period of time by various investigators (Table 1). Nearly all counts were done by R. Brugam and C. Patterson. Care was taken to coordinate identifications so that counts would be comparable.

Diatom identification and taxonomy

Diatom identification was based on Patrick and Reimer (1966, 1975), Hustedt (1930, 1977a, 1977b, 1977c) and Cleve-Euler (1968, 1969). Assignment of species to environmental preference groups followed Lowe (1974).

The small *Stephanodiscus* species in the data set have caused the most taxonomic difficulties. Because of the changes in taxonomy of these species over the past 10 years (Håkansson and Locker, 1981; Håkansson and Stoermer, 1984), the names applied to the small *Stephanodiscus* species used in this data set were deemed unreliable. For the statistical analyses, the species are lumped: the taxa included in this grouping are *Stephanodiscus minutula, S. dubius, S. hantzschii, S. parvus, S. alpinus,* and *S. subtilus.*

Chemical analysis and environmental variables

Chemical analyses followed *Standard Methods for the Examination of Water and Wastewater* (American Public Health Association, 1985). Total phosphorus was measured on unfiltered samples using persulfate digestion followed by analysis using the phosphomolybdate blue method. Alkalinity was measured by titration of the sample with 0.02 N H_2SO_4. A glass electrode pH meter was used to measure pH within 24 hours of sample collection. Specific conductivity was measured using a conductivity meter standardized with KCl solution.

Secchi disk transparency data was taken from the Minnesota Secchi Disk Project (Shapiro and others, 1975). Averages were made of summer data.

Statistical analysis

DECORANA ordination (Hill, 1979; Gauch, 1982) was performed using 31 species of the surface sample data set (Table 2). A subset of species was chosen because the data set contained more than 400 species, too many to use in a statistical analysis. The species chosen were the ones that were most abundant and that had shown strong environmental preferences in other investigations (Brugam and others, 1988; Brugam and Patterson, 1983; Brugam, 1979). No transformation of the original percentage data was made before ordination.

The SPSSx computer package (SPSS, Inc., 1983) was used to perform Spearman rank correlations, cluster analysis, and Mann-Whitney U tests.

Cluster analysis was performed using Ward's method (SPSS, Inc., 1983) on untransformed percentage data from the Elk Lake core and the 174 surface samples. The same 31 species used in the ordination were used for cluster analysis.

RESULTS

Geographical distributions of diatom species

Maps of diatom abundances across Minnesota are helpful in identifying environmental optima for particular species. With maps, the distributions of species can be compared with known lake characteristics.

TABLE 1. SAMPLING METHODS AND INVESTIGATORS FOR THE SURFACE SAMPLE DATA SET

Number of Samples	Location	Sample Method	Investigator	Reference
100	Throughout Minnesota	Hongve Corer	Brugam	Brugam, 1983
22	Shallow lakes in Minnesota and South Dakota	Ekman Dredge	Brugam	Birks, 1973; Synerholm, 1979; Brugam, 1983
28	Eastern Minnesota	Hongve Corer	Patterson	Brugam, 1981; Brugam and Patterson, 1983
8	Throughout Minnesota	Ekman Dredge	Florin	Bright, 1968
16	Northeast Minnesota	Hongve Corer	Patterson	Brugam and others, 1988

TABLE 2. SPECIES USED IN THE DECORANA ANALYSIS*

	Name	pH Preference
1.	Aulacoseira ambigua (Grun.) O. Müller	Alkaliphil
2.	Sum of Aulacoseira qranulata (E) Ralfs and A. granulata angustissima Müller	Alkaliphil
3.	Aulacoseira italica (Ehr.) Küntz.	Alkaliphil
4.	Aulacoseira distans (Ehr.) Küntz.	Acidophil
5.	Cyclotella comta (Ehr.) Kütz.	Alkaliphil
6.	Cyclotella meneghiniana Kütz.	Alkaliphil
7.	Cyclotella michiganiana Skv.	Unknown
8.	Cyclotella stelligera Cl. and Grun.	Indifferent
9.	Cyclotella ocellata Pant.	Indifferent
10.	Stephanodiscus niagarae Ehr.	Unknown
11.	Sum of Small Stephanodiscus	Alkaliphil
12.	Tabellaria fenestrata (Lyngb.) Küntz.	Acidophil
13.	Tabellaria flocculosa (Roth) Küntz.	Acidophil
14.	Fragilaria crotonensis Kitton	Alkaliphil
15.	Fragilaria construens (Ehr.) Grun.	Alkaliphil
16.	Sum of Fragilaria construens venter (Ehr.) Grun. and Fragilaria pinnata Ehr.	Alkaliphil
17.	Fragilaria brevistriata Grun.	Alkaliphil
18.	Anomoeoneis serians (Brèb. ex Kütz.) Cl.	Acidophil
19.	Synedra ulna (Nitz.) Ehr.	Alkaliphil
20.	Synedra acus Kütz.	Alkaliphil
21.	Synedra nana Meist.	Unknown
22.	Asterionella formosa (Hantz.) Grun.	Alkaliphil
23.	Eunotia incisa W. Sm. ex Greg.	Unknown
24.	Eunotia flexuosa Brèb. ex Kütz.	Acidophil
25.	Cocconeis placentula Ehr.	Alkaliphil
26.	Mastogloia smithii Thwaites ex W. Sm.	Alkaliphil
27.	Frustulia rhomboides (Ehr.) DeT.	Acidophil
28.	Pinnularia biceps Greg.	Acidophil
29.	Amphora ovalis (Kütz.) and varieties	Alkaliphil
30.	Rhopalodia gibba (Ehr.) O. Müller	Alkalibiontic
31.	Sum of Gomphonemas	Unknown

*Numbered according to their order of appearance in the program.

The small *Stephanodiscus* species (Fig. 2) are most abundant (>25%) in the prairie lakes of southwestern Minnesota, in the lakes of Hennepin and Ramsey counties, and in eastern Minnesota, where eutrophication problems are important.

Fragilaria crotonensis (Fig. 3) is not as abundant as the small *Stephanodiscus* species. Its main centers of abundance (>10%) are in the lakes of eastern Minnesota, Hennepin County, and the Itasca transect. It is not abundant in the prairie southwest or in the northeast.

Aulacoseira ambigua (Fig. 4) is a very abundant and widespread species throughout Minnesota, and shows no obvious geographical preferences.

The *Aulacoseira granulata* mapped in Figure 5 includes both the typical form and the more lightly constructed variety, *Aulacoseira granulata v. angustissima.* It is most abundant in the lakes of the prairie part of the state, but it is not strictly limited to this region.

High numbers of *Stephanodiscus niagarae* (>10%) are limited to the southwestern prairie lakes (Fig. 6). This species only reaches these abundances at Elk Lake between 1000 and 3000 varve yr B.P. Lower abundances of the species are found in many of the calcareous lakes of southern and southwestern Minnesota.

Cyclotella michiganiana is very rare in surface samples from Minnesota (Fig. 7). It also shows no strong geographical trends. Only a few modern lakes have abundances that approach those from the fossil record of Elk Lake: Sarah Lake and McIntyre Lake (Table 3), both in northeastern Minnesota. Two other lakes with high abundances of *C. michiganiana* are Fish Lake in the southwest and Arco Lake in the Itasca transect (Table 3).

Cyclotella ocellata, like *C. michiganiana,* is rare in modern Minnesota lakes (Fig. 8). It occurs in abundances greater than 1% in only a few lakes. It has also been very rare in Elk Lake except at the very beginning of the fossil record.

Cyclotella comta is not as rare as the last two *Cyclotella* species (Fig. 9), but it seldom reaches the levels seen in the Elk Lake core. The species is widespread across the state, but major centers of abundance exist on the Itasca transect and in the northeast. The highest abundance of this species occurs at Lake of the Valley (Table 3), a clear (Sechi disc >7 m) lake on the Itasca transect.

Environmental optima

It is possible to use a surface sediment data set like the Minnesota samples to understand environmental optima of diatom species. Although single chemical samples cannot be expected to show the seasonal changes in water chemistry that control diatom succession, they are adequate to distinguish between alkaline hard-water and acid soft-water lakes. Single water

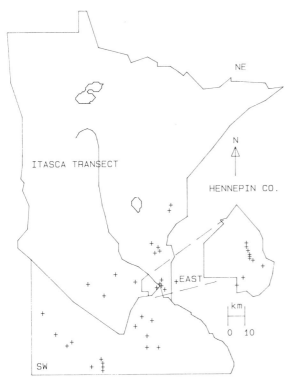

Figure 2. Locations of lakes with percentages of small *Stephanodiscus* species greater than 25%.

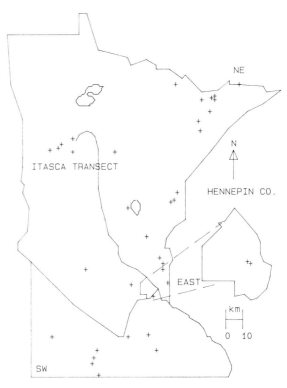

Figure 4. Locations of lakes with percentages of *Aulacoseira ambigua* greater than 25%.

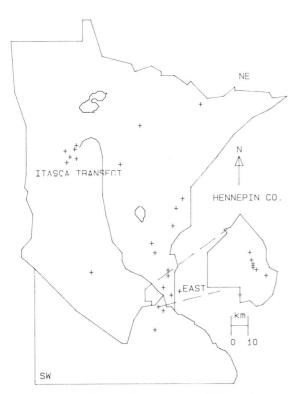

Figure 3. Locations of lakes with percentages of *Fragilaria crotonensis* greater than 10%.

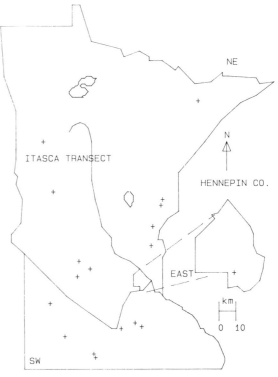

Figure 5. Locations of lakes with percentages of *Aulacoseira granulata* greater than 20%.

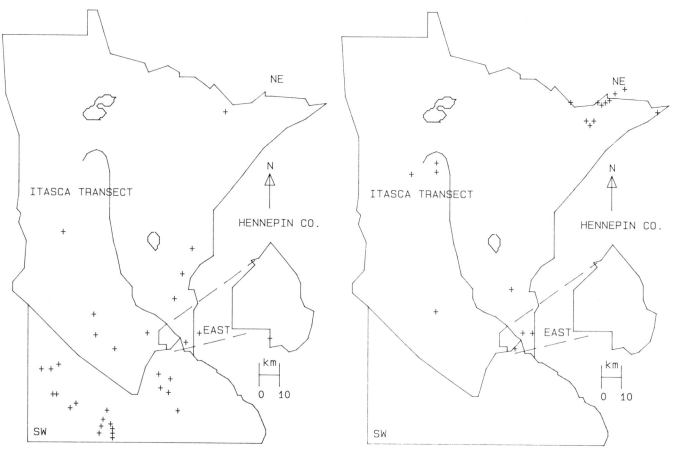

Figure 6. Locations of lakes with percentages of *Stephanodiscus niagarae* greater than 5%.

Figure 7. Locations of lakes with percentages of *Cyclotella michiganiana* greater than 1%.

chemistry samples should also allow us to measure trophic status of the study lakes.

The Spearman rank correlation coefficient can be a useful index of the relation between diatom abundance and environmental variable. Because it is a nonparametric statistic (Sokal and Rohlf, 1981), it does not force nutrient relations of diatoms into a linear model, as does linear regression analysis. The Spearman coefficient is organized in the same way as the more familiar Pearson product moment correlation coefficient. Positive correlation results in $0 < r_s < 1.0$, and negative correlation yields $-1.0 < r_s < 0$.

Correlation coefficients must always be compared with scatter diagrams of the original data. Scatter diagrams of the abundances of selected diatom taxa versus water chemistry (Figs. 10, 11, 12, and 13) show a large amount of variance. The original data are presented in Table 4 (environmental variables and lake locations) and Table 3 (diatom percentages). It is evident from the scatter diagrams that there is no linear relation between diatom abundances and environmental variables. The high variability of the data probably comes from the diversity of factors that limit the abundance of particular species. For example, *Stephanodiscus niagarae* seems to have an environmental optimum in shallow lakes (Fig. 10), because its highest abundances occur in lakes less than 20 m deep.

The dominant relation between the abundances of diatom taxa and environmental variables seems to be a step function in which a species is rare below (or above) a particular cut-off level of an environmental variable, but abundant above (or below) it. A step function is the relation that might be expected if resource-based competition as described by Tilman (1977) is important in determining diatom distributions. This theory predicts a "smoothed" step function relation between the abundance of particular species and ratios of limiting nutrients (Reynolds, 1984).

The small *Stephanodiscus* (Table 5) species reach their maximum percentages in high conductivity ($r_s = 0.56$), high total

Elk Lake fossil diatom assemblages

TABLE 3. PERCENT ABUNDANCES OF DIATOMS USED IN THE DECORANA ANALYSIS*

Lake	2	10	11	6	19	20	25	29	26	30	17	31	1	15	22	9
Woodworth 1	0	0	0	0	0	0	0	0	0	9.7	0	1.7	0	0	0	0
Woodworth 2	0	0	0	0	0	0	4.6	0	0	2.1	0	3.7	0	26.7	0	0
Telegraph	0	0	2	0	0	0	0.1	0.3	0.4	0	26	0.9	0	0	0	0
Reichow	0	0	0	0	0.4	0.4	0	5.7	0	0.7	1.4	10.1	0	1.7	0	0
Hen Pond	0	0	0	0	0	0.5	3.2	2.4	1.6	1.9	20.3	8.6	0	4.1	0	0
Waubun Pr.	0	0	0	0	0	0	5.4	1.2	0.7	0	9.3	31.9	0	0	0	0
Ballard	0	0	0	0.4	0.4	0	2.6	3.2	0.6	0.9	2.4	12.4	0	3	0	0
Quallen	0	0	0	0	0.1	6.8	1.7	1.4	0	0	3.8	3.1	0.6	1.7	0	0
Squirrel	0	0	0	0.1	0.1	0.1	0.5	0.4	1.7	1.6	67	0.5	0.7	1.2	0	0
Terhell	2.8	0	0.3	0	0.1	0	8.7	6.4	0.6	0	23.3	2.6	0	1.3	0	0
Wyoming	0	0	0.8	0.1	0	0	25.7	0.8	0.4	0	0	3.7	0	0	0	0
Woodworth 4	0	0	0	0.2	0.4	0	12.1	23.3	0	0	0	8.7	0	0	0	0
Cucumber	0	0	4.5	0	0.2	0	38.7	2.8	0	0.9	7.9	2.3	0	0	0	0
Thompson	2.1	0	8.3	0	0	0	48.2	2.5	0	0	0	3.7	0.2	0	0	0
Mays	1.5	0	2.2	0	0	0	8.8	0.6	0	0	0	1.7	0.2	0.9	0	0
Cedar Bog	2.9	0	1.9	0	4	2.3	18.4	1.5	0	1.2	0	12.5	0	0	0	0
Cokato	0.7	0.3	40.9	0	0	0	0.5	0.3	0	0	3	0.4	0.2	33.2	0	0
Tetonka	7.8	3	11	0	0.1	0	1.3	0.6	0	0	4.5	1.2	0.2	36.1	0	0
School Grove	26.5	6.5	8	0.2	0	3.7	2.6	1.4	0	0	2.6	0.2	4.9	24.8	1.2	0
Wood	6.9	7.8	6.7	5.4	0	1.1	6.1	8.7	0.2	1.1	6.3	1.4	2.8	0	0	0
Wagonga	47.8	0	10.1	3.7	0	0	4.3	1.3	0.6	0.2	8	3.1	4.7	5.6	0	0
Andrew	38.7	3.1	0.3	0	0	0	1.1	8.3	0.8	0.3	5.3	2.2	15.6	2.5	0	0.3
Willmar	20.2	0.2	1.1	0.1	0	0.9	2.8	2.7	1.4	0.1	14	11.9	21.3	10.9	0	0
Minnewashta	5.2	6	9.9	0	0.4	0	2.5	0.3	0.4	0.1	0	0.2	30.1	4.8	1	0.1
Hunt	20.9	6.2	6	0.2	0	0	0	0	0.2	0	2.1	0.5	20.8	9.6	6.5	0
Beaver	3.4	6.1	41.4	0	0	0.1	0	0.3	0	0	0.9	0.2	25.7	2.3	0.6	0
Pokegamma	24.6	15.8	11.1	0.7	0.2	0.2	0.7	0.5	0	0	5.4	0	15.3	2.3	0	0
Sarah	11.9	8.7	1.2	0.7	0	0.5	1	2.2	0	0	0	0	62.8	0	0	0
Cottonwood	16.6	0.3	23.4	25.6	1.8	1.4	0.8	3.8	0	0	0	1.3	0.5	0	0	0
Waubun	27.3	0	0	23	0.5	0.2	3.7	2.2	0	0	0	0	27.8	0	0	0
Buffalo	18	24	6.8	1.6	0.1	0.1	1.9	0.7	0	0.3	0.4	0	31.8	1.4	0	0
Cedar	19.7	18.9	10	1.5	0	0.5	0.2	0.9	0	0	0	0	41.4	0.3	0	0
S. Silver	3.8	15.7	54	2.1	0	0	1.7	0.9	0	0	1.7	0.2	0.2	1.5	0	0
Charlotte	26	17.5	31.1	8.6	0.1	0	0	0.3	0	0	0	0.9	8.9	2	0	0
Dead Coon	5.1	13.3	65.2	2.9	0.3	0	0.2	0.3	0	0	0	0.4	3.7	0	0	0
Shetek	22.2	24.7	37	1.2	0.2	0	0.2	0	0	0	0	0.2	8.8	0.8	0.4	0
Oaks	29.8	5.8	36.7	7.1	0.2	0	0	1.3	0	0	0	0.2	12.3	0.2	0	0
Talcott	13.3	20.6	42.4	6.7	0.1	0.3	0.4	0.1	0	0	0.1	0.1	10.6	0	1.8	0
Allie	12.2	6.7	43	6.7	1.6	0	0	5	0	0	0.2	1.2	0	0	0	0
Traverse	42	26.5	4.2	2.8	0.2	0	0.3	2	0	0	0	0	10.7	0	0	0
Fox	13.2	11.5	51.8	0.2	0	0	0.9	0.6	0	0	0	0	5.6	0.6	2.4	0
Budd	21.1	19.2	48.3	1.2	0	0	0.1	0.1	0	0	0.3	0.6	2.3	0.5	0	0
Hall	26.9	12.7	37.7	1	0.2	0	0.2	2.2	0.2	0	1.4	0	5.2	1 4	0	0
Bryant	22.2	2.2	38.6	0	0.2	0	1.4	0.6	0	0.1	0.1	0.2	17	1	0.6	0
Riley	16.9	6.6	36.2	1.2	0	0	0.5	0.2	6	0	0.6	0.4	0	0	2.2	0
Mazaska	32.5	12.8	2.4	0.8	0	0	1.7	0.5	0	0.1	0	0	25.3	1.9	5.2	0
Kandyohi	5.2	11.3	49.2	0	0	0	1.4	2.4	0	0	0.2	0	0.2	2.4	3.6	0
Skogman	7.9	5.1	37.6	0	0.2	0	1.3	0.2	0	0	0	0.8	19.6	0	0.9	0
Big Horseshoe	27.7	2.9	29.3	0	0	0	1.6	0	0	0	0	0.3	11.5	1	1.7	0
Big Pine	22.1	5.6	27.3	0.6	0.9	0	4.2	0.2	0	0	0	1.1	13.6	0.2	1.6	0
Carver	4.8	5.9	65.1	0	0	0	2.7	0.4	0	0	0.7	0.4	0	0.2	2.9	0
Bone	14.2	2.9	22.6	0	0	0	0.3	0	0	0	0	0	26.3	1	0	0
Shields	17.4	9.8	31.3	0	0.3	0	0.6	1.3	0.3	0	0	0.6	5.1	8.2	0.6	0
Madison	21.8	4.3	23.1	0.3	0	0	1.1	0	0	0	0	0	29.1	0	3.3	0
St. Olaf	13.1	2.9	46.4	0	0.1	0	2	0.1	0	0	0.3	0	11.5	1	0	0
Nest	24.2	11	8	0	0	0.2	1.4	0.5	0.7	0	0.6	1	27.8	0	1.1	0.1
Pine	23.3	4	27.3	0.4	0	0.1	4.2	0.4	0	0	0	1.8	15.2	0.1	1.4	0
Tamarack	20.7	0	13.1	1.9	0	0.5	0.7	0.2	0	0	0	0.2	0.2	0	15.6	0
Long	3	0	4.1	0.4	0	0.9	0.4	0	0	0	0	3.5	16.8	0	1.9	0
Shields	4.6	1.3	33.8	16.3	1.1	0	2.4	0	0	0	0	0.7	2	0	0	0

TABLE 3. PERCENT ABUNDANCES OF DIATOMS USED IN THE DECORANA ANALYSIS* (continued)

Lake	14	5	16	7	21	3	12	24	18	23	27	28	4	8	13
Woodworth 1	0	0	0	0	0	0	0	0	0	0	0	0	0	0	0
Woodworth 2	0	0	16.7	0	0	0	0.1	0	0.2	0	0	0	0.3	0	0
Telegraph	0	0	0	0	0	0	0	0	0	0	0	0	0	0	0
Reichow	0	0	14	0	0	0	0	0	1	0	0	0.2	0	0	0
Hen Pond	0	0	3.2	0.4	0	0	0	0	1	0	0	0	0	0	0
Waubun Pr.	0.3	0	4.9	0	0	0	0	0	1.3	0	0	0	0	0	0
Ballard	0	0	8.3	0	0	0	0	0	0.4	0	0	0.2	0	0	0
Quallen	0	0	60.5	0	0	0	0	0	0	0	0	0	0	0	0
Squirrel	0	0	6.2	0	0	0	0	0	0	0	0	0	0	0	0
Terhell	0	0	49.5	0	0	0	0	0	0	0	0	0	0	0	0
Wyoming	1.3	0	4	0	0	0	0	0.5	0	0	0	0	0	0	0
Woodworth 4	0.8	0	0	0	0	0	0	0	0	0	0	0	0	0	0
Cucumber	0	0	3.2	0	0	0	0	0	0	0	0	0	0	0	0
Thompson	0	0	0.9	0	0	0	0	0	0	0	0	2.8	0	0	0
Mays	3.2	0	25.3	0	0	0	0	0	0	0	0	0	0	0.3	0
Cedar Bog	0	0	11.5	0.1	0	0	0	1.9	0	0	0	0	0	0.4	0
Cokato	0	0.3	2.3	0	0	0	0	0	0	0	0	0	0	0	0
Tetonka	0	0	10.3	0	0	0	0	0	0	0	0	0	0	0	0
School Grove	0.2	0	7.5	0	0	0	0	0	0	0	0	0	0	0	0
Wood	2.2	0.4	4.5	0	0	0	0	0	0	0	0	0	0	0	0
Wagonga	0.4	0	4.9	0	0	0	0	0	0	0	0	0	0	0	0
Andrew	1.4	0	7.2	0	0	0	0	0	0	0	0	0	0	0	0
Willmar	0.9	0.1	2	0	0	0.1	0	0	0	0	0	0	0	0	0
Minnewashta	2.3	4.6	8.1	0	0	0.1	0	0.4	0	0	0	0	0	0	0.8
Hunt	2.8	3.1	1.1	0	0	0.1	0	0	0	0	0.2	0	0	0	0
Beaver	2.6	2.1	6.2	0	0	0	0	0	0	0	0	0	0	0	0
Pokegamma	4.5	0	7.9	0	0	0	0	0	0	0	0	0	0	0.7	0
Sarah	0	0	0	0	0	0.5	0.2	0	0	0.8	0	0	0	0	0
Cottonwood	0	0	0	0	0	0	0	0	0	0	0	0	0	0	0
Waubun	0	0	0	0	0	0	0	0	0	0	0	0	0	0	0
Buffalo	0	0	6.2	0	0	0.1	0	0	0	0	0	0	0	0	0
Cedar	0	0	0.2	0	1.4	0.7	0	0	0	0	0	0	0	0	0
S. Silver	0	0	1.7	0	0	0	0	0	0	0	0	0	0	0	0
Charlotte	0	0	1.3	0	0	0	0	0	0	0	0	0	0.6	0	0
Dead Coon	0	0	0	0	0	0	0	0	0	0	0	0	0	0.3	0
Shetek	0	0	0.6	0	0	0.2	0	0	0	0	0	0	0	0	0
Oaks	0	0	0.7	0	0.2	0	0	0	0	0	0	0	0	0	0
Talcott	0	0	0	0	0	0.4	0	0.1	0	0	0	0	0	0	0
Allie	0	0.2	0	0	0	0	0	0	0	0	0	0	0	0	0
Traverse	0	0	0	0	0	0	0	0	0	0	0	0	0	0	0
Fox	0.9	0	1.5	0	0	0	0	0	0	0	0	0	0	0	0
Budd	0.4	0	0.5	0	0	0	0.4	0	0	0	0	0	0	0.1	0
Hall	0.8	0	3.4	0.2	0	0	0	0	0	0	0	0	0	0	0
Bryant	0.8	0	4.7	0	0	0	0	0	0	0	0	0	0	0	0
Riley	16.9	0	8.5	0	0	0	0	0	0	0	0	0	0	0	0
Mazaska	8.9	0	2.2	0	0	0	0	0	0	0	0	0	0	0	0
Kandyohi	8.7	0	0.4	0	0	0	0	0	0	0	0	0	0	0	0
Skogman	4.6	0	0.9	0	0	0	0.2	0.4	0	0	0	0	0.4	0	0
Big Horseshoe	6.8	0.3	1.3	0	0	0	0	0.3	0	0	0	0	0	0	0
Big Pine	4.9	0	5.1	0	0	0	2.5	0	0	0	0	0	0.3	0.3	0.2
Carver	2.2	0	0.9	0	0	0	0	0	0	0	0	0.2	0	0.2	0
Bone	1.5	0	11	0	0	0	0	0	0	0	0	0	0	0.2	0
Shields	4.7	0	4.4	0	0	0	0	0	0	0	0	0	0	0	0
Madison	6.6	0	1.2	0	0	0	0	0	0	0	0	0	0	0.7	0
St. Olaf	1.3	8.6	1.8	0	0	0	0	0	0	0	0	0	0	0	0
West	7.3	0	0.4	0	0	0.1	0.1	0	0	0	0	0	0	0	0
Pine	4.4	0	4.5	0	0	0	2	0.1	0	0	0	0	0.4	0.5	0
Tamarack	5.5	0	0	0	0.7	0	0	0	0	0	0	0	0	18.9	0
Long	2.1	0	56.5	0	0	0	0	0	0	0	0	0	0	0.7	0
Shields	9.6	0	0.8	0	0	0	0.4	0	0	0	0	0	0	2.4	0

TABLE 3. PERCENT ABUNDANCES OF DIATOMS USED IN THE DECORANA ANALYSIS* (continued)

Lake	2	10	11	6	19	20	25	29	26	30	17	31	1	15	22	9
Sakatah	3.4	0	61.6	1.1	0.3	0	2.4	1.4	0	0	0	0.6	1.1	11.3	0	0
Fish Lake	7.5	7.8	29.3	0.5	0.2	0	5.5	0.7	0	0	1.2	0.2	0	0.3	2.5	0
Dudley	0.3	0	28.7	0.3	0.7	0	5.7	0	0	1	0	0.6	15.9	0.1	8.6	0
Wabasso	0	0	48.1	0	0.4	0	3.2	0	0	0.2	0	1.1	0	0	6.2	0
Little Horse	9.4	4.8	13.7	0	0.2	0	3.1	0	0	0	0	0.5	21.4	0	18	0
Grove	44.7	2.5	8.2	0	0.8	0	1.7	0	0	0	0	0	0.4	0	9.1	0
Hiawatha	2.6	0.4	25.6	0	0.2	1	3.2	0	0	0	0.8	0.6	8.9	0	11.7	0
Wirth	0.6	1.6	63.8	0.9	0.1	1.9	1.7	0.2	0	0	1.4	1.1	3	0.2	0	0
Twin	0.3	0.1	62.3	0	0.6	0.6	0.7	0.1	0.1	0	0	0.5	0.4	0	9.3	0
Lake of Isles	2	0.8	42.4	2	0.1	3.3	1	1	0	0	0	0	6.4	0	8.1	0
Twin	2.4	2	49.5	0	0.2	3.4	2.8	0	0	0	0	0.4	1.3	0	10	0
Clear	0.8	0	19.3	0.4	0	6.8	2.5	0	0	0	1.4	2.9	0.8	0	10.4	0
Brownie	6.7	0.6	33	2.2	0.9	0.6	0.4	0	0	0.1	0	2.9	2.3	0	12.8	0
McCarron	0.3	3	24.9	2.1	0.2	0	3.5	0	0	0	0	0.2	0	0	0.5	0
Adams	0	0	70.9	0	0	0	0.1	0	0	0	0.2	0.7	0	0	0	0
Fox	0.2	9.3	25.3	0	0.1	0	1.3	0	0	0	0	0	0	0.1	3.8	0
Owasso	3.1	0	45	0	0	0	0.2	0	0	0	0.2	1.2	11.6	0.7	3.6	0
Gervais	2.3	1.7	57.2	0.5	0	0	0.5	0	0	0	0	1.1	4.9	0.4	4.9	0
Cloverdale	0.8	0	0	0	0	1.8	0	0	0	0	0	0.6	14.4	0	0	0
Fish 2	0.6	0	12.3	0	0	2.6	0	0	0	0	0	1	0	0	3.7	0
Holland	0	0	7.6	0	0.7	0.2	2.1	0	0	0	0	0	4	3.6	5	0
Powers	2.4	0.2	11.6	0.7	0	0.7	1.1	0.4	0	0	0	1.1	6.9	0	0	0
McCraney	18.7	0	0	0	2.5	0	0.8	0	0	0	0.2	0.2	52.8	0.2	10.2	0
Long	9.6	9.9	7.8	0	0	0	0.4	2	0	0	0.2	0.6	34.1	3.9	7	0
Island	4.3	1	1.1	0	0.3	0	1	0	0	0	0	0.3	40.1	0	3.6	0
Fall	37	1.9	7.9	0.1	0	0	0.5	0.1	0	0	0	0.3	11.6	1.6	2.2	0
Whitefish	4.1	3.4	1.4	0	2.1	0.2	0.3	1.2	1	0	2.5	6.8	30.8	0	1.4	0.9
Cedar	1.2	1.2	49.5	0.4	0	0.4	0.5	0.2	0	0	0.1	0.7	11.6	0	0	1.5
Harriet	0.2	2.3	31.5	0	0	0.4	1.3	0.3	0.2	0.2	0.2	1.2	25.6	0.3	6.2	0.1
Calhoun	9.5	2.6	33.3	0	0	0.2	1.9	0.3	0	0	0	0.6	25.7	0	3.7	0
Nokomis	3.2	2.7	1.5	0.2	0	0.2	1.2	0.4	0	0	2.7	3	20.2	1.2	1	0
Washburn	6.2	1.5	1.5	0.1	0.5	0.2	0.1	0.1	0	0	0.1	0.5	64.3	0	1.7	0.1
Leavitt	6.1	0.3	16.1	0	2.2	0	3.1	0.3	0	0.6	1.2	1.8	11.6	1.5	0	0
Forest	12.5	1.4	3.3	0.5	0	1.4	2.8	3	0	0	0.7	0.7	29.9	1.5	5.9	0
Green	6.7	1.2	6.3	0	0.5	0.3	2.5	1.8	3	1.2	1.5	11.2	8.7	0	4.2	1.5
Comfort	9.4	1.6	8.5	0.2	0	0	0.9	0	0	0	0	0.9	38.2	4.3	4.3	0
Goose	9.4	1.9	35.9	0.2	0.2	0	0	0	0	0	0.2	1.1	24.6	2.1	2.3	0
Fish 1	18.6	3.2	1.1	0	0	0	1.1	0	0	0	0.2	1.4	30.9	0	1.2	0
Devils	5.8	2.9	0.3	0	0	0	1.4	0.8	0	0	0.3	1.4	31.7	12.1	0	0
Kroon	8.8	0	1.5	0	0	0.2	3.5	0	0	0	0	1.7	17.8	0.4	0	0
Clear	8.7	2.9	12.6	0	0	0	0	1.6	0	0	1.2	1.7	23.9	3.7	0.2	0
Square	0	4.3	2.2	0	0.2	0	0	1	0	0	0.2	0	25.4	0.4	2	0
Maple	8.5	2	8.6	0	0.3	0	1.8	0.5	0	0	3.3	0	26.6	0	8.3	0
Bright	4	0.2	1.1	0.1	0	0.1	0	1.7	0	0	2.1	0.3	51.8	16.9	0	0
Shagawa	9.9	7.4	22.6	0	0	0	0.6	0.2	0	0	2.9	0.2	17.3	1.6	1	0
Wisc. Pine	0	6.1	28.5	0	0.3	0.1	0.4	0.4	0	0.4	2.1	0	0.9	6.3	2.9	11.3
Christmas	0	4.6	29.2	0	0	0	1.3	1.9	0.8	0	2.5	1.6	8.6	6.5	0	0
Pickerel	0	1.4	0	0	0	0	0	0.1	0	0.5	1.7	0.1	52.6	10.9	3.8	0
Lake of Valley	0	0	0	0	0.2	0	0.2	2.2	0	0	1.8	0	30.2	5.7	0.4	0
Sturgeon	4.8	0.5	0	0	0.5	0	2.2	2.9	0	0	0	0.2	13	2.6	4.3	0.5
Long	0.4	0.4	0	0	0.2	0	0.7	0.4	0	0	0	0.4	59.5	0.4	1.4	0
Pike	0.6	2.5	5.4	0.1	1.1	0	0.3	0	0.1	0	0	0.9	16.9	0.4	0	1
Nemec	2.7	0.5	0	0	0	3.9	0	0.7	0	0	0.8	1.4	18.2	7.7	0.2	0
Elk	0	0.5	0	0	1.6	0	3.9	3.6	0.3	0	0.7	0.2	0	0.5	0	0
Gull	0.5	0	0	0	11.6	3.7	5.9	0.9	0	0	0	33.5	2.7	0.7	4.3	0.9
Fenske	0	0	0	0	0	0	0	0.2	0	0	10.5	0	12.5	0	0.9	0.2
Saganaga	0.4	0.8	0.5	0.1	0.2	0.1	0	0.2	0	0	0	0	4	0.7	6.7	0
Trout Bay	0.6	0.7	0.7	0	0	0.2	0	0	0	0	0	0.5	19.3	2.5	2.8	0
L.S. Burnside	0.3	0.3	0.3	0.3	0	0.1	0.1	0	0	0	0.4	0.4	32.6	0	2.5	0.3
Burntside	0	0.6	0	0.3	0.1	0.3	0.2	0.1	0	0	2.1	0.7	2.6	2.6	0	4.6

TABLE 3. PERCENT ABUNDANCES OF DIATOMS USED IN THE DECORANA ANALYSIS* (continued)

Lake	14	5	16	7	21	3	12	24	18	23	27	28	4	8	13
Sakatah	2	0	3.1	0	0	0	0	0	0	0	0	0	0	0.3	0
Fish Lake	16.3	0	0.5	16.3	0	0	0	0	0	0	0	0	0	0	0
Dudley	0.5	10.4	2.1	0	0	0	0	0	0	0	0	0	0	0	0
Wabasso	6.9	2.5	0.2	0.2	0.2	0	0	0	0	0	0	0	0	0	0
Little Horse	4.9	2.8	0.4	0	0	0	0	0	0	0	0	0	0	0.2	0
Grove	0	8.7	0	2.7	0	0	0	0	0	0	0	0	0	0	0
Hiawatha	19.7	0	0	0	0	0	0.6	0	0	0	0	0	0	0	0
Wirth	7.4	0.4	0.3	0	0	0.1	0	0	0	0	0	0	0	0	0
Twin	14.9	0	0	0	0	0.1	3.2	0	0	0	0	0	0	0	0
Lake of Isles	23.8	0.1	0	0	0	0.1	0	0	0	0	0	0	0	0	0
Twin	5.6	0.2	0	1	0.5	0	0	0.3	0	0	0	0	0	0.8	0
Clear	1.9	0	0.8	0.4	0	0	0	0	0	0	0	0	0	0	0
Brownie	29.1	0	0.1	0	0	0	0	0	0	0	0	0	0.4	0	0
McCarron	35.8	1.8	7.2	0	0	0	0.2	0	0	0	0	0	0	0.2	0
Adams	28.5	0	19.2	0.1	0	0	0	0	0	0	0	0	0	0	0
Fox	11.8	0	0.2	2	0	0	0	0	0	0	0	0	0	0	0
Owasso	6.2	0.5	5.8	0.2	0	0	0	0	0	0.2	0	0	0	0.7	0
Gervais	2.5	0.2	4.1	0	0	0	0	0	0	0	0	0	0.4	2.3	0
Cloverdale	47.4	0	4	0	0	0	0	0	0	0	0	0	0	1.8	0
Fish 2	30	0	0.4	0	0	0	0	0	0	0	0	0	0	0.6	0
Holland	33	0	1.2	0	0	0	0	0	0	0.9	0	0	0	4.1	0
Powers	8.2	0	0.3	0.2	0	0	0	0.4	0	0	0	0	0	30.6	0
McCraney	0.2	1.9	0	0	0	0	6.8	0	0	0	0	0	0	0	0
Long	0.4	0.2	9.5	0	0	0	0	0	0	0	0	0	0	0	0
Island	16.8	2	2.7	0	0	0.3	4.8	0	0	0.3	0	0.3	0	0.8	0
Fall	4	0.5	4.3	0.5	0	11.4	5	0	0	0	0	0	3	0.8	0
Whitefish	7.2	1.5	6.7	0.9	0	0.2	5.2	0.3	0	0	0	0	0	1	0
Cedar	14.3	1.1	0.3	0	0	0	0.5	0	0	0	0	0	0	0.2	0
Harriet	18.3	1.1	0.1	0.1	0	0.3	1.1	0	0	0	0	0	0	0	0
Calhoun	12.9	2	0	0	0	0.1	3.1	0	0	0.3	0	0	0.2	0	0
Nokomis	0.7	0.5	5.3	0.2	0	0	0	0	0	0	0	0	0	0	0
Washburn	8.8	1.1	0	0.9	0	3.3	5	0	0	0	0	0	0	0.7	0
Leavitt	11.7	8.6	2.8	0	0	0	0	0.1	0	0	0	0	0	0.1	0
Forest	16.3	4.4	3.3	0	0	0	0.1	0	0	0	0	0	0	0.4	0
Green	18.6	1.3	7.4	6.5	0	0	3.7	0	0	0	0	0	0	0.5	0
Comfort	10.4	1.2	2.5	0	0	0	0	0	0	0	0	0	0	0	0
Goose	0	0.6	6	0	0	0	0.2	0	0	0	0	0	0	0.4	0
Fish 1	6.2	4.6	0.4	0	0	0	12.9	0	0	0	0	0	0	0	0
Devils	14.8	5.5	3.6	0.3	0	0	5.2	0	0	0	0	0	0	0	0
Kroon	11.2	0	10.9	0	0	1	3.3	0	0	0	0	0	0	0	0
Clear	3.2	3.2	8.4	0	0	0	0	0	0	0	0	0	0	0.2	0
Square	0.8	6.3	3.5	3.7	0	0	1.2	0	0	0	0	0	0	2.3	0
Maple	9.1	1.3	0.8	7.3	0	0	0	0	0	0	0	0	0	0	0
Bright	3.3	0.1	8.6	0.2	0	0.6	0	0	0	0	0	0	0	0	0
Shagawa	12.4	0.2	14.5	0	0	2.4	1.1	0.1	0	0	0	0	0	0.2	0.1
Wisc. Pine	11.1	5.4	9.6	0.5	0	0	0	0	0	0	0	0	0	0	0
Christmas	4.8	9	11.6	0.2	0	0	7.9	0.2	0	0	0	0	0	0	0
Pickerel	19.4	5.8	0	0	0	0	2.3	0	0	0	0	0	0	0	0
Lake of Valle	14.3	22.5	0.2	0	0	0	4	0	0	0	0	0	0	0	0
Sturgeon	8.4	6.5	23.6	0.2	0	0	2.9	0	0	0.5	0	0	0	0	0
Long	15.2	6.5	1.9	0	0	1.9	4.4	0	0	0	0	0	0	0	0
Pike	35.9	6	0	3.3	0	0	9.3	0	0	0	0	0	0	0	0
Nemec	40	0	0	0	0	0	0.2	0	0	0	0	0	0	0	0
Elk	48.4	19.8	0	0.2	0	0	0	0	0	0	0	0	0	0	0
Gull	4.6	6.6	0.7	0	0	0	4.6	0	0	0	0	0	0	0	0
Fenske	0.2	8.9	1.4	0	0	36.1	5.2	0.5	0	0.7	0	0	0.9	0.9	1.8
Saganaga	0	3.8	1	4	0	35.6	14.7	0	0.2	0.2	0.1	0	4.9	11.2	0.2
Trout Bay	1	3.8	0.5	1.7	0	34.9	12.7	0	0.1	0.1	0	0.1	4.4	6.2	0
L.S. Burnsid	0.5	3.4	2.7	0	0	16.4	15.1	0.3	0.1	0.3	0	0.1	1.8	12	0
Burntside	2.1	2.5	1.9	4.6	0	9	23.3	0.1	0	0.1	0	0.1	1.4	22.6	0

TABLE 3. PERCENT ABUNDANCES OF DIATOMS USED IN THE DECORANA ANALYSIS* (continued)

Lake	2	10	11	6	19	20	25	29	26	30	17	31	1	15	22	9
First Lake	2.4	0.3	0	0	0	0	1.4	0.2	0	0	0	1.5	69.6	0.2	0.2	0
Bald Eagle	1.2	0.7	1.4	0	0.1	0	0.7	0.2	0.2	0	2.5	0	26.7	1.2	0.1	0
Bassett	3.5	0.2	0.2	0	0.2	0	0	0.2	0	0	0.2	0.4	35.4	0	0	0
Gabbro	3.6	0.2	0.2	0	0.4	0.2	0	0.2	0	0	0	0	32.3	0.2	1.1	0
Wilson	0	1.3	0	0	0	0	0	0	0	0	2	0.2	53.2	2.9	0	0
Nickel	0.3	0	0	0	0	0	0	0.6	0	0	0	0	19.9	0.8	0.4	0
Hornby	0	0	0	0	0	0	0	0	0	0	0	0.2	37	0.9	9.9	0
Coffee	0	0.5	0	0	0	0	3.7	1	0	0	1	0.8	13.7	7	0	0.3
Lords	0	0	0	0	0	0	0	0.5	0	0	0	1	0.3	0.3	0.5	0
Railroad	1	0	0	0	0	0	0	0.9	0	0	0	0.4	2	1.6	0	0
Road Lake	0	0	0	0	0.2	0	0.1	0	0	0	1.4	2	7.9	1	0.1	0
Tony Lake	0	0	0	0	0	0	0	0.4	0	0	2.6	0.1	5.9	8.1	0.2	0
Wampus	0	0	0	0	0.1	0	0.2	0	0	0	0.1	0.5	1.5	2.5	0.4	0
Jane	0	0	0	1.7	0	0	1.2	0.2	0	0	1.7	1.7	0	3.1	0	0
Dogfish	0	0	0	0	0	0	0	0	0	0	0.7	0	5.7	0.2	2.5	0
Burnside	1.5	0.8	0	0	0	0	0.5	1	0	0	0.8	0	6.3	0.5	2	0
Caribou	0	0	0	0	0	0	0	0.9	0	0	3.4	0.3	0	0.8	3.9	0.3
Big Lake	0	0	0	0	0	0	0	0	0	0	1.9	0	17	0	0	0
Squaw	0.6	0	0.3	0	0.3	0.3	0.6	1.4	0	0.3	0	1.3	33.5	14.8	0	0
Bass	0.4	0	3.3	0	0.2	0	0	0.2	0	0	0	1.2	0.2	0	0	0
Grindstone	0.3	0	2.1	0	0.2	0	0.5	0	0.2	0	0	0.3	1.6	0.3	1.6	0
Moody	1.1	0.3	3.4	1	0.3	0	0.2	0	0	0	0	1.4	1.8	0	8.7	0
Peterson	4.2	0	3.5	0	0	0	1.7	0	0	0	1.7	1.9	4.2	0	0	0
Sand	0	0	0	0	0	0	0.7	0	0	0	0	1.1	8.7	0	4.5	0
Kremer	0.1	0	0	0	0	0	0	0.1	0	0	0	0.6	0	0	0.9	0
Surprize	0.5	0	0	0	0	0	0	0	0	0	0	0.3	0.8	0	12.2	0
Woods	0	0.2	0	0	0	0	0	1.3	0	0	0	0	0	0	1.5	0
Margaret	0	0	0	0	0	0	0	0	0	0	0	1.7	0	0	0	0
Beckman	0	0	0	0	0	0	0	0	0	0	0	5.3	0	0	0.1	0
Beaver Lodge	0	0	0	0	0	0	0	0	0	0	0	0.9	0	0	0	0
Observation	0	0	0	0	0	0	0	0	0.2	0	0	2.8	0	0	0	0
Kylen Lake	0	0	0	0	0	0	0	0	0	0	0	0.1	0	0	0	0
West Boot	12.7	0	0.6	0	0	0	1.7	0.4	0	0.4	0	1.2	16.5	0	7.5	0
East Boot	1.2	0	0	0	0	0	4	0	0	0.8	0	2.4	0.4	0	5.4	0
Lower Good	0	0.2	0.2	0	0	0	0.3	0.2	0	0	0.6	0.2	54.3	6.8	0.8	0
Upper Good	0	0.2	1.4	0	0	0	0	0.2	0	0	0	0.5	32.7	1.3	7.3	0
Bass Ed	0	1	0.2	0	0	0	0	0.2	0	0.2	0.5	0	16.2	3.8	12.1	0
Low	0.6	0.4	0.4	0	0	0	0.2	0	0	0.2	1.2	0.2	29.5	4	5.5	0
Sunfish	0	0	1.7	0	0	2	0.2	0.7	0	0	0	0	31.9	0	0	0
Kimbal	0	0	0	0	0	0	0.9	0	0	0	4.7	0.4	10.6	21.3	0	0
Sarah Ed	0	0	0	0	0	0	0	0	0	0	0.2	0	2.4	1.8	1.3	0
High	0	0	0	0	0	0	0	0.2	0	0	0.6	0	12.3	1.4	3.3	0
Nest Ed	0	0	0	0	0	0	0	0	0	0	0.7	0	4.6	1.8	5.1	0
McIntyre	0	0	0	0	0	0	0	0	0	0	0	0	1.2	0.6	3.7	0
Dry	0	0	0.2	0	0	0	0	0	0	0	0.6	0	2.6	7.3	7.8	0
Indiana	0	0	0	0	0	0	0	0	0	0	0.2	0	0.9	2.4	0	0
Meander	0.3	0	0	0	0	0	0	0	0	0	0	0	32.4	0.3	1.3	0
Trout	0	0	0	0	0	0	0	0	0	0	7.7	0	0	7	0.3	0
Gardner	0	0	0	0	1.3	0	0.2	0	0	0.2	0.2	0.2	0	0	0	0
Greenwater	0	0	1.9	0	0.1	0	0.1	0.1	0.6	0	0.8	0.1	0	0.6	5.1	0
Arco	0	0	0	0	0	0	0	0	0	0	0	0	0	0	0	0
Josephine	0	0	0	0	0	0	0	0	0.1	0	0	0.1	0	0	0.1	0
Deming	0	0	0.2	0	0	0	0	0	0	0	0	0.8	0	0	0	0
Carlson	0	0	0	27.6	0	0	1.2	0	0	0	0	4.1	0	0	0	0

TABLE 3. PERCENT ABUNDANCES OF DIATOMS USED IN THE DECORANA ANALYSIS* (continued)

Lake	14	5	16	7	21	3	12	24	18	23	27	28	4	8	13
First Lake	3.3	2.6	0.4	0	0	2.2	1.9	0.5	0	0.3	0	0	0.9	0.2	0
Bald Eagle	0.3	1.9	2.6	0	0	8	3.9	0.5	0	3.2	0.3	0	35.5	3	0
Bassett	2.9	0.9	0.5	0	0	17.7	0	0.3	0	0.5	0	0	20.6	3.1	0.2
Gabbro	2.3	0.9	2.1	0	0	13.2	4.2	1.3	0	0.6	0.2	0	27.6	3.2	0.2
Wilson	0.1	1.6	11.1	0	0	9.3	0.9	0	0	0	0	0	1.1	0.9	0.1
Nickel	0.7	0.2	7.1	0	0	0	3.8	0.9	0	0.6	1.3	1	16.6	5.2	0.4
Hornby L	0	0	2.3	0	0	1	2.1	0.1	0.3	0.9	0.2	0.1	3.2	9.3	0.6
Coffee	3	3.7	10.4	0	0	3.5	4.7	1.5	0	3.5	0	0	0	0	0
Lords	3.1	0.3	52.1	0	0	0	0.3	0	0	0.5	0	0	0	1.8	1.6
Railroad	0	0	31.7	0	0	3.9	7.7	2.3	0.7	1.9	2	1.7	6.2	0	0
Road Lake	0	0	23.8	0	0	0.4	2	1.1	0.1	0.4	0.4	0	0.5	0.5	1.6
Tony Lake	0.1	0	63.4	0	0	0.9	0.5	0.2	0.3	0	0	0.5	1.8	0.7	0
Wampus	0	0.2	48.7	0	0	0	0.8	0.1	1.5	0.1	0.3	0.7	0	0.2	1.1
Jane	1.8	5.6	29.6	0	0	0	0.5	4.4	0	1.4	0.2	0	0	0	0
Dogfish	1.1	8.4	14.5	0.6	0	0	13.2	0.7	0	0.7	0	2.3	0.2	13	0
Burnside	0.3	5.3	13	0	0	0.5	0.8	0.5	0	0	0.2	1	0.8	24.6	0
Caribou	1.7	6	11.2	0	0	0	12.3	0.2	0	0.5	0	6.6	0	34.2	0.2
Big Lake	1.2	5.7	18.7	0	0	0	21.2	0	0.4	0	0	1.5	0	12.6	1.9
Squaw	12.1	2.3	3.4	0	0	0.3	2	0.6	0	0.3	0	0	0	0.6	0.3
Bass	5.7	0.4	21.9	0	0	0	26.4	0	0	0.2	0	1	0	3.9	1.6
Grindstone	10.8	2.8	2.4	0.5	0	2.8	54.4	0	0	0	0	0	0	2.9	0
Moody	20.9	0	25.2	0	0	0	0	0.3	0	0	0	0	0	7.6	0.2
Peterson	0.7	0	25.6	0	0	0.6	2.1	0.9	0	2.6	0	0	0.7	5.3	0.7
Sand	19.9	0	15.2	0	0	0	1.1	0.2	0	0.4	0	0	0.5	2.5	2.2
Kremer	0.5	5.2	8.3	0	0	0	13.7	0	0	0	0.1	9.4	0	53.1	0.4
Surprize	12.2	0	4.6	0	0	0	5.6	3.6	0	3.8	2.5	14.7	0	1.3	5.3
Woods	1.5	0.2	3.2	0	0	0	27.9	0	0	0.9	0	35	0	3.2	0.2
Margaret	0	1	5.4	0	0	0	2.7	0.5	0	18.5	6.3	0	0	0	3.9
Beckman	0	0	9.1	0	0	0	1.2	1.3	4	36.2	0	0	0	0	2.1
Beaver Lodge	0	0	2.8	0	0	0	2.4	3.2	1.1	0.9	3	3.3	0	0	4.5
Observation	0.3	0	0	0	0	0	2.2	3.1	0	6.1	0	6.1	0	0	1.7
Kylen Lake	0	0	0.4	0	0	0	0	0.5	9.8	0.5	46.8	2.6	0	3.9	0.8
West Boot	4.8	0	0	0.8	0	0	1	0	0	0	0	0	0	5.4	0.6
East Boot	0.4	0	23.3	0.2	0	0.2	0	0.2	0	0	0	0	0	0	0
Lower Good	0.2	1.9	8	0	0.2	1.1	0	0.2	0	0	0	0	0.5	1	3.6
Upper Good	0	4.5	1.8	0.7	1.6	1.6	0	0	0	0	0	0	0.9	3.1	19.3
Bass Ed	3.3	3.8	11.2	6.2	0	0.3	0	0.2	0	0	0	0	0.2	3.9	11.2
Low	1	1.4	6.7	0	0	1.6	0	0	0	0.2	0	0	0.4	7.7	8.2
Sunfish	0	0.2	5.9	3.9	0	0	0	0.2	0	0	0	0	0	18.3	19.6
Kimbal	1.2	3	18.6	4.5	0	1.7	2.6	0	0	0	0	0	0.2	2.4	0.5
Sarah Ed	0	5.5	1.1	14.2	0.2	0.4	0	0	0	0	0	0	1.1	44.6	7.5
High	0	2.3	2	4.1	0.2	0	0	0	0	0	0	0	1.8	27.4	10.9
Nest Ed	0	1.3	7.5	2	0.5	5	0	0	0	0	0	0	0.6	31.6	21.8
McIntyre	0	7.3	3.8	27.6	1.2	0	0	0	0	0	0	0	0	20.9	9.5
Dry	0.6	1.7	16	3.9	0	4.5	0	0	0	0	0	0	2.3	9.7	12.9
Indiana	0	0.4	58.7	0.6	0	0	0	0	0	0	0	0	0	5.9	5
Meander	0.5	2.1	8	0.5	0	0.8	0.4	0	0	0.5	0.1	0.3	3	31.5	3.3
Trout	5.2	5.1	6.6	0.4	0	12	0	0	0.3	0.3	0	0	0.6	34.1	0
Gardner	0	0	8	0	42.1	0	0	0	0	0	0	0	0	29.8	0
Greenwater	3.1	1.2	7.2	9.1	49.7	0	0	0.1	0	0	0	0.1	0	6.4	0
Arco	0.8	0	0.8	3.8	41.9	0.2	0.6	0	0	0	0	0	0	40.9	0.8
Josephine	0	0	0.5	0.2	19.4	0	0.1	0	0	0	0	0	0	75	0.6
Deming	3	1.9	0.6	1.3	39	0	3	0.2	0	1.1	0	0	0	25.8	5.1
Carlson	0	0	0	0	0	0	0	0	0	0	0	0	0	27.6	0

*The 31 most abundant species were used in the analysis. Percentages were calculated on the basis of all diatoms in the sample. Table rearrangement was done using TWINSPAN (Gauch, 1982). Numbers at the heads of columns indicate taxon numbers assigned on Table 2.

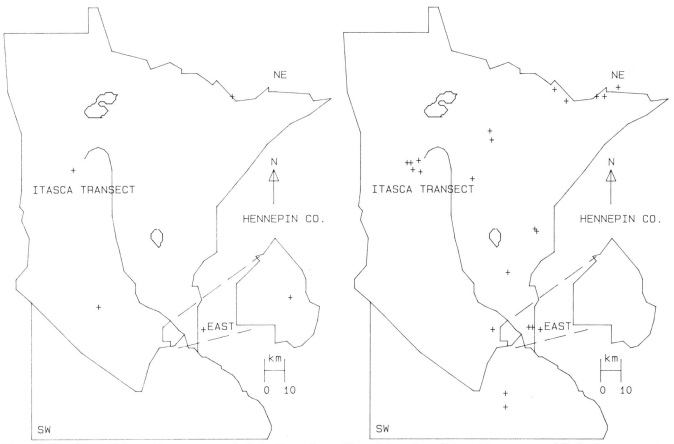

Figure 8. Locations of lakes with percentages of *Cyclotella ocellata* greater than 1%.

Figure 9. Locations of lakes with percentages of *Cyclotella comta* greater than 5%.

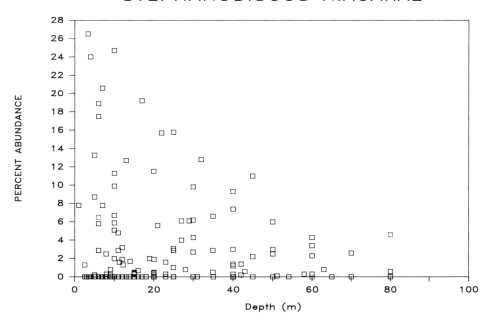

Figure 10. Plot of percentages of *Stephanodiscus niagarae* vs. maximum lake depth.

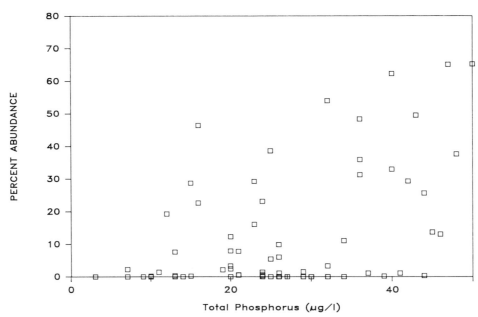

Figure 11. Plot of percentages of small *Stephanodiscus* species vs. lake total phosphorus concentrations.

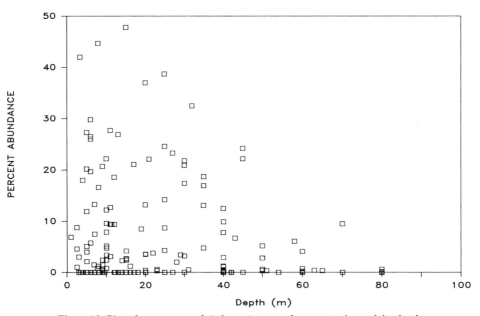

Figure 12. Plot of percentages of *Aulacoseira granulata* vs. maximum lake depth.

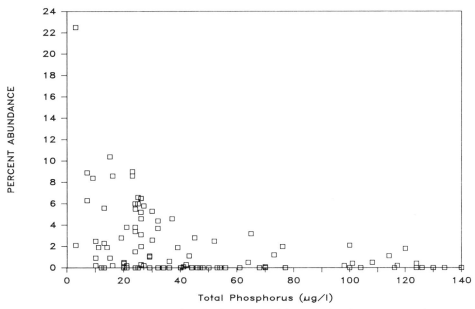

Figure 13. Plot of percentages of *Cyclotella comta* vs. lake total phosphorus concentrations.

phosphorus (r_s = 0.37, Fig. 11), and high alkalinity (r_s = 0.50) lakes. They show no particular optimal lake depths or pHs.

Fragilaria crotonensis and *Aulacoseira ambigua* show no clear optima in the Minnesota data set (Table 5). *Aulacoseira granulata* shows a decided optimum in shallow (r_s = 0.31 on depth, Fig. 12), high total phosphorus (r_s = 0.30), high pH (r_s = 0.30), and high alkalinity (r_s = 0.45) lakes. *Stephanodiscus niagarae* shows the same optimum (Fig. 10, Table 5). *Cyclotella comta* (Fig. 13) reaches its highest abundances in low total phosphorus (r_s = –0.56) lakes.

Ordination of diatom samples

An ordination of the diatom data can summarize the environmental preferences of individual species more efficiently than the graphs presented above. Detrended correspondence analysis (performed by the program DECORANA) was chosen as the ordination technique because it assumes that species are distributed in bell-shaped (Gaussian) curves with respect to environmental variables (Gauch, 1982). Our examinations of species distributions (above) show that the diatoms are distributed according to step functions with a high scatter of points. These are approximated by truncated Gaussian curves in the DECORANA ordination.

Although four axes were derived from the data by the program DECORANA, axes 1 and 2 summarize the major portion of the variance in the data set (Table 6). In the ordination of species, the taxa are spread along axis 1 according to pH and alkalinity preferences (Fig. 14). The species with high values on axis 1 are all acidophilic or indifferent in the Hustedt (1937a, 1937b, 1937c, 1937d, 1939a, 1939b) spectrum. The species that plotted to the low end of axis 1 are alkaliphils, which show distributions across Minnesota that are consistent with their environmental preference. For example, *Aulacoseira granulata* (2, Fig. 14) and *Stephanodiscus niagarae* (10, Fig. 14) are alkaliphils that reach their highest abundances in the southwest. In contrast, *Pinnularia biceps* (24, Fig. 14) is an acidophil that is abundant in the acidic lakes of the northeast (Fig. 15).

Axis 2 seems to separate species according to depth preference. Species that are located high on axis 2 are benthic and epiphytic. The axis 1 division between alkaliphils and acidophils is preserved among these benthic species. *Rhopalodia gibba* (30, Fig. 14), a benthic alkaliphil, is located high on axis 2 but low on

TABLE 4. LOCATIONS AND ENVIRONMENTAL VARIABLES FOR LAKES IN THE MINNESOTA DATA-SET*

S. Name	County	Lat.	Long.	Sample Depth (m)	Maximum Depth (m)	Specific Conducitvity Micromho	pH	Total P (ug/l)	SiO$_2$ (mg/l)	Alkalinity (mg/l CaCO$_3$)	SECCHI Disk Transparency (m)
Woodworth 1	North Dakota	47°88'	99°18'	3	3	1,600	-1	-1	-1	-1	-1
Woodworth 2	North Dakota	47°08'	99°18'	4	4	1,100	-1	-1	-1	-1	-1
Telegraph	Mahnomen	47°10'	96°40'	2.5	2.5	350	9.3	-1	-1	118	-1
Reichow	Mahnomen	47°10'	96°40'	2	2	417	8.1	-1	-1	104	-1
Hen Pond	Mahnomen	47°10'	96°40'	1.8	1.8	324	8.6	-1	-1	63	-1
Wabun Pr.	Mahnomen	47°10'	96°40	2.5	2.5	1,230	8.3	-1	-1	147	-1
Ballard	Mahnomen	47°10'	96°40'	2	20	306	8.8	-1	-1	60	-1
Quallen	Mahnomen	47°12'	95°47'	8	8	401	8.4	215	16	136	0.5
Squirrel	Mahnomen	47°11'	92°52'	7	17	658	8.5	190	22	184	0.9
Terhell	Mahnomen	47°10'	96°40'	5	5	361	8.2	-1	-1	75	-1
Wyoming	Anoka	45°30'	93°00'	3	3	180	-1	-1	-1	-1	-1
Woodworth 4	North Dakota	47°08'	99°18'	2	2	2,200	-1	-1	-1	-1	-1
Cucumber	Mahnomen	47°10'	96°00'	2	2	722	8.2	-1	-1	146	-1
Thompson	Mahnomen	47°10'	98°00'	5	5	324	8	-1	-1	105	-1
Mays	Washington	45°10'	92°48'	6.8	6.8	197	7.9	56	4	92	1.2
Cedar Bog	Wright	45°20'	94°10'	40	40	-1	-1	-1	-1	-1	-1
Cokato	Wright	45°07'	94°10'	33	60	578	8.3	156	20	211	2.8
Tetonka	LeSeur	44°14'	93°36'	18	40	416	7.9	140	18	141	1.5
School Grove	Lyon	44°33'	95°37'	6	6	2,414	8.2	70	12	292	-1
Wood	Lyon	44°20'	96°02'	6	1	1,711	7.7	101	13	200	-1
Wagonga	Kandyohi	45°04'	94°59'	6	15	1,216	8.4	2,100	80	292	0.2
Andrew	Kandyohi	45°19'	95°03'	22	25	438	8.4	44	16	134	1
Willmar	Kandyohi	45°05'	94°58'	5	5	380	8.1	170	17	202	-1
Minnewashta	Carver	44°52'	93°37'	27	50	374	7.9	26	4.5	107	2.3
Hunt	Rice	44°20'	93°27'	22	30	385	8.1	26	1	119	0.6
Beaver	Waseca	43°54'	93°22'	29	29	383	7.8	100	4	156	1.7
Pokegamma	Pine	45°52'	93°02'	13	25	241	7.2	34	7.2	148	0.9
Sarah	Murray	44°09'	95°46'	5	5	1,146	8	54	16	172	0.4
Cottonwood	Lyon	44°37'	95°41'	5	8	3,086	8.2	104	20	173	0.8
Waubun	Mahnomen	47°12'	95°53'	5	5	869	8.7	400	19	336	0.9
Buffalo	Renville	44°35'	94°34'	2	4	803	7.8	161	10	262	-1
Cedar	Martin	43°47'	94°44'	4	6	1,105	8.1	130	32.5	256	0.3
S. Silver	Martin	43°31'	94°28'	22	22	421	8.3	32	1.8	161	0.5
Charlotte	Martin	43°39'	94°28'	4	6	916	8	134	29	283	-1
Dead Coon	Lyon	44°22'	96°06'	5	5	978	8.2	50	13	126	0.3
Shetek	Murray	44°06'	95°42'	9	10	817	7.9	233	19	106	0.8
Oaks	Cottonwood	43°58'	95°30'	6	6	447	8.8	520	13	93	-1
Talcott	Cottonwood	43°52'	95°27'	7	7	422	7.6	590	10	76	0.4
Allie	Renville	44°40'	94°31'	10	10	548	8.2	165	7.5	145	-1
Traverse	Traverse	45°40'	96°30'	3.36	3.36	1,050	8.45	-1	30.8	163	1.1
Fox	Martin	43°41'	94°42'	13	20	626	8.1	178	19	203	0.8
Budd	Martin	43°38'	94°28'	10	17	557	8.2	36	8.3	174	1
Hall	Martin	43°37'	94°28'	5	13	658	8.4	130	23.5	207	1.4
Bryant	Rice	44°53'	93°26'	45	45	578	7.9	25	6.5	156	1.8
Riley	Rice	44°50'	93°31'	35	35	372	7.8	124	15	106	0.8
Mazaska	Rice	44°20'	93°24'	32	32	301	7.7	20	0.5	116	0.6
Kandyohi	Kandyohi	44°59'	94°58'	10	10	775	8	126	14.5	197	-1
Skogman	Isanti	45°34'	93°09'	10	10	258	6.8	48	1.9	117	0.9
Big Horseshoe	Chisago	45°35'	93°04'	11	11	258	7.1	42	1.8	119	0.9
Big Pine	Pine	46°11'	93°05'	21	21	191	7.1	68	12.4	79	1.2
Carver	Washington	44°54'	92°58'	10	-1	363	7.4	47	0.6	119	-1
Bone	Washington	45°17'	92°52'	25	25	-1	8.2	77	-1	71	1.5
Shields	Rice	44°20'	93°26'	20	30	366	8	36	22	131	1.3
Madison	Blue Earth	44°11'	93°49'	30	30	390	7.7	24	15	153	0.7
St. Olaf	Waseca	43°53'	93°25'	32	35	349	8	16	3	129	2
Nest	Kandyohi	45°15'	94°58'	32	45	395	8.2	116	0.5	175	1.3
Pine	Pine	46°12'	93°03'	22	27	194	7.6	150	9.6	88	1.8
Tamarack	Anoka	45°12'	93°4.5'	6	6	406	7	46	4.7	152	0.7

TABLE 4. LOCATIONS AND ENVIRONMENTAL VARIABLES FOR LAKES IN THE MINNESOTA DATA SET* (continued)

S. Name	County	Lat.	Long.	Sample Depth (m)	Maximum Depth (m)	Specific Conducitvity Micromho	pH	Total P (ug/l)	SiO$_2$ (mg/l)	Alkalinity (mg/l CaCO$_3$)	SECCHI Disk Transparency (m)
Long	Washington	45°11'	92°51'	3	3	197	7.1	61	5.6	90	0.6
Shields	Washington	45°15'	92°57'	2.5	2.5	-1	7.3	375	-1	82	1.7
Sakatah	LeSeur	44°15'	93°31'	7	10	456	7.5	206	23	156	-1
Fish Lake	Jackson	43°55'	95°00	7.02	7.02	360	8.55	-1	7.6	131.6	0.7
Dudley	Rice	44°21'	93°22'	65	65	205	7.8	15	0.5	89	3.4
Wabasso	Ramsey	45°02'	93°07'	13	13	393	6.7	52	10.1	151	2.5
Little Horse	Chisago	45°35.5'	93°04.3'	11	11	270	7.2	45	0.7	127	0.9
Grove	Pope	45°31'	95°10'	7.94	7.94	370	8.25	-1	13.2	168	1.7
Hiawatha	Hennepin	44°55'	93°14'	13	15	583	7.8	44	1.6	145	1.8
Wirth	Hennepin	44°59'	93°20'	23	23	646	7.6	124	6.5	150	0.8
Twin	Hennepin	45°00'	93°20'	51	51	660	7.6	40	2.1	152	3.2
Lake of Isles	Hennepin	44°57'	93°19'	28	28	573	7.8	70	1.3	225	0.7
Twin	Ramsey	45°02'	93°05'	10	10	209	6.4	880	1.1	64	4.2
Clear	Washington	45°10'	93°48.3'	7.7	7.7	252	8	12	1	120	3.4
Brownie	Hennepin	44°58'	93°19'	43	43	729	7.5	40	8.3	160	1.4
McCarron	Ramsey	45°00'	93°07'	25	50	441	7.4	120	0.5	114	0.9
Adams	Isanti	45°39.5'	93°14.5'	15	15	461	7.3	375	22.3	232	0.9
Fox	Rice	44°24'	93°20'	40	40	364	7.7	100	10	121	4
Owasso	Ramsey	45°02'	93°07'	11	11	307	6.6	64	0.7	115	1.8
Gervais	Ramsey	45°01'	93°07'	14	14	467	7	117	3	129	1.2
Cloverdale	Washington	45°01'	92°51.5'	10	10	49	6.7	53	1.9	11	0.6
Fish 2	Dakota	44°49'	93°01'	9	9	258	6.5	20	0.7	96	3.4
Holland	Dakota	44°47'	93°09'	15	15	190	6.2	13	1	77	2.1
Powers	Washington	44°56'	92°54'	9	-1	98	6.5	68	1.8	43	-1
McCraney	Mahnomen	47°09'	95°43'	40	35	432	7.9	14	18.7	240	4.6
Long	Watonwan	43°54'	94°40'	10	10	571	8	21	4	159	0.5
Island	Pine	46°25'	92°45'	24	25	92	-1	26	3	26	1.4
Fall	St. Louis	47°57'	91°44'	7	20	82	6.7	20	4.5	21	1.5
Whitefish	Crow Wing	46°12'	93°38'	41	60	229	8	24	11	102	-1
Cedar	Hennepin	44°58'	93°19'	40	40	481	7.7	43	1.8	121	1.9
Harriet	Hennepin	44°55'	93°18'	55	60	513	7.7	114	1.3	119	3.1
Calhoun	Hennepin	44°56'	93°51'	63	70	586	7.8	76	1.3	118	1.8
Nokomis	Hennepin	44°54'	93°14'	15	30	540	7.9	108	9.4	133	0.8
Washburn	Cass	46°52'	93°59'	90	105	716	7.8	29	7	44	3.2
Leavitt	Cass	46°50'	94°00'	58	58	36	8.9	23	18.7	93	0.3
Forest	Chisago	45°17'	92°58'	34	40	319	7.7	32	13	143	0.6
Green	Kandyohi	45°15'	94°55'	135	110	167	7.9	144	11	167	2.3
Comfort	Chisago	45°19'	92°57'	11	11	369	7.2	73	8.6	157	1.2
Goose	Chisago	45°37.5'	93°05'	12	12	269	7.7	36	1.3	132	1
Fish 1	Chisago	45°35'	93°04'	12	12	246	7.5	37	4	110	1.7
Devils	Kanabec	45°18'	93°19.5'	6	6	234	6.0	24	13.9	108	2.3
Kroon	Chisago	45°21'	92°51'	2.5	2.5	-1	8	77	-1	52	0.8
Clear	Chisago	45°16'	93°00'	25	25	-1	8.7	65	-1	11	1.1
Square	Washington	45°08'	92°48'	60	60	-1	8.6	7	-1	111	6.2
Maple	Douglas	45°45'	95°10'	18.91	18.91	400	8.25	-1	5	180.5	3
Bright	Martin	43°35'	94°35'	3	5	690	8	41	40	198	-1
Shagawa	St. Louis	47°55'	91°52'	21	40	76	6.6	16	3.5	18	0.9
Wisc. Pine	Unknown	?	?	27	27	378	-1	-1	-1	113	-1
Christmas	Carver	44°54'	93°33'	54	80	341	7.8	23	1	131	4.3
Pickerel	Hubbard	47°09'	95°27'	42	42	238	8.2	27	17	116	2.9
Lake of Valley	Becker	47°07'	95°24'	50	70	274	8.3	3	6	134	6.9
Sturgeon	Pine	46°23'	92°46'	34	35	103	7.5	26	3	26	2.9
Lone	Pine	46°17'	92°52'	19	20	112	7.5	26	1.5	60	2.9
Pike	Mahnomen	47°08'	95°31'	44	50	396	8.1	25	17	198	2.6
Nemec	Mahnomen	47°04'	95°55'	15	15	369	8.5	145	7.3	137	1.5
Elk	Mahnomen	47°10'	96°30'	-1	-1	-1	-1	-1	-1	-1	-1
Gull	Mahnomen	47°11'	95°41'	36	40	324	8.1	25	19.3	156	2.5
Fenske	St. Louis	47°59'	91°54'	20	40	35	6.1	7	2.5	6	5.8

TABLE 4. LOCATIONS AND ENVIRONMENTAL VARIABLES FOR LAKES IN THE MINNESOTA DATA SET* (continued)

S. Name	County	Lat.	Long.	Sample Depth (m)	Maximum Depth (m)	Specific Conducitvity Micromho	pH	Total P (ug/l)	SiO$_2$ (mg/l)	Alkalinity (mg/l CaCO$_3$)	SECCHI Disk Transparency (m)
Saganaga	St. Louis	48°09'	90°55'	63	63	45	6.2	21	-1	7	-1
Trout Bay	Ontario, Canada	48°10'	90°44'	-1	-1	42	6.1	24	-1	4	-1
L.S. Burnside	Ontario, Canada	48°09'	90°45'	23	23	45	6.6	24	-1	18	-1
Burntside	St. Louis	47°56'	91°58'	40	80	36	6.3	10	2.7	7	3.8
First Lake	Pine	46°19'	92°49'	13	15	84	7.6	30	13.5	44	2.2
Bald Eagle	Lake	47°50'	91°33'	16	16	82	7.4	11	8.5	25	1.8
Bassett	St. Louis	47°50'	91°33'	19	20	87	7.4	10	16	28	2
Gabbro	Lake	47°51'	91°36'	12	20	79	7.3	15	14.5	26	1.5
Wilson	Lake	47°44'	91°01'	12.2	12.2	46.4	6.9	-1	14.4	17	4.4
Nickel	St. Louis	47°50'	91°38'	9	9	53	6.1	10	5	10	0.9
Hornby L.	St. Louis	47°20'	92°00'	9.5	9.5	32	6.9	-1	-1	-1	-1
Coffee	Pine	46°26'	92°45'	18	20	101	7.4	32	1.3	37	-1
Lords	Pine	46°24'	92°47'	8	8	58	6.9	26	1.3	32	2.2
Railroad	St. Louis	47°20'	92°00'	2.5	2.5	35	6.5	-1	-1	-1	-1
Road Lake	Cook	48°00'	91°00'	12	12	50	-1	-1	-1	-1	-1
Tony Lake	St. Louis	47°20'	92°00'	7	7	66	7.5	-1	-1	-1	-1
Wampus	Cook	48°00'	91°00'	3.5	3.5	40	-1	-1	-1	-1	-1
Jane	Washington	45°01'	92°55.5'	9	-1	190	7.3	13	1	74	4.3
Dogfish	St. Louis	48°12'	92°11'	18	18	16	6	9	-1	8	2.7
Burnside	Ontario, Canada	48°10'	90°45'	9	9	37	6.3	30	-1	16	-1
Caribou	Itasca	47°32'	93°38'	80	80	50	6.4	24	15	16	6.8
Big Lake	St. Louis	48°07'	92°00'	5.18	5.18	23.8	6	-1	2.2	5.5	4.3
Squaw	Clearwater	47°14'	95°16'	54	80	134	7.4	13	3	71	4.7
Bass	Pine	46°09'	93°03'	15	20	47	6.4	20	1.4	9	1.5
Grindstone	Pine	46°07'	93°01'	57	106	132	7.5	19	6.2	49	3.5
Moody	Washington	45°18'	92°52'	40	40	-1	7.1	301	-1	32	-1
Peterson	Chisago	45°22'	92°47'	15	15	-1	6.9	332	-1	36	0.6
Sand	Washington	45°14'	92°48'	15	15	-1	6.5	187	-1	11	0.9
Kremer	Itasca	47°28'	93°38'	54	80	35	6.3	26	1.5	8	-1
Surprize	Itasca	47°29'	93°38'	31	31	53	6.6	20	2	8	-1
Woods	Cass	46°49'	94°00'	42	42	42	6.9	27	1.3	12	1.5
Margaret	Cass	46°50'	94°00'	31	40	28	6.5	29	1.5	1	-1
Beckman	Mahnomen	47°10'	96°40'	3	3	90	-1	-1	-1	-1	-1
Beaver Lodge	Mahnomen	47°10'	96°30'	7	7	33	7.3	-1	-1	6	-1
Observation	Mahnomen	47°10'	96°30'	6	6	31	6.8	-1	-1	7	-1
Kylen Lake	St. Louis	47°20'	92°00'	4	4	31	6.7	-1	-1	-1	-1
West Boot	Washington	45°09'	92°51'	11	11	234	7.1	21	3.7	118	2.8
East Boot	Washington	45°09'	92°49'	8	-1	184	7.1	34	1.5	87	1.8
Lower Good	Cook	48°00'	91°37'	-1	-1	57	8.1	-1	-1	-1	1.3
Upper Good	Cook	48°00'	91°37'	-1	-1	58	7.8	-1	-1	-1	1.4
Bass Ed	St. Louis	47°58'	91°51'	-1	-1	113	8.1	-1	-1	-1	1.8
Low	St. Louis	47°58'	91°50'	-1	-1	69	7.4	-1	-1	-1	1.4
Sunfish	Dakota	44°52.5'	93°06'	9	9	246	7	98	1.9	91	0.9
Kimbal	Cook	47°50'	90°14'	4.88	4.88	56.3	7.4	-1	5.3	22	4.1
Sarah Ed	Unknown	?	?	-1	-1	20	8.2	-1	-1	-1	2.5
High	St. Louis	47°58'	91°53'	-1	-1	34	7.6	-1	-1	-1	2.3
Nest Ed	Unknown	?	?	-1	-1	24	7.6	-1	-1	-1	1.8
McIntyre	Cook	?	?	-1	-1	20	6.8	-1	-1	-1	2.7
Dry	St. Louis	47°58'	91°50'	-1	-1	62	7.6	-1	-1	-1	1.9
Indiana	Cook	48°01'	91°36'	-1	-1	35	7.8	-1	-1	-1	2.3
Meander	St. Louis	48°08'	92°09'	23	23	17	5.6	3	-1	5	3.1
Trout	Cook	47°50'	90°13'	13.73	13.73	41.4	6.6	-1	3	13	8.8
Gardner	Mahnomen	47°18'	95°53'	40	40	236	7.7	254	1.5	116	1.1
Greenwater	Unknown	?	?	50	60	-1	7.1	-1	-1	212	-1
Arco	Hubbard	47°10'	95°10'	35	40	89	7.6	32	1.5	39	-1
Josephine	Hubbard	47°10'	95°10'	38	40	74	7.6	29	1.5	27	-1
Deming	Hubbard	47°10'	95°10'	54	54	80	7.5	39	2	75	-1
Carlson	Dakota	?	?	16	16	244	8.1	143	6	92	-1

*-1 = Value undetermined. Latitude and longitude give approximate locations.

TABLE 5. SPEARMAN RANK CORRELATION COEFFICIENTS BETWEEN DIATOM SPECIES, PERCENTAGES, AND ENVIRONMENTAL VARIABLES

Species	Max. Depth	Specific Conductivity	pH	Total Phosphorus	Alkalinity
Aulacoseira ambigua	-0.01	-0.02	0.04	-0.19†	0.05
A. granulata	-0.21†	0.49§	0.30§	0.30§	0.45†
A. italica	-0.01	-0.31§	-0.22†	-0.21†	-0.33§
A. distans	-0.05	-0.46§	-0.34§	-0.22†	-0.40§
Cyclotella comta	0.37§	-0.41§	-0.20†	-0.56§	-0.27†
C. kutzingiana	-0.25§	0.49§	0.23†	0.39§	0.36§
C. michiganiana	0.15*	-0.26§	-0.11	-0.22†	-0.09
C. stelligera	0.14*	-0.68§	-0.54§	-0.27§	-0.59§
C. oscellata	0.34§	-0.02	-0.04	-0.18†	-0.06
Stephanodiscus niagarae	-0.002	0.45§	0.29§	0.22†	0.38§
Small Stephanodiscus	-0.0007	0.56§	0.19†	0.37§	0.50§
Tabellaria fenestrata	0.27§	-0.46§	-0.40§	-0.47§	-0.52§
T. floculosa	0.003	-0.63§	-0.38§	-0.25§	-0.56§
Fragilaria crotonensis	0.33§	0.03	-0.22§	-0.12	-0.03
F. construens	-0.04	-0.06	0.21§	-0.13	0.13
F. construens venter	0.05	0.08	-0.43§	-0.18†	-0.30§
F. brevistriata	0.06	0.08	0.31§	-0.02	0.18†
Anomoeoneis serians	-0.15*	-0.27§	-0.09	-0.14	0.24§
Asterionella formosa	0.23§	-0.12	-0.25§	-0.24§	0.02
Cocconeis placentula	-0.03	0.43§	0.29§	0.26§	0.36§
Mastogloia smithii	0.17†	0.17†	0.30§	0.04	0.17†
Frustulia rhomboides	-0.09	-0.39§	-0.32§	-0.31§	-0.36§
Pinnularia biceps	-0.07	-0.46§	-0.39§	-0.34§	-0.44§
Rhopalodia gibba	0.19	0.20†	0.19†	0.02	0.17†

*0.05 <P <0.10
†0.01 <P <0.05
§P <0.01

TABLE 6. EIGENVALUES OF THE DECORANA ORDINATION OF 174 MINNESOTA DIATOM SURFACE SAMPLES

Axis	Eigenvalue	Length (Standard Deviation Units)
1	0.660	5.39
2	0.453	4.40
3	0.326	4.53
4	0.234	2.57

axis 1. In contrast, *Frustulia rhomboides* (27, Fig. 14), a benthic acidophil, is located high on both axes.

The ordination of lakes clearly differentiates among lake types following the species ordination (Fig. 16). The group of lakes with coordinates low on axis 1 are alkaline eutrophic prairie lakes that contain alkaliphilic species. Those high on axis 1 are northeastern acidic lakes with acidophilic or indifferent diatoms. Axis 2 divides deep lakes in the northeast (low on axis 2) from shallow lakes (high on axis 2). The shallow lakes form two groups. One group is acid (high on axis 1) and the other is alkaline (low on axis 1).

The axis 1 coordinates of the lake samples can be plotted on the map of Minnesota to show the geographical distribution of lakes with high (>200) values (Fig. 17). It is apparent that lakes in the northeast and east and along the Itasca transect have the highest axis 1 values. Lakes with axis 1 <200 are, in general, located in the prairie southwest.

The lakes with large values on axis 2 (>200) are not located in any particular geographic region (Fig. 18). These are differentiated not by water chemistry (which varies geographically), but rather by depth (which does not vary geographically).

Spearman rank correlation coefficients between ordination axes and environmental variables can reveal factors that control diatom distributions (TerBraak and Prentice, 1988). Axis 1 has a strong negative correlation with specific conductivity, pH, total phosphorus, dissolved silica, alkalinity, and Secchi depth (Table 7). The majority of low conductivity, oligotrophic, soft-water lakes have high axis 1 coordinates. Axis 2 correlates with few environmental variables. Axis 3 is positively correlated with most chemical measurements, but not Secchi depths. Axis 4 shows few strong correlations with any axis.

COMPARISON OF SURFACE SAMPLES WITH THE ELK LAKE CORE

When the down-core samples from the Elk Lake core are plotted on the ordination plane (Fig. 19) they form a tight cluster.

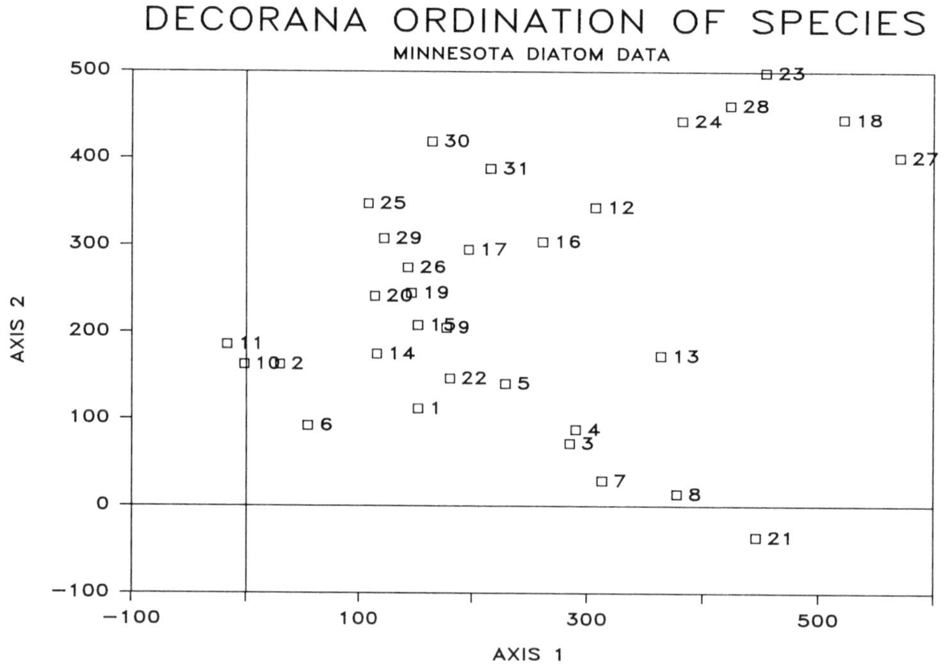

DECORANA ORDINATION OF SPECIES
MINNESOTA DIATOM DATA

Figure 14. Plot of species on axis 1 and axis 2 of the DECORANA ordination. Species codes are listed in Table 2.

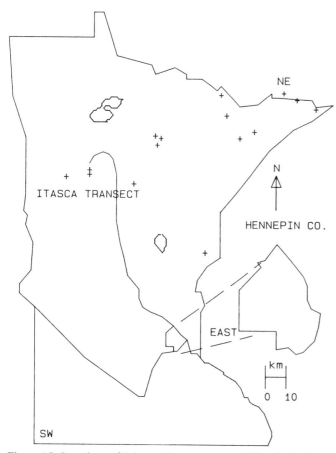

Figure 15. Locations of lakes with percentages of *Pinnularia biceps* greater than 0.5%.

This result means that there has been little change in the species composition of Elk Lake fossil diatom assemblages through time. It is apparent (Bradbury and Dietrich-Rurup, Chapter 15) that *Fragilaria crotonensis* and small *Stephanodiscus* spp. have dominated the lake for a long time: these species cause the Elk Lake assemblages to be among the high alkalinity lakes at the low end of axis 1 in the ordination.

The cluster analysis of surface samples and down-core samples reveals the modern lakes whose diatom assemblages are most similar to those of Elk Lake (Fig. 20). Table 8 shows the species that are most abundant in particular clusters.

Only a few clusters seem to be limited to particular time periods. Cluster 4 contains samples older than 3250 yr, and clusters 1 and 6 contain samples that are younger. These distributions are related to the distribution of diatom species in the cores. Samples older than 3250 yr contain *Cyclotella michiganiana*, whereas younger samples contain *Asterionella formosa*. The *C. michiganiana* cluster has few surface samples (only Fish and Pike lakes).

Because *C. michiganiana* is so rare in Minnesota samples, it controls the cluster analysis, causing the Elk Lake assemblages to group together in cluster 4. Neither *Aulacoseira ambigua* nor *Stephanodiscus niagarae* formed the basis of clusters, even though they have quite limited depth distributions in the Elk Lake core.

Differences between the environmental variables associated with the surface samples in each cluster were tested for significance using the Mann-Whitney U test. No significant differences were found. Because tests of significance are difficult with the small numbers of surface samples included in the clusters, it

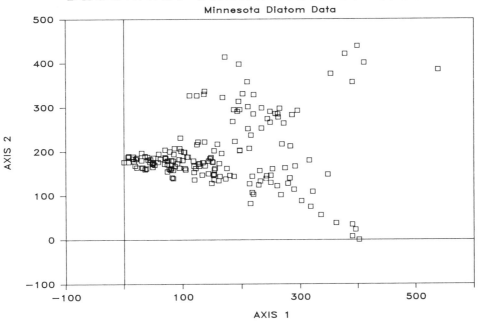

Figure 16. Plot of DECORANA ordination of diatom samples.

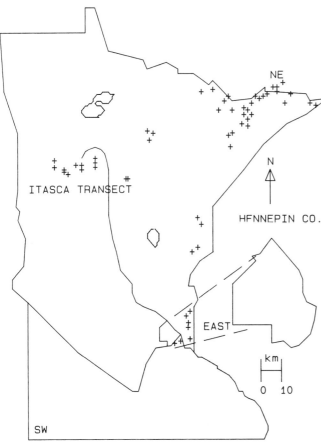

Figure 17. Locations of lakes with values of axis 1 greater than 200 in the DECORANA ordination.

should not be assumed that there have been no changes in the Elk Lake environment over the past 10,000 yr. There are too few surface samples in the clusters to draw firm conclusions about past environments based on the cluster analysis alone.

DISCUSSION

The high abundance of *Stephanodiscus minutulus* throughout the Elk Lake diatom diagram (Bradbury and Dieterick-Rurup, Chapter 15) indicates that total phosphorus levels have usually been high in the lake. Other Minnesota lakes with large numbers of small *Stephanodiscus* are eutrophic. This interpretation is consistent with the physiological information on phosphorus limitation of growth in this species (Tilman and others, 1982).

The distribution of *Fragilaria crotonensis* in Minnesota does not shed much light on its behavior in Elk Lake; in Minnesota, it has no clear environmental preferences. In low-alkalinity lakes, however, it is an indicator of eutrophic conditions (Brugam, 1988; Brugam and Patterson, 1983; Stockner and Benson, 1967). The physiological data of Tilman and others (1982) suggest that *F. crotonensis* is not as good a competitor as *Stephanodiscus* in the high-phosphorus and low-silica environments of eutrophic lakes.

Bradbury (1989) suggested that the dominance of *Fragilaria crotonensis* in some lakes may be related to the timing of ice melt in spring and its effect on nutrient cycling. Such a relation is possible, but too complex to test using the simple chemical data available. However, Bradbury (1989) used the geographic distri-

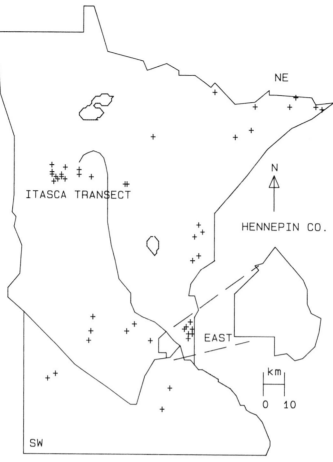

Figure 18. Locations of lakes with values of axis 2 greater than 200 in the DECORANA ordination.

TABLE 7. SPEARMAN RANK CORRELATION COEFFICIENTS OF DECORANA AXES COMPARED WITH ENVIRONMENTAL VARIABLES

Axes	1	2	3	4	N
Maximum depth	0.13	-0.14	0.08	0.24§	156
Specific conductivity	-0.74§	0.13	0.31§	0.23†	163
pH	-0.38§	0.08*	0.31§	-0.03	164
Total phosphorus	-0.46§	0.18†	0.29†	-0.19†	133
Dissolved silica	-0.22†	-0.01	0.28†	-0.03	125
Alkalinity	-0.64§	-0.01	0.31§	0.02	151
Secchi disc transparency	0.38§	-0.03	-0.14	0.12	123

*$0.05 < P < 0.10$
†$0.01 < P < 0.05$
§$P < 0.01$

bution of *F. crotonensis* from part of the Minnesota data set to support his argument.

The increases in *Aulacoseira* species during the prairie period at Elk Lake are especially interesting. *Aulacoseira granulata* is a species of moderately shallow, alkaline, prairie lakes of southwestern Minnesota. The presence of this species indicates that during the prairie period Elk Lake became more like modern-day prairie lakes.

The large increase of *Aulacoseira ambigua* during the prairie period (to nearly 40% of the fossil assemblage) is difficult to interpret because this species, like *Fragilaria crotonensis*, shows no particular optima among the environmental variables we tested. *A. ambigua* is a ubiquitous species in Minnesota that lives in many different lake types. Lund (1954, 1955) showed that *Aulacoseira italica* in the English Lake District depends for survival on resting stages that endure adverse seasons in the sedi-

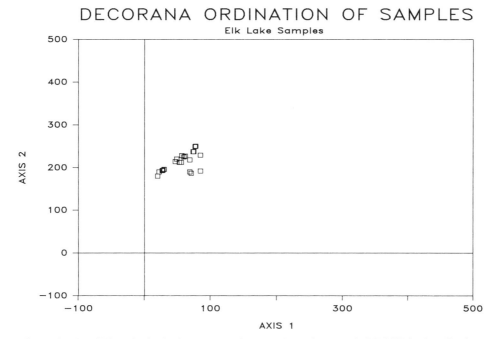

Figure 19. Plot of Elk Lake fossil diatom assemblages on the surface sample DECORANA ordination. Only samples at 500 yr intervals are plotted.

ment, but are recirculated into the water column during periods of water turbulence. It is likely that other species of *Aulacoseira* behave similarly. I suggest that the abundance of *Aulacoseira* that characterizes the fossil record of Elk Lake from 6000 to 4000 yr B.P. represents a time of lowered lake levels and/or of increased turbulence. Either of these alternatives would allow the dominance of heavily silicified species like *Aulacoseira*. These conditions also allow the release of *Aulacoseira* resting cells from the enclosing sediment. In a shallower lake the probability of resuspension would be high enough to allow *Aulacoseira* to survive.

The ordination of surface samples shows that diatoms follow the general northeast to southwest trend in lake water chemistry. The first axis created by ordination of the surface sample data set is correlated with lake pH and alkalinity, northeastern lakes being at the high end of the axis and southwestern lakes being at the lower end. The second axis is correlated with lake depth.

The fossil assemblages of Elk Lake fall on the ordination plane among the high-alkalinity, high-productivity lakes from areas of the state with carbonate-rich tills. When they are plotted on the ordination axes, the fossil assemblages do not change much over time, suggesting that gross environmental conditions at Elk Lake have been stable over the past 10,000 yr.

Cluster analysis supports the ordination results because the analogue lakes chosen by the cluster analysis are all high-alkalinity, moderately eutrophic lakes. Unfortunately, there are too few surface sample analogues in the cluster analysis to make firm statements about changes in past environmental conditions over time. The dominance of *C. michiganiana* reduces the number of analogues among the surface samples.

CONCLUSIONS

The diatoms of Minnesota show a strong relation with water chemistry across the state. *Aulacoseira italica, Aulacoseira distans, Cyclotella michiganiana,* and *Cyclotella stelligera* are species of deep, low-alkalinity lakes in the northeast. The shallow, alkaline, eutrophic prairie lakes of the southwest contain small *Stephanodiscus* spp., *Stephanodiscus niagarae,* and *Aulacoseira granulata.*

Elk Lake has always been alkaline and somewhat eutrophic. The fluctuations in *Stephanodiscus minutulus* and *Fragilaria crotonensis* that characterize most of the lake's history result from environmental variations that are too subtle to be reconstructed from the data we have for the surface samples. Similarly, the increase of *Aulacoseira ambigua, A. granulata,* and *Stephanodiscus niagarae* that occurred during the prairie period probably resulted from decreased depth or increased water turbulence, but the surface sample analysis was not sensitive enough to detect changes.

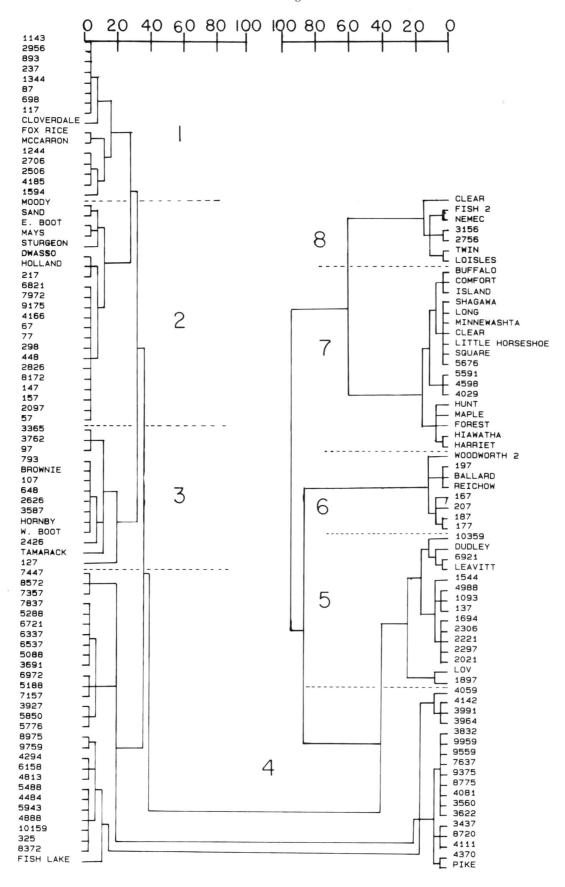

Figure 20. Cluster diagram for Elk Lake fossil diatom assemblages and the Minnesota surface sample diatom data set. Only clusters containing Elk Lake samples are shown. The cluster diagram has been folded to save space. See Table 8 for cluster number definitions.

TABLE 8. DOMINANT SPECIES IN THE ELK LAKE CLUSTER ANALYSIS

Cluster	Species
1	*Synedra acus, Asterionella formosa*
2	*Aulacoseira ambigua, A. granulata,* and Small *Stephanodiscus* spp.
3	*Gomphonemas*
4	*Cyclotella comta*
5	*Cyclotella michiganiana*
6	*Asterionella formosa*
7	*Fragilaria crotonensis*
8	*Stephanodiscus niagarae*

ACKNOWLEDGMENTS

I thank H. E. Wright, Jr., for encouraging me in this work. His unfailing support has allowed me to develop my ideas on diatom distributions. I also thank the many investigators who have been at the University of Minnesota Limnological Research Center for their help on this project. They include Claire Keister, Edward Swain, and Thomas Crisman.

REFERENCES CITED

American Public Health Association, 1985, Standard methods for the examination of water and wastewater: Washington, D.C., American Public Health Association, 874 p.

Birks, H. H., 1973, Modern macrofossil assemblages in lake sediments in Minnesota, *in* Birks, H.J.B., and West, R. G., eds., Quaternary plant ecology: Oxford, Blackwell, p. 173–189.

Bradbury, J. P., 1989, A climatic-limnologic model of diatom successions for paleolimnological interpretation of varved sediments at Elk Lake, Minnesota: Journal of Paleolimnology, v. 1, p. 115–133.

Bright, R. C., 1968, Surface-water chemistry of some Minnesota lakes, with preliminary notes on diatoms: University of Minnesota Limnological Research Center, Interim Report 3, 57 p.

Brugam, R. B., 1979, A re-evaluation of the Araphidineae/Centrales index as an indicator of lake trophic status: Freshwater Biology, v. 9, p. 451–460.

—— , 1981, Chemistry of lake water and groundwater in areas of contrasting glacial drifts in eastern Minnesota: Hydrobiologia, v. 80, p. 47–62.

—— , 1983, The relationship between fossil diatom assemblages and limnological conditions: Hydrobiologia, v. 98, p. 223–235.

—— , 1988, Long-term history of eutrophication in Washington lakes, *in* Adams, W. J., Chapman, G. A., and Landis, W. G., eds., Aquatic toxicology and hazard assessment, Volume 10: Philadelphia, American Society for Testing and Materials, p. 63–70.

Brugam, R. B., and Patterson, C., 1983, The A/C (Araphidineae/Centrales) ratio in high and low alkalinity lakes in eastern Minnesota: Freshwater Biology, v. 13, p. 47–55.

Brugam, R. B., and Speziale, B. J., 1983, Human disturbance and the paleolimnological record of change in the zooplankton community of Lake Harriet, Minnesota: Ecology, v. 64, p. 578–591.

Brugam, R. B., Grimm, E. C., and Eyster-Smith, N., 1988, Holocene environmental changes in Lily Lake, Minnesota, inferred from fossil diatom and pollen assemblages: Quaternary Research, v. 30, p. 53–66.

Charles, D. F., and Norton, S. A., 1986, Paleolimnological evidence for trends in atmospheric deposition of acids and metals, *in* Acid deposition: Long-term trends: Washington, D.C., National Academy Press, p. 335–506.

Cleve-Euler, A., 1968, Die diatomeen von Schweden und Finnland, *in* Bibliotheca Physologia Band 5: Lehre, German, J. Cramer.

Dixit, S., 1986, Diatom-inferred pH calibration of lakes near Wawa, Ontario: Canadian Journal of Botany, v. 64, p. 1129–1133.

Gasse, F., 1986, East African diatoms and water pH, *in* Smol, J. P., Battarbee, R. W., Davis, R. B., and Meriläinen, J., eds., Diatoms and lake acidity: Dordrecht, Dr. W. Junk Publishers, p. 149–168.

Gauch, H. G., Jr., 1982, Multivariate analysis in community ecology: New York, Cambridge University Press, 298 p.

Gorham, E., Dean, W. E., and Sanger, S. E., 1983, The chemical composition of lakes in the north-central United States: Limnology and Oceanography, v. 28, p. 287–301.

Håkansson, H., and Locker, S., 1981, *Stephanodiscus* Ehrenberg 1846, A revision of the species described by Ehrenberg: Nova Hedwigia, v. 85, p. 117–150.

Håkansson, H., and Stoermer, E., 1984, Observations on the type material of *Stephanodiscus hantzschii* Grunow in Cleve and Grunow: Nova Hedwigia, v. 39, p. 477–495.

Hill, M. O., 1979, DECORANA: A FORTRAN program for detrended correspondence analysis and reciprocal averaging: Ithaca, New York, Cornell University Section of Ecology and Systematics, 52 p.

Hustedt, F., 1930, Die Süsswasserflora Mitteleuropas: *Bacillariophyta (Diatomeae),* Heft 10, 466 p.

Hustedt, F., 1937a, Systematische und ökologische Untersuchungen über die Diatomeen Flora von Java, Bali, und Sumatra: Archiv für Hydrobiologie (Supplement), v. 15, p. 131–177.

—— , 1937b, Systematische und ökologische Untersuchungen über die Diatomeen Flora von Java, Bali, und Sumatra: Archiv für Hydrobiologie (Supplement), v. 15, p. 187–295.

—— , 1937c, Systematische und ökologische Untersuchungen über die Diatomeen Flora von Java, Bali, und Sumatra: Archiv für hydrobiologie (Supplement), v. 15, p. 393–506.

—— , 1937d, Systematische und ökologische Untersuchungen über die Diatomeen Flora von Java, Bali, und Sumatra: Archiv für Hydrobiologie (Supplement), v. 15, p. 790–836.

—— , 1939a, Systematische und ökologische Untersuchungen über die Diatomeen Flora von Java, Bali, und Sumatra: Archiv für Hydrobiologie (Supplement), v. 16, p. 1–155.

—— , 1939b, Systematische und ökologische Untersuchungen über die Diatomeen Flora von Java, Bali, und Sumatra: Archiv für Hydrobiologie (Supplement), v. 16, p. 274–394.

—— , 1977a, Die Kieselalgen Deutschlands, Osterreichs und der Schweiz. Teil 1, Band VII, *in* Rabenhorst, L., ed., Kryptogamen Flora Deutschlands, Osterreich und der Schweiz: Königstein, Germany, Otto Koeltz, 920 p.

—— , 1977b, Die Kieselalgen Deutschlands, Osterreichs und der Schweiz. Teil 2, Band VII, *in* Rabenhorst, L., ed., Kryptogamen Flora Deutschlands, Osterriech und der Schweiz: Königstein, Germany, Otto Koeltz, 845 p.

—— , 1977c, Die Kieselalgen Deutschlands, Osterreichs und der Schweiz. Teil 3, Band VII, *in* Rabenhorst, L., ed., Kryptogamen Flora Deutschlands, Osterreich und der Schweiz: Königstein, Germany, Otto Koeltz, 816 p.

Huttunen, P., and Meriläinen, J., 1986, Application of multivariate techniques to infer limnological conditions from diatom assemblages, *in* Smol, J. P., Battarbee, R. W., Davis, R. E., and Meriläinen, J., eds., Diatoms and lake acidity: Dordrecht, Dr. W. Junk Publishers, p. 201–211.

Lowe, R., 1974, Environmental requirements and pollution tolerance of freshwater diatoms: U.S. Environmental Protection Agency, Environmental Monitoring Series EPA-67014-74-005, 333 p.

Lund, J. W.G., 1954, The seasonal cycle of the planktonic diatom *Melosira italica* (Ehr.) Kutz. subsp. *subarctica* O. Muller: Journal of Ecology, v. 42, p. 151–179.

—— , 1955, Further observations on the seasonal cycle of *Melosira italica* (Ehr.) Kutz. subsp. *subarctica* O. Muller: Journal of Ecology, v. 43, p. 90–102.

McAndrews, J., 1966, Postglacial history of prairie, savanna, and forest in northwestern Minnesota: Torrey Botanical Club Memoirs, v. 22, p. 1–72.

Patrick, R., and Reimer, C. W., 1966, The diatoms of the United States 1: Philadelphia Academy of Natural Sciences Monograph 13, 668 p.

—— , 1975, The diatoms of the United States 2, Part 1: Philadelphia Monograph 13, Academy of Natural Sciences, 213 p.

Reynolds, C. S., 1984, The ecology of freshwater phytoplankton: New York, Cambridge University Press, 384 p.

Shapiro, J., Lundquist, J. B., and Carlson, R. E., 1975, Involving the public in limnology. An approach to communication: Verhandlungen der International Vereinigung für Theoretische und Angewandte Limnologie, v. 19, p. 866–874.

Sokal, R. B., and Rohlf, F. J., 1981, Biometry: Principles and practice of statistics in biological research: San Francisco, W.H. Freeman & Co., 875 p.

SPSS, Inc., 1983, SPSSx user's guide: New York, McGraw-Hill, 806 p.

Stockner, J. G., and Benson, W. W., 1967, The succession of diatom assemblages in the recent sediments of Lake Washington: Limnology and Oceanography, v. 12, p. 513–532.

Synerholm, C. C., 1979, The chydorid cladocera from surface lake sediments in Minnesota and South Dakota: Archiv für Hydrobiologie, v. 86, p. 137–151.

TerBraak, C.J.F., and Prentice, I. C., 1988, A theory of gradient analysis: Advances in Ecological Research, . 18, p. 272–313.

Tilman, D., 1977, Resource competition between planktonic algae: An experimental and theoretical approach: Ecology, v. 58, p. 338–348.

Tilman, D., Kilham, S. S., and Kilham, P., 1982, Phytoplankton community ecology: The role of limiting nutrients: Annual Review of Ecology and Systematics, v. 13, p. 349–372.

MANUSCRIPT ACCEPTED BY THE SOCIETY JULY 27, 1992

Holocene diatom paleolimnology of Elk Lake, Minnesota

J. Platt Bradbury and K. V. Dieterich-Rurup
U.S. Geological Survey, MS 919, Box 25046, Federal Center, Denver, Colorado 80225

ABSTRACT

Planktonic diatoms dominate the Holocene varved-sediment record of Elk Lake, Minnesota. For the past ~10,400 yr, the lake never became shallow enough to allow large numbers of benthic and epiphytic diatoms to become deposited in the center of the lake. The relatively great depth of Elk Lake throughout this time is consistent with the continuous presence of varves in the record and the predominantly autochthonous character of sediment in the profundal part of the lake.

The planktonic diatom assemblages are dominated by two species, *Fragilaria crotonensis* and *Stephanodiscus minutulus*. They alternate in dominance on scales of hundreds to thousands of years and indicate shifting limnological conditions under subtle climatic control. *Fragilaria crotonensis* is typically a summer and early-fall diatom that prospers when supplies of silicon are high compared to those of phosphorus. *Stephanodiscus minutulus* blooms in the early spring when circulation provides abundant phosphorus. The alternation of these two diatoms reflects principally the climatic conditions that drive spring circulation and summer stagnation and thereby control the fluxes of silicon and phosphorus to and within the lake. Cold and dry climates in late spring and early summer promote blooms of *S. minutulus,* and hot summers with some frontal storm activity provide conditions suitable for *F. crotonensis*.

The mid-Holocene prairie period (8.2 to 4.0 ka) is characterized by a greatly increased diatom accumulation rate and a general dominance of *S. minutulus.* Between 6.4 and 4.0 ka *Aulacoseira ambigua* became prominent, implying increased nutrient fluxes and summer turbulence, probably related to winds and storms in that season. Lake levels were probably lower at times during this period. This part of the Holocene, however, was interrupted by a 600 yr interval of moister and warmer climates (5.4 to 4.8 ka), with low diatom influx and a dominance of *F. crotonensis*.

After 4.0 ka diatom productivity fell, and *F. crotonensis* tended to dominate in response to reduced spring circulation and probably increased precipitation. The Little Ice Age, between A.D. 1450 and 1850, is documented by increased abundances of *S. minutulus,* indicating cooler late spring conditions. Logging activities in the vicinity of Elk Lake in the early twentieth century allowed *Aulacoseira ambigua* to return by increasing turbulence and nutrient fluxes to the lake.

INTRODUCTION: GEOLOGICAL AND ENVIRONMENTAL SETTING

Elk Lake lies within a terrane of glacial deposits formed when the Wadena ice lobe advanced to produce the Itasca moraine about 20,000 ^{14}C yr B.P. (Wright, 1972, and Chapter 2). During deglaciation, between 18,000 and 14,000 ^{14}C yr B.P., meltwater carved subglacial tunnel valleys and partially filled them with gravelly sands and silts that buried large blocks of stagnant ice detached from the shrinking glacier. Subsequent melting of the ice blocks during the final stages of the Wisconsin, 11,000 ^{14}C yr B.P., produced a series of depressions aligned

---

Bradbury, J. P., and Dieterich-Rurup, K. V., 1993, Holocene diatom paleolimnology of Elk Lake, Minnesota, *in* Bradbury, J. P., and Dean, W. E., eds., Elk Lake, Minnesota: Evidence for Rapid Climate Change in the North-Central United States: Boulder, Colorado, Geological Society of America Special Paper 276.

roughly north-south and complexly interconnected. Lake Itasca, Elk Lake, and numerous smaller basins trace the complex pattern of these ancient drainage channels (Wright, 1972; Stark, 1976).

Elk Lake is located in the mixed conifer-deciduous forest of Itasca State Park about 50 km east of the prairie-forest border (Fig. 1). The border marks a major climatically determined vegetation boundary between forest communities to the east that prosper under an excess of precipitation over evaporation, and prairie and savanna communities to the west that survive seasonal moisture stress and periodic fire (e.g., Grimm, 1983). Proximity to this climatic-vegetation boundary places Elk Lake in a suitable position to monitor environmental and climatic changes associated with moisture thresholds that determine the location and productivity of major vegetation types.

Elk Lake is essentially a seepage lake. It receives much of its water from a complex of marshes and wetlands to the south (Figs. 2 and 4 in Wright, Chapter 2), and from the permeable gravelly and sandy tunnel-valley fill that links the basin to the ground-

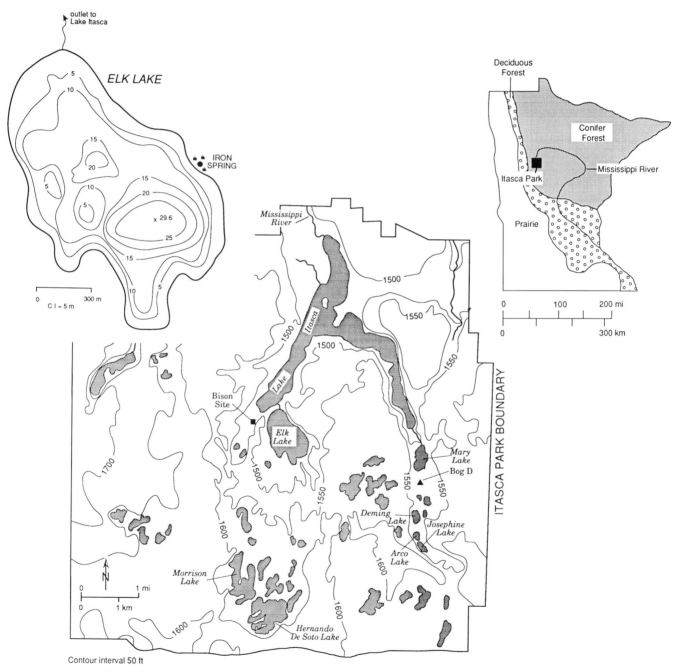

Figure 1. Index map showing the location and bathymetry of Elk Lake, Itasca State Park, bison-kill site of Shay (1971), and the general vegetation of Minnesota.

water system (Oaks and Bidwell, 1968). Because of this, water and nutrient input to the lake is moderated and regulated to some extent by hydrogeological conditions.

The morphometry of Elk Lake (Fig. 1) contributes significantly to the value of its sedimentary record for recording past environmental change. Elk Lake is deep relative to its area, presumably a consequence of the size and shape of the ice blocks it replaced. Anoxic conditions in the deep hole in the southeastern third of the basin preserve diatoms, pollen, and other microfossils and exclude burrowing organisms that disturb and mix the sediments. Seasonal fluxes of sediments to the deep hole are therefore preserved intact as 10,400 annual laminations or varves (Dean and others, 1984) and provide a means of timing precisely paleolimnologic changes. Focusing of sediments to the deep hole has produced comparatively high sediment accumulation rates (2 mm/yr) for the 21.3 m core and thereby provides adequate material for multiple analyses at small time intervals.

Because the location of the deep hole is well away from near-shore regions of Elk Lake (Fig. 1), planktonic diatom communities dominate the core record. Their succession is finely tuned to seasonal fluxes of light, nutrients, and turbulence, which are interrelated themselves and directly or indirectly related to seasonal climatic, hydrologic, and biological environments both within and outside the lake. Land-water interactions are the major focus of this paleolimnological study, because it is through them that Elk Lake's lacustrine history can be extended and interpreted at regional paleoenvironmental scales.

MATERIALS AND METHODS

The 22 m section of lake sediments was cored through 30 m of water with a Livingstone-type square-rod piston sampler (Wright, 1967) near the approximate center of the southeastern deep hole of Elk Lake (Fig. 1) in the winters of 1978 and 1982. The uppermost sediments and sediment-water interface were obtained by an 83 cm frozen core (Wright, 1980). Additional diatom analyses of the uppermost sediments have come from frozen cores collected in 1969 and 1971.

After varve counting, samples containing 50 varve yr of time were cut from the cores and homogenized for diatom and geochemical analysis. A sample of known dry weight at 25 °C was processed for diatoms by digestion in hot (100 °C) concentrated nitric acid. After digestion, acid and solutes were removed from the diatom suspension by centrifugation and decanting. All diatoms from the processed sample were suspended in distilled water, and an aliquot was settled onto cover slips (Battarbee, 1973) and mounted in Hyrax. At least 500 diatoms from each level were identified (e.g., Bradbury, 1975) and counted. Because the area of the count traverse was recorded, the number of diatoms per milligram dry weight (diatom concentration) could be calculated. Diatom net accumulation rate (referred to here as "influx" = diatoms/cm^2/yr) was derived from bulk density, water content, sample varve count, and diatom concentration data (Dean, Chapter 10). Relative frequency of major diatom taxa (occurrence in at least three consecutive levels and minimum frequency of 0.4%) was calculated from the diatom count. Chrysophyte cysts, scales, sponge spicules, phytoliths, and reworked Cretaceous diatoms were tabulated outside the diatom count. The diatom slides are part of the permanent diatom collection at the U.S. Geological Survey, Denver, Colorado.

Samples from the surficial frozen core were collected for 10 varve yr intervals and were processed and counted as above. The diatom biostratigraphy is based on analysis of 108 10 or 50 yr samples; the maximum time separation between samples is 200 yr.

A surface-sample transect from the shallow arm of the lake to the central profundal area (Fig. 2) was taken by Ekman grab sampler to provide information on distribution of modern diatoms relative to lake morphometry. These samples were processed by the same techniques, although without quantitative calibration.

Varve counts, verified by pollen-zone correlations and radiometric dates (Anderson and others, Chapter 4; Dean and Stuiver, Chapter 12), provide the chronology for the core. Varve

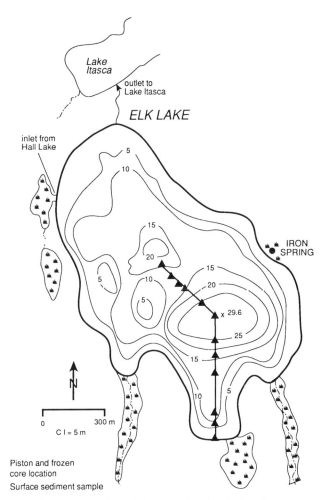

Figure 2. The morphometry of Elk Lake, showing the location of core (x) and the transect of surface sediment samples.

counts on cores taken at a later date (Sprowl, Chapter 6) confirm the general accuracy of the chronology. The age of the surficial frozen cores has been determined by varve counts and the stratigrahic position of the *Ambrosia* rise, an increase in the pollen of ragweed attributable to historic cultivation in the region beginning about A.D. 1890 (Swain, 1973; Birks and others, 1976; Bradbury, 1986). The effects of logging in the drainage area of Elk Lake at the beginning of the twentieth century (Hansen and others, 1974) is also represented by a disturbed, nonvarved layer in the stratigraphic record of the frozen cores.

RESULTS

Relative diatom frequency

Holocene fossil diatom assemblages from Elk Lake are dominated by two planktonic species, *Stephanodiscus minutulus* and *Fragilaria crotonensis* (Fig. 3). Throughout the 10,400 yr varved record of Elk Lake, these two species account for an average of 62% of the total diatom assemblage. Other planktonic diatoms of the genera *Stephanodiscus, Cyclotella, Aulacoseira, Asterionella, Fragilaria, Tabellaria,* and *Synedra* attain significant percentages locally in the core, but their individual frequencies seldom rise above 20% of the assemblage in any sample (Figs. 3 and 4).

Taken together, planktonic diatoms average 82% of the entire diatom assemblage; the only benthic taxon exceeding 5% of the assemblage at any level is *Amphora ovalis*, and most are less than 1%.

The subdominant planktonic diatom species of the Elk Lake core can be grouped according to patterns of stratigraphic distribution. For example, *Cyclotella ocellata* and *C. kuetzingiana* have their highest percentages at the base of the core, and *C. kuetzingiana, Stephanodiscus alpinus,* and *Cyclotella michiganiana* have their principal distribution at levels in the core that are greater than 3.0 ka (Fig. 3). The *Aulacoseira* species and *Stephanodiscus parvus* characterize the central part of the core, particularly between 4.0 and 6.4 ka, whereas *S. niagarae* has a somewhat broader stratigraphic distribution (Fig. 3). *Synedra acus, Asterionella formosa, Cyclotella comta,* and *Tabellaria fenestrata* characterize the past 4000 yr although the latter three species have fairly broad distributions (Fig. 4).

In general, the stratigraphic distribution of benthic diatoms is sporadic, although some species seem to characterize specific zones of the core and may be paleolimnologically significant (Figs. 5, 6, and 7). For example, *Mastogloia, Amphora perpusilla, Rhopalodia gibba,* and several other akaliphilous benthic diatoms are more common before 4.0 ka (Fig. 5), whereas species of *Cymbella, Gomphonema, Eunotia,* and others characterize sediments deposited after 4.0 ka (Fig. 7).

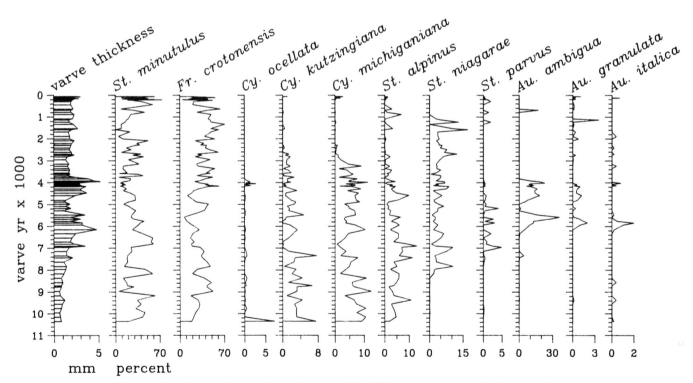

Figure 3. Dominant planktonic diatoms (percent) of the Elk Lake core plotted with the varve thickness chronology.

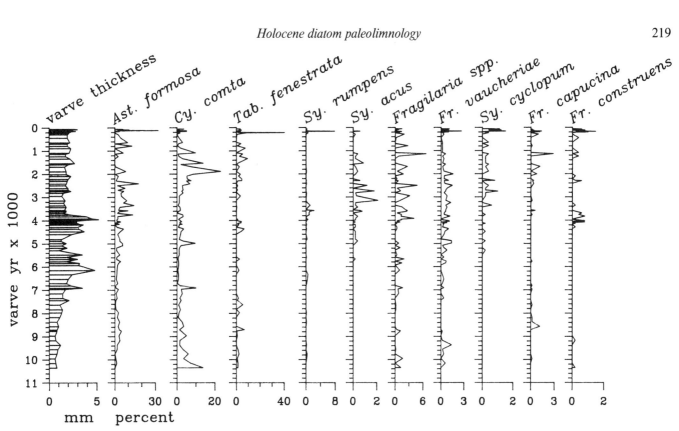

Figure 4. Minor planktonic and tychoplanktonic diatoms (percent) of the Elk Lake core plotted with the varve thickness chronology.

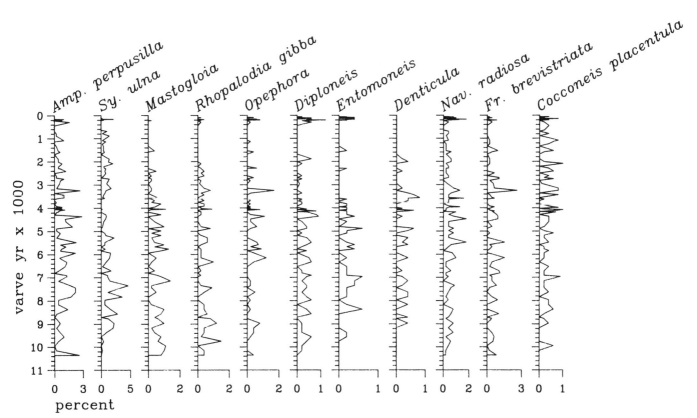

Figure 5. Benthic diatoms (percent) from the Elk Lake core that characterize the Elk Lake core before 4.0 ka.

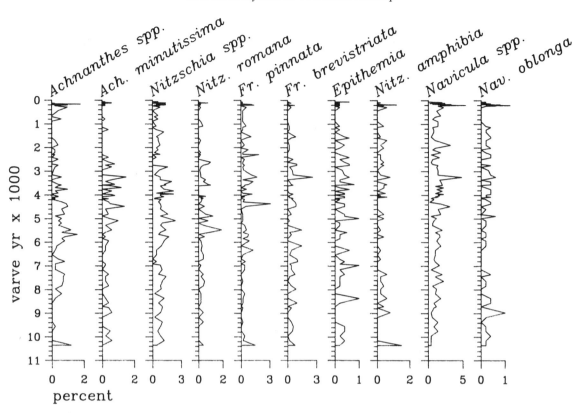

Figure 6. Benthic and tychoplanktonic diatoms (percent) that characterize the Elk Lake core after 4.0 ka.

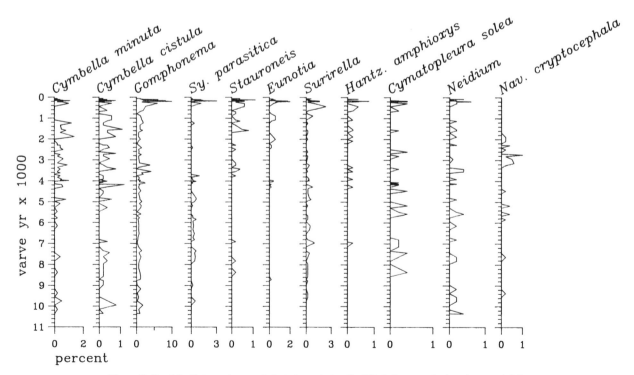

Figure 7. Benthic diatoms (percent) that characterize the Elk Lake core during the past 1.0 ka.

Abundance of other siliceous microfossils

The comparative abundances of reworked Cretaceous marine diatoms, phytoliths, chrysophyte cysts and scales, and sponge spicules (Fig. 8) were determined separately from the diatom count to provide additional paleolimnological information. Sporadic values of Cretaceous diatoms are highest at the base of the core. Large numbers of phytoliths, chrysophyte cysts and scales, and sponge spicules characterize levels deposited in the past 2000 yr although chrysophyte scales are also abundant around 9.0 and 5.0 ka.

Diatom concentration and accumulation rate

Diatom concentration (diatoms/g) fluctuates between high and low values, but in general high concentrations occur between 4.0 and 8.0 ka (Fig. 8). The net accumulation rate (influx) of diatoms (diatoms/cm^2/yr) generally correlates with diatom concentration and varve thickness (Figs. 8 and 9), because diatoms contribute significantly to the thickness of some varves. High influxes of individual diatom taxa (Figs. 9 and 10) show some stratigraphic variation. For example, the influx of *Aulacoseira* (Fig. 10) is stratigraphically restricted to periods between 4.0–4.8 and 5.3–5.7 ka, whereas *Cyclotella michiganiana*, *Tabellaria flocculosa*, *Fragilaria crotonensis*, and *Stephanodiscus niagarae* all have peak influxes shortly before 4.0 ka (Figs. 9 and 10).

Influx peaks of other planktonic diatoms occur at 8.2, 6.8, 5.7, and 4.2 ka, and *Synedra acus* peaks are highest between 2.0 and 4.0 ka.

Surface frozen-core studies

Paleolimnological changes during the past 200 yr, and especially during the interval when Europeans began lumbering in the Itasca Park region at the end of the nineteenth century, have been investigated by detailed diatom and pollen analysis of frozen cores of the upper meter of sediment. In the 1978 frozen core, diatom counts extend to the base of the "disturbed layer" (Anderson and others, Chapter 4) that represents flooding of Elk Lake by logging activities in A.D. 1903 (Stark, 1971; Bradbury, 1975).

Profiles of major diatom taxa for the 223 yr between A.D. 1680 and A.D. 1903 in the 1978 frozen core (Figs. 11 and 12) illustrate limnological changes related to climate change and fire history in Itasca State Park. The dominant planktonic diatoms of Elk Lake during this period remain *Fragilaria crotonensis* and *Stephanodiscus minutulus*. Together they make up about 80% of the assemblage with a reciprocating dominance. *S. minutulus* has periods of dominance between A.D. 1690–1730, 1750–1795, and 1810–1865 (Fig. 11). *Gomphonema* species and phytoliths show a dramatic increase and decline between A.D. 1700 and 1740, as do some of the minor components of the diatom assemblage, such as *Amphora ovalis* and species of *Surirella* and

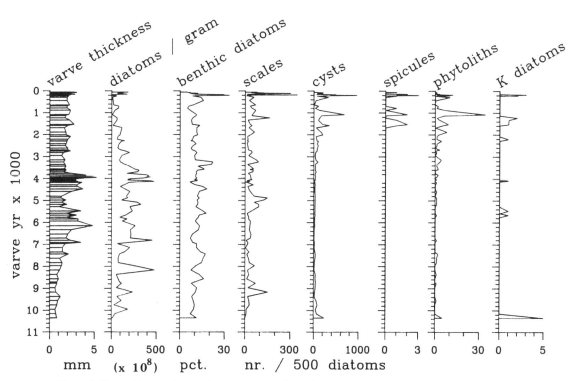

Figure 8. Varve thickness (mm), diatom concentration (diatoms/gram), and the comparative abundance of benthic diatoms, chrysophyte scales and cysts, sponge spicules, phytoliths, and Cretaceous (K) marine diatoms in the Elk Lake varved core.

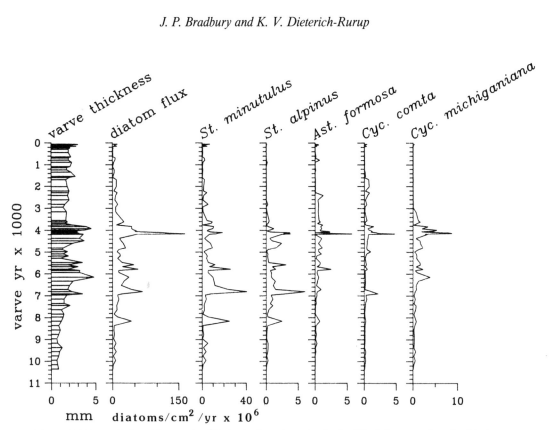

Figure 9. Diatom influx (diatoms/cm²/yr) of selected species in the Elk Lake core plotted with the varve thickness chronology.

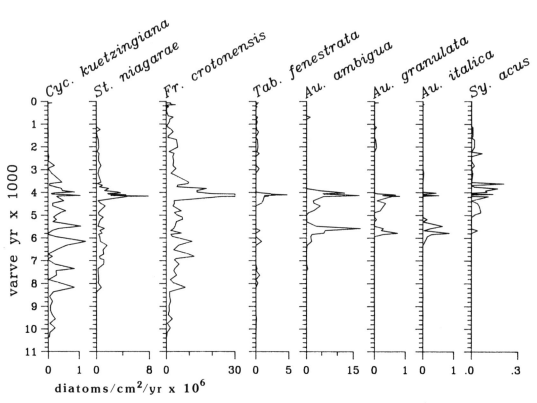

Figure 10. Influx of characteristic mid-Holocene planktonic diatoms in the Elk Lake core.

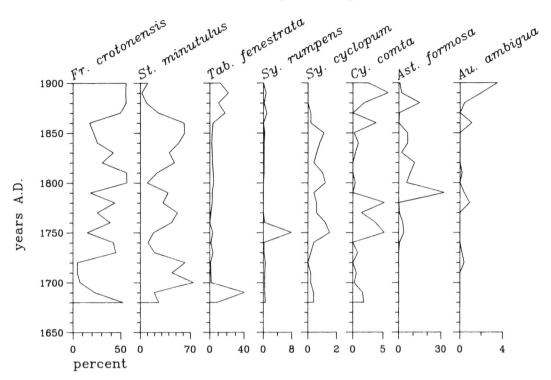

Figure 11. Percentage of characteristic diatoms in the Elk Lake 1978 frozen core between A.D. 1690 and A.D. 1900.

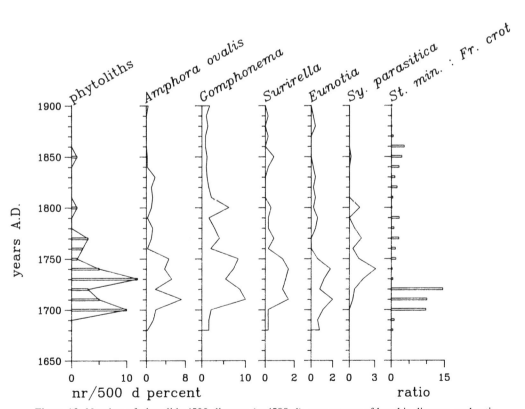

Figure 12. Number of phytoliths/500 diatoms (nr/500 d), percentages of benthic diatoms, and ratio between *Stephanodiscus minutulus* and *Fragilaria crotonensis (St. min: Fr. crot)* in the Elk Lake 1978 frozen core between A.D. 1690 and 1900.

Eunotia (Fig. 12). Other planktonic diatoms appear at or after A.D. 1730 in a successional progression that leads to an increase in *Aulacoseira ambigua* at the top of the laminated section.

The diatom stratigraphy of the upper 64 cm of sediment in the deep hole of Elk Lake records lumbering activities in Itasca Park (A.D. 1902–1917; Spurr, 1954). The diatoms (Stark, 1971; Bradbury, 1975) (Fig. 13) are characterized initially by high values of *Stephanodiscus minutulus*. A subsequent increase in *Fragilaria crotonensis, Asterionella formosa*, and *Tabellaira fenestrata* coincides with the *Ambrosia* rise (A.D. 1890) at a depth of 56 cm. Weakly laminated sediments of the "disturbed zone" extend from 53 to 14 cm and encompass an increase in *Aulacoseira ambigua*.

Pollen and diatom analyses (H. E. Wright, 1984, written commun.) from another frozen core of deep-basin surface sediment taken in 1971 can be correlated with Stark's (1971) observations. The *Ambrosia* rise in the 1971 core begins at 47 cm and the first increase in *Aulacoseira ambigua* occurs in the "disturbed zone" at 44 cm (Fig. 14).

Visual correlation of distinctive varves in the 1971 and 1978 frozen core (H. E. Wright, 1984, written commun.) locates the position of the *Ambrosia* rise just below the "disturbed zone" in the 1978 core, and the initial increase of *Aulacoseira ambigua* at this level confirms the assignment of the base of the "disturbed zone" to the beginning of the twentieth century. The distinctive laminated and disturbed horizons in the upper meter of sediment are of irregular thickness and thin toward the margin of the deep hole; laminated sediments are absent at water depths less than 21 m (H. E. Wright, 1984, written commun.). This circumstance and potential disturbance of the sediment record by coring activi-

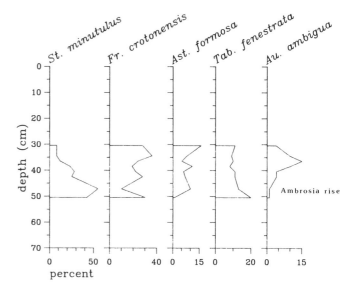

Figure 14. Percentage of characteristic diatoms spanning the *Ambrosia* rise (31–51 cm) in an Elk Lake frozen core taken in 1971. Analyst: B. J. Speziale.

ties may account for the different depths of the "disturbed zone" in frozen cores and the missing sediment at the top of the 1978 frozen core.

Surface sediment transect

A transect of surface sediment samples, beginning at 1 m depth near the shore of the southern embayment of the lake and extending northward through the deep hole and into the central region of the lake (Fig. 2), illustrates how planktonic and benthic diatoms accumulate on the lake bottom. Planktonic diatoms dominate all offshore sediments, whereas benthic diatoms such as *Cocconeis placentula* and species of *Fragilaria* attain significant frequencies only in the nearshore sample (Figs. 15 and 16). The percentage composition of the plankton diatom community shows some variation across the lake bottom, but it is small and not obviously related to depth. The major planktonic species represented in the uppermost samples of the frozen surface core studied by Stark (1971) (Fig. 13) are present in the surface sediments in similar frequencies (Fig. 15), suggesting that the average limnological character of the lake has not changed appreciably since the early part of the twentieth century. The restriction of benthic diatoms to shallow water depths, may suggest that sediment focusing is not a very active process in Elk Lake.

INTERPRETATION AND DISCUSSION

Nutrient dynamics of Elk Lake and diatom ecology

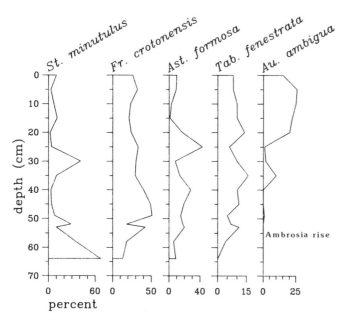

Figure 13. Percentage of characteristic diatoms spanning the *Ambrosia* rise and logging period at Elk Lake. Counts are from Stark (1971) on the 64-cm-long Elk Lake frozen core taken in 1969.

The dominant diatoms of the Elk Lake cores are planktonic species, the abundance and succession of which is controlled by seasonal changes in nutrients (chiefly silicon and phosphorus),

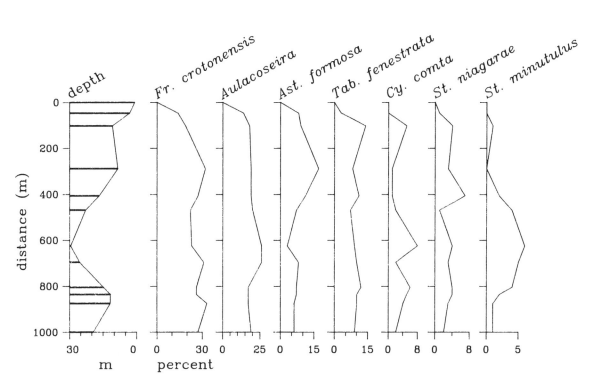

Figure 15. Dominant planktonic diatoms in surface sediments along a transect from shallow (south end of lake) to deep water (Fig. 2).

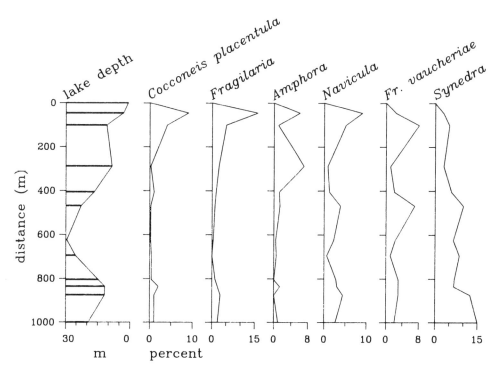

Figure 16. Characteristic benthic diatoms in surface sediments along a transect from shallow (south end of lake) to deep water (Fig. 2).

light levels, and turbulence. Elk Lake behaves as a more or less typical dimictic lake, generally thawing in April or early May and circulating until stratification becomes fully developed by June. Stratification weakens and deepens with cooler temperatures in September and October, and the lake circulates again in early November and freezes shortly afterward (Megard and others, Chapter 3).

During the periods of summer and winter stagnation, hypolimnetic oxygen is depleted, and anoxic conditions occur beneath the thermocline (Megard and others, Chapter 3). Phosphorus (P) and nitrogen (N) accumulate in the hypolimnion during stratification and become distributed to the epilimnion to be utilized by phytoplankton during periods of circulation. The turbulence accompanying circulation suspends the diatoms in the photic zone, where photosynthesis occurs. Increasing light levels during spring circulation often produces larger diatom blooms than those in the fall. Silica for diatom growth is derived from the hypolimnion, where sedimented diatom frustules partially dissolve, and from ground water entering Elk Lake (Megard and others, Chapter 3).

In Elk Lake, and probably in many other temperate dimictic lakes, diatom productivity and succession are closely governed by climatic factors that control length and depth of circulation and nutrient recharge from ground water. Nutrient buildup in the hypolimnion, partially controlled by the length of periods of stratification, may also be important, especially during the summer. Supply ratios of Si/P (Kilham and others, 1986) appear to exert the major influence on diatom productivity and succession in this region, and the climatic and limnologic factors that control these nutrients are the most critical for paleolimnologic and paleoclimatic interpretation of fossil diatom floras.

At Elk Lake today the dominant planktonic diatoms are *Tabellaria fenestrata, Synedra acus, Asterionella formosa, Fragilaria crotonensis,* and *Stephanodiscus* spp., particularly *S. minutulus.* Both distributional and laboratory studies of these diatoms indicate that their dominance depends upon their relative abilities to compete for Si and P (Kilham and others, 1986; Tilman and others, 1982; Mechling and Kilham, 1982). *Tabellaria fenestrata* and *Synedra* species are the best competitors for phosphorus and relatively poor competitors for silicon: consequently, these diatoms grow best with ample silicon supplies. The reverse is true for *Stephanodiscus species*; they compete well for silicon and grow best when phosphorus is abundant. *Fragilaria crotonensis* and *Asterionella formosa* are intermediate in their silicon and phosphorus requirements.

Sediment-trap studies at Elk Lake support this general model of diatom ecology by relating diatom productivity with observed seasonal weather conditions, which control circulation and stratification of Elk Lake (Nuhfer and others, Chapter 7). In 1979, ice out did not occur until 7 May, and the lake was stratified by mid-May. Consequently, the period of spring circulation was only nine days long in 1979, and probably large quantities of soluble phosphorus were not generated from the hypolimnion or from the sediments. The presumed low phosphorus fluxes were apparently not adequate to support a large bloom of *Stephanodiscus minutulus*; this species is not recorded in the sediment traps during that time interval. However, because Si/P ratios were apparently high, *Synedra* species were favored and they, along with smaller numbers of *Fragilaria crotonensis* and *Tabellaria fenestrata,* dominated the sediment-trap record of the spring 1979 diatom bloom (Bradbury, 1988).

In 1984, however, ice out occurred on April 19 and was followed by about 25 days of cool to cold weather, which apparently allowed an extended period of deep circulation in Elk Lake and a proportionally large flux of soluble phosphorus from the hypolimnion and/or from dissolution of phosphorus-rich minerals in the anoxic bottom sediments. A pure layer of *Stephanodiscus minutulus* was deposited in the sediment trap at this time, reflecting the presumed lower Si/P nutrient supply ratio favorable for this species (e.g., Kilham and others, 1986).

In a similar fashion, Stauffer and Armstrong (1984, 1986) demonstrated that large epilimnetic increases in phosphorus in Shagawa Lake (northeastern Minnesota) occurred in response to sudden downward migration of the epilimnion into the nutrient-rich metalimnion and hypolimnion by wind-generated eddy diffusion during summer cold fronts. A notable event of this nature occurred in late July, 1974, when epilimnetic phosphorus increased 130% in Shagawa Lake and caused a bloom of *Stephanodiscus* (Bradbury, 1978).

Although mean daily temperatures and wind energy are important in determining the nature of spring and fall circulation in dimictic lakes, other processes, such as the amount and timing of precipitation prior to and during the open-water season, can play an important role in nutrient fluxes. This may be particularly true for ground-water silicon fluxes, as mentioned earlier. Precipitation, in combination with seasonal warming and cooling, could diversify and complicate greatly the character of nutrient fluxes to Elk Lake. It is also evident that a given Si/P loading ratio might result from different combinations of seasonal climate. For example, long periods of spring circulation (high phosphorus loading) could result from early ice out related to warm winter weather, or from cold, windy late spring and early summer weather.

Nevertheless, some hypothetical climate-diatom relations can be suggested for the two dominant planktonic diatoms in the Elk Lake core, *Stephanodiscus minutulus* and *Fragilaria crotonensis*. High influxes or percentages of *Stephanodiscus minutulus* should reflect climatic and limnologic factors that promote low Si/P ratios. Anything that reduces silicon and increases phosphorus would tend to favor this diatom. Extended and deep circulation (to provide a high phosphorus flux) due to early ice out and/or late spring stratification is one possibility. Reduction of late winter and spring precipitation and a concomitant reduction of ground-water silicon fluxes could similarly change the Si/P ratio. Such factors are not mutually exclusive.

Conversely, limnologic and climatic factors that either increase the flux of silicon and/or reduce that of phosphorus would tend to favor diatoms that are good competitors for phosphorus: *Fragilaria crotonensis, Asterionella formosa, Synedra* species, and *Tabellaria fenestrata.*

Other kinds of evidence for diatom ecology come from the distribution of diatoms in the sediment of Minnesota lakes. Samples of surface sediment can be related to the limnology, geography, and climate of the lakes. For example, *Stephanodiscus* species, particularly *S. hantzschii* (a close relative of *S. minutulus*), dominate in lakes in the prairie regions of southwestern Minnesota, where phosphorus loading of lakes tends to be high and precipitation is low. At the other extreme, *Synedra nana* characterizes dilute lakes in central and northern Minnesota that often fail to circulate every year (Brugam, 1983). When this happens, nutrients, especially phosphorus, remain in the bottom waters and anoxic sediments and are unavailable for diatom growth. In all probability, the loading ratio of Si/P is high in such lakes.

The diatom stratigraphy of sediment cores from (culturally) enriched lakes also confirms the role of soluble phosphorus in supporting large blooms of *Stephanodiscus minutulus* and *S. hantzschii* (e.g., Bradbury, 1975).

Two other diatom genera, *Cyclotella* and *Aulacoseira*, are important in the recent history of Elk Lake. Considerably less is known about the autecology of these genera, even though they are common in many temperate lakes. The *Cyclotella* species *C. comta*, *C. michiganiana*, and *C. kuetzingiana* characterize oligotrophic to mesotrophic lakes of circumneutral pH (Brugam, 1983, and unpublished data; Charles, 1985). *Cyclotella comta* and *C. michiganiana* were common in sediment-trap samples during the spring of 1979, implying that they prosper when Si/P ratios are high, like *Synedra* spp., *Tabellaria fenestrata*, and *Fragilaria crotonensis*. The distribution of these *Cyclotella* species in a sediment core from Harvey's Lake, Vermont, suggests that they prosper under conditions of modest nutrient enrichment (Engstrom and others, 1985).

The species of *Aulacoseira* all appear to have high growth requirements for silicon coupled with active turbulence to keep these heavy, thread-like diatom colonies suspended within the photic zone of a lake (Kilham and others, 1986). Phosphorus requirements for the different *Aulacoseira* species are not known specifically, but distributional data suggest that *A. granulata* requires the highest levels of phosphorus, *A. ambigua* is intermediate in its needs for this nutrient, and *A. italica* has the lowest needs (Brugam, 1983, and unpublished data; Kilham and others, 1986). These species may also have different light requirements; *A. granulata* characterizes low-light environments (typically shallow, turbid, and eutrophic lakes) and *A. ambigua* and probably *A. italica* predominate in more transparent (i.e., high-light) lake systems (e.g., Kilham and others, 1986).

Paleolimnology of Elk Lake

Boreal and coniferous forest environments: 10.4–8.2 ka. The diatom record at Elk Lake begins at a core depth of 21.3 m below the sediment-water interface with the appearance of laminated (varved) sediments. According to the varve chronology, these sediments were first deposited 10,360 yr ago. Sediments cored below that depth, to a depth of 22.5 m, were barren of diatoms, although *Phacotus*, a green calcareous algae, is abundant throughout the basal 1.2 m and suggests that limnic environments existed prior to the preservation of varved sediments. The unvarved basal sediments are characterized by *Picea* pollen with variable amounts of woody, forest floor trash (e.g., Shay, 1971; Whitlock and others, Chapter 12; Fig. 17). At this time, Elk Lake may have been a shallow, sag-pond environment overlying a stagnant block of melting glacial ice. The lack of diatoms probably results from their destruction in this shallow-water, possibly somewhat alkaline system.

The earliest horizon with adequate diatom remains for analysis is dominated by planktonic species of *Cyclotella, Fragilaria crotonensis,* and *Stephanodiscus minutulus* (Fig. 3). Diatom concentration is low, and there are comparatively large numbers of the reworked Cretaceous marine diatoms (Fig. 8), presumably

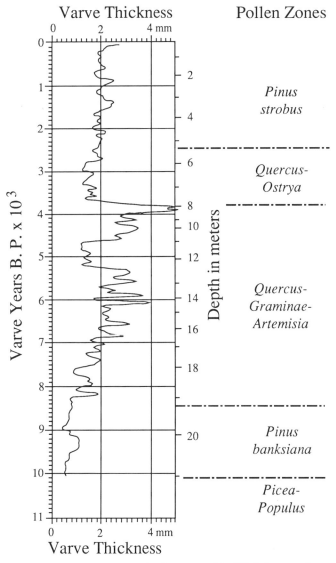

Figure 17. Varve thickness and pollen zones of the Elk Lake varved core.

indicating erosion of glacial till derived from Cretaceous deposits to the northwest (e.g., Florin, 1970; Wright, Chapter 2).

The diatom dominants of this assemblage, *Stephanodiscus minutulus* and *Fragilaria crotonensis,* are the same species that characterize Elk Lake today. In fact, the stratigraphic distribution of these two planktonic diatoms persists with alternations in dominance throughout the record. This indicates that the limnology of Elk Lake, at least in terms of seasonality of nutrient fluxes and general hydrochemistry, has remained remarkably constant for at least 10,000 yr.

Cyclotella ocellata has its greatest representation in the first diatom sample (Fig. 3). This diatom characterizes the least-disturbed regions of the Great Lakes, where it is an inhabitant of cool, chlorophyll-rich metalimnetic waters (Stoermer and others, 1985). In general, the species is considered characteristic of cool, oligotrophic, oxygen-rich lakes with a pH range of 8.6–10.9 (Gasse and Tekaia, 1983). *Cyclotella kuetzingiana* and *C. michiganiana* are also well represented in the early levels of the core. These diatoms appear to reflect somewhat warmer temperatures in Lake Michigan (15–20 °C) (Stoermer and Ladewski, 1976) and are common in, but not restricted to, circumneutral oligotrophic lakes (Bradbury, 1978). Cool oligotrophic waters in Elk Lake is consistent with pollen evidence of coniferous forests in the region at that time (Whitlock and others, Chapter 17; Fig. 17) and probably the representation of *C. ocellata, C. kuetzingiana,* and *C. michiganiana* record the waning phases of cool climate responsible for the boreal vegetation. Low plant-pigment concentrations in the core at about 10.0 ka (Sanger and Hay, Chapter 13) also imply oligotrophic conditions with poor pigment preservation at Elk Lake.

A faunal paleolimnologic study of Elk Lake (Stark, 1976) documents the presence of chironomids (*Tanytarsus*) and ostracods (*Cytherissa lacustris*) in the *Picea* and *Pinus banksiana* pollen zones of mid-depth lake cores. These organisms require cool, oxygen-rich water, and their presence in the deeper zones of Elk Lake during that time also supports the interpretation of comparative oligotrophy 10.0 to about 8.0 ka.

Small percentages of benthic diatoms such as *Amphora perpusilla, A. ovalis, Nitzschia amphibia, N. romana, Mastogloia* spp. (Figs. 5 and 6), and *Gyrosigma attenuatum* in the earliest part of the Elk Lake record indicate that periodically higher alkalinities existed in the basin.

Diatom productivity as measured by the annual influx of diatoms to the sediments of Elk Lake (Fig. 9) increased slightly by 10.0 ka, but remained generally low until 8.2 ka. The diatom influx record correlates with the record of thin varves throughout this interval and with the dominance of *Pinus banksiana* pollen (Whitlock and others, Chapter 17). *Stephanodiscus minutulus* dominates until about 9.0 ka, and *Fragilaria crotonensis* is subdominant. This suggests, according to the model established earlier, that Elk Lake experienced early mild springs and/or cooler summers and therefore extended periods of circulation between 10.0 and 9.0 ka. Late winter and early spring precipitation may also have been less than modern averages during this time.

Between 9.0 and 8.3 ka, the relative dominance of *Stephanodiscus minutulus* and *Fragilaria crotonensis* reversed (Fig. 3). Accordingly, periods of spring circulation during this time may have been insufficient to provide phosphorus to support large blooms of *S. minutulus*. Precipitation may have increased to provide the additional ground-water supplies of silicon that allowed *F. crotonensis* to prosper. Nevertheless, plant-pigment data for this interval in the core (Sanger and Hay, Chapter 13) indicate increased production of cyanobacteria (bluegreen algae). It is possible that *S. minutulus* was limited by silicon during this period, because cyanobacteria in such circumstances are equally efficient competitors for phosphorus (Tarapchak and Herche, 1987). Lack of turbulence during strong summer stratification may have selectively allowed diatoms to sink below the productive photic zone, whereas cyanobacteria can float (e.g., Horne and Goldman, 1972). In addition, cyanobacteria such as *Oscillatoria* may selectively inhibit centric diatom production through allelopathy (Engstrom and others, 1985).

The two distinct limnologic episodes during this time interval (10.0 to 8.3 ka) probably represent climatic changes. *Pinus banksiana* dominates throughout this interval, although between 9.6 and 9.1 ka both its percentages and influx decreased markedly (Whitlock and others, Chapter 17). This decrease correlates with a slight increase in varve thickness (Fig. 3) and with increases in quartz, feldspar, calcite, and other carbonate minerals (Dean, Chapter 10). The mineral, geochemical, and varve-thickness parameters that increase between 9.6 and 9.1 ka reflect the presence of turbidites, which are relatively thick sediment layers that originated on the margins of Elk Lake and slumped or flowed to the profundal regions of this deep lake. The turbidites, because of their origin on the margins of Elk Lake, have low pollen concentrations and probably account for the low values of *Pinus banksiana* in this part of the core. Minor increases in benthic diatoms such as *Fragilaria brevistriata* and *F. vaucheriae* (Fig. 4) in this interval may reflect the influx of turbidites from the shallow parts of Elk Lake, although in general the diatom stratigraphy does not indicate major limnologic change. Perhaps the peak of *Stephanodiscus minutulus* percentages at about 9.2 ka represents a period of exceptionally vigorous circulation, which promoted the formation of turbidite layers as documented by the quartz and feldspar mineralogy in that part of the core (Dean, Chapter 10). Throughout the interval between 10.0 and 8.3 ka, low concentrations of plant pigments (Sanger and Hay, Chapter 13) and low values of diatom influx indicate reduced algal productivity.

Prairie period: 8.5–4.0 ka. The prairie period begins with the pollen-zone transition from *Pinus banksiana* to *Quercus*-Gramineae-*Artemisia* about 8.5 ka (Whitlock and others, Chapter 17; Fig. 17). The pollen transition is regarded as reflecting the beginning of a major change to drier climates that supported an oak savanna with prairie vegetation in areas occupied earlier by tall pine forests (McAndrews, 1966; H. E. Wright, 1968, 1976; Whitlock and others, Chapter 17). As the forests became less dense, westerly winds became more effective and sand dune deposition occurred 70 km east-northeast of Elk Lake between 8000 and 5000 ^{14}C yr B.P. (Grigal and others, 1976).

At Elk Lake the diatom record of the prairie period is more clearly defined by changes in diatom influx than by changes of species assemblages (Fig. 9). *Stephanodiscus minutulus* and *Fragilaria crotonensis* continue to dominate the record, alternating in dominance in a roughly cyclic style with approximate periods between 1000 and 300 yr (Fig. 3). The diatom record indicates that drastic chemical or physical limnologic changes at Elk Lake did not occur during the prairie period. Salinity probably remained below 1000 ppm, and lake chemistry remained dominated by Ca^{++} and HCO_3^- ions. Probably variations in phosphorus loadings were the major cause of shifts in dominance between *S. minutulus* and *F. crotonensis,* although variation in silicon loadings may have been important. Increased productivity with minimal change in species assemblages at Elk Lake during the prairie period may parallel the effects of fertilization on other oligotrophic or mesotrophic lakes with well-integrated primary-producer communities (Stoermer and others, 1985).

A dramatic increase of diatom influx to the Elk Lake sediments occurred about 8.2 ka. The accompanying increase in varve thickness (Fig. 9) is due both to the increase in diatom flux and in clastic sedimentation, as manifested by the stratigraphic profiles of quartz, feldspar, and sodium (Dean, Chapter 10). *Stephanodiscus minutulus* dominates over *Fragilaria crotonensis* in this comparatively brief transitional period (Fig. 3), suggesting that extended periods of spring circulation predominated. Perhaps such periods of spring circulation also accounted for the increases in clastic influx at this time.

The transition from pine-dominated to prairie-savanna vegetation recorded in the Elk Lake core is also marked by strong changes in several geochemical parameters. Si, Al, Na, Mg, and K all increase as a result of the increase in clastic sedimentation and increased influx of diatoms. Fe, P, and Mn, however, decrease rapidly after 8.2 ka. Some of the phosphorus was probably recycled by long periods of circulation and utilized by diatoms and other algae to increase the productivity of Elk Lake. Nevertheless, the mass accumulation rate of phosphorus remained comparatively constant (Dean, Chapter 10).

Stephanodiscus minutulus was particularly abundant from 8.4 to 7.8 ka and between 7.0 and 6.2 ka. After 6.2 ka, this species follows an oscillating diminishing trend until the close of the prairie period, 4.0 ka. Distribution of *Fragilaria crotonensis* more or less mirrors that of *S. minutulus,* but also decreases in the mid-prairie period as *Aulacoseira ambigua* becomes an important part of the diatom assemblage (Fig. 3).

Although the dominant diatoms persisted into the prairie period, some noteworthy species changes occurred during this interval. *Stephanodiscus niagarae* first becomes abundant about 8.0 ka (Fig. 3) and appears to record increasing eutrophy and turbulence of Elk Lake. Elsewhere in Minnesota this diatom is found predominantly in shallow, eutrophic, prairie lakes with low transparencies and comparatively high levels of dissolved sulfate (Brugam, 1983). Morphological studies on *Stephanodiscus niagarae* (Theriot and Stoermer, 1984) indicate that fine-structured valves of this diatom are characteristic of very eutrophic lakes, whereas coarse-structured valves are found under mesotrophic to oligotrophic conditions. The valves of *S. niagarae* in Elk Lake are typically coarse-structured, indicating that although the lake became more eutrophic than it was previously, it was not limnologically similar to typical prairie lakes of southwestern Minnesota today.

Another *Stephanodiscus* species, *S. parvus,* has an irregular stratigraphic distribution in the Elk Lake core that coincides closely with the prairie period, although it also appears sporadically in the upper part of the record as well (Fig. 3). *Stephanodiscus parvus* can dominate the spring diatom bloom in highly eutrophic lakes in both Europe and North America (Stoermer and Hakansson, 1984). In a core from Lake Ontario this species has its greatest development in sediments that represent the periods of maximal nutrient loading, particularly of phosphorus (Stoermer and others, 1985). Its presence in the Elk Lake core may document periods during the prairie period, when phosphorus loading rapidly increased, perhaps as a result of wind entrainment of nutrient-rich metalimnetic water (e.g., Stauffer and Armstrong, 1984, 1986).

Stephanodiscus alpinus maintains comparatively high percentages throughout the prairie period, although it was sporadically abundant before that time (Fig. 3). Like other *Stephanodiscus* species, *S. alpinus* lives under a wide range of mildly eutrophic conditions (Theriot and Stoermer, 1982). In the Great Lakes and other areas it is typical of cold water (<2 °C) and is restricted to winter or subthermocline growth (Stoermer and others, 1985). *Fragilaria crotonensis* is the dominant diatom between 6.8 and 7.6 ka, implying lake stratification early in the season and low phosphorus levels in the epilimnion. However, the highest percentage abundance of *Stephanodiscus alpinus* also lies between 6.8 and 7.6 ka, where it appears to correlate with the distribution of *Limnocythere herricki,* an ostracode indicative of cold prairie environments in central Canada (Forester and others, 1987). The Canadian prairies have colder and drier fall seasons than are characteristic of Minnesota today, and perhaps these climatic conditions promoted the development of *Stephanodiscus alpinus* in the early part of the prairie period by providing cold water, abundant nutrients, and falling light levels during and immediately following fall circulation.

Minor but significant percentages of *Entomoneis ornata, Synedra parasitica, Amphora perpusilla,* and species of *Achnanthes* are also consistently present throughout the prairie period (Figs. 5, 6, and 7) although all have sporadic distributions in other parts of the core. *Entomoneis ornata* is a large diatom that lives on muddy bottoms in the shallow parts of lakes, and its prevalence in the prairie period sediments may testify to increased turbulence and transport of diatoms and other materials from the littoral regions of the lake. *Synedra parasitica* lives epiphytically on large littoral diatoms like *Surirella* (Hustedt, 1930). Its increased distribution in the profundal region of Elk Lake, like that of the other notable benthic diatoms, *Amphora perpusilla* and

Achnanthes species, relates to increased littoral transport, suggesting occasional lower lake levels and increased wind-driven turbulence. A palynologic study of a transect of cores from Elk Lake (Stark, 1976) suggests that a possible 6 m lowering of Elk Lake occurred during some part of the prairie period.

Aulacoseira ambigua and minor numbers of *A. granulata* are also characteristic of the prairie period in Elk Lake. Their principal distribution is restricted to two intervals, 6.4 to 5.3 and 4.9 to 4.0 ka (Fig. 3), although small percentages are found elsewhere in the core. Maximum influx of *Aulacoseira* occurs between 5.4 and 5.9 ka and at 4.2 ka (Fig. 10).

The dominant *A. ambigua*, one of the most widespread diatoms in Minnesota lakes today (Brugam, 1983), requires high levels of silicon, moderate loadings of phosphorus, and sufficient turbulence to keep it suspended in the photic zone, where intermediate light levels exist for photosynthesis (Kilham and others, 1986). The mean specific conductance of seven Minnesota lakes where this diatom composes greater than 50% of the surface sediment assemblage is about 500 μmho, the mean silica concentration is 16 mg/l, and the mean phosphorus concentration is about 32 μg/l. *A. ambigua* bloomed from August to October, 1974, in Shagawa Lake, northeastern Minnesota, when both silicon and phosphorus were transported to the epilimnion as a result of wind mixing and entrainment of nutrient-rich metalimnetic water (Stauffer and Armstrong, 1984; Bradbury, 1978). Moderate nutrient fluxes, turbulence, and high light conditions are required to allow *A. ambigua* to prosper. Low light conditions during spring and fall circulation probably account for its generally low dominance during these times, even though nutrient levels may be adequate. However, wind-generated turbulence in the high-light summer season can provide the necessary nutrient levels for this diatom, even in comparatively deep stratified lakes such as Elk Lake, because the metalimnion contains adequate supplies of silicon and phosphorus (e.g., Stauffer and Armstrong, 1984, 1986).

Aulacoseira species form resting cells that remain in the sediment during times of unfavorable conditions (Sicko-Goad and others, 1989). Once resuspended by turbulent circulation, the resting cells rejuvenate rapidly to form large populations capable of exploiting the return of favorable environments.

The *Aulacoseira ambigua* distribution in the Elk Lake core can be interpreted as representing environments of increased wind stress on the lake during the summer and early fall, probably as a result of increased frontal storm activity and a more open (less forested) vegetation of short oaks and prairie herbs. The distribution of *A. ambigua* correlates closely with the abundance of clastic materials (e.g., Na, Al, and quartz) to Elk Lake (Dean and others, 1984); the increases in these and related parameters also probably reflect turbulence in the lake and drier conditions surrounding it.

High influxes of *Stephanodiscus minutulus*, *S. alpinus*, and *S. niagarae* and other diatoms (Figs. 9 and 10) during these two periods (5.4 to 5.9 ka and at 4.2 ka) indicate that spring and fall circulation was also important in contributing diatoms and sediment to the profundal regions of Elk Lake.

Between 5.4 and 4.8 ka *Aulacoseira ambigua* disappears and *Fragilaria crotonensis* percentages increase. Diatom influx is low in this interval, and thin varves and reduced quantities of clastic materials (Dean and others, 1984) indicate that Elk Lake became less eutrophic and perhaps deeper than it had been previously. Phosphorus and iron increase during this time (Dean, Chapter 10), suggesting that circulation at Elk Lake was often incomplete and that clastic sedimentation did not dilute these autochthonous components. This 600 yr interval of presumed moister and less windy climate began and terminated abruptly, within a decade, according to varve measurements (Anderson, Chapter 5), and represents a major short-term climatic-limnologic reversal in the prairie period. The pollen record of surrounding vegetation does not clearly record this climate change, although pollen influx is generally low at this time, perhaps for sedimentological reasons.

Typical prairie period limnological conditions returned to Elk Lake 4.8 ka and persisted until about 4.0 ka. *Aulacoseira ambigua* and *Stephanodiscus minutulus* are the dominant diatoms, and comparatively high percentages of *S. alpinus* characterizes the first part of this period. By 4.2 ka, *Fragilaria crotonensis* dominates and diatom influx is at an all-time high (Figs. 3 and 10), indicating the maximum productivity of Elk Lake in the Holocene. Pigment ratios indicate that cyanobacteria production and input of allochthonous plant pigments were probably also high at this time. Chemical and mineralogical indicators of clastic influx are high or at a maximum at 4.2 ka, although maximum varve thickness occurs at about 3.9 ka. These events mark the end of the prairie period. According to geochemical, mineralogical, and diatom data, the end came rapidly. Ostracode stratigraphy of a shallow water core shows similar abrupt changes at this time (Forester and others, 1987).

Mixed deciduous and coniferous forest environments: 4.0–0.5 ka. After 4.0 ka, varves become consistently thin (~2 mm; Fig. 3), and pollen of mesic deciduous trees (*Betula, Ulmus, Ostrya*) increases in the record dominated by *Quercus* (Fig. 17). Pollen of this mixed deciduous forest reaches percentage and influx maxima at about 3.0 ka (Whitlock and others, Chapter 17). The arrival of deciduous forest communities around Elk Lake signals the arrival of moister climates (McAndrews, 1966; Bartlein and Whitlock, Chapter 18). Iron, manganese, phosphorus, and organic carbon concentrations in the sediment begin to increase at this time (Dean, Chapter 10).

Fragilaria crotonensis dominates with strong fluctuations between 4.2 and 3.2 ka. *Asterionella formosa* and *Synedra acus* also have significant representation at this time (Figs. 3 and 4). This diatom assemblage indicates generally low phosphorus fluxes, short periods of circulation, and ample supplies of silicon, possibly derived from increased spring and summer precipitation. However, pigment studies indicate continued high production of cyanobacteria throughout this and higher intervals in the strata

(Sanger and Hay, Chapter 13). These algae compete effectively with the diatoms for phosphorus, and because they are able to float in the epilimnion during periods of lake stratification (summer) and thereby raise the Si/P ratio, they coexist with the *Fragilaria-Asterionella-Synedra* diatom community, which thrives under such nutrient conditions.

A number of benthic and tychoplanktonic diatom species and phytoliths (Figs. 5, 6, 7, and 8) increase to small percentages after 4.0 ka. These diatoms typically inhabit the littoral regions of lakes, and their minor increase in the core may reflect the reduced production of planktonic species and increased transportation of diatom frustules from shallow areas. Increases in phytoliths probably reflect the same process.

Shortly before 3.0 ka *Synedra cyclopum* becomes more or less consistently represented in the Elk Lake core stratigraphy. This diatom is not planktonic, but is often found in the limnetic regions of lakes, because it lives attached to planktonic crustaceans (Huber-Pestalozzi, 1942). Its presence in the core correlates closely with the increased diversity of cladocerans and the appearance of *Daphnia* at the same levels (Boucherle, 1982).

Around 3.0 ka *Stephanodiscus minutulus* becomes sporadically abundant and coincides with sharp increases in pollen influx, particularly of deciduous tree species (Whitlock and others, Chapter 17). Perhaps the two events are related. The long and vigorous spring circulation implied by *S. minutulus* may also have been responsible for winnowing pollen from the littoral sediments and redepositing it in the profundal region of Elk Lake.

After 3.0 ka the deciduous forest is modified progressively by the migration of *Pinus strobus* into the region. Arrival of this species at Elk Lake apparently occurred about 2.7 ka (Whitlock and others, Chapter 17). By 2.0 ka *P. strobus* had largely replaced the mixed deciduous forest at Elk Lake, probably as a result of yet cooler and moister climates (Jacobsen, 1979).

Diatom influx remains low throughout the transition to conifer forests. *Fragilaria crotonensis* dominates about 2.8 ka, but *Stephanodiscus minutulus* replaces it by 2.0 ka (Fig. 3). *Fragilaria crotonensis* then returns to dominate until about 1.0 ka, when *S. minutulus* increases with fluctuations throughout the past millennium. The broad alternations between these species are probably keyed to mean spring and fall temperatures and wind variations that control lake circulation and stratification in the manner previously discussed. In addition to a short-lived increase in *S. minutulus*, the transition period to *Pinus strobus* forests (3.0 to 2.0 ka) is characterized by an irregular decrease in *Asterionella formosa* and an increase in *Cyclotella comta*, which replaces *S. minutulus* by 1.8 ka. *Stephanodiscus alpinus, Cyclotella michiganiana*, and *C. kuetzingiana* largely disappear from the Elk Lake record at or shortly after 3.0 ka.

The cooler and moister climate that promoted the immigration of *Pinus strobus* (e.g., Bryson and Wendland, 1967) may have both directly and indirectly reduced the trophic status of Elk Lake. As pines occupied and closed the open oak-deciduous forest, Elk Lake became increasingly protected from winds that partially govern spring and fall circulation. The consequence of reduced circulation is the reduction of nutrient fluxes to the epilimnion and the eventual loss of nutrients to the varved profundal sediments. Moister climate probably also differentially increased ground-water silicon supplies over phosphorus supplies as increased amounts of vegetation utilized the latter nutrient at the beginning of the spring growing season (e.g., R. F. Wright, 1976).

Transportation of mollusk shells from shallow to deeper littoral environments in Elk Lake was reduced as *Pinus strobus* pollen increased in the record, also suggesting a reduction of wave-induced currents and turbulence in Elk Lake at this time (Stark, 1976).

Cyclotella kuetzingiana and *C. michiganiana* are oligotrophic to mesotrophic warm-water (summer) species (Stoermer and Ladewski, 1976), and therefore require modest nutrient deliveries in that season for growth. Perhaps these nutrients are derived from the metalimnion by wind-generated eddy diffusion. If so, the closing of the forests around Elk Lake would reduce wind stress and thereby reduce or eliminate this nutrient supply. The same process operating in the late fall or winter would reduce the duration and effectiveness of fall circulation and might account for the elimination of *Stephanodiscus alpinus*.

Cyclotella comta is a fairly common diatom in oligotrophic to mesotrophic woodland lakes, and in Minnesota it is found in moderately deep transparent lakes in the deciduous and coniferous forests (Brugam, 1983). In Lake Michigan the diatom is considered to be a warm-water form (Stoermer and Ladewski, 1976), and in Burntside Lake, northeastern Minnesota, it appears in the middle to late summer (Bradbury, 1978). Its geographical and successional distribution suggests that *Cyclotella comta* is adapted to moderately low nutrient levels, particularly phosphorus (Stoermer and others, 1985), but its occurrence in many lake types implies that this species is reasonably opportunistic.

In general, *Cyclotella comta* is comparatively heavily silicified (Einsele and Grimm, 1938), implying that high levels of silicon might be required for rapid growth of this diatom. In a Lake Ontario core *Cyclotella comta* was reduced drastically in abundance as the lake water became depleted in silicon (Stoermer and others, 1985). In a core from a small lake in southern Ontario, *C. comta* showed a sharp rise in numbers coincident with pollen changes that indicated increased regional precipitation (Delorme and others, 1986), much as in the case of the Elk Lake core at about 2.0 ka. These circumstances are probably related. Increased precipitation at Elk Lake may have provided sufficient silicon via ground-water input to allow *C. comta* to become a significant part of the diatom assemblage. Similar changes in diatom assemblages are noted in the paleolimnologic record of Harvey's Lake, Vermont, although the source of increased nutrients, which we presume to have included silicon, was cultural; specifically, postlogging soil erosion and early agriculture (Engstrom and others, 1985).

Perhaps the diatom and pollen changes that suggest a cooler and wetter climate in northwestern Minnesota are related in some way to the roughly correlative "Neoglacial" period of arctic and alpine glacier expansion (Denton and Karlén, 1973) or to the "Subatlantic" climatic episode of Bryson and Wendland (1967).

After 1.8 ka, *Cyclotella comta* drops out and is replaced by increasing numbers of *Asterionella formosa* (Figs. 3 and 4) and *Stephanodiscus minutulus*. *Fragilaria crotonensis* also diminished in importance, although there is a notable but short-lived reversal about 800 yr ago that coincides with a sharp increase in *Aulacoseira ambigua* and *A. granulata*.

Minor increases in *Stephanodiscus alpinus* and *S. niagarae* after 1.6 ka precede the shift to dominance of *S. minutulus* by 1.2 ka and reflect increasing productivity of fall circulation periods. In general this change suggests colder fall seasons and possibly drier conditions. Cretaceous diatoms and sponge spicules also increase at this time, and shortly afterward, by 1.1 ka, phytoliths reach a maximum. All of these imply increased transport of materials from the lake margin or littoral areas, or possibly, in the case of fragments of Cretaceous diatoms, eolian deposition into Elk Lake. The large peak of chrysophyte scales (principally *Mallomonas pseudocoronata*) and cysts at 1.1 ka probably relates to increased lake productivity in the spring and early summer (Tippett, 1964; Carney and Sandgren, 1983). Silty turbidites occur in Elk Lake at about 0.9 ka (Anderson and others, Chapter 4; Dean, Chapter 10), and may represent a rapid or especially large lake-level fall.

The sudden increase of *Aulacoseira ambigua* and a strong influx of quartz about 0.7 ka (Dean, Chapter 10) appear to record yet drier and windier summer climates that may correspond to drought conditions documented by archaeology and pollen changes in northwestern Iowa (Bryson and others, 1970).

The pollen record of Elk Lake documents significant vegetation changes after 0.9 ka by steeply falling percentages of *Pinus strobus* and rising percentages of *P. banksiana*/or *P. resinosa* (Whitlock and others, Chapter 17). Perhaps this reflects increased fire frequencies (cf. McAndrews, 1966) due to widespread drought conditions that began about this time (cf. Swain, 1978).

The Little Ice Age: 0.5–0.1 ka. The diatom record of the past 500 yr at Elk Lake is characterized by very rapid fluctuations in dominance between *Fragilaria crotonensis* and *Stephanodiscus minutulus* (Fig. 3). In part this is the result of closer sampling and intervals of smaller temporal duration, although it may reflect real short-term limnologic variability to some degree. This period of time encompasses the Little Ice Age (A.D. 1450–1850) (Denton and Karlén, 1973) as well as the time when Europeans first arrived in the area. Because the history of fire in the Itasca Park area has been documented for the past ~300 yr, there is an opportunity to evaluate the limnological effects of fire at Elk Lake.

Several benthic diatoms, such as species of *Surirella* and *Gomphonema*, show comparatively large increases at about 0.5 ka, and other benthic species and phytoliths (Figs. 5, 6, 7, and 8) increase afterward. These changes occur when *Stephanodiscus minutulus* is very abundant, and the presence of benthic diatoms in the profundal region of Elk Lake may relate to the strong and persistent spring circulation implied by *S. minutulus*. It is possible that lake levels were sporadically lower during this interval of increased benthic diatom input. Peaks of chrysophyte cysts and scales imply that high productivity persisted into the summer, perhaps as a result of wind entrainment of metalimnetic nutrients.

The diatom record between A.D. 1680 and 1740 is also characterized by large percentages of benthic diatoms and numbers of phytoliths that correlate with high percentages of *Stephanodiscus minutulus* during the first quarter of the eighteenth century (Figs. 11 and 12). The record of drought conditions around A.D. 1700 (Blasing and Duvick, 1984; DeWitt and Ames, 1978) may be related to the diatom record, which suggests somewhat drier conditions around Elk Lake.

The Little Ice Age climate is not especially well known in the north-central United States. Mesic deciduous forests developed in south-central Minnesota about 400 yr ago in response to moister climates that might correlate with the main phase of the Little Ice Age (Grimm, 1983). At Lake of the Clouds in northeastern Minnesota, lower influx values of charcoal between A.D. 1550 and 1850 may record cooler and moister climates with less fire frequency (Swain, 1973). Farther east, in north-central Wisconsin, Bernabo (1981) calculated 1 °C lower growing-season temperatures around A.D. 1700; these were preceded by a cooling trend beginning at A.D. 1200 and were followed by a warming trend ending around A.D. 1850. Moister conditions are also implied for north-central Wisconsin during this time (Swain, 1978).

Sediments of Dogfish Lake in northeastern Minnesota that span the Little Ice Age record shallower and more alkaline lake conditions at this time (Bradbury, 1986), and a similar interpretation applies to Mirror Lake, east-central Wisconsin (Farris, 1981). Pollen studies from varved lakes in northwestern Wisconsin show decreased percentages and influxes of *Pinus strobus* between A.D. 1700 and 1850 (Gajewski and others, 1985) that might be interpreted as recording the effects of drier summers (e.g., Bryson and Wendland, 1967). Tree-ring records from central Iowa (Blasing and Duvick, 1984) show cyclic changes in precipitation, but no consistently lower or higher values during the past 300 yr. However, precipitation around A.D. 1700 was very low. A chronology of tree-ring width indices from Itasca State Park (Fig. 18) apparently documents drought conditions between A.D. 1691 and 1703 by the presence of abnormally narrow tree rings during this time interval (DeWitt and Ames, 1978), and a charcoal chronology from neighboring varved lakes in the park shows high charcoal influx (from forest fires) during this time (Clark, Chapter 19).

It is apparent that climate during the Little Ice Age was not consistently cool and wet, but included episodes of drought. It is also apparent that paleoclimatic records of the Little Ice Age show spatial variability, although temperatures may have been generally cooler throughout this period (Lamb, 1982). If the Little Ice Age included both periods of cooler and moister climates, as well as periods of drought, it is possible that the lakes responded more sensitively to the combination of cool weather and

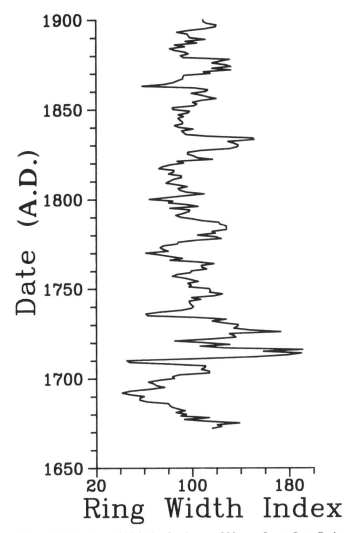

Figure 18. Tree ring width index for the past 300 yr at Itasca State Park. Data from DeWitt and Ames (1978).

dry conditions, while terrestrial vegetation recorded summer moisture stress and the incidence of forest fires, also partly under climatic control (e.g., Clark, 1989).

The fire history of Itasca State Park has been determined by tree-ring analysis of fire-scarred trees, by the distribution of even-aged stands of trees that became established after forests were largely burned, and by analysis of charcoal in varved lake sediments elsewhere in the park (Spurr, 1954; Frissell, 1973; Clark, Chapter 19). Major fires that included all or part of the Elk Lake shoreline occurred in the following years: A.D. 1727, 1759, 1772, 1803, 1811, 1864, 1875, and 1885. These years do not have a clear relation to the diatom stratigraphy of Elk Lake (Figs. 11 and 12), suggesting that individual fires did not affect the limnology of the lake as measured by 10 yr integrated diatom samples (e.g., Bradbury, 1986). Even the tree-ring chronology of Itasca (Fig. 18) is not related obviously to the sequence of major fires in the park (Frissell, 1973). Probably many independent factors, including human activity, have controlled fire history during the past 200 years, and these factors are not detected adequately by tree-ring or varved lacustrine records.

The general distributions of *Asterionella formosa, Cyclotella comta, Tabellaria fenestrata,* and species of *Synedra* after A.D. 1740 may relate to variable silicon fluxes to Elk Lake. All these diatoms probably prosper when silicon is high, and the increased incidence of forest fires near Elk Lake after A.D. 1742 (Frissell, 1973) may have increased rates of silicon input to the lake by increasing runoff (cf. R. F. Wright, 1976). For example, *Asterionella formosa* dominated the phytoplankton (35%) of Meander Lake in northeastern Minnesota the fall following the 1971 fire that encircled this lake (Tarapchak and Wright, 1986). Up to that time, this species was only a minor component (3%) of the phytoplankton assemblage of the lake (Bradbury, 1986). A similar case can be made for *Cyclotella comta.* Although discrete fires may have been responsible for the increases in some of these species, a cause and effect relation requires information about past fire seasonality and patterns of diatom succession that are not yet available for Itasca Park.

The final dominance of *Stephanodiscus minutulus* in Elk Lake occurred between A.D. 1830 and 1870. This dominance may again reflect drier than average conditions in northwestern Minnesota. Tree rings document a severe drought in North Dakota at this time (Will, 1946), and the diatom records from Lake Sallie, 65 km southwest of Elk Lake, indicate low lake levels that probably correspond to this dry period (Bradbury, 1975). Narrow tree rings are also recorded in the chronology of Itasca Park (Fig. 18) between A.D. 1840 and 1860, and A.D. 1863 and 1864 were fire years with large water-balance deficits (Clark, 1989, and Chapter 19).

European settlement and logging in the twentieth century. Although the Itasca region had been visited by European settlers by the middle of the nineteenth century, their impact on the area and on Elk Lake specifically was limited to a possible increase in forest-fire frequency (Hansen and others, 1974). Rapid development of farming activities at the end of the nineteenth century in the Red River Valley 130 km west of Itasca is recorded in the rise of *Ambrosia* pollen in most Minnesota lakes and provides a time datum (A.D. 1890) to judge paleolimnological changes in Elk Lake related to lumbering (e.g., Birks and others, 1976).

The impact of lumbering on lakes in Itasca Park resulted from several activities (cf. Hansen and others, 1974): (1) timber removal, including road building and log hauling; (2) damming of lakes to raise water levels to float and transport logs; and (3) storage of logs in lakes prior to transport down the Mississippi River.

The first two activities would undoubtedly increase nutrient loading by increasing runoff and soil erosion, but the impact of log storage in a lake is difficult to assess. Typically logs were stacked on the ice in the winter, and floated off with the spring melt. Depending on how long the logs stayed in the lake water, there might be some opportunity for leaching of tannins and lignins from the wood and bark; in addition, the epilimnion might be shaded by the logs. Otherwise, the presence of logs on the ice would probably hasten ice out in the spring by submerging the ice into the water and by providing a more heat-adsorptive surface to trap spring insolation. Floating the logs in a lake would probably affect water turbulence as well.

After A.D. 1860 *Fragilaria crotonensis* replaces *Stephanodiscus minutulus* as the dominant planktonic diatom in Elk Lake, suggesting an end of the dry and extended spring-season climate that existed previously (Fig. 11). In A.D. 1902, a dam was built on the Mississippi River at the outflow of Lake Itasca to raise water levels for floating logs downstream to lumber mills (Dobie, 1959). By 1904 back-water flooding had created severe damage to shoreline vegetation in both Itasca and Elk lakes (Brower, 1904). The nutrients that entered Elk Lake in response to these floods may have produced the short-lived increase in *Aulacoseira ambigua* near the base of the zone of disturbed sediments in frozen cores from the deep hole (Figs. 13 and 14).

Actual logging within the immediate drainage of Elk Lake and log storage in the lake did not occur until A.D. 1918 and 1919 (Aaseng, 1976). The disturbed-zone sediments in the frozen cores probably represent this activity at least in part. *Stephanodiscus minutulus* becomes very common in this part of the 1969 core (Stark, 1971) and probably represents logging activities on and around Elk Lake at this time. Certainly higher nutrient fluxes and prolonged circulation in the spring could be a result of timber cutting and log storage in the lake.

Following logging, *Aulacoseira ambigua* and smaller numbers of *A. granulata* become significant components of the phytoplankton of Elk Lake. The implied high levels of nutrients and turbulence appear to correlate with elevated numbers of fungal spores and of *Chydorus sphaericus* in a shallow-water core from this time period (Birks and others, 1976), taken to indicate inwash of terrestrial materials (soils) and eutrophication, respectively (e.g., Bradbury and Waddington, 1973).

Perhaps some limnologic changes at Elk Lake came in response to drought conditions of the 1930s, although human impact on Elk Lake did not end with the cessation of logging. A permanent dam on the outflow of the lake was constructed in 1935, and may have affected the lake. Modern transect samples of surface sediment show that comparatively high percentages of *Aulacoseira ambigua* persist in Elk Lake, and this species commonly occurs in fall sediment-trap samples (Nuhfer and others, Chapter 7). Although it is unlikely that recreational and scientific activities on Elk Lake drastically affect its present limnology, the possibility has never been critically examined.

It is possible that Elk Lake is becoming shallow enough through sediment accumulation during the past 10,000 yr that its status as an incompletely circulating lake is finally changing. As more of the bottom rises to levels that eddy currents from spring, fall, or even mid-summer turbulence can reach, more nutrients can be transported to the epilimnion to increase productivity of the lake (e.g., Megard and others, Chapter 3). Alternatively, general logging in Itasca Park during the first 20 yr of the twentieth century may have reduced forest density sufficiently to allow a significant increase of wind stress on the lake, which could account for episodes of increase depth and duration of circulation.

SUMMARY

The diatom study of Elk Lake cores has generated several interrelated conclusions about the limnology of the lake, its paleolimnological history, and local paleoclimatic and paleoenvironmental conditions during the Holocene.

1. Although Elk Lake behaves essentially as a dimictic lake, the comparatively great depth of part of its profundal zone seasonally maintains an anoxic environment that serves to store nutrients for periodic release to the epilimnion during times of vigorous circulation. This means that Elk Lake is largely buffered from the variability of external sources and fluxes of nutrients. Because Elk Lake is predominantly a seepage lake, its hydrologic balance is also buffered from short-term changes in precipitation and evaporation. The buffered character of Elk Lake ensures that diatom communities have remained comparatively stable, i.e., the same species have coexisted although in different proportions for the past 10,000 yr.

2. Climatic and environmental changes around Elk Lake generally have not been drastic enough to overcome the buffered limnological thresholds and significantly modify the diatom communities of Elk Lake during the Holocene. Neither did lake chemistry change sufficiently to alter the diatom community, although salinity may have increased slightly during the prairie period. Nevertheless, Elk Lake remained comparatively fresh, with probably less than 1000 ppm total dissolved solids in the water throughout the Holocene.

The lake never became so shallow that large benthic diatom communities were recorded at the core site. Transects of cores studied for pollen and ostracodes (Stark, 1976) suggest a 6 m fall in lake level during the middle Holocene. Abundant ostracodes inhabited the ridge separating the two deep holes in Elk Lake, in water that at that time may have been about 10 m deep. However, this sublittoral environment was still 250 m away from the core site in the deep hole where contemporaneous water depths were 33–40 m. If benthic diatoms lived on the ridge at this time, they were not transported to the core site in the deep hole in large numbers.

3. Elk Lake responded to climatic change in the middle Holocene by substantial increases in diatom productivity, which resulted from a more rapid cycling of the nutrient pool maintained in the anoxic basins of the lake and from addition of nutrients from terrestrial sources. The specific climate changes that provided increased circulation of nutrients were related to factors that lengthen the spring and fall seasons and to the frequency and strength of summer frontal storms. Throughout the middle Holocene, the specific climatic factors that enhanced productivity may not have always occurred together.

4. *Stephanodiscus minutulus* responds positively to low

Si:P ratios and low light conditions. Such environments would occur when spring circulation was vigorous, long, and in the context of dry winters. Vigorous circulation would extend deep into the nutrient-rich bottom waters of the lake to provide phosphorus for this diatom, and deep circulation implies turbidity and therefore low light levels. Lake ice is thinner and ice out is earlier in dry winters, which provides for a long period of circulation that allows for massive development of *S. minutulus*. Reduced runoff and ground-water recharge in dry winters could reduce silicon influx and thereby help insure low Si:P ratios. The correlation of peaks of *S. minutulus* with increased numbers of benthic diatoms, increased levels of clastic sedimentation in the cores, and with narrow tree rings in the same time periods supports its association with dry spring weather. The abundance of *S. minutulus* in prairie lakes of shallow to moderate depth implies that high percentages and influx values of this diatom in Elk Lake indicate a shift toward prairie-type climates.

5. *Fragilaria crotonensis* responds positively to high Si:P ratios and probably to high light conditions. It prospers during the summer after phosphorus has been reduced in the epilimnion by previous phytoplankton blooms. High percentages and influxes of this diatom suggest short, wet springs that produce minimal phosphorus during circulation yet ample silicon via ground-water influx. Summer and fall blooms of *F. crotonensis* may also relate to frontal storm activity with winds that deepen the thermocline and circulate nutrients from the metalimnion to the photic zone.

6. *Aulacoseira ambigua* is a summer diatom that requires moderate to high light levels, high Si levels, and strong turbulence to support it in the epilimnion. It probably increases in abundance when strong winds from frontal storms and influx of silicon-rich water from rain provide the necessary conditions for its growth. In the Elk Lake core, *A. ambigua* increases at the expense of *F. crotonensis*. When there are nutrient fluxes and the turbulence increases, the former diatom appears to be the better competitor.

High abundances of *F. crotonensis* and/or *A. ambigua* imply essentially modern climate conditions at Elk Lake.

Alternations in abundance of these diatoms (Fig. 19) show in a general way how climate changed around Elk Lake during the past 10,000 yr. The ratio of *Stephanodiscus minutulus* to *Fragilaria crotonensis* is a relative measure of prairie-like climates with dry, long springs (high ratios) versus warm, stormy summers, probably with increased moisture. The percentage plot of *Aulacoseira ambigua* depicts exceptional frequencies of frontal storms in the summer and early fall.

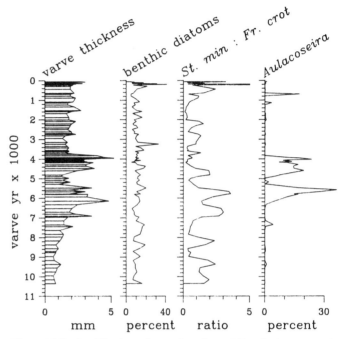

Figure 19. Ratio of *Stephanodiscus minutulus* and *Fragilaria crotonensis* and percentage of *Aulacoseira* species and benthic diatoms from the Elk Lake varved core.

ACKNOWLEDGMENTS

Many people have generously shared data, ideas, and material relevant to this study of Elk Lake. We thank the following for their contributions to the many phases of this project: N. E. Aaseng for the history of logging in Itasca State Park; John Almendinger for correlation of frozen cores and determination of the *Ambrosia* rise in Elk Lake; A. L. Baker for limnological data from Elk and nearby lakes; D. G. Baker for climate data and insights for Itasca State Park; David Bosanko for lake sampling, chlorophyll analyses, and temperature measurements from Elk Lake; John King for core material and discussions on the correlation of Elk Lake cores; Vera Markgraf for pollen analysis to determine pre- and post-*Ambrosia* rise sediments; Jon Ross for information on ice up and ice out for Elk and Itasca lakes and field assistance at Elk Lake; Barbara Speziale for diatom analyses of a frozen core from Elk Lake; and H. E. Wright for help in coring Elk Lake, facilities for research, and many helpful discussions and relevant data.

REFERENCES CITED

Aaseng, N. E., 1976, The history, nature, and extent of the major logging operations in Itasca State Park (1901–1919) [M.S. thesis]: St. Paul, University of Minnesota, 48 p.

Battarbee, R. W., 1973, A new method for the estimation of absolute micofossil numbers, with reference especially to diatoms: Limnology and Oceanography, v. 18, p. 647–653.

Bernabo, J. C., 1981, Quantitative estimates of temperature changes over the last 2700 years in Michigan based on pollen data: Quaternary Research, v. 15, p. 143–159.

Birks, H. H., Whiteside, M. C., Stark, D. M., and Bright, R. C., 1976, Recent paleolimnology of three lakes in northwestern Minnesota: Quaternary Research, v. 6, p. 249–272.

Blasing, T. J., and Duvick, D., 1984, Reconstruction of precipitation history in North American corn belt using tree rings: Nature, v. 307, p. 143–145.

Boucherle, M. M., 1982, An ecological history of Elk Lake, Clearwater County, Minnesota, based on Cladocera remains [Ph.D. thesis]: Bloomington, University of Indiana, 135 p.

Bradbury, J. P., 1975, Diatom stratigraphy and human settlement in Minnesota: Geological Society of America Special Paper 171, 74 p.

—— , 1978, A paleolimnological comparison of Burntside and Shagawa lakes, northeastern Minnesota: Environmental Protection Agency Ecological Research Series EPA-600/3-78-004, 50 p.

—— , 1986, Effects of forest fire and other disturbances on wilderness lakes in northeastern Minnesota II. Paleolimnology: Archiv für Hydrobiologie, v. 106, p. 203–217.

—— , 1988, A climatic-limnologic model of diatom succession for paleolimnological interpretation of varved sediments at Elk Lake, Minnesota: Journal of Paleolimnology, v. 1, p. 115–131.

Bradbury, J. P., and Waddington, J.C.B., 1973, The impact of European settlement on Shagawa Lake, northeastern Minnesota, U.S.A., in Briks, H.J.B., and West, R. G., eds., Quaternary plant ecology: Oxford, Blackwell, p. 289–307.

Brower, J. V., 1904, Itasca State Park, an illustrated history: Minnesota Historical Society Collections, Volume XI: St. Paul, Minnesota Historical Society, 285 p.

Brugam, R. B., 1983, The relationship between fossil diatom assemblages and limnological conditions: Hydrobiologia, v. 98, p. 223–235.

Bryson, R. A., and Wendland, W. M., 1967, Tentative climatic patterns for some late-glacial and post-glacial episodes in central North America, in Mayer-Oaks, W. J., ed., Life, land and water: Proceedings of the 1966 Conference on Environmental Studies of the Glacial Lake Agassiz Region: Winnepeg, University of Manitoba, p. 271–298.

Bryson, R. A., Baerreis, D. A., and Wendland, W. M., 1970, The character of late-glacial and post-glacial climate changes, in Dort, W., and Jones, J. K., eds., Pleistocene and recent environments of the central Great Plains: Lawrence, University of Kansas Press, p. 53–74.

Carney, H. J., and Sandgren, C. D., 1983, Chrysophycean cysts: Indicators of eutrophication in the recent sediments of Frains Lake, Michigan, U.S.A.: Hydrobiologia, v. 101, p. 195–202.

Charles, D. F., 1985, Relationships between surface sediment diatom assemblages and lakewater characteristics in Adirondack lakes: Ecology, v. 66, p. 994–1011.

Clark, J. S., 1989, Effects of long-term water balances on fire regime, northwestern Minnesota: Journal of Ecology, v. 77, p. 989–1004.

Dean, W. E., Bradbury, J. P., Anderson, R. Y., and Barnowsky, C. W., 1984, The variability of Holocene climate change: Evidence from varved lake sediments: Science, v. 226, p. 1191–1194.

Delorme, L. D., Duthie, H. C., Esterby, S. R., Smith, S. M., and Harper, N. S., 1986, Prehistoric pH changes in Batchawana Lake, Ontario, from sedimentary diatom assemblages: Archiv für Hydrobiologie, v. 108, p. 1–22.

Denton, G. H., and Karlén, W., 1973, Holocene climatic variations—Their pattern and possible cause: Quaternary Research, v. 3, p. 155–205.

DeWitt, E., and Ames, M., 1978, Tree ring chronologies of eastern North America: Chronology Series IV, Volume 1: Tucson, University of Arizona Laboratory of Tree Ring Research, 42 p.

Dobie, J., 1959, The Itasca story: Minneapolis, Minnesota, Ross and Haines, 202 p.

Einsele, W., and Grim, J., 1938, Über den Kieselsäuregehalt planktischer Diatomeen und dessen Bedeutung für einige Fragen ihrer Ökologie: Zeitschrift für Botanik, band 32, p. 545–590.

Engstrom, D. R., Swain, E. B., and Kingston, J. C., 1985, A paleolimnological record from Harvey's Lake, Vermont: Geochemistry, pigments and diatoms: Freshwater Biology, v. 15, p. 261–288.

Farris, D. P., 1981, The recent historical limnology of Mirror Lake, Waupaca County, Wisconsin (U.S.A.) evidenced by the diatom stratigraphy [M.S. thesis]: Ann Arbor, University of Michigan, 117 p.

Florin, M. B., 1970, Late glacial diatoms of Kirchner Marsh, southeastern Minnesota: Nova Hedwigia, v. 31, p. 667–756.

Forester, R. M., Delorme, L. D., and Bradbury, J. P., 1987, Mid-Holocene climate in northern Minnesota: Quaternary Research, v. 28, p. 263–273.

Frissell, S. S., 1973, The importance of fire as a natural ecological factor in Itasca State Park, Minnesota: Quaternary Research, v. 3, p. 397–407.

Gajewski, K., Winkler, M. G., and Swain, A. M., 1985, Vegetation and fire history from three lakes with varved sediments in northwestern Wisconsin (U.S.A.): Review of Paleobotany and Palynology, v. 44, p. 277–292.

Gasse, F., and Tekaia, F., 1983, Transfer functions for estimating paleoecological conditions (pH) from East African diatoms: Hydrobiologia, v. 103, p. 85–90.

Grigal, D. F., Severson, R. C., and Goltz, G. E., 1976, Evidence of eolian activity in north-central Minnesota 8,000 to 5,000 yr ago: Geological Society of America Bulletin, v. 87, p. 1251–1254.

Grimm, E. C., 1983, Chronology and dynamics of vegetation change in the prairie-woodland region of southern Minnesota, U.S.A.: New Phytologist, v. 93, p. 311–350.

Hansen, H. L., Kurmis, V., and Ness, D. D., 1974, The ecology of upland forest communities and implications for management in Itasca State Park, Minnesota (Forestry Series 16): St. Paul, University of Minnesota Agricultural Experiment Station, Technical Bulletin 298, 43 p.

Horne, A. J., and Goldman, C. R., 1972, Nitrogen fixation in Clear Lake, California. I. Seasonal variation and the role of heterocysts: Limnology and Oceanography, v. 17, p. 678–692.

Huber-Pestalozzi, G., 1942, Das Phytoplankton des Süsswassers, Teil 2, 2 Hälfte, in Thienemann, A., ed., Diatomeen: Die Binnengewässer, v. 16, p. 366–549.

Hustedt, F., 1930, Bacillariophyta (Diatomeae), in Pascher, A., ed., Die Süsswasser-Flora Mitteleuropas, Heft 10: Jena, Gustav Fischer, 464 p.

Jacobson, G. L., 1979, The paleoecology of white pine (*Pinus strobus*) in Minnesota: Journal of Ecology, v. 67, p. 697–726.

Kilham, P., Kilham, S. S., and Hecky, R. E., 1986, Hypothesized resource relationships among African planktonic diatoms: Limnology and Oceanography, v. 31, p. 1169–1181.

Lamb, H. H., 1982, Climate history and the modern world: London and New York, Methuen, 387 p.

McAndrews, J. H., 1966, Postglacial history of prairie, savanna and forest in northwestern Minnesota: Torrey Botanical Club Memoirs, v. 22, p. 1–72.

Mechling, J. A., and Kilham, S. S., 1982, Temperature effects on silicon limited growth of the Lake Michigan diatom *Stephanodiscus minutus* (Bacillariophyceae): Journal of Phycology, v. 18, p. 199–205.

Oaks, E. L., and Bidwell, L. E., 1968, Water resources of the Mississippi headwaters watershed, north-central Minnesota: U.S. Geological Survey Hydrologic Investigations Atlas HA-278, 4 map sheets.

Shay, C. T., 1971, The Itasca bison kill site—An ecological analysis: St. Paul, Minnesota Historical Society, 133 p.

Sicko-Goad, L., Stoermer, E. F., and Kociolek, J. P., 1989, Diatom resting cell rejuvenation and formation: Time course, species records, and distribution: Journal of Plankton Research, v. 11, p. 375–389.

Spurr, S. H., 1954, The forests of Itasca in the nineteenth century as related to fire: Ecology, v. 35, p. 21–25.

Stark, D. M., 1971, A paleolimnological study of Elk Lake in Itasca State Park, Clearwater County, Minnesota [Ph.D. thesis]: Minneapolis, University of Minnesota, 178 p.

——, 1976, Paleolimnology of Elk Lake, Itasca State Park, northwestern Minnesota: Archiv für Hydrobiologie, Supplement 50 (Monographische Beiträge), p. 208–274.

Stauffer, R. E., and Armstrong, D. E., 1984, Lake mixing and its relationship to epilimnetic phosphorus in Shagawa Lake, Minnesota: Canadian Journal of Fisheries and Aquatic Sciences, v. 41, p. 57–69.

——, 1986, Cycling of iron, manganese, silica, phosphorus, calcium and potassium in two stratified basins of Shagawa Lake, Minnesota: Geochimica et Cosmochimica Acta, v. 50, p. 215–229.

Stoermer, E. F., and Hakansson, H., 1984, *Stephanodiscus parvus:* Validation of an enigmatic and widely misconstrued taxon: Nova Hedwigia, v. 39, p. 497–511.

Stoermer, E. F., and Ladewski, T. B., 1976, Apparent optimal temperatures for the occurrence of some common phytoplankton species in southern Lake Michigan: Ann Arbor, University of Michigan Great Lakes Research Division Publication 18, 49 p.

Stoermer, E. F., Wolin, J. A., Schelske, C. L., and Conley, D. J., 1985, An assessment of ecological changes during the recent history of Lake Ontario based on siliceous algal microfossils preserved in the sediments: Journal of Phycology, v. 21, p. 257–276.

Swain, A. M., 1973, A history of fire and vegetation in northeastern Minnesota as recorded in lake sediments: Quaternary Research, v. 3, p. 383–396.

——, 1978, Environmental changes during the past 2000 years in north-central Wisconsin: Analysis of pollen, charcoal and seeds from varved lake sediments: Quaternary Research, v. 10, p. 55–68.

Tarapchak, S. J., and Herche, L. R., 1987, Lake Michigan phosphorus-phytoplankton dynamics. A test of the Schelske-Stoermer hypothesis and community structure analysis [abs.] (American Society of Limnology and Oceanography, 1987 Annual Meeting Proceedings): Madison, Wisconsin, American Society of Limnology and Oceanography, p. 78.

Tarapchak, S. J., and Wright, H. E., 1986, Effects of forest fire and other disturbances on wilderness lakes in northeastern Minnesota I. Limnology: Archiv für Hydrobiologie, v. 106, p. 177–202.

Theriot, E. C., and Stoermer, E. F., 1982, Observations on North American populations of *Stephanodiscus* (Bacillariophyceae) species attributed to Friederich Hustedt: American Microscopial Society Transactions, v. 101, p. 368–374.

——, 1984, Principal components analysis in character variation in *Stephanodiscus niagarae* Ehrenb.: Morphological variation related to trophic status: Proceedings of the 7th International Diatom Symposium: Koenigstein, Otto Koeltz, p. 97–111.

Tilman, D., Kilham, S. S., and Kilham, P., 1982, Phytoplankton community ecology: The role of limiting nutrients: Annual Review of Ecology and Systematics, v. 13, p. 349–372.

Tippett, R., 1964, An investigation into the nature of the layering of deep-water sediments in two eastern Ontario lakes: Canadian Journal of Botany, v. 42, p. 1693–1709.

Will, G. F., 1946, Tree ring studies in North Dakota: North Dakota Agriculture Experiment Station Bulletin 338, 24 p.

Wright, H. E., Jr., 1967, A square-rod piston sampler for lake sediments: Journal of Sedimentary Petrology, v. 37, p. 975–976.

——, 1968, The history of the prairie peninsula, *in* The Quaternary of Illinois: University of Illinois College of Agriculture Special Publication 14, p. 78–88.

——, 1972, Quaternary history of Minnesota, *in* Sims, P. K., and Morey, G. B., eds., Geology of Minnesota: A centennial volume: St. Paul, Minnesota Geological Survey, 632 p.

——, 1976, The dynamic nature of Holocene vegetation. A problem in paleoclimatology, biogeography, and stratigraphic nomenclature: Quaternary Research, v. 6, p. 581–596.

——, 1980, Cores of soft lake sediments: Boreas, v. 9, p. 107–114.

Wright, R. F., 1976, The impact of forest fire on the nutrient influxes to small lakes in northeastern Minnesota: Ecology, v. 57, p. 649–663.

MANUSCRIPT ACCEPTED BY THE SOCIETY JULY 27, 1992

Postglacial chrysophycean cyst record from Elk Lake, Minnesota

Barbara A. Zeeb and John P. Smol
Paleoecological Environmental Assessment and Research Lab, Department of Biology, Queen's University, Kingston, Ontario K7L 3N6, Canada

ABSTRACT

We have identified a total of 56 chrysophycean stomatocyst morphotypes from the postglacial sediments of Elk Lake, Minnesota. Cysts were well preserved and abundant throughout the lake's history. Stratigraphic changes in the 22 dominant cysts were correlated with other paleoecological information available for this core. In general, shifts in chrysophyte assemblages coincided with inferred changes in past climate. The most striking change in chrysophytes occurred about 8.5 ka, and coincided with the shift to prairie-vegetation dominance in the lake's drainage basin. The relative proportion of chrysophycean stomatocysts to diatom frustules also decreased at that time. A second major shift occurred about 5.3 ka, with the return of a chrysophyte assemblage resembling, in some respects, the early postglacial flora. Stratigraphic analyses at decade resolution recorded more recent changes in cyst assemblages that presumably track climatic variables. The relative abundance of chrysophytes appeared to be exceptionally high (values of about 40%–80%) during the Little Ice Age.

INTRODUCTION

Chrysophycean algae (Classes Synurophyceae and Chrysophyceae; Anderson, 1987) are a diverse group of more than 1000 documented species. Chrysophytes are common in many freshwater habitats, but are typically euplanktonic (Sandgren, 1988). Evidence that these algae have well-defined optima and narrow tolerances to a number of limnological variables has been accumulating in recent years, and suggests that chrysophytes should be valuable biomonitors (Christie and others, 1988; Cumming and others, 1991; Dixit and others, 1989a, 1989b, 1990; Siver, 1988, 1989a, 1989b; Siver and Chock, 1986; Smol, 1986; Zeeb and Smol, 1991).

A characteristic feature of all Chrysophyceae and Synurophyceae is their ability to form endogenous resting stages called statospores or stomatocysts (Cronberg, 1986; Sandgren, 1991). Cysts provide chrysophytes with a competitive advantage over many other algae because they allow survival during periods when the physical or chemical conditions of the planktonic habitat are beyond the physiological tolerances of the vegetative cells (Sandgren, 1983).

Most stomatocysts are thickly silicified, resistant to dissolution and fragmentation, and are often well preserved in lake sediments (Fig. 1) (Smol, 1986, 1988). They are spherical to oval or subtriangular in shape, and have diameters that range between 2.5 μm and >30 μm (Adam and Mahood, 1981). All cysts have a single pore through which the germinating cell emerges. An assortment of silicious thickenings, referred to as the collar complex, may surround the pore (Fig. 1). The wall of a mature stomatocyst is composed of a single layer of silica, overlain by one or more subsequent layers of silica. The external cyst wall may be smooth or ornamented. This ornamentation and any collar structure adds even more silica to the structure. Chrysophyte cyst morphotypes possessing distinctive collar and ornamentation features appear to be species specific (Cronberg, 1986) and can be distinguished by size, collar structure, and surface ornamentation.

Chrysophytes can provide additional and complementary information concerning lake conditions to that provided by diatoms alone. Chrysophytes are a group of predominantly flagellated and planktonic algae (Kristiansen and Takahashi, 1982) that occur most abundantly in circumneutral-pH, low-alkalinity,

Zeeb, B. A., and Smol, J. P., 1993, Postglacial chrysophycean cyst record from Elk Lake, Minnesota, *in* Bradbury, J. P., and Dean, W. E., eds., Elk Lake, Minnesota: Evidence for Rapid Climate Change in the North-Central United States: Boulder, Colorado, Geological Society of America Special Paper 276.

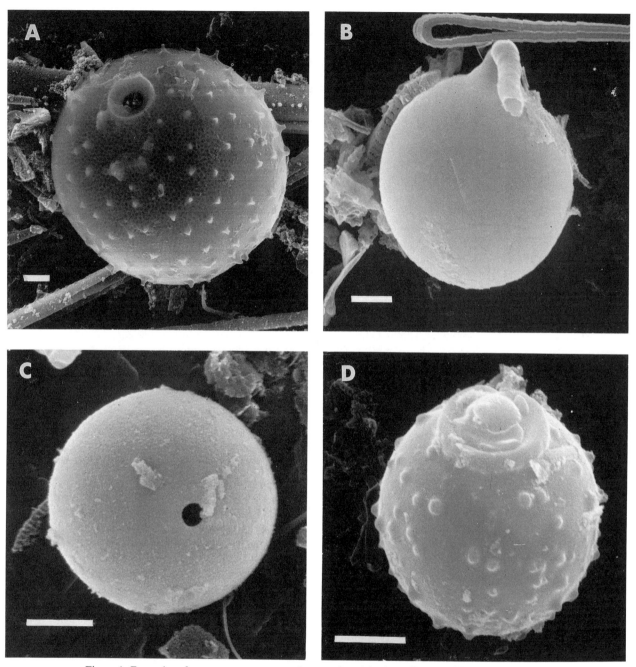

Figure 1. Examples of some stomatocyst morphotypes found in the sediments of Elk Lake (scanning electron microscope). A: cf. *Mallomonas crassisquama*; B: stomatocyst 41; C: stomatocyst 9, D: stomatocyst 158. Scale bars = 2 μm.

low-phosphorus, low-productivity lakes (Sandgren, 1988). Diatoms, however, also occur in more eutrophic lakes, and include both planktonic and benthic forms (Battarbee, 1986). As a group, cysts have been analyzed in relation to diatoms (e.g., Smol, 1983; Stoermer and others, 1985); the ratio of cysts to diatoms has been proposed as a rough measure of lake production in temperate lakes (Smol, 1985), and may provide insights on long-term trends in algal succession.

Although the above ratio may provide a general paleolimnological index of trends in lake trophic status, much information is inevitably lost when such an ecologically diverse group of organisms is lumped together. Additional paleoenvironmental information could be gleaned from stratigraphic analyses of individual cyst morphotypes. For example, shifts in cyst assemblages have been correlated with known changes in environmental conditions such as trophic status (e.g., Carney, 1982; Carney and Sandgren,

1983; Rybak, 1986, 1987; Zeeb and others, 1990), pH (e.g., Rybak and others, 1987; Duff and Smol, 1991), and metal concentrations (Elner and Happey-Wood, 1978). Until now, investigators have rarely linked cyst morphotypes to the species that produce them. Fortunately, surface-sediment calibration sets can be used to describe the distribution of cysts in relation to environmental variables, and will likely yield valuable quantitative estimates of the ecological optima and tolerances of the morphotypes. We anticipate that quantitative reconstruction of past lake environments will eventually be possible using the stratigraphic analysis of fossil cyst morphotypes.

Our study has two major objectives: (1) to expand our knowledge of stomatocysts by describing the cysts found in the sediments of Elk Lake, using established guidelines (Cronberg and Sandgren, 1986); and (2) to explore the extent to which chrysophycean stomatocysts can be used to track past lake conditions. Elk Lake is an ideal site for such an investigation because it has a well-studied paleolimnological record (chapters in this volume). Our study is the first to use International Statospore Working Group (ISWG) standards to describe changes in cyst assemblages for the entire postglacial sequence of a lake. It also marks the first attempt to use chrysophycean cysts as indicators of long-term (ca. 10,000 yr) climatic change. For these reasons, our interpretation should be considered exploratory and still somewhat speculative.

METHODS

Sediment collection and preparation

A 20-m-long sediment core (hereafter referred to as the "long core") was removed from Elk Lake's profundal zone in 1978. The sediments are divided into annual laminations (varves) and 10,400 varves have been counted and measured to form a 10,400 yr time series. All time references are relative to the varve time scale in thousands of years (ka) (Dean and others, 1984). The core was sectioned at about 50 yr intervals, according to the varve chronology. A frozen-finger core of the recent sediments (hereafter referred to as the "short core") was also obtained, which spans the 230 yr history of lake ontogeny from about A.D. 1630 to A.D. 1864. This short core was sectioned at finer intervals (about 10 yr).

Chrysophyte cysts were prepared and studied using the same digestive techniques used for fossil diatoms. Sediment was digested with acid, rinsed, and the cleaned silicious material was resuspended and evaporated onto glass cover slips, which were then mounted in Hyrax on glass slides. We used the identical microscope slides prepared by Bradbury and Dieterich-Rurup (Chapter 15).

For scanning electron microscopy (SEM), an aliquot of the resuspended silicious material was evaporated onto a smooth piece of aluminum foil. Double-sided tape was used to affix the foil to aluminum SEM stubs, and then each stub was sputter-coated with gold and examined at 20 kV at a working distance of 15–20 mm.

Morphological descriptions

Cysts were measured and described from scanning electron micrographs according to the ISWG guidelines (Cronberg and Sandgren, 1986). Light micrographs were matched to electron micrographs and each stomatocyst was compared with numerous published descriptions, including: Gritten (1977), Adam and Mahood (1981), Nicholls (1981), Sandgren and Carney (1983), Carney and Sandgren (1983), Smol (1984), Smith and White (1985), Sandgren and Flanagin (1986), Takahashi and others (1986), Rybak (1986, 1987), Rybak and others (1987, 1991), Cronberg (1988, 1989), Duff and Smol (1988, 1989, 1991), Zeeb and others (1990), Carney and others (1992), Duff and others (1992), Pienitz and others (1993), and many others. Previously undescribed stomatocysts were assigned a reference number following the ISWG recommendations; these numbers continue our established sequence from Pienitz and others (1993). The Elk Lake morphotypes are described in Zeeb and Smol (1993).

Collective categories (CC) consist of two cysts that were morphologically similar and could not be distinguished reliably using light microscopy (LM). Morphotypic differences were, however, clearly visible using SEM, because SEM resolution is three to four orders of magnitude greater than that of LM. CC1 consists of smooth, spherical cysts with a shallow, concave pore. CC2 consists of smooth to psilate cysts, with a short cylindrical collar and a distinct annulus. CC3 includes small, spherical cysts with a regular pore. Full descriptions of these collective categories are found in Zeeb and Smol (1993).

Data compilation and analyses

At least 300 stomatocysts were counted along parallel transects at each of our 105 sediment levels using a Leitz Dialux 20 light microscope. Bradbury provided total diatom and cyst counts, which we recalculated to show the relative proportions, expressed as a percentage of chrysophyte cysts to diatom frustules (see Smol, 1985). Relative abundances of individual cyst types were also calculated and the percentage data graphed relative to depth (i.e., age) in the core. A principal components analysis (PCA) was performed on the cyst assemblage data from all depths in the long and short cores to ordinate the samples with respect to their species assemblages. A preliminary correspondence analysis (CA) revealed that the first axis was less than 1.0 standard deviation unit from the mean sample score. This suggested that cyst distributions along the first axis were linear, and therefore a PCA would be a more appropriate model (Jongman and others, 1987).

RESULTS

The ratio of chrysophycean stomatocysts to diatom frustules was high in the immediate postglacial and remained relatively high throughout the early postglacial (until ca. 8.5 ka) (Fig. 2A). The period from 8.5 to 4.0 ky is characterized by a relatively

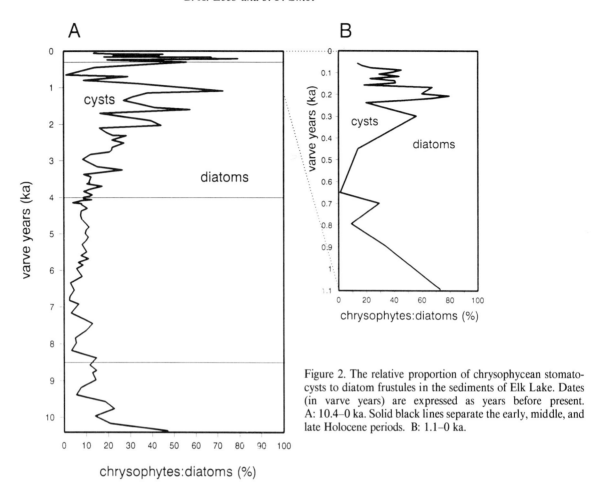

Figure 2. The relative proportion of chrysophycean stomatocysts to diatom frustules in the sediments of Elk Lake. Dates (in varve years) are expressed as years before present. A: 10.4–0 ka. Solid black lines separate the early, middle, and late Holocene periods. B: 1.1–0 ka.

lower abundance of chrysophytes. Chrysophyte abundances generally increase from this point to ca. 1.1 ka and then decrease dramatically to trace levels at 0.6 ka. This was followed by another increase in cyst abundance peaking at 0.2 ka, then decreasing again toward the surface of the core.

We identified and described 56 stomatocyst morphotypes in the Elk Lake core. Our taxonomic descriptions are presented elsewhere (Zeeb and Smol, 1993), and include SEM and LM micrographs. Ecological interpretations from the cyst data were confounded by our finding that ~25% of the stomatocysts from Elk Lake had not been previously described. Moreover, based on our taxonomic study of this flora, we had to refine the taxonomic scheme described by Duff and Smol (1991). For example, we have now divided a large collective category (CC1: unornamented cysts lacking collars) into two distinct groups, based on careful LM observations of slight differences in pore morphology. The first group includes cysts whose pores are flush with the cyst surface, and the second group includes cysts whose pores lie in a slight depression. Taxonomic details are discussed fully in Zeeb and Smol (1993).

Nineteen individual stomatocysts, and the three collective categories, were found in relative abundances greater than 2% in at least half the sediment samples, and only these are presented in Figures 3 and 4. Unidentified cysts were never found in relative abundances of greater than 1.5%. Both cores were dominated by stomatocysts 41 and 9, which composed between 15% and 50% of the total cyst assemblage at any given level in the cores. Distinct shifts in cyst assemblage are described below.

Long core

The long core covers a span of time from about 10.4 to 0.3 ky. At the base of this core, and throughout the early history of Elk Lake, 20 of the 22 dominant stomatocysts can be separated stratigraphically into two distinct groups (Table 1). Group 1 cysts were common at the bottom of the core and decreased with varying degrees of abruptness at about 8.5 ka. Group 2 cysts generally replaced group 1 cysts after 8.5 ka.

Figure 3. Stratigraphic distributions of common chrysophycean stomatocysts in Elk Lake (10.4 to 0.3 ka), expressed as relative (%) abundances. CC = collective category. Dates (in varve years) are expressed as years before present. Solid black lines separate the early, middle, and late Holocene periods. Dashed lines indicate an interval during the middle Holocene when conditions changed suddenly.

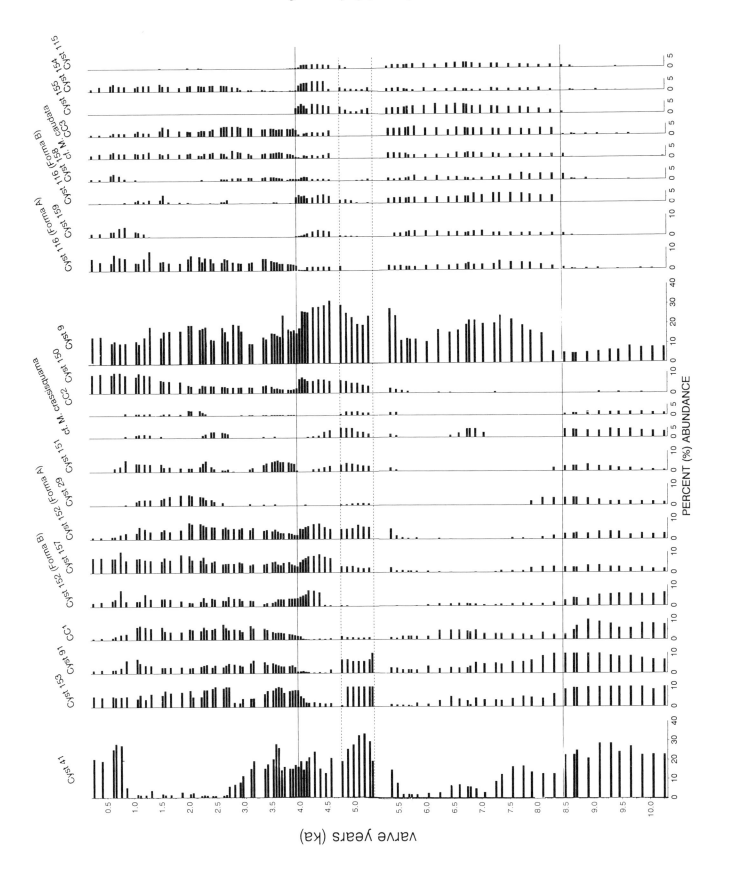

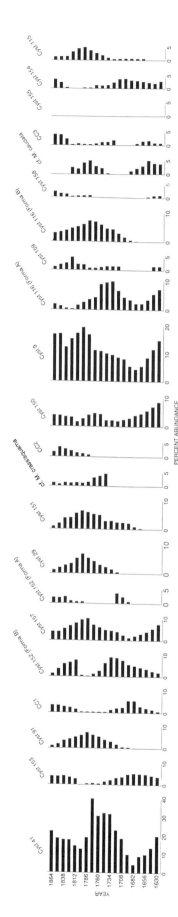

Figure 4. Stratigraphic distributions of common chrysophycean stomatocysts in the recent sediments (A.D. 1630–1864) of Elk Lake, expressed as relative (%) abundances. CC = collective category.

Most of the group 1 stomatocysts increased again in relative abundance for about 500 yr beginning at about 5.3 ka (Fig. 3). During this period, these cysts regained, or even surpassed, their pre–8.5 ka abundances, whereas all the group 2 stomatocysts decreased dramatically in abundance, and the cyst assemblage resembled the one recorded during the early postglacial period.

Cyst assemblages are less-clearly defined between 4.0 and 0.3 ky. Stomatocyst 41 decreased drastically for a short interval between 2.8 and 0.8 ky, but then regained its former maximum abundance by 0.8 ka. Stomatocysts 153, 152 (Forma B), 157, 9, 116 (Forma A), cf. *Mallomonas caudata*, and 154 remain relatively stable throughout this period. In contrast, cysts 91, CC1, 152 (Forma A), 29, 151, cf. *M. crassisquama*, CC2, and CC3 decreased in abundance, whereas cyst 150 increased in abundance.

Short core

Our close interval sediment analyses of the short core recorded distinct shifts in cyst assemblages (Fig. 3), and the group 1 and 2 designations we used in the long core could not be applied. Both dominant cysts (41 and 9), and stomatocysts 157, 150, 116 (Forma A), cf. *M. caudata*, and CC3 declined in abundance from

TABLE 1. STRATIGRAPHIC GROUPING OF DOMINANT STOMATOCYSTS IN ELK LAKE

Group 1	Group 2
Stomatocyst 41 (Duff and Smol, 1989)	
Stomatocyst 153	Stomatocyst 116 (Forma A) (Zeeb and others, 1990)
Stomatocyst 91 (Duff and Smol, 1991)	
CC1; includes Duff and Smol stomatocyst 120 (Duff and others, 1992) and stomatocyst 149	Stomatocyst 159
	Stomatocyst 116 (Forma B) (Zeeb and others, 1990)
Stomatocyst 152 (Forma B)	Stomatocyst 158
Stomatocyst 157	cf. *Mallomonas caudata*
Stomatocyst 152 (Forma A)	CC3; includes stomatocyst 1 (Duff and Smol, 1988) and stomatocyst 148
Stomatocyst 29 (Duff and Smol, 1989)	
Stomatocyst 151	Stomatocyst 155
cf. *Mallomonas crassisquama*	Stomatocyst 154
CC2; includes Zeeb and Smol stomatocyst 146 (Pienitz and others, 1993), and stomatocyst 156	Stomatocyst 115 (Zeeb and others, 1990)

A.D. 1630 to A.D. 1690. In contrast, stomatocysts CC1 and 152 (Forma B) increase in abundance, whereas cysts 91, 152 (Forma A), 29, 151, cf. *M. crassisquama,* CC2, and 116 (Forma B) are rare. Circa A.D. 1690 appears to have been a turning point; previously declining populations stabilized or regained abundance during the subsequent century. Four stomatocysts that were previously present in very low relative abundances (cysts 91, 152 [Forma A], 116 [Forma B], and 115) increase around this time.

A second major shift in cyst assemblages occurs at about A.D. 1720, when stomatocysts 29 and 159 reappear, stomatocysts 91, 157, 29, 151, 116 (Forma A and B), 159, and 115 start to increase, and stomatocysts 153, CC1, 152 (Forma A), and 154 decrease. The final major shift occurred at about A.D. 1810, when previously declining taxa recovered (with one exception—stomatocyst 159).

Principle components analysis

The first and second axes of the PCA analysis explained 40% and 18% of the total variance, respectively. Elk Lake stomatocysts were clustered into several similar assemblages (Fig. 5). First, the early postglacial (10.4 to 8.5 ky) cyst assemblages most closely resemble those during the Little Ice Age (ca. A.D. 1860–1750). Second, cyst assemblages from 6.3 to 5.7 years ago were almost identical to those of 2.8 ka to A.D. 1022. It is interesting that cyst assemblages from 5.5 to 4.1 ky are unique, and have no other analogues during the lake's history.

DISCUSSION

Our interpretations should be considered exploratory, because little is yet known about cyst taxonomy and ecology. In this study, for example, only two of the dominant cysts could be linked to known algal species (i.e., cf. *Mallomonas crassisquama* and *M. caudata*). Ecological calibrations of cyst distributions are underway, but as yet unfinished, in several regions. For this reason, it is most useful to compare our results to those found for other environmental indicators described for this core, as presented in other papers in this volume.

Early postglacial (10.4 to 8.5 ky)

The relatively high abundance of chrysophytes in the immediate postglacial sediments (~48%) has been recorded in many other studies (Fig. 2A). This response is comparable to those recorded from high arctic lakes (Smol, 1983), and is probably related to the oligotrophic, cold water conditions present.

Although the ratio of chrysophytes to diatoms was high, stomatocysts (along with diatom valves) were relatively uncommon in the lake's early history. Nevertheless, the assemblage was diverse, and remained largely unchanged throughout the early postglacial period. Of the 14 cysts that dominated during this time, most (i.e., 12) were smooth, unornamented cysts differing only in size and collar morphology. The dominant cyst (i.e., stomatocyst 41) possesses a distinct tubular collar and is probably produced by *Dinobryon cylindricum* Imhof (Sandgren, 1980, 1983; Donaldson and Stein, 1984), which is reportedly more abundant in cool oligotrophic to mildly eutrophic waters (Rybak, 1986; Rybak and others, 1987). Carney and Sandgren (1983) noted that it occurs almost exclusively in presettlement sediments, although a more recent study showed that this cyst also occurs in postsettlement sediments and under eutrophic conditions (Zeeb and others, 1990). In a survey of the genus *Dinobryon* in Finnish lakes, Eloranta (1989) found the highest frequency of *D. cylindricum* in oligotrophic clear-water lakes, although this species was also found less frequently in mesotrophic and eutrophic lakes, at water temperatures up to 21.9 °C.

Other common stomatocysts (maintaining an abundance of ~10%) at the base of the long core include stomatocysts 153, 91, CC1, 152 (Forma B), and 9. CC1 is composed of smooth, spherical cysts with a shallow, concave pore. This is a very common cyst morphotype that has been described by many researchers in many different habitats (Adam and Mahood, 1979; Adam and Mehringer, 1980; Rybak and others, 1987; Duff and Smol, 1989, 1991). Stomatocyst 9 is also a very common morphotype, with no ornamentation or collar structure. It is distinguished from CC1 cysts entirely on the basis of pore morphology. This cyst has been described from the sediments of a high arctic lake (Duff and Smol, 1988) and high arctic ponds (Duff and others, 1992). In addition, cysts identical in morphology, but varying in size, have been described from more temperate habitats (Zeeb and others, 1990; Rybak and others, 1991). It is likely that both stomatocyst morphotypes CC1 and cyst 9 are produced by several closely related chrysophyte species. Stomatocysts 152 (Forma B) and 153 also possess no surface ornamentation, but are distinguished by their collars; 152 has a low and wide collar, 153 has a high and cylindrical collar. Of the dominant cysts at the base of the core, only cyst 91 is ornamented with a number of ridges.

Several stomatocysts were present at the base of the core at about 5% abundances (157, 152 [Forma A], 29, 151, cf. *M. crassisquama*). Stomatocyst 29 has previously been described from the sediments of a mid-arctic lake (Duff and Smol, 1989) and from a small, shallow, high-elevation pond (Adam and Mehringer, 1980). *M. crassisquama* is one of the most widely distributed *Mallomonas* species in the world, and generally exhibits wide ecological tolerances, with the exception of low pH waters (e.g., Hartmann and Steinberg, 1986; Smol, 1986; Dixit and others, 1988). We acknowledge that other chrysophyte species produce cysts that are very similar to those of *M. crassisquama,* and it is possible, although unlikely, that these counts represent several different taxa (Cronberg, 1991, personal commun.).

It is not possible to comment on the ecological affinities of cysts that have not been previously described (i.e., those numbering 150 and above), and interpretations are always vague where collective categories are involved. Our PCA analyses show a remarkable similarity in cyst assemblages throughout the early postglacial (Fig. 5). It is difficult to make any conclusive interpretations, except to say that the cysts found in the early postgla-

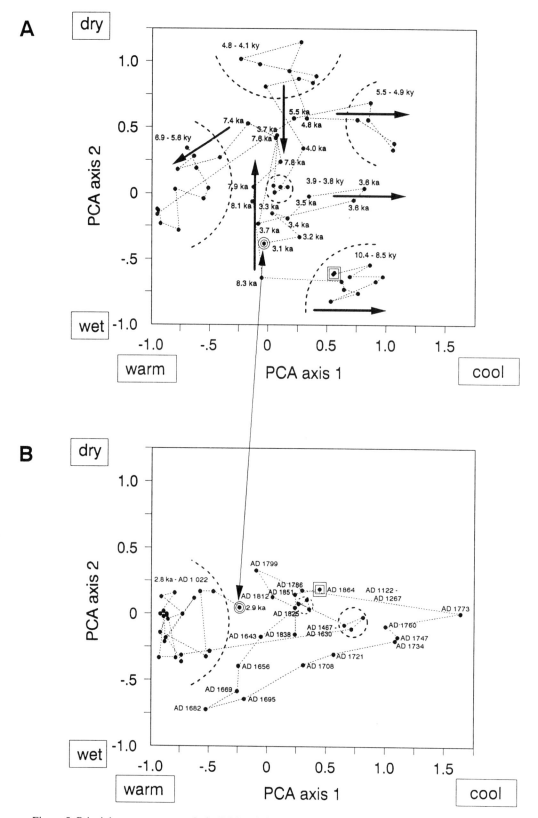

Figure 5. Principle components analysis (PCA) of site scores, derived from the relative percentages of the chrysophyte morphotypes, at intervals analyzed in the sediment cores. For reasons of clarity, the figure is chronologically presented in two panels. A: Long core: site scores from about 10.4 to 3.2 ka. B: Top of long core and short core: site scores from 3.0 ka to A.D. 1864. Arrows indicate major trends.

cial sediments of Elk Lake may be generalists, although found often in cold waters (Duff and Smol, 1988, 1989, 1991).

Mid-Holocene (Hypsithermal) (8.5 to 4.0 ky)

Climate models for the midwestern United States, based on past vegetational patterns, suggest that temperatures and aridity increased from about 8.5 ka to a maximum at about 7.0 ka (Wright, 1976). The presence of sagebrush in the basin of Elk Lake (Dean and others, 1984; Whitlock and others, Chapter 17) supports an interpretation of a dry prairie between 8.5 and 4.0 ky. Dean and others (1984) determined that Elk Lake became shallower and slightly more saline during the Hypsithermal, as a result of drier prairie conditions. During this time period, there was an increase in the transport of unweathered clastic material into the deep part of the lake. This model for climate change is supported by the relative decrease in abundance of chrysophytes ca. 8.5 ka (Fig. 2A). Sandgren (1988) suggested that chrysophytes might be sensitive to elevated alkalinity, pH, and productivity, all of which may have been higher during this prairie period. Our unpublished observations also suggest that chrysophytes appear to be less competitive in waters of higher salinity. Our chrysophyte to diatom ratio indicates the persistence of these conditions for ~4500 yr (Fig. 2A). Using ostracods and diatoms, Forester and others (1987) have shown that in northern Minnesota, part of the mid-Holocene (8.5 to 4.0 ky) consisted of three distinct, dry climatic periods in which temperature and moisture changed rapidly. These changes correspond to two dry pulses, lasting from about 8.5 to 5.4 ky and from 4.8 to 4.0 ky, separated by a moister interval (Dean and others, 1984).

The cyst assemblages show a sudden increase in species richness at about 8.5 ka, when group 2 cysts are recorded in larger numbers, with compensatory declines in group 1 cysts. Group 2 cysts may be intolerant of the cold conditions immediately following deglaciation, or are more competitive in waters of higher salinity and productivity. There is less information in the literature about these cysts; only four have been described previously. Cysts in CC3 are similar to but larger than cyst 9, and have previously been described from both arctic and temperate regions (Duff and Smol, 1988, 1991; Zeeb and others, 1990; Rybak and others, 1991). Stomatocysts 115 and 116 (Forma B) were believed to be indicators of eutrophic conditions in Little Round Lake, Ontario (Zeeb and others, 1990). *Mallomonas caudata* is a common and widely distributed chrysophyte species (Nicholls, 1982; Smol, 1986) that has been found in lakes with high conductivities (Siver and Hamer, 1989) and high chloride concentrations (Zeeb and Smol, 1991).

Our cyst-assemblage data indicate that limnological conditions during the mid-Holocene prairie period shifted for a period of about 500 yr between 5.3 and 4.8 ky. Dominant cysts indicate a striking return to early postglacial conditions (Fig. 3): it is interesting that our PCA diagram shows this less clearly (Fig. 5), because assemblage scores appear relatively similar with respect to the first PCA axis, but are widely separated on the second axis.

This suggests environmental conditions were similar in some respects, perhaps temperature, but differed in others, perhaps salinity (i.e., precipitation-evaporation changes). The PCA analysis indicates that during the mid-Holocene, cyst assemblages frequently shifted with respect to the first and second axes.

Late Holocene (4.0 to 0.3 ky)

Relative numbers of chrysophytes again increased ca. 4.0 ka (peaking ca. 1.1 ka), suggesting a return to cooler and oligotrophic nutrient conditions (Fig. 2A). The fluctuations in the profiles are probably related to several shorter-term oscillations in lake trophic status between 4.0 and 1.1 ka. The dramatic decrease in relative abundance of chrysophytes that followed and the resultant low abundances that persisted for ~400 yr (0.8 to 0.4 ky) (Fig. 2B) suggest higher trophic status and/or conductivity than between 8.5 and 4.0 ky. This time period corresponds well with the Medieval Warm Period, a period of maximum temperature between 0.9 and 0.5 ky (Lamb, 1981). The increase and resulting dominance in relative abundances of chrysophytes between 0.4 and 0.1 ky are again comparable to those recorded in high arctic lakes. This time period coincides with the Little Ice Age and suggests that the climate was similar to that of the early postglacial (Fig. 2B).

With the exception of stomatocysts 41 and 115, the dominant cysts maintained fairly stable abundances, with some oscillations from 4.0 to 0.3 ky (Figs. 3 and 5). At this time we cannot offer an explanation for the complete disappearance of cyst 115 at about 4.0 ka. The dramatic decrease of cyst 41 at about 2.6 ka appears to be related to lake nutrient levels, which were probably relatively high during this period (see other chapters in this volume).

Recent times (A.D. 1630 to A.D. 1864)

Chrysophytes decreased in relative abundances with respect to diatoms near the top of the core, suggesting a return to higher nutrient conditions and probably higher temperatures. Bradbury (1988) used diatom and phytolith data to suggest that dry winters and early springs were more common between A.D. 1690 and A.D. 1720. Conditions between A.D. 1720 and A.D. 1810 were much like those of today, with wetter winters and later springs. The diatom data indicate a drier climate with earlier springs ca. A.D. 1810.

Distinctive shifts in cyst assemblages occurred in response to known climatic changes (Fig. 4). Based on Bradbury's (1988) interpretations, we tentatively suggest that stomatocyst 41, 153, 157, 150, 9, 116 (Forma A), *Mallomonas caudata* and CC3 may be more common in lakes with later than average ice-out dates and heavy snowfalls. Stomatocysts 152 (Forma A and B) and 154 may indicate drier winters and earlier springs; the abundance of these cysts is coincident with the abundance of benthic diatoms and phytoliths (Bradbury, 1988).

CONCLUSIONS

Our exploratory stomatocyst work on Elk Lake sediments has shown that chrysophytes respond to environmental changes on a long-term perspective in the range of centuries to millennia. Our ecological interpretations are currently still largely speculative. However, based on the other paleoecological information we have available for this core, we speculate that the first PCA axis may be temperature dependent, going left to right, from warm to cool. This would place the early prairie period and part of the more recent deciduous and coniferous forest period (2.8 to 0.3 ky) at one extreme, and the early postglacial and the period believed to represent the Little Ice Age (mid 1700s [A.D.]) at the other extreme. We speculate that the second PCA axis may be precipitation-evaporation dependent (i.e., salinity), going from top to bottom, from drier to wetter. This would place the early postglacial and Little Ice Age in the wettest periods and the early and late prairie periods in the driest periods. In the near future, as surface-sediment calibration sets become available for cysts, and we can define the ecological optima and tolerances of morphotypes, we hope to refine our paleoecological interpretations.

ACKNOWLEDGMENTS

This study was supported by a Natural Sciences and Engineering Research Council of Canada operating grant to Smol. We thank J. P. Bradbury for providing the material, and the cyst to diatom data, used in this study. We thank C. D. Sandgren, P. A. Siver, W. E. Dean, and P. Leavitt and scientists at PEARL for their excellent suggestions and constructive comments.

REFERENCES CITED

Adam, D. P., and Mahood, A. D., 1979, Modern and Holocene chrysomonad cysts from Alta Morris Lake, Eldorado County, California: U.S. Geological Survey Open-File Report 79-1461, 13 p.

——, 1981, Chrysophyte cysts as potential environmental indicators: Geological Society of America Bulletin, v. 92, p. 839–844.

Adam, D. P., and Mehringer, P. J., Jr., 1980, Scanning electron micrographs of modern and Holocene chrysomonad cysts from Fish Lake, Steens Mountains, Oregon: U.S. Geological Survey Open-File Report 80-1249, 18 p.

Anderson, R. A., 1987, Synurophyceae classis nov., a new class of algae: American Journal of Botany, v. 74, p. 337–353.

Battarbee, R. W., 1986, Diatom analysis, in Berglund, B. E., ed., Handbook of palaeoecology and palaeohydrology: Toronto, John Wiley and Sons, p. 527–570.

Bradbury, J. P., 1988, A climatic-limnologic model of diatom succession for paleolimnological interpretation of varved sediments at Elk Lake, Minnesota: Journal of Paleolimnology, v. 1, p. 115–131.

Carney, H. J., 1982, Algal dynamics and trophic interactions in the recent history of Frains Lake, Michigan: Ecology, v. 63, p. 1814–1826.

Carney, H. J., and Sandgren, C. D., 1983, Chrysophycean cysts: Indicators of eutrophication in the recent sediments of Frains Lake, Michigan, U.S.A.: Hydrobiologia, v. 101, p. 195–202.

Carney, H. J., Whiting, M. C., Duff, K. E., and Whitehead, D. R., 1992, Chrysophycean cysts in Sierra Nevada (California) lake sediments: Paleoecological potential: Journal of Paleolimnology, v. 7, p. 73–94.

Christie, C. E., Smol, J. P., Huttunen, P., and Meriläinen, J., 1988, Chrysophyte scales recorded in lake sediments from eastern Finland: Hydrobiologia, v. 161, p. 237–243.

Cronberg, G., 1986, Chrysophycean cysts and scales in lake sediments: A review, in Kristiansen, J., and Andersen, R. A., eds., Chrysophytes: Aspects and problems: Cambridge, Cambridge University Press, p. 281–315.

——, 1988, Variability in size and ultrastructure of the statospore of *Mallomonas caudata*: Hydrobiologia, v. 161, p. 31–39.

——, 1989, Stomatocysts of *Mallomonas hamata* and *M. heterospina* (Mallomonadacea, Synurophyceae) from south Swedish lakes: Nordic Journal of Botany, v. 8, p. 683–692.

Cronberg, G., and Sandgren, C. D., 1986, A proposal for the development of standardized nomenclature and terminology for chrysophycean statospores, in Kristiansen, J., and Andersen, R. A., eds., Chrysophytes: Aspects and problems: Cambridge, Cambridge University Press, p. 312–328.

Cumming, B. F., Smol, J. P., and Birks, H.J.B., 1991, The relationship between sedimentary chrysophyte scales (Chrysophyceae and Synurophyceae) and limnological characteristics in 25 Norwegian lakes: Nordic Journal of Botany, v. 11, p. 231–242.

Dean, W. E., Bradbury, J. P., Anderson, R. Y., and Barnosky, C. W., 1984, The variability of Holocene climate change: Evidence from varved lake sediments: Science, v. 226, p. 1191–1194.

Dixit, S. S., Dixit, A. S., and Evans, R. B., 1988, Chrysophyte scales in lake sediments provide evidence of recent acidification in two Quebec (Canada) lakes: Water, Air, and Soil Pollution, v. 38, p. 97–104.

Dixit, S. S., Dixit, A. S., and Smol, J. P., 1989a, Relationship between chrysophyte assemblages and environmental variables in seventy-two Sudbury lakes as examined by canonical correspondence analyses (CCA): Canadian Journal of Fisheries and Aquatic Sciences, v. 46, p. 1667–1676.

Dixit, S. S., Dixit, A. S., and Evans, R. D., 1989b, Paleolimnological evidence for trace-metal sensitivity in scaled chrysophytes: Environmental Science and Technology, v. 23, p. 110–115.

Dixit, S. S., Smol, J. P., Anderson, D. S., and Davis, R. B., 1990, Utility of scaled chrysophytes for inferring lakewater pH in northern New England lakes: Journal of Paleolimnology, v. 3, p. 269–286.

Donaldson, D. R., and Stein, J. R., 1984, Identification of planktonic Mallomonadaceae and other Chrysophyceae from selected lakes in the lower Fraser Valley, British Columbia, Canada: Canadian Journal of Botany, v. 62, p. 525–539.

Duff, K. E., and Smol, J. P., 1988, Chrysophycean stomatocysts from the postglacial sediments of a High Arctic lake: Canadian Journal of Botany, v. 66, p. 1117–1128.

——, 1989, Chrysophycean stomatocysts from the postglacial sediments of Tasikutaaq Lake, Baffin Island, N.W.T.: Canadian Journal of Botany, v. 67, p. 1649–1656.

——, 1991, Morphological descriptions and stratigraphic distributions of the chrysophycean stomatocysts from a recently acidified lake (Adirondack Park, N.Y.): Journal of Paleolimnology, v. 5, p. 73–113.

Duff, K. E., Douglas, M.S.V., and Smol, J. P., 1992, Chrysophyte cysts in 36 Canadian High Arctic ponds: Nordic Journal of Botany, v. 12, p. 471–499.

Elner, J. K., and Happey-Wood, C. M., 1978, Diatom and chrysophycean cyst profiles in sediment cores from two linked but contrasting Welsh lakes: British Phycological Journal, v. 13, p. 341–360.

Eloranta, P., 1989, On the ecology of the genus *Dinobryon* in Finnish lakes: Beiheft zur Nova Hedwigia, v. 95, p. 99–109.

Forester, R. M., Delorme, D. L., and Bradbury, J. P., 1987, Mid-Holocene climate in northern Minnesota: Quaternary Research, v. 28, p. 263–273.

Gritten, M. M., 1977, On the fine structure of the Chrysophycean cysts: Hydrobiologia, v. 53, p. 239–252.

Hartman, H., and Steinberg, C., 1986, Mallomonadacean (Chrysophyceae) scales: Early biotic paleoindicators of lake acidification: Hydrobiologia, v. 143, p. 87–91.

Jongman, R.H.G., ter Braak, C.J.F., and Van Tongeren, O.F.R., 1987, Data analysis in community and landscape ecology: Wageningen, The Netherlands, Pudoc, 299 p.

Kristiansen, J., and Takahashi, E., 1982, Chrysophyceae: Introduction and bibliography, *in* Rowowski, J. R., and Parker, B. C., eds., Selected papers in phycology, 2: Lawrence, Kansas, Phycological Society of America, Inc., p. 698–704.

Lamb, H. H., 1981, An approach to the study of the development of climate and its impact in human affairs, *in* Wigley, T.M.L., Ingram, M. J., and Farmer, G., eds., Climate and history: Studies in past climates and their impact on Man: Cambridge, Cambridge University Press, p. 291–309.

Nicholls, K. H., 1981, *Chrysococcus furcatus* (Dolg.) comb. nov.: A new name for *Chrysastrella furcata* (Dolg.) Defl. based on the discovery of the vegetative stage: Phycologia, v. 20, p. 16–21.

—— , 1982, *Mallomonas* species (Chrysophyceae) from Ontario, Canada, including descriptions of two new species: Nova Hedwigia, v. 34, p. 80–124.

Pienitz, R., Walker, I., Zeeb, B. A., Smol, J. P., and Leavitt, P. R., 1993, Biomonitoring past salinity changes in an athalassic sub-arctic lake: International Journal of Salt Lake Research (in press).

Rybak, M., 1986, The chrysophycean paleocyst flora of the bottom sediments of Kortowskie Lake (Poland) and its ecological significance: Hydrobiologia, v. 140, p. 67–84.

—— , 1987, Fossil chrysophycean cyst flora of Racze Lake, Wolin Island (Poland) in relation to paleoenvironmental conditions: Hydrobiologia, v. 150, p. 257–272.

Rybak, M., Rybak, I., and Dickman, M., 1987, Fossil chrysophycean cyst flora in a small meromictic lake in southern Ontario, and its paleoecological interpretation: Canadian Journal of Botany, v. 65, p. 2425–2440.

Rybak, M., Rybak, I., and Nicholls, K., 1991, Sedimentary chrysophycean cyst assemblages as paleoindicators in acid sensitive lakes: Journal of Paleolimnology, v. 5, p. 19–72.

Sandgren, C. D., 1980, Resting cyst formation in selected chrysophyte flagellates: An ultrastructural survey including a proposal for the phylogenetic significance of interspecific variations in the encystment process: Protistologica, t. XVI, fasc. 2, p. 289–303.

—— , 1983, Survival strategies of chrysophycean flagellates: Reproduction and the formation of resistant resting cysts, *in* Fryxell, G. A., ed., Survival strategies of the algae: Cambridge, Cambridge University Press, p. 23–48.

—— , 1988, The ecology of chrysophyte flagellates: Their growth and perennation strategies as freshwater phytoplankton, *in* Sandgren, C. D., ed., Growth and reproductive strategies of freshwater phytoplankton: Cambridge University Press, p. 9–104.

—— , 1991, Chrysophyte reproduction and resting cysts: A paleolimnologists primer: Journal of Paleolimnology, v. 5, p. 1–9.

Sandgren, C. D., and Carney, H. J., 1983, A flora of fossil chrysophycean cysts from the recent sediments of Frains Lake, Michigan, U.S.A.: Nova Hedwigia, v. 38, p. 129–163.

Sandgren, C. D., and Flanagin, J., 1986, Heterothallic sexuality and density dependent encystment in the chrysophycean alga *Synura petersenii* Korsh: Journal of Phycology, v. 22, p. 206–216.

—— , 1988, Distribution of scaled chrysophytes in 17 Adirondack (New York) lakes with special reference to pH: Canadian Journal of Botany, v. 66, p. 1391–1403.

Siver, P. A., 1989a, The distribution of scaled chrysophytes along a pH gradient: Canadian Journal of Botany, v. 67, p. 2120–2130.

—— , 1989b, The separation of *Mallomonas acaroides* v. *acaroides* and v. *muskokana* (Synurophycea) along a pH gradient: Beihefte zur Nova Hedwigia, v. 95, p. 111–117.

Siver, P. A., and Chock, J. S., 1986, Phytoplankton dynamics in a chrysophycean lake, *in* Kristiansen, J., and Anderson, R. A., eds., Chrysophytes: Aspects and problems: Cambridge, Cambridge University Press, p. 165–183.

Siver, P. A., and Hamer, J. S., 1989, Multivariate statistical analysis of the factors controlling the distribution of scaled chrysophytes: Limnology and Oceanography, v. 34, p. 368–381.

Smith, M. A., and White, M. J., 1985, Observations on lakes near Mount St. Helens: Phytoplankton: Archiv für Hydrobiologie, v. 104, p. 345–362.

Smol, J. P., 1983, Paleophycology of a high arctic lake near Cape Herschel, Ellesmere Island: Canadian Journal of Botany, v. 61, p. 2195–2204.

—— , 1984, The statospore of *Mallomonas pseudocoronata* (Mallomonadacea, Chrysophyceae): Nordic Journal of Botany, v. 4, p. 827–831.

—— , 1985, The ratio of diatom frustules to chrysophycean statospores: A useful paleolimnological index: Hydrobiologia, v. 123, p. 199–208.

—— , 1986, Chrysophycean microfossils as indicators of lakewater pH, *in* Smol, J. P., Battarbee, R. W., Davis, R. B., and Meriläinen, J., eds., Diatoms and lake acidity: Dordrecht, The Netherlands, Dr. W. Junk, p. 275–287.

—— , 1988, Chrysophycean microfossils in paleolimnological studies: Palaeogeography, Palaeoclimatology, Palaeoecology, v. 62, p. 287–297.

Stoermer, E. F., Wolin, J. A., Schelske, C. L., and Conley, D. J., 1985, An assessment of ecological changes during the recent history of Lake Ontario based on silicious algal microfossils preserved in the sediments: Journal of Phycology, v. 21, p. 257–276.

Takahashi, E., Watanabe, K., and Satoh, H., 1986, Silicious cysts from Kita-No-Seto Strait, north of Syowa Station, Antarctica: National Institute for Polar Research Memoirs, Special Issue, v. 40, p. 84–95.

Wright, H. E., Jr., 1976, The dynamic nature of Holocene vegetation: A problem in paleoclimatology, biogeography, and stratigraphic nomenclature: Quaternary Research, v. 6, p. 581–596.

Zeeb, B. A., and Smol, J. P., 1991, Paleolimnological investigation of the effects of road salt seepage on scaled chrysophytes in Fonda Lake, Michigan: Journal of Paleolimnology, v. 5, p. 263–266.

Zeeb, B. A., Duff, K. E., and Smol, J. P., 1990, Morphological descriptions and stratigraphic profiles of chrysophycean stomatocysts from the recent sediments of Little Round Lake, Ontario: Nova Hedwigia, v. 51, p. 361–380.

Zeeb, B. A., and Smol, J. P., 1993, Chrysophycean cyst flora from the postglacial sediments of Elk Lake, Minnesota: Canadian Journal of Botany, v. 71 (in press).

MANUSCRIPT ACCEPTED BY THE SOCIETY JULY 27, 1992

Vegetation history of Elk Lake

Cathy Whitlock and Patrick J. Bartlein
Department of Geography, University of Oregon, Eugene, Oregon 97403
William A. Watts
Trinity College, Dublin 2, Ireland

ABSTRACT

A pollen record from Elk Lake reveals the character and timing of major vegetation changes in northwestern Minnesota for the past 11.6 ka, the past 10.4 ka of which are recorded by varves. Fossil pollen spectra are compared with modern pollen data to identify the closest analogues and thereby to infer past climatic changes. The late glacial *Picea* assemblage (ca. 11,638–10,000 varve yr) lacks an exact analogue in the modern vegetation; initially it compares most closely with modern samples from Manitoba and Saskatchewan and later it is most similar to samples from northeastern Canada. The *Picea* decline at Elk Lake occurs between 10,234 and 9984 varve yr. Within this interval, percentages of *Larix, Juniperus, Betula, Quercus, Ulmus,* and *Fraxinus* increase, but the pollen-accumulation rates of these and other taxa decline. The *Pinus banksiana-resinosa* assemblage (10,000–8500 varve yr) has its closest modern analogues in northern Wisconsin and implies warmer slightly drier conditions than before. The prairie period, with high percentages of *Quercus,* Gramineae, and *Artemisia* (8500–4400 varve yr), is first matched by surface samples from central and southern Minnesota, then from southern Saskatchewan, and later from southern and central Wisconsin and Minnesota. The change in the location of the analogues suggests gradually wetter conditions after 5723 varve yr. A *Quercus-Ostrya* assemblage (4400–3000 varve yr) has it modern counterparts in the conifer-hardwood forest of the southern Great Lakes region, where the climate is wetter than that of the prairie period. The *Pinus strobus* assemblage (3000 varve yr to present) marks the development of cooler moister conditions in the late Holocene.

The Elk Lake chronology was converted to radiocarbon years to compare it with other pollen records from the midwestern United States. The *Picea* decline is registered as a time-transgressive event, although it occurred 1000 years earlier in the west than in the east. The late glacial *Ulmus* maxima and the middle Holocene prairie period appear to be synchronous across the region. Discrepancies in the timing of these events are attributed to radiocarbon dating errors, which are particularly severe in the western part of the region.

INTRODUCTION

Minnesota has been the setting for some of the most important advances in Quaternary paleoecology of recent decades, in terms of understanding the postglacial development of vegetation and the response of plant communities to specific changes in environment. Detailed pollen studies from the state have revealed a vegetational history sensitive to a wide spectrum of disturbances, ranging from deglaciation and its attendant modification of the late glacial landscape (Wright, 1968; Birks, 1976, 1981), to Holocene climatic fluctuations (Cushing, 1967; McAndrews, 1966; Watts and Winter, 1966; Birks, 1981; Webb and others, 1983; Grimm, 1983; Keen and Shane, 1990), changes in fire frequency (Amundson and Wright, 1979; Clark, 1988a, 1988b), and human disturbances of the environment (Grimm, 1984).

The northwestern part of the state and in particular the

Itasca Park region have played a central role in paleoecologic investigations. A transect of sites, including Bog D from within the park, first disclosed the eastward movement of the prairie-forest boundary during a period of middle Holocene warmth and aridity (McAndrews, 1966). Pollen and phytosociological studies at the northern edge of the park delineated the relative contribution of local versus regional pollen rain in a small basin, Stevens Pond (Janssen, 1967). A comparison of several microfossil groups in a transect of cores across Elk Lake revealed the timing and extent of lake shallowing during the middle Holocene (Stark, 1976). Archeologic and paleoecologic studies near the shores of Elk Lake led to an understanding of prehistoric utilization of the prairie region during times of vegetational change (Shay, 1971). Detailed studies of fossil charcoal and pollen within the park have led to refinements in the use of paleoecologic data to reconstruct fire history (Clark, 1988a, 1988b, this volume).

In this chapter we describe a ca. 11.6 ka pollen record from Elk Lake in Itasca Park that builds upon this body of literature in three respects. First, a varve chronology for the past 10.4 ka at Elk Lake provides a time scale by which paleoenvironmental events in the region are dated in calendar years. The timing of major vegetational changes are addressed, including (1) the late glacial transition from *Picea*-dominated forest to one largely composed of *Pinus*; (2) the eastward expansion of prairie in the middle Holocene; and (3) the late Holocene spread of pine and hardwood species. Second, modern analogues are selected quantitatively from the large data base of surface samples that now exists from eastern North America; these analogues are used to reconstruct the past vegetation and environment at Elk Lake. Third, the Elk Lake varve chronology is converted to radiocarbon years and compared with other pollen records from the midwestern United States (Midwest). An analysis of important pollen events along north-south and east-west transects helps clarify the regional patterns of vegetational change and assess the variability of radiometric age determinations.

MODERN SETTING AND GLACIAL HISTORY

Elk Lake is located within the pine-hardwood forest, about 80 km east of the eastern limit of prairie (Fig. 1). A belt of deciduous forest dominated by *Quercus macrocarpa* (bur oak) and *Populus tremuloides* (quaking aspen) lies between conifer forest and prairie in the Itasca region. The pine-hardwood forest includes elements from both the boreal forest and the eastern hardwood forest. On droughty, nutrient-poor soils are extensive stands of *Pinus banksiana* (jack pine), *P. resinosa* (red pine), *Q. macrocarpa* (bur oak), *Q. ellipsoidalis* (northern pine oak), *Q. borealis* (red oak), and *Q. alba* (white oak). On widespread till soils, *P. strobus* (white pine) is common on fine-textured soil, whereas *P. resinosa* grows on coarse-textured soil. *Abies balsamifera* (balsam fir), *Picea glauca* (white spruce), *P. mariana* (black spruce), and *Larix laricina* (tamarack) are widespread but are not common in either pine or hardwood forest (McAndrews, 1966). Important hardwood elements from the boreal forest, particularly in burned areas, are *Betula papyrifera* (white birch), *Populus balsamifera* (balsam poplar), and *P. tremuloides* (quaking aspen). Hardwoods typical of the eastern hardwood forest grow on fine-textured soils and include *Acer rubrum* (red maple), *A. saccharum* (sugar maple), *Betula lutea* (yellow birch), *Ulmus* spp. (elm), *Fraxinus* spp. (ash), *Tilia americana* (American basswood), *Ostrya virginiana* (eastern hophornbeam), and *Alnus rugosa* (speckled alder). Common upland shrubs include *Corylus* (hazel), *Cornus* (dogwood), *Amelanchier* (serviceberry), *Prunus* (cherry), *Viburnum*, and *Ribes*. Less common shrubs are *Alnus crispa* (green alder), *Taxus canadensis* (American yew), *Juniperus communis* (common juniper), and *Carpinus* (hornbeam). Many of the slopes around the site were extensively logged from A.D. 1901 to 1921 (Dobie, 1959), and today the vegetation at the lake margin is balsam fir and red pine, while aspen and birch are more abundant upslope.

Itasca Park lies in the middle-latitudinal belt of prevailing westerly winds. Air masses originate from three source areas: from the south comes warm moist subtropical air, from the north comes cold dry continental polar air, and from the west comes dry continental air (Bryson, 1966). The character of the weather and climate reflects the frequency of cyclonic storms associated with different air-mass systems. Mean temperatures are 20 °C in July and –7.8 °C in January. About 670 mm of precipitation is received yearly, most between April and September (U.S. Department of Commerce Weather Bureau, 1960).

Two lobes of the Laurentide ice sheet covered the Elk Lake region in late Wisconsin time (Wright, 1972, and Chapter 2). The Wadena lobe formed the Itasca moraine at ca. 20 ka. The Elk Lake depression is part of a subglacial tunnel-valley system that formed beneath the Wadena lobe. Later the Wadena till and stagnant ice were buried by outwash from the St. Louis sublobe of the Des Moines lobe, which reached its maximum ca. 13 ka. The oldest sediments at Elk Lake date to ca. 11.6 ka, suggesting that the basin formed from the melting of a buried ice complex ca. 1500 yr after retreat of the Des Moines lobe.

METHODS

Only the palynological procedures are discussed here, inasmuch as the field methods and the varve chronology are presented in Anderson and others (Chapter 4). The varve chronology is discussed as varve years ago (varve yr); radiocarbon years are listed as ka (1000 yr before present).

Pollen samples were extracted from the core with a 0.5 cm^3 rectangular metal sampler at intervals of 50 yr; 90 samples were analyzed. The 50th yr lamina was in the middle of each sample and the number of varve years per sample was recorded. Samples processed for pollen analysis were subjected to standard treatments with HCl, KOH, HF, acetolysis, and alcohol washes outlined by Faegri and Iversen (1975), and stored in silicon oil.

Pollen was examined under magnifications of 400× and 1000×. Identifications were based on the reference collection at the Limnological Research Center, University of Minnesota, and

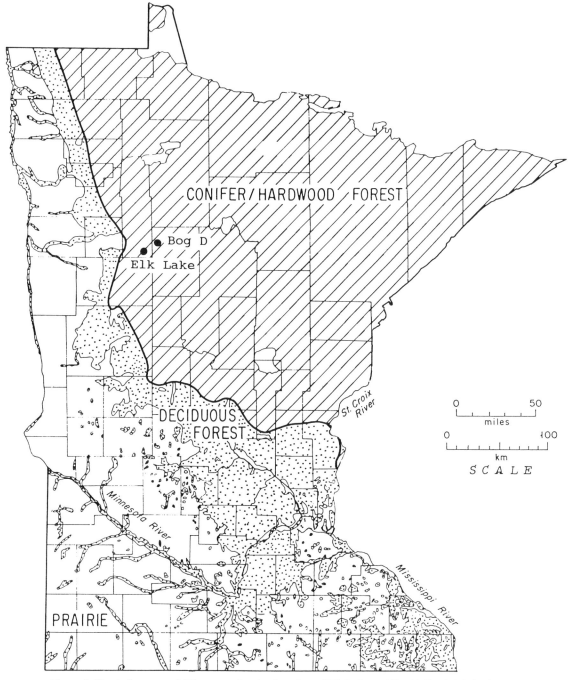

Figure 1. Vegetation map of Minnesota showing location of Elk Lake and Bog D Pond (McAndrews, 1966).

published atlases (Faegri and Iversen, 1975; McAndrews and others, 1973; Moore and Webb, 1978). Grains that could not be identified were tallied as "unknown"; grains that were deteriorated, corroded, hidden, or broken beyond recognition were considered "indeterminate." Diploxylon-haploxylon determinations were made on *Pinus* pollen with intact distal membranes; other *Pinus* were tallied as undifferentiated.

Percentage data were the major tools for reconstructing past vegetation and climate. The sum of the terrestrial pollen and spores was the denominator in calculating the percentages of trees, shrubs, herbs, and terrestrial pteridophytes. The sum of all palynomorphs was used to calculate the percentages of aquatic and wetland taxa. A *Eucalyptus* tracer was added to each sample to permit calculation of pollen concentration (grains/cm^3).

When divided by the deposition time (varve yr/cm), the concentration values were converted to pollen-accumulation rates (grains/cm^2/varve yr). The deposition time in varved sequences can be calculated with greater accuracy than that achieved from radiocarbon-dated nonlaminated sediments, and therefore some of the imprecision inherent in calculations of pollen-accumulation rates is mitigated at Elk Lake.

The interpretation of the fossil percentage data was facilitated by identifying quantitatively the closest modern analogues for each fossil spectrum. This analysis involved selecting a set of modern pollen data from eastern North America and calculating a dissimilarity value between a fossil spectrum and the surface samples. These dissimilarity values were mapped to identify the locations of the modern spectra that are similar to each fossil spectrum.

The data set consisted of 1200 modern spectra, including some selected from the set of modern pollen data at Brown University (e.g., Overpeck and others, 1985; Bartlein and others, 1984), supplemented by additional samples from Lichti-Federovich and Ritchie (1965, 1968) and MacDonald and Ritchie (1986) for the central and western interior of Canada. The pollen sum consists of the 44 most abundant pollen types, including herb taxa with the notable exception of *Ambrosia* and *Iva,* which are generally overrepresented in modern samples as a result of agriculture and other anthropogenic disturbances. Some of the types in the sum never appear at Elk Lake (e.g., *Magnolia*) but are included to avoid fortuitous analogues. For example, the *Pinus* and *Quercus* percentages are such that, if southeastern hardwood taxa were omitted from the sum, some modern spectra from the southeastern United States would be similar to some of the fossil values at Elk Lake, even though they are clearly from a nonanalogous vegetation.

Following Overpeck and others (1985), the squared chord distance was used as the dissimilarity measure, or

$$d_{ij} = \Sigma_k (p_{ik}^{1/2} - p_{jk}^{1/2})^2,$$

where d_{ij} is the squared chord distance between two pollen spectra i and j, and p_{ik} is the proportion (between 0.0 and 1.0) of pollen type k in pollen spectrum i.

Anderson and others (1989) listed the desirable properties of the squared chord distance for identifying modern analogues, including (1) the ability of this measure to differentiate vegetation types at the scale of the formation and even the forest type; (2) the values of this measure that indicate that "good analogues" are robust with respect to the choice of the pollen sum; and (3) at the large geographic scale considered here, this measure emphasizes the large-scale patterns in the data and suppresses the local variability, or noise, in the modern pollen samples. Interpretation of the squared chord was facilitated by determining its value for each unique pairing of the 1200 modern samples (yielding 119,400 values). The first, fifth, and tenth percentile (0.092, 0.205, and 0.305, respectively) of these values are provisionally selected as indicators of very good, good, and fair analogues (see Anderson and others, 1989). The modern spectraa that are similar to the fossil spectra at Elk Lake lie within a limited region in eastern North America (i.e., no spectra from the southeastern United States are similar to any of the fossil spectra at Elk Lake).

POLLEN RECORD

It is not surprising that the pollen record from Elk Lake compares closely with that from nearby Bog D. McAndrews (1966) identified five assemblages that provide a useful basis for discussion: *Picea* assemblage; *Pinus banksiana-P. resinosa-Pteridium* assemblage; *Quercus*-Gramineae-*Artemisia* assemblage; and *Pinus strobus* assemblage, with a *P. banksiana-resinosa* subassemblage. Pollen percentages for the entire record are presented in Figure 2; the late glacial and early Holocene portions are plotted on a larger scale in Figure 3. The pollen-accumulation rates are in Figure 4.

Pollen values of the best modern analogue for each spectrum are plotted stratigraphically and compared to a fossil-pollen diagram that uses the same pollen sum (Fig. 5). Maps display the location of the modern pollen data with variably shaded symbols to reflect the degree of similarity between a particular fossil spectrum and each modern spectrum (Fig. 6). These maps identify the location of modern counterparts and illustrate the strength of the analogue. Throughout the record, the most similar modern spectrum for each fossil spectrum has a squared chord distance less than the tenth percentile of the squared chord distances among the modern samples. However, some variations exist in the strength of the similarities through time. Four episodes can be identified: (1) the interval during which the *Picea* assemblage prevailed, when the squared chord distances are relatively large, indicating relatively poor modern analogues for these spectra; (2) the interval spanning the time when *Pinus banksiana-P. resinosa-Pteridium* and *Quercus*-Gramineae-*Artemisia* assemblages occurred, when the squared chord distances are relatively small; (3) the interval when the *Quercus-Ostrya* assemblage prevailed, when the dissimilarities are again large; and (4) the interval from 3000 varve years to present (the *Pinus strobus* assemblage), when the squared chord distances are small, indicating good modern analogues for the fossil spectra.

Picea assemblage (11,638–10,000 varve yr)

This assemblage is characterized by high pollen percentages of *Picea* (8%–52%) and significant amounts of *Pinus, Larix,* Gramineae, Cyperaceae, and *Artemisia.* Low combined percentages (<8%) of *Quercus, Ulmus,* and *Fraxinus* are also present. Indeterminate pollen is abundant and includes degraded Cretaceous taxa derived secondarily from the bedrock. Rates of pollen accumulation of *Picea, Larix, Juniperus,* and *Betula* are variable, and the overall pollen-accumulation rates in the varved sediments are less than 8000 grains/cm^2/yr, typical of modern forest tundra and boreal forest (Ritchie and Lichti-Federovich, 1967; M. B. Davis and others, 1973; R. B. Davis and others, 1975).

The six bottommost pollen samples are from the unvarved part of the core—the interval from 10,400 varve yr and extrapolated to 11.638 ka. The period probably records the initial formation of Elk Lake during ice melting. The sediments contain lots of woody plant debris derived from forest that may have grown on a buried ice block prior to the formation of the lake. Pollen-accumulation rates are lowest (<18,900 grains/cm^2/yr) above and below two sand layers estimated to be 11.20 and 11.4 ka. Ostracodes from these sand units suggest a lake at least a few meters deep with irregular deposition patterns (Smith, 1991). This hypothesis is consistent with the evidence for variable deposition times implied by pollen-accumulation rates.

The pollen data suggest that before 10,000 varve yr the vegetation was a spruce forest with local hardwoods, as well as openings that supported *Artemisia,* Gramineae, and other herbs. The landscape during the spruce period was probably characterized by meltwater sand plains, unstable slopes, and shallow soils. On well-drained soil, mixtures of white spruce, paper birch, grass, and *Artemisia* may have occurred. Tamarack may also have grown on the uplands, as it does today in Canada. Poorly drained substrates would have also sustained tamarack, black spruce, and sedges. Ash and elm would have grown on seasonally moist soils.

In comparison to the Bog D record, the spruce period at Elk Lake features more *Juniperus* and less *Populus* pollen. Juniper favors open areas with high snow accumulation, and it is possible that the slopes around Elk Lake were more suited to its growth than at Bog D. *Populus* (probably both *P. tremuloides* and *P. balsamifera* types) pollen at Bog D was abundant enough for McAndrews (1966) to designate a *Picea-Populus* assemblage. Its poor representation at Elk Lake suggests either that the sedimentary matrix was less suitable for the preservation of this delicate grain or that the tree was less abundant in the pollen-recruitment area.

Modern analogues for the spruce assemblage are not exact and change midway. The analogue map for 11,638 yr shows the best match in modern samples from Manitoba and Saskatchewan, where *Picea* and herb percentages are high and *Pinus* values are relatively low (Fig. 6). Even the best modern analogues tend to overrepresent *Pinus* and *Betula* and underrepresent *Picea* and *Juniperus* in comparison with the late glacial spectra. The reconstructed climate for this period was cold and relatively dry (see Bartlein and Whitlock, Chapter 18, for specific climatic estimates).

The end of the spruce period occurs between 10,234 and 9984 varve yr, when *Picea* pollen declines from 51% to 10%. During this interval *Pinus banksiana-resinosa* pollen rises from 6% to 28%. A decline in nonarboreal percentages at Elk Lake begins at 10,334 varve yr. Prior to the rise in *Pinus* percentages, there is a brief interval in which percentages of *Abies, Larix, Juniperus, Betula, Quercus, Ulmus,* and *Fraxinus* are higher than before or after. This increase is not matched by a change in pollen-accumulation rates at Elk Lake. The intervening interval is more conspicuous farther south in Minnesota (Amundson and Wright, 1979) and has been interpreted as a time when various conifer and hardwood taxa grew on a landscape in which spruce populations were declining, but before the expansion of pine. The low accumulation rates of Elk Lake, however, imply that these taxa were only minor elements of an open spruce forest in the Itasca region, if present at all. It seems likely that the climatic conditions that limited spruce suppressed all species and the period between 10,200 and 10,000 varve yr was sparsely vegetated.

Modern analogues for the 10,084 varve yr spectra can be found west of Elk Lake, where *Picea* accounts for >35% and *Pinus* percentages are high (Fig. 6). Warming and possibly drier conditions may have characterized the Elk Lake climate at that time. The analogues, however, do not contain significant percentages of *Quercus, Ulmus,* and *Fraxinus.*

Pinus banksiana-resinosa-Pteridium assemblage (10,000-8500 varve yr)

An increase in *P. banksiana-resinosa* percentages and *Betula* (10%-23%) occurs between 10,034 and 9050 varve yr. *Pteridium* is represented by low values (<4%). *Ulmus* (4%-8%), Gramineae (4%-8%), Cyperaceae (<5%), and *Artemisia* are also present in small percentages. *Picea, Larix, Juniperus,* and *Abies* values are insignificant. Percentages of *Artemisia* and Gramineae increase about 8450 varve yr at the end of this assemblage. The total pollen-accumulation rates during this interval range from 5740 to 20,090 grains/cm^2/yr, typical of those from modern boreal forest (Ritchie and Lichti-Federovich, 1967; M. B. Davis and others, 1973; R. B. Davis and others, 1975).

The pollen record suggests development of an upland forest composed largely of *Pinus banksiana, Betula papyrifera,* and possibly *P. resinosa. Picea* and *Larix* were probably restricted to boggy sites. McAndrews (1966) suggested that *Ulmus* was an important tree on fine-textured and moist soils. High levels of charcoal at other sites, along with an abundance of *Pteridium*-type spores, suggest that frequent fires allowed bracken fern to become an important understory species in a forest dominated by jack pine. The pine period is registered at Bog D between 11 and 8.56 ka, although Webb and others (1983) decreased these dates by 1000 yr, which matches more closely the Elk Lake record. The anomalously old dates at Bog D are on calcareous gyttja and are attributed to a carbonate error. Radiocarbon dates on the Elk Lake record (Anderson and others, Chapter 4) also imply a carbonate error of about 1000 yr in this interval. Dates of 9.36 and 9.69 ka on *Larix* wood within the spruce zone from the nearby Nicollet Valley are younger and considered more reasonable (Shay, 1971; Webb and others, 1983).

Paleoecologic data point to the Appalachian region and the Atlantic coastal plain as the glacial refugia for jack pine (Davis, 1983; Watts, 1979). The species migrated to Minnesota from the south and east, colonizing areas disturbed by fire and wind throw within a spruce forest that was declining as a result of warmer, effectively drier summers and winters (Amundson and Wright, 1979). The establishment of pine about 10,000 yr ago may be attributed to climatic warming and its effect on the disturbance regime. The closest modern analogues are found in northern Wis-

ELK LAKE, Clearwater County, Minnesota
Pollen Percentages

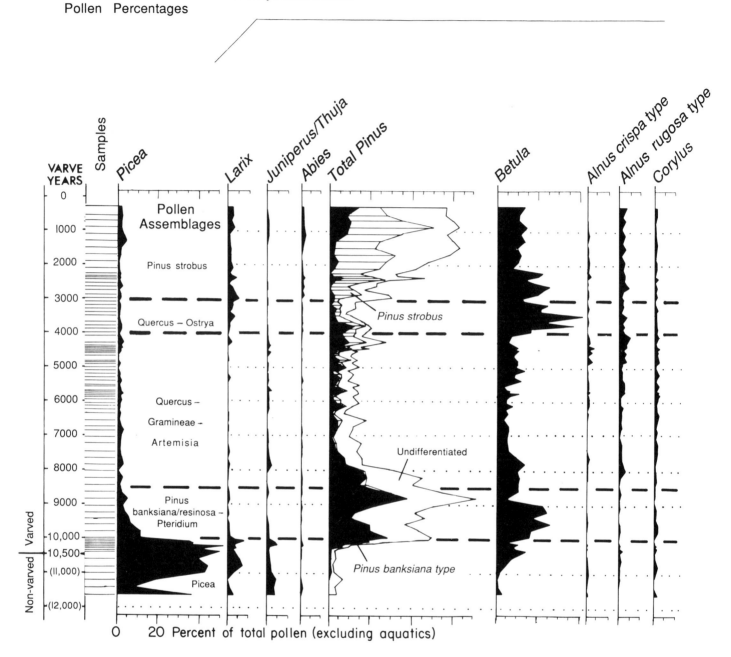

Figure 2 (on this and following three pages). Pollen percentage diagram.

consin and southern Alberta at the southern edge of the boreal forest (Fig. 6, analogue maps for 9984 and 9584 varve yr).

Quercus-Gramineae-*Artemisia* assemblage (8500–4000 varve yr)

Percentages of *Quercus*, Gramineae, and *Artemisia* increase after 8500 varve yr. These pollen types are most abundant after 8000 varve yr, when they occur with *Ambrosia,* Tubuliflorae, *Iva ciliata*–type, Chenopodiineae, and other herbaceous taxa. *Salix, Corylus, Fraxinus,* and *Alnus* are represented steadily throughout the assemblage. The pollen-accumulation rates for this period fluctuate widely between 1870 and 48,050 grains/cm^2/yr.

The assemblage suggests a gradual transition from jack pine forest to oak savanna and prairie in the middle Holocene. McAndrews (1966) suggested that *Quercus macrocarpa* was the dominant tree during the prairie period, but that *Q. ellipsoidalis*

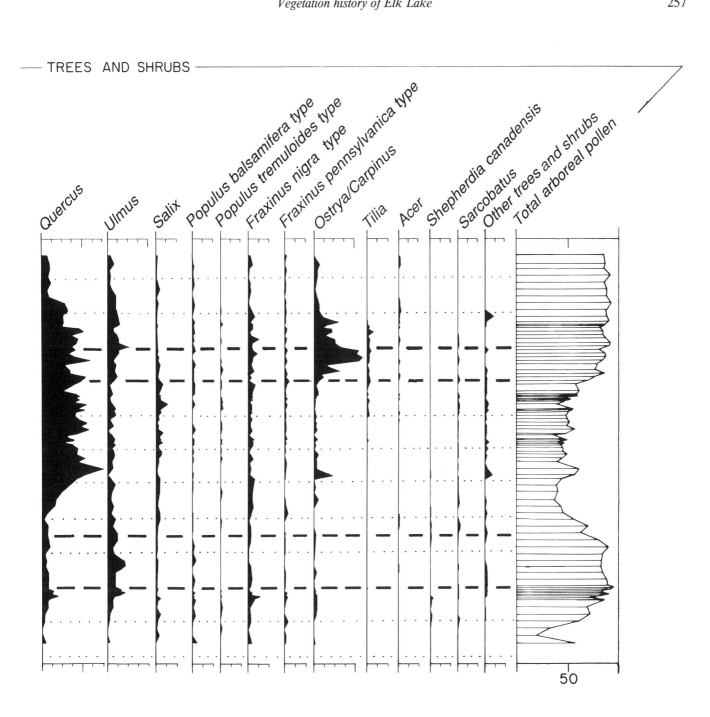

probably grew on outwash soils. *Artemisia* pollen is attributed to *A. ludoviciana, A. frigida,* and *A. glauca. Corylus, Salix, Alnus rugosa,* and *Fraxinus* may have been common on fine-textured soils.

Modern analogues for the prairie period first lie in northern Wisconsin and Minnesota and after 8247 varve yr in central and southern Minnesota (Fig. 6). This shift suggests a significant drying trend (See Bartlein and Whitlock, Chapter 18). The analogues for 7862 varve yr are found along the United States–Canada border, where modern Gramineae values are high and suggest drier conditions than before. By 7232 varve yr, the best modern analogues are from southwest of Elk Lake, where the *Artemisia* and Chenopodiineae percentages are high and conditions are even drier. The best modern analogues for 5723 varve yr come from southern and central Wisconsin and Minnesota, where *Betula* pollen is well represented today and the climate is wetter. It is noteworthy that this shift toward wetter conditions

258   C. Whitlock and Others

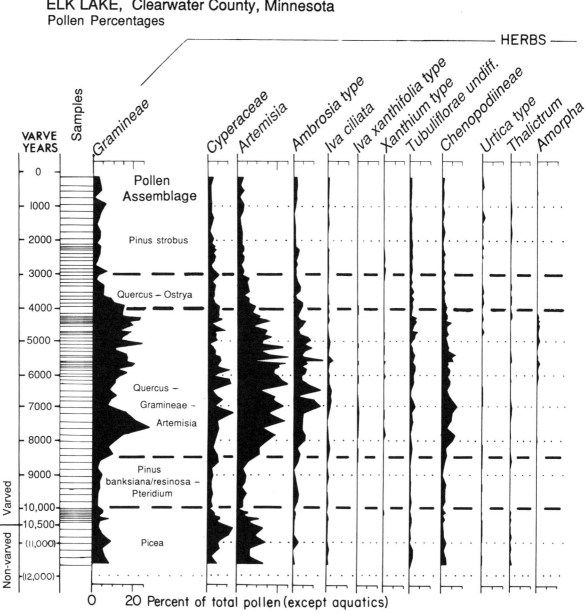

coincides with renewed alluviation of flood plains in southwestern Wisconsin, which is also attributed to increased precipitation at the end of the prairie period (Knox, 1983, 1985; McDowell, 1983). It also coincides with an abatement of eolian activity and evidence of higher precipitation in east-central Minnesota (Keen and Shane, 1992).

At Elk Lake the prairie period has been divided into three climatic phases on the basis of diatom, sedimentologic, and geochemical data: an early xeric phase between 8500 and 5400 varve yr, a somewhat wetter phase from 5400 to 4800 varve yr, and a dry phase between 4800 and 4000 varve yr (Dean and others, 1984; Dean, Chapter 10; Bradbury and Dieterich-Rurup, Chapter 15). The pollen record does not show this subdivision clearly, although subtle fluctuations in pollen percentages do occur for several taxa and considered together these may reflect climatic changes (see Bartlein and Whitlock, Chapter 18). *Artemisia* percentages, for example, increase at 8447 varve yr and decline beginning at 4439 varve yr. *Quercus* percentages increase signifi-

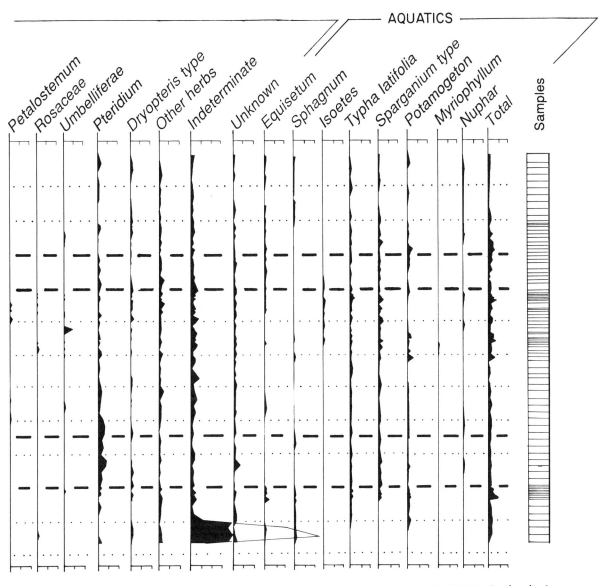

C. Whitlock, Analyst

cantly at 7232 varve yr as *Pinus* and *Betula* decrease, and they remain high until 1600 varve yr. Some herb pollen types show minor trends during the prairie period. Gramineae pollen, an indicator of relative wetness, shows two peaks centering at 7660 and 4675 varve yr. *Ambrosia* percentages are high between 8047 and 4573 varve yr. Chenopodiineae is best represented between 8047 and 5534 varve yr, and values decline gradually after that. Pollen of *Petalostemum purpureum*–type and *P. candidum*–type (combined together as *Petalostemum* in pollen diagrams) and *Amorpha*-type are present consistently after 6000 varve yr. Furthermore, pollen-accumulation rates are lowest for the period from 5400 to 4800 varve yr, which may represent a time when slopes were less vegetated than before or after.

The prairie period is generally attributed to unidirectional warming and drying in the middle Holocene, caused by the reduced frequency of arctic air masses and more common occurrence of warmer and drier Pacific air masses (Bartlein and others, 1984). However, Forester and others (1987) on the basis of os-

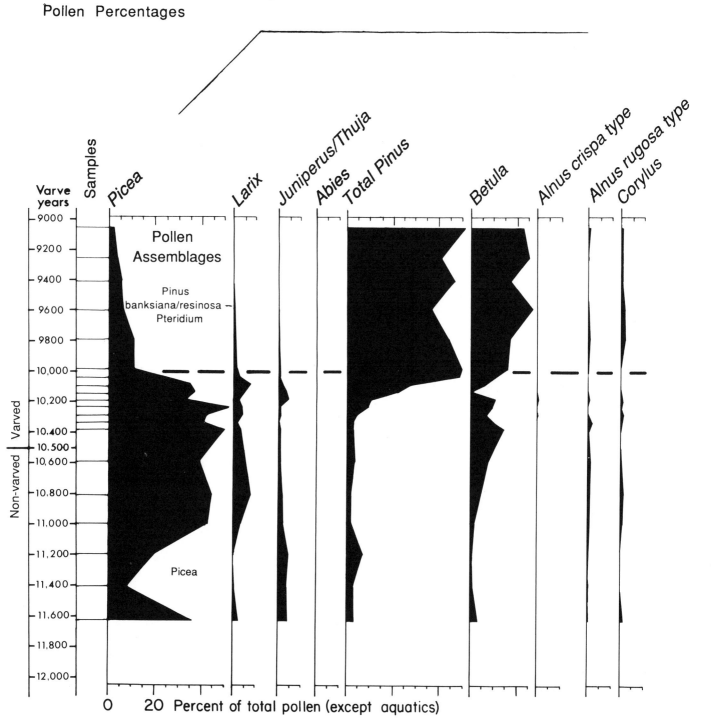

Figure 3. Expanded pollen percentage diagram showing the late glacial and early Holocene periods.

Vegetation history of Elk Lake

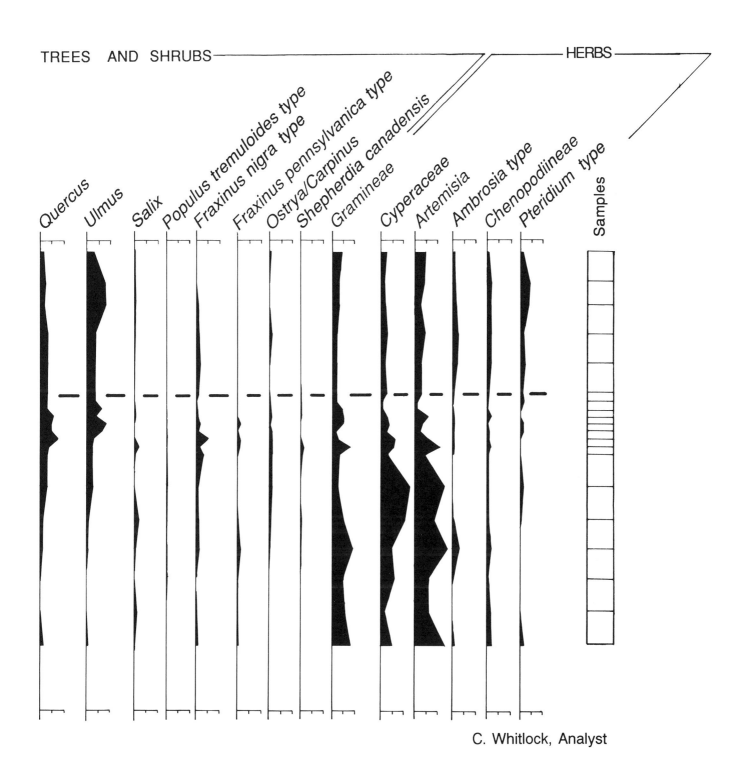

C. Whitlock, Analyst

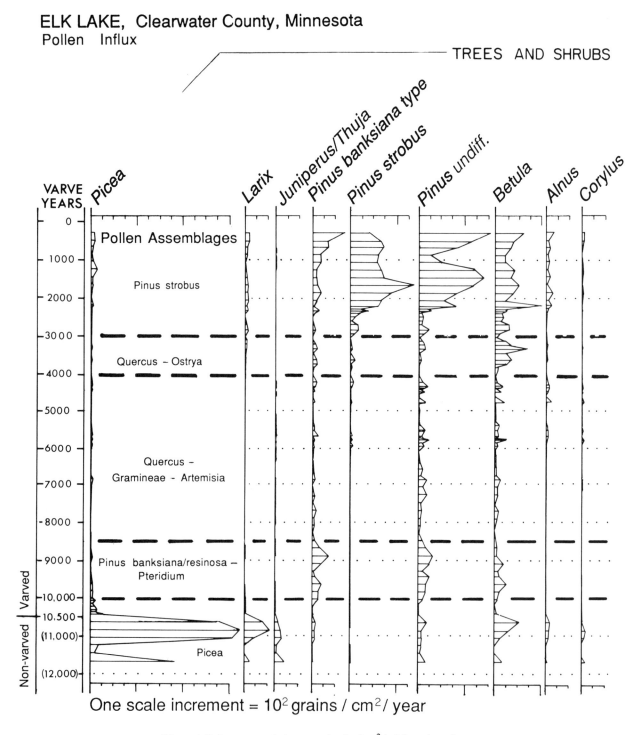

Figure 4. Pollen-accumulation rates (grains/cm²/yr) for selected taxa.

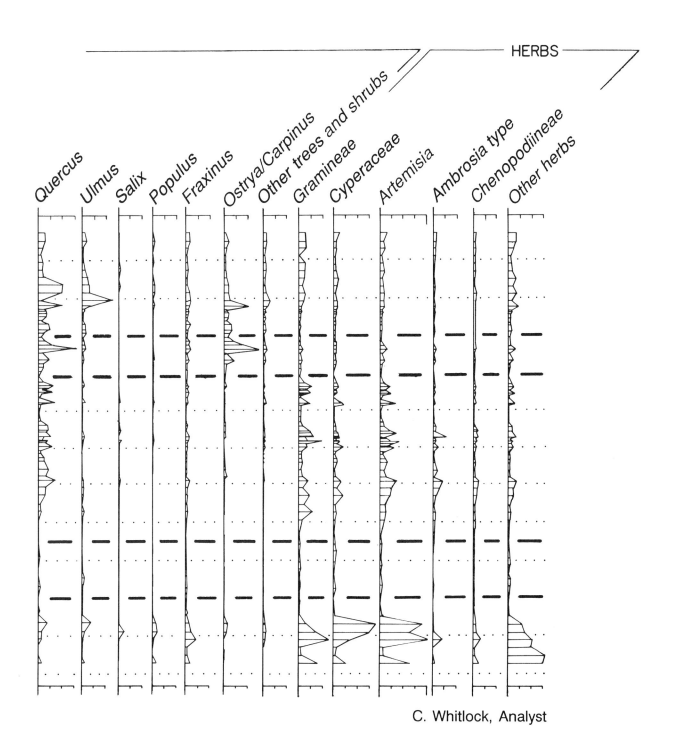

C. Whitlock, Analyst

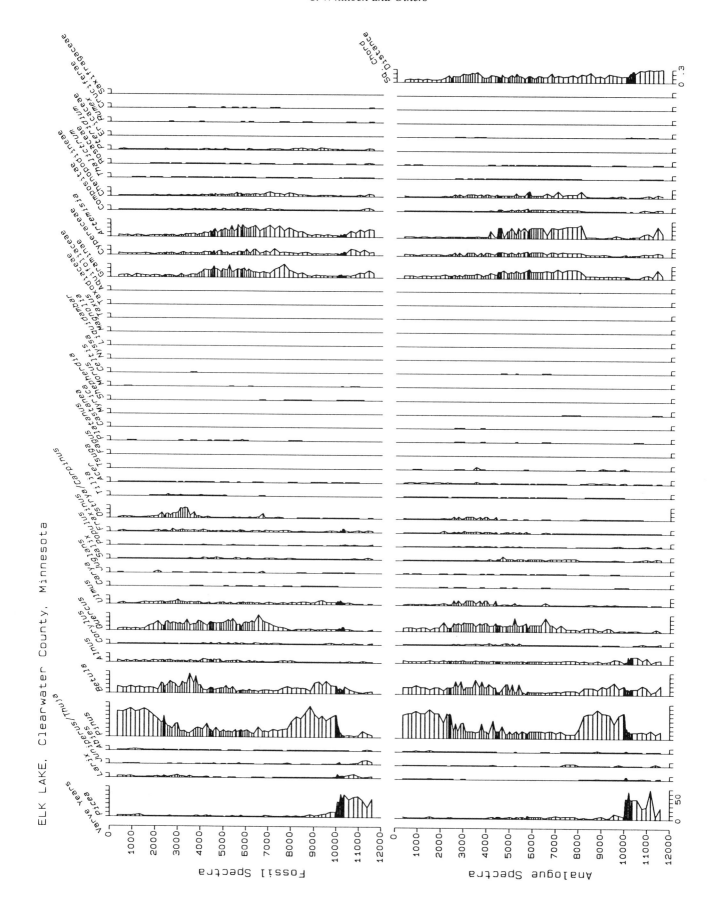

tracode data, suggested that winters and summers were cooler and summers were driest between 7800 and 6700 varve yr. From 6700 to 4000 varve yr, the ostracode record indicates that the climate was transitional leading to the modern climate. Summers were initially warmer than at present, and precipitation was highly variable. This reconstruction has some support in the pollen record, which has its best analogues for 7800 to 6600 varve yr in southern Saskatchewan (Fig. 6). The analogues lie west of the modern prairie forest border for the spectra between 7232 and 5723 varve yr and then shift eastward for spectra between 4848 and 4509 varve yr. The pattern is interrupted slightly at 4848 varve yr, when the analogues occur farther east in southern Wisconsin and imply greater precipitation.

Quercus-Ostrya assemblage (4000–3000 varve yr)

Nonarboreal percentages decline to <25% after 4000 varve yr, whereas percentages of *Quercus* and *Betula* increase. In addition, pollen of other hardwood taxa become significant in the record, namely *Ostrya-Carpinus, Ulmus, Tilia, Fraxinus,* and *Acer saccharum*. Percentage increases in *Abies, Picea, Larix,* and *Pinus strobus*–type also occur. Pollen-accumulation rates of these taxa increase, and the overall values compare with those from the modern mixed conifer-hardwood forest (M. B. Davis and others, 1973; R. B. Davis and others, 1975).

This assemblage indicates decline of prairie vegetation and oak savannah around Elk Lake and the spread of mesic deciduous forest, dominated by oak and birch. Oak, both *Q. macrocarpa* and *Q. rubra*, according to McAndrews (1966), continued to grow on the dry upland slopes and outwash gravels, but the mesic lowlands were invaded by birch, hophornbeam, elm, basswood, and maple. Tamarack, spruce, and fir probably occupied peatlands, which were developing at this time (Webb and others, 1983). The increase in *Pinus strobus*–type pollen ca. 3000 varve yr records the westward migration of white pine into the region.

The pollen assemblage at 3880 varve yr typifies the first part of the assemblage. The best modern analogues lie in the mixed conifer-hardwood forest boundary of the southern Great Lakes region (Fig. 6), which implies a wetter warmer climate than before. The *Ostrya-Carpinus* maximum at 3390 varve yr has fewer analogues, but the inferred climate is similar to or slightly warmer than before.

Pinus strobus assemblage (3000 varve yr to A.D. 1890)

During the past 3000 varve yr pine is once again the dominant pollen type at Elk Lake, but its source is *P. strobus* as well as *P. banksiana* and *P. resinosa*. Jacobson (1979) described the Holocene migration of white pine in the Great Lakes region from glacial refugia in the Appalachian highlands. White pine reached northeastern Minnesota by 7.2 ka but was unable to expand farther west during the prairie period. Its appearance at Elk Lake coincides with the introduction of cooler, moister conditions in the late Holocene. A second rise in the pollen-accumulation rates of white pine probably indicates its local prominence ca. 2320 varve yr. Pollen percentages and accumulation rates of *Pinus banksiana* and *P. resinosa* increase in the past millennium. Jack pine would have occupied the sandy substrates of drift deposits, and both pines would have been favored by frequent fires (Almendinger, 1992).

At 2990 varve yr the climate reconstruction suggests cooler and slightly drier conditions. Modern spectra from central Minnesota and northern Wisconsin match the assemblage at 2651 varve yr, suggesting the type of vegetation that was present just before pollen percentages of white pine increase. After the immigration of white pine, the best analogues center on northern Wisconsin and northern Minnesota (Fig. 6). At the white pine maximum, ca. 1319 varve yr, the analogues imply that winters became warmer than before. Thus the exclusion of *Pinus strobus* from Minnesota during the preceding prairie period could have been effected by cold winters as well as by dry conditions. In the past 4000 yr, the climate has become drier, as evidenced by the fact that the best analogues lie progressively westward, and from 1319 to 320 varve yr they occur in the Itasca region.

The varved record does not include a rise in the percentages of *Ambrosia*-type and other weedy taxa at the top of the core that is conspicuous in latest Holocene records from Minnesota. This increase marks the beginning of extensive agriculture in the region after 1890. Frozen cores taken from Elk Lake, however, register the *Ambrosia* rise (Bradbury and Dieterich-Rurup, Chapter 15; Anderson and others, Chapter 4).

REGIONAL CONTEXT OF THE ELK LAKE POLLEN STRATIGRAPHY

The vegetation history in the vicinity of Elk Lake is a reflection of larger changes occurring throughout the Midwest. Cushing (1967) examined these regional patterns along a north-south transect of pollen records in eastern Minnesota. In our study three major pollen events are compared: the *Picea* decline–*Pinus* rise, the late glacial *Ulmus* maximum, and the maximum of prairie forbs in the prairie period. The comparison is made for pollen records along three transects: south to north from western Iowa to Manitoba, south to north from eastern Iowa to northeastern Minnesota, and west to east from northeastern South Dakota to northeastern Minnesota (Figs. 7 and 8). The time scale for all the pollen profiles is in radiocarbon years, and accordingly the varved records at Elk Lake and Lake of the Clouds were converted with the calibration equations of Stuiver and others (1986).

To interpret pollen profiles in terms of regional vegetational changes required mapping at close time intervals and an assessment of the magnitude of dating errors. Variation in the timing of

◂

Figure 5. Observed and analogue pollen spectra for Elk Lake. The upper pollen diagram shows the observed fossil-pollen spectra. The bottom diagram was constructed by plotting the most similar modern pollen spectrum for each fossil spectrum; the right column shows the squared chord distance between each fossil spectrum and the closest modern spectrum.

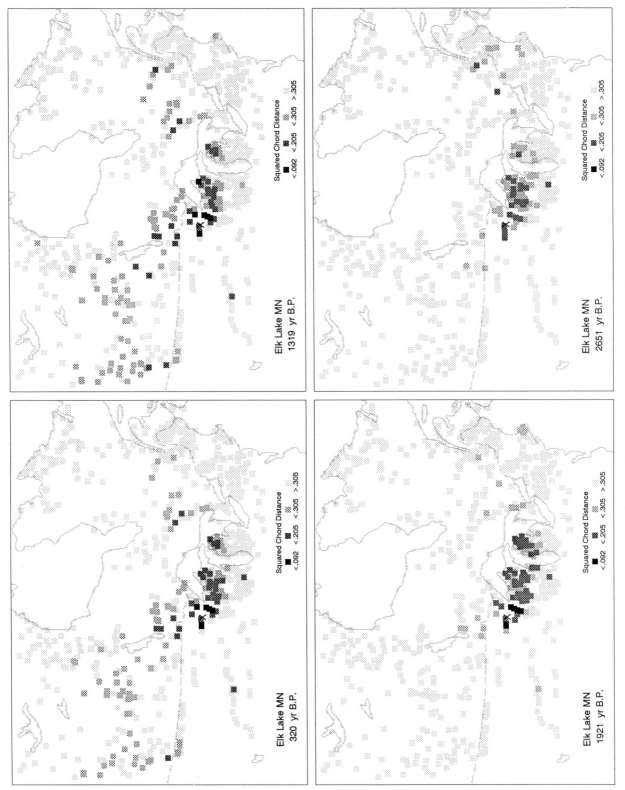

Figure 6 (on this and following four pages). Analogue maps showing the pattern of modern analogues for a particular fossil spectrum. The location of each modern pollen spectrum is shown by a shaded square, and the degree of similarity to the particular fossil spectrum indicated by the intensity of the shading. The location of Elk Lake is indicated by an X.

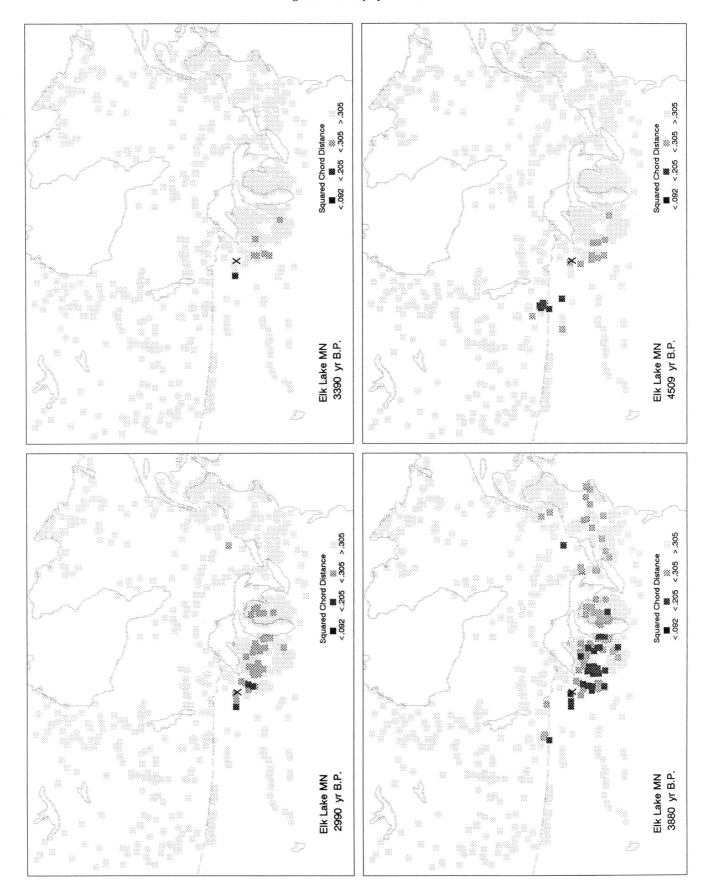

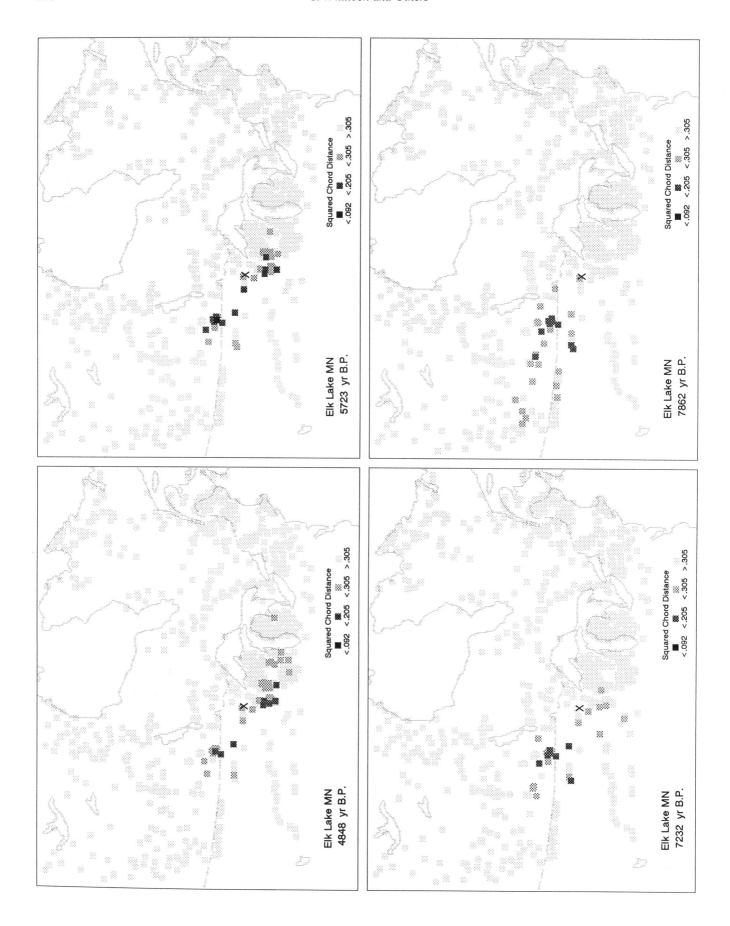

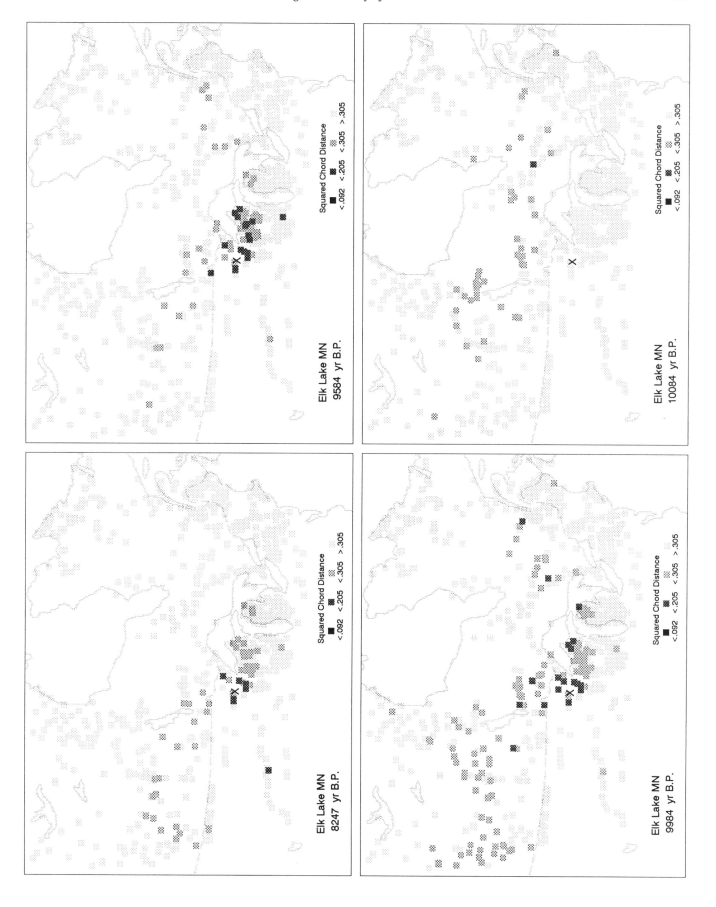

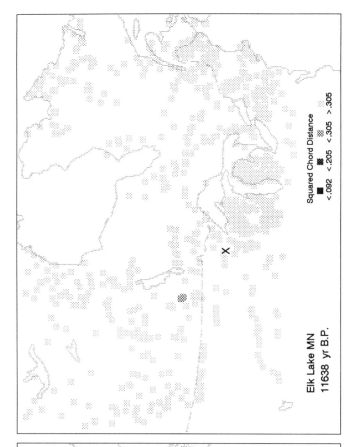

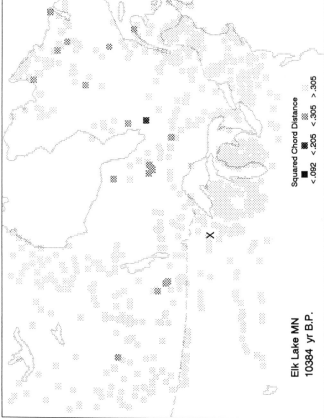

important pollen stratigraphic events may have several explanations. First, the event or episode may be a response to a time-transgressive change in climate or to a migration lag. It is difficult, perhaps impossible, to separate these two possibilities (Prentice, 1983). Second, local habitat variability creates landscape heterogeneity that may result in differences in the timing of pollen events from one site to another. Third, local geology, including the composition of the surficial deposits, may lead to variability in dating. Sites along the western transect, for example, are more likely to have sediments containing old carbon from limestone- and shale-rich tills than sites from the eastern transect; thus the western pollen records may register a synchronous event earlier than those from farther east. Finally, the type of material being dated creates variability in radiocarbon ages, and unfortunately Midwest pollen chronologies have been developed from a combination of radiocarbon dates on wood, peat, and lake sediments.

Picea fall–Pinus rise

The abrupt decline of *Picea* and subsequent spread of *Pinus* is perhaps the most striking feature in the late glacial pollen record of eastern North America. This transition is commonly perceived as a time-transgressive decline in spruce forest from south to north followed by the spread of pine (Ritchie and MacDonald, 1986). In the Midwest spruce spread from Illinois, Missouri, and northeast Kansas into the Great Lakes region after 15 ka. Spruce formed an open woodland at Wolf Creek Lake at 13 ka (Birks, 1976), but it took another 1500 yr before it reached Kylen Lake, 350 km to the northeast (Birks, 1981).

The south to north transect in eastern Minnesota shows the time transgression of the *Picea* decline (Figs. 7 and 8). *Picea* percentages decrease between 10.2 and 9 ka from south to north; *Pinus* percentages increase between 10.8 and 9 ka with no obvious spatial trend. A south to north transect of western sites shows the *Picea* decline before 10.0 ka at Lake West Okoboji, Medicine Lake, Pickeral Lake, Glenboro, and Bog D, and after 10.0 ka at Elk Lake and Lake E (Riding Mountain). That Elk Lake and Bog D differ from each other by 1200 yr points to the likelihood of a dating problem in the western sites, where the radiocarbon ages for the *Picea* decline are about 1000 yr older than the age determined by the Elk Lake varve chronology.

A west to east transect of sites from Pickerel Lake to Weber Lake shows considerable spatial variability in the timing of the *Picea* decline and in the *Pinus* rise, but the transition occurred between 10.8 and 9.2 ka.

The overall pattern is consistent with the hypothesis that spruce declined in response to environmental changes accompanying deglaciation, when the frontal position of arctic air shifted northward. The subsequent spread of pine was more variable. Superimposed on this general pattern are west-east variations related to radiocarbon dating errors.

Late glacial Ulmus maximum

The profiles for *Ulmus* place its greatest abundance in western Minnesota between 10.8 and 9 ka and in eastern Minnesota

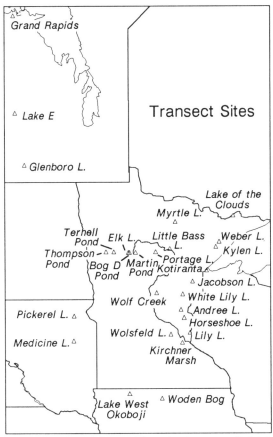

Figure 7. Location of sites used in regional transects (L is lake).

between 10 and 8.6 ka. The west to east transect has the *Ulmus* maximum at ca. 10 ka. Allowing for the uncertainty in dating, the apparent synchroneity in the timing suggests that the event was not time transgressive or related to deglaciation. It may be a region-wide environmental change related to the amplification of the seasonal cycle of radiation, which was a hemispheric-scale climatic control (Kutzbach and Guetter, 1986; Wright, 1992).

Prairie period

A comparison of prairie forb percentages along the west to east transect shows some of the patterns presented as isopoll maps in Webb and others (1983). The prairie moved eastward in the Great Lakes region from 9 to 7 ka and then retreated westward to its present position. Between 8 and 7 ka the prairie reached its easternmost limit in Minnesota and Wisconsin, although farther south in Illinois the border began to shift westward. By 3 ka the prairie border shifted westward in Minnesota; in Illinois it moved eastward again (Webb and others, 1983).

The sites in the western north to south transect, again with the exception of Elk Lake, record a rise in forbs at 9 ka that reaches maximum values before 8 ka. The eastern sites from Martin Pond to Weber Lake, and Elk Lake, show a forb rise later, ca. 8 ka. This same conclusion is reached when the south to north transects are compared for eastern and western Minnesota. The prairie period begins about 1000 radiocarbon yr earlier in the west than in the east, which might be used as evidence that the vegetational changes were time transgressive from west to east. Alternatively, if radiocarbon dates in the west were too old because of a carbonate error, they would create a false impression of time transgression. The radiocarbon conversion of the Elk Lake record places the beginning of the prairie period at the same time as in the east, which suggests that the transition was a synchronous event across Minnesota.

CONCLUSIONS

Pollen diagrams from varved sediments offer an opportunity to define important stratigraphical boundaries in terms of calendar years. At Elk Lake, the spruce decline occurred during a 250 yr interval from 10,234 to 9984 varve yr ago. Although higher percentages of hardwoods (*Betula, Quercus, Ulmus,* and *Fraxinus*) and other conifers (*Abies, Larix, Juniperus*) occurred during the spruce decline, they are not matched by a decrease in pollen-accumulation rates, and the ecological significance of their percentage rise is suspect. The increase in *Pinus* pollen began at 10,000 varve yr. The prairie period, defined by high percentages of *Quercus, Artemisia,* and Gramineae, occurred between 8500 and 4000 varve yr. Fluctuations within this interval suggest that the climate was variable. The period began with a drying trend, which culminated about 5723 varve yr, followed by increased precipitation and warmer January and continued warm July temperatures. The analogues during this interval shift from southern Saskatchewan southeastward to southern Wisconsin. During the *Quercus-Ostrya* assemblage, from 4400 to 3000 varve yr, the best analogues are from the southern Great Lakes region. The *Pinus strobus* assemblage after 3000 varve yr occurs with the onset of cooler and slightly drier conditions. The expansion of white pine across the Great Lakes region may be related to cooler winters and wetter conditions than those of the prairie period.

Our ability to reconstruct patterns of vegetational change through time depends upon the accuracy of the chronology at each of the sites considered. The radiocarbon chronology of Midwest pollen sites and the varve chronology at Elk Lake, converted to radiocarbon years, suggest that the decline of *Picea* pollen and the increase of *Pinus* pollen in the latest Pleistocene was a time-transgressive event from south to north. The likely explanation for the spatial pattern is the environmental changes that occurred with the progressive retreat of Laurentide ice northward. The late glacial increase in *Ulmus* occurred across the region synchronously and may have been a response to environmental changes that affected the entire region (such as the seasonal cycle of radiation). The prairie period at Elk Lake is dated 1000 yr later than at adjacent sites, suggesting that the dates at other sites are erroneously old. The period at Elk Lake, however, occurs concurrently with sites farther east, where the possibility of a dating error is less. Thus the onset of the prairie period may appear to be time transgressive, when in fact it was a regionally synchronous event.

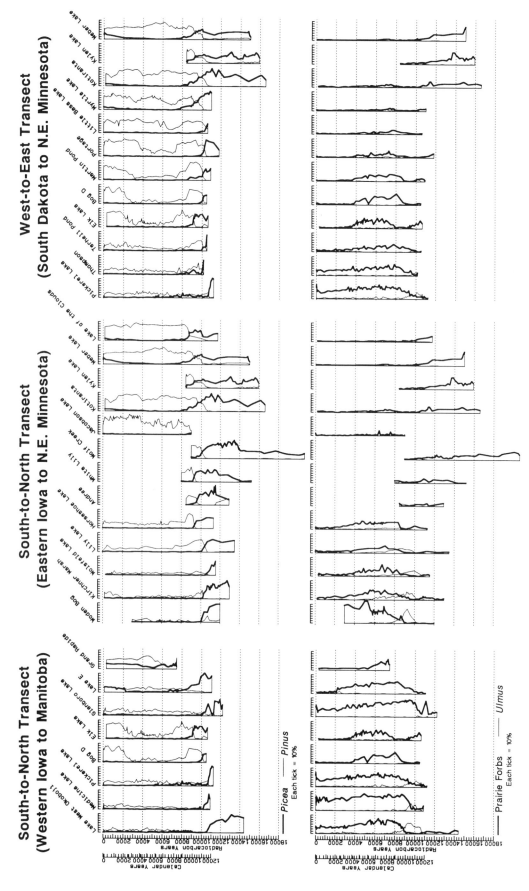

Figure 8. Three transects of pollen sites, comparing the timing of three pollen stratigraphical events in the Midwest: the *Picea* fall–*Pinus* rise, the late glacial *Ulmus* maximum, and the "prairie period." The chronology at all sites is in terms of radiocarbon years and varve years. Data are from Webb and others (1983), who used the set of fossil pollen data at Brown University.

ACKNOWLEDGMENTS

We thank J. P. Bradbury and H. E. Wright for their help during all phases of this research, and V. Markgraf and R. S. Thompson for thoughtful reviews of the manuscript. L. K. Shane and G. Hannon provided valuable assistance with laboratory and data compilation. E. J. Cushing aided in critical pollen identifications. R. Karchefsky, M. Davies, and B. B. Lipsitz prepared illustrations, and J. Crandlemire Sacco helped prepare the manuscript. The research was funded by grants from the U.S. Geological Survey and the National Science Foundation, Climate Dynamics Program (ATM-873980 and ATM-8815170).

REFERENCES CITED

Almendinger, J. C., 1992, The late Holocene history of prairie, brush-prairie, and jack pine (*Pinus banksiana*) forest on outwash plains, north-central Minnesota, USA: The Holocene, v. 2, p. 37–50.

Amundson, D. C., and Wright, H. E., Jr., 1979, Forest changes in Minnesota at the end of the Pleistocene: Ecological Monographs, v. 49, p. 1–16.

Anderson, P. M., Bartlein, P. J., Brubaker, L. B., Gajewski, K., and Ritchie, J. C., 1989, Modern analogues of late-Quaternary pollen spectra from the western interior of North America: Journal of Biogeography, v. 16, p. 573–596.

Bartlein, P. J., Webb, T., III, and Fleri, E., 1984, Holocene climatic change in the northern Midwest: Pollen derived estimates: Quaternary Research, v. 22, p. 361–374.

Birks, H.J.B., 1976, Late Wisconsin vegetational history at Wolf Creek, central Minnesota: Ecological Monographs, v. 46, p. 395–429.

——, 1981, Late Wisconsin vegetational history at Kylen Lake, northeastern Minnesota: Quaternary Research, v. 16, p. 322–355.

Bryson, R. A., 1966, Air masses, streamlines, and the boreal forest: Geographical Bulletin, v. 8, p. 228–269.

Clark, J. S., 1988a, Particle motion and the theory of charcoal analysis: Source area, transport deposition, and sampling: Quaternary Research, v. 30, p. 67–80.

——, 1988b, Stratigraphic charcoal analysis on petrographic thin sections: Application to fire history in northwestern Minnesota: Quaternary Research, v. 30, p. 81–91.

Cushing, E. J., 1967, Late Wisconsin pollen stratigraphy and the glacial sequence in Minnesota, *in* Cushing, E. J., and Wright, H. E., Jr., eds., Quaternary paleoecology: New Haven, Connecticut, Yale University Press, p. 59–88.

Davis, M. B., 1983, Holocene vegetational history of the eastern United States, *in* Wright, H. E., Jr., ed., Late Quaternary environments of the United States, Volume 2, The Holocene: Minneapolis, University of Minnesota Press, p. 166–181.

Davis, M. B., Brubaker, L. B., and Webb, T., III, 1973, Calibration of pollen influx, *in* Birks, H.J.B., and West, R. G., eds., Quaternary plant ecology: Oxford, Blackwell Scientific Publications, p. 9–25.

Davis, R. B., Bradstreet, J. E., Stuckenrath, R., Jr., and Borns, H. W., Jr., 1975, Vegetation and associated environments during the past 14,000 years near Moulton Pond, Maine: Quaternary Research, v. 5, p. 435–465.

Dean, W. E., Bradbury, J. P., Anderson, R. Y., and Barnosky, C. W., 1984, The variability of Holocene climate change: Evidence from varved lake sediments: Science, v. 226, p. 1191–1194.

Dobie, J., 1959, The Itasca story: Minneapolis, Minnesota, Ross and Haines, Inc., 202 p.

Faegri, K., and Iversen, J., 1975, Textbook of pollen analysis: Oxford, Blackwell Scientific Publications, 295 p.

Forester, R. M., Delorme, L. D., and Bradbury, J. P., 1987, Mid-Holocene climate in northern Minnesota: Quaternary Research, v. 28, p. 263–273.

Grimm, E. C., 1983, Chronology and dynamics of vegetation change in the prairie-woodland region of southern Minnesota, U.S.A.: New Phytologist, v. 93, p. 311–350.

——, 1984, Fire and other factors controlling the Big Woods vegetation of Minnesota in the mid-nineteenth century: Ecological Monographs, v. 54, p. 291–311.

Jacobson, G. L., Jr., 1979, The paleoecology of the white pine (*Pinus strobus*) in Minnesota: Journal of Ecology, v. 67, p. 697–726.

Janssen, C. R., 1967, A postglacial pollen diagram from a small typha swamp in northwestern Minnesota: Archiv für Hydrobiologie, Supplement 50 (Monographische Beitrage), p. 208–274.

Keen, K. L., and Shane, L.C.K., 1990, A continuous record of Holocene eolian activity and vegetation change at Lake Ann, east-central Minnesota: Geological Society of America Bulletin, v. 102, p. 1646–1657.

Knox, J. C., 1983, Responses of river systems to Holocene climates, *in* Wright, H. E., Jr., ed., Late Quaternary environments of the United States, Volume 2, The Holocene: Minneapolis, University of Minnesota Press, p. 26–41.

——, 1985, Responses of floods to Holocene climatic change in the upper Mississippi Valley: Quaternary Research, v. 23, p. 287–300.

Kutzbach, J. E., and Guetter, P. J., 1986, The influence of changing orbital parameters and surface boundary conditions on climate simulations for the past 18,000 years: Journal of Atmospheric Sciences, v. 43, p. 1726–1759.

Lichti-Federovich, S., and Ritchie, J. C., 1965, Contemporary pollen spectra in central Canada, II. The forest-grassland transition in Manitoba: Pollen et Spores, v. 7, p. 63–87.

——, 1968, Recent pollen assemblages from the western interior of Canada: Review of Palaeobotany and Palynology, v. 7, p. 297–344.

MacDonald, G. M., and Ritchie, J. C., 1986, Recent pollen spectra from the western interior of Canada and the interpretation of the late Quaternary vegetation development: New Phytologist, v. 103, p. 245–268.

McAndrews, J. H., 1966, Postglacial history of prairie, savanna, and forest in northwestern Minnesota: Torrey Botanical Club Memoirs, v. 22, p. 1–72.

McAndrews, J. H., Berti, A. H., and Norris, G., 1973, Key to the Quaternary pollen and spores of the Great Lakes region: Royal Ontario Museum Life Sciences Miscellaneous Publications, v. 61, 61 p.

McDowell, P. F., 1983, Evidence of stream response to Holocene climatic change in a small Wisconsin watershed: Quaternary Research, v. 19, p. 100–116.

Moore, P. D., and Webb, J. A., 1978, An illustrated guide to pollen analysis: New York, John Wiley and Sons, 133 p.

Overpeck, J. T., Prentice, I. C., and Webb, T., III, 1985, Quantitative interpretation of fossil pollen spectra: Dissimilarity coefficients and the method of modern analogs: Quaternary Research, v. 23, p. 87–108.

Prentice, I. C., 1983, Postglacial climatic change: Vegetation dynamics and the pollen record: Progress in Physical Geography, v. 3, p. 273–286.

Ritchie, J. C., and Lichti-Federovich, S., 1967, Pollen dispersal phenomena in arctic-subarctic Canada: Review of Palaeobotany and Palynology, v. 3, p. 255–266.

Ritchie, J. C., and MacDonald, G. M., 1986, The patterns of post-glacial spread of white spruce: Journal of Biogeography, v. 13, p. 527–540.

Shay, C. T., 1971, The Itasca Bison Kill Site—An ecological analysis: St. Paul, Minnesota Historical Society, 133 p.

Smith, A. J., 1991, Lacustrine ostracodes as paleohydrological indicators in Holocene lake records of the north-central United States [Ph.D. thesis]: Providence, Rhode Island, Brown University, 306 p.

Stark, D. M., 1976, Paleolimnology of Elk Lake, Itasca State Park, northwestern Minnesota: Archiv für Hydrobiologie, Supplement 50 (Monographische Beitrage), p. 208–274.

Stuiver, M., Kromer, B., Becker, B., and Ferguson, C. W., 1986, Radiocarbon age calibration back to 13,300 years BP and the ^{14}C age matching of the German oak and US bristlecone pine chronologies: Radiocarbon, v. 29, p. 1022–1030.

U.S. Department of Commerce Weather Bureau, 1960, Climatic summary of the United States—Supplement for 1931–1952: Climatography of the United States, no. 11–39.

Watts, W. A., 1979, Late Quaternary vegetation of central Appalachia and the New Jersey Coastal Plain: Ecological Monographs, v. 49, p. 427–469.

Watts, W. A., and Winter, T., 1966, Plant macrofossils from Kirchner Marsh, Minnesota—A paleoecological study: Geological Society of America Bulletin, v. 77, p. 1339–1359.

Webb, T., III, Cushing, E. J., and Wright, H. E., Jr., 1983, Holocene changes in vegetation of the Midwest, *in* Wright, H. E., Jr., ed., Late Quaternary environments of the United States, Volume 2, The Holocene: Minneapolis, University of Minnesota Press, p. 142–165.

Wright, H. E., Jr., 1968, The roles of pine and spruce in the forest history of Minnesota and adjacent areas: Ecology, v. 49, p. 937–955.

—— , 1972, Quaternary history of Minnesota, *in* Sims, P. K., and Morey, G. B., eds., Geology of Minnesota: A centennial volume: Minneapolis, Minnesota Geological Survey, p. 515–548.

—— , 1992, Patterns of Holocene climatic change in the midwestern United States: Quaternary Research, v. 38, p. 129–134.

MANUSCRIPT ACCEPTED BY THE SOCIETY JULY 27, 1992

Printed in U.S.A.

Paleoclimatic interpretation of the Elk Lake pollen record

Patrick J. Bartlein and Cathy Whitlock
Department of Geography, University of Oregon, Eugene, Oregon 97403

ABSTRACT

The pollen record from Elk Lake is interpreted in climatic terms by three different numerical approaches. The paleoclimatic record inferred for Elk Lake can be described as a sequence of climatic zones, separated by short transitional intervals: (1) the cold and dry late glacial zone (11,600–11,000 varve yr), (2) the cool and moist early Holocene zone (10,000–8500 varve yr), (3) the warm and dry middle Holocene zone (7800–4500 varve yr), and (4) the warm and moist late Holocene zone (3500 varve yr to present).

Two large-scale controls of this climatic sequence can be inferred from paleoclimatic model experiments. The first is the effect of the Laurentide ice sheet on surface winds and temperatures; this influence was strongest prior to 9 ka. This control was replaced by the amplified seasonal cycle of solar radiation between 12 and 6 ka that increased summer temperature and net radiation and decreased effective moisture. Mesoscale controls on the climate of the Itasca region possibly include a lake effect during the various stages of Lake Agassiz and subtle changes in atmospheric circulation during the prairie period (about 8000 to 4000 varve yr).

INTRODUCTION

In several respects, the paleoecological record of Elk Lake offers a unique opportunity to study climatic variations during the past 11,600 years and their influence on the lake and its watershed. First, the varve chronology provides a temporal resolution not common in late Quaternary lake records, and thus it is possible to document timing and rates of change with considerable precision. Furthermore, the lake is located in a region of steep vegetational gradients controlled by present-day climate; changes in past climate should be evident in the fossil-pollen record and the vegetational changes inferred therein. Finally, the availability of multiple paleoenvironmental indicators at Elk Lake permits a comparison of their relative sensitivities to climate change (Dean and others, 1984).

In Chapter 17, we examined in detail the fossil-pollen record of Elk Lake and emphasized the reconstruction of the past vegetation in the Itasca region. In this chapter we focus on the interpretation of the fossil-pollen data in climatic terms and use the inferred climate to discuss the paleoenvironmental history of the region and its potential causes. Three different numerical approaches—regression, analogue, and response-surface analyses—were used to reconstruct climatic variations. Each method utilizes the relationship between modern climate and pollen abundance in eastern North America to infer past climate from fossil-pollen data. The comparison of the three methods allows us to identify and minimize uncertainty in the reconstructions. The specific climatic history at Elk Lake is compared with simulations of large-scale paleoclimatic patterns (Kutzbach and Guetter, 1986; Webb and others, 1987) to elucidate the potential large-scale and mesoscale controls of regional climatic variations.

DATA

Modern pollen and climate data

The modern pollen data set consisted of 1200 modern spectra, extracted from the data set at Brown University (see Overpeck and others, 1985; Bartlein and Webb, 1985) and supplemented by additional samples from Lichti-Federovich and Ritchie (1965, 1968) and MacDonald and Ritchie (1986) for the central and western interior of Canada. In the analogue and response surface approaches, the entire data set was analyzed simultaneously with a pollen sum that included the 44 most abundant pollen types in the modern data set. Some taxa that never appear at Elk Lake were included in this sum to reduce the possibility of obtaining fortuitous analogues (see Whitlock and others, Chapter 17, for the specific list of types). In the regression approach, subsets of the entire data set were used (see below), each with its own pollen sum. The climate values assigned to each modern pollen spectrum were obtained from the closest climatological station to each sample (see Huntley and others, 1989, for discussion of various approaches for assigning modern climate values to individual pollen sampling sites).

Climate-model output

We used output from Kutzbach and Guetter's (1986) experiments with the National Center for Atmospheric Research Community Climate Model (NCAR CCM) to examine how changes in the large-scale controls of regional climates may have influenced the sequence of paleoclimatic variations (see Barnosky and others, 1987; Webb and others, 1987; COHMAP Members, 1988). We used "areal averages" of individual variables obtained by averaging together the simulations for individual model grid points (see Webb and others, 1993, for specific details on the averaging procedure and the specific gridpoints involved).

METHODS

General approach

Whether pollen data are interpreted qualitatively or numerically, their interpretation in climatic terms is a two-step procedure. First, a relationship is constructed between modern pollen and climate, and second, this relationship is applied to the fossil data. In this study, the reconstructed climate variables are mean January temperature, mean July temperature, and mean annual precipitation. These variables were selected to represent the large-scale moisture and temperature controls of plant distribution (see Bartlein and others, 1986).

Individual reconstruction approaches differ mainly in the way in which the relationship between pollen and climate is constructed. For example, the relationship can take the form of a specific prediction equation, as in the regression approach (e.g., Webb, 1980; Bartlein and others, 1984), in which the relative abundance values of particular pollen types acting as predictors are plugged-in to yield predictions of an individual climate variable, acting as the dependent variable or predictand. The relationship can also exist in the form of a tabulation (e.g., Klimanov, 1984) that illustrates the association between particular categories of pollen abundance and ranges of climate variables, or as response surfaces (e.g., Bartlein and others, 1986) that show how the abundance of a particular pollen type varies as a nonlinear function of a small number of climate variables. Most simply, the relationship can take the form of an associated list of pollen spectra and climate values (i.e., the analogue approach).

In this chapter we use the following three specific approaches.

1. In the *regression* approach, regression equations (also referred to as "transfer functions") are established that relate a specific climate variable to a set of pollen predictors (Bartlein and others, 1984). The relationships are then applied to fossil-pollen data to produce estimates of the climate variables.

2. In the *analogue* approach (see Whitlock and others, Chapter 17; Overpeck and others, 1985), a data set of paired observations of modern pollen and climate are searched for the best analogues of a particular fossil spectrum. The modern climate values associated with these analogues are used as the reconstructed values.

3. In the *response-surface* approach (see Bartlein and others, 1986; Webb and others, 1993), response surfaces that describe the relationship between the abundance of a particular pollen type and a small set of climate variables are used as the source of the analogues.

One problem common to all methods, including subjective approaches, is the occasional lack of a modern analogue for a particular fossil spectrum. This situation does not automatically preclude the production of an estimate; however, that estimate may be quite unreliable. A "no-analogue" condition is probably most serious in the regression approach, where it could produce greatly misleading extrapolations (Weisberg, 1985). It is possible to monitor the performance of the reconstruction methods as they proceed and recognize when the problem arises.

A second problem in all reconstruction methods is the potential violation of certain underlying ecological assumptions (Howe and Webb, 1983): (1) that the modern vegetation, as represented by pollen data, is in equilibrium with the modern climate on the temporal and spatial scale of the reconstruction; (2) that variations in the fossil record under consideration are ultimately attributable to climatic changes; and (3) that the vegetation has responded to climatic changes during the course of the record without significant temporal lags (Webb, 1986). Only when the response time of the vegetation is short relative to the time scale of the climatic changes can the vegetation variations be interpreted in climatic terms (Prentice, 1986; Webb, 1986). At Elk Lake, multiple paleoenvironmental indicators are available to test the equilibrium hypothesis by examining the synchroneity of the response of several independent systems to past climatic variations. Another way to minimize violations of these assumptions is to focus attention on longer temporal and larger spatial scales, concentrating, for example, on millenial-scale variations that are coherent across the upper Midwest (Prentice, 1983, 1986; Bartlein and others, 1984).

Specific approaches

The three numerical approaches differ in their relationship-building procedures, and in the measures of goodness of fit, uncertainty in the reconstructed values, and extent of extrapolation in the reconstructed values. Measures of goodness of fit, the extent to which a method can reproduce the observed modern climate from modern pollen data, help judge the overall ability of a procedure to reconstruct past climate. Measures of uncertainty in the reconstructed values and of the extent of extrapolation help to evaluate the reliability of the reconstructed values.

Regression approach. The regression or transfer function approach involves fitting a regression equation that expresses the values of a particular climate variable as a function of the abundances of several pollen types. This equation can then be applied to fossil data in order to interpret the fossil data in climatic terms. The calibration procedure of Bartlein and others (1984) and Bartlein and Webb (1985) was used because it examines explicitly the statistical assumptions that underlie the regression procedure and uses a sequence of steps to minimize any violations of those assumptions.

Bartlein and Webb (1985) defined a number of calibration regions for eastern North America, within which the relationships between a particular climatic variable and most pollen types are linear or monotonic. The reconstruction of climatic variations from the fossil data thus involves first the selection of the appropriate calibration region and second the application of its regression equation to each fossil pollen spectrum. As described by Bartlein and Webb (1985), a criterion for calibration-region selection is the Mahalanobis distance, which provides a measure of the dissimilarity between a particular pollen spectrum and the mean of the pollen spectra within a particular region. The Mahalanobis distance is also an approximate measure of the extent of extrapolation (Weisberg, 1985). In effect, this measure helps to identify the calibration region with pollen analogues for each individual fossil spectrum (and hence the appropriate equation to use), and to assess the extent of extrapolation. Mahalanobis distances are expressed here as probability values, in which values greater than 0.95 indicate little similarity between the fossil spectrum and the modern data set of a calibration region.

Uncertainty in the predicted values generated by a particular regression equation arises from two sources: (1) variations of the values of the dependent (i.e., climatic) variable about the regression line, and (2) uncertainty in the location of the line. Prediction confidence limits that reflect these two components were constructed for the climatic reconstructions in the usual way (Neter and others, 1983, section 7.7). Goodness of fit in the regression approach is measured by the R^2 value for the individual equations.

Analogue approach. Of the three reconstruction approaches used here, the analogue approach is perhaps the most intuitive. The paleoclimatic reconstructions are the modern climate values associated with the best modern pollen analogues of individual fossil spectra. A key assumption of this approach is therefore that the analogue data set (the modern pollen spectra) contains examples of all of the fossil spectra being analyzed. Analogues are identified using the squared-chord distance—a numerical measure of dissimilarity with certain desirable properties for comparing pollen spectra (Whitlock and others, Chapter 17; Anderson and others, 1989; Overpeck and others, 1985).

Rather than taking the climate values of the single best modern analogue, we used a weighted average of the climate values of the ten closest modern analogues (see also Guiot, 1987), where the weights were taken as the inverse of the squared-chord distance. This approach has the advantage of providing a degree of interpolation among modern samples that can compensate for the uneven distribution of those samples in geographic as well as climatic space.

In the analogue approach, uncertainty in the reconstructed climatic values was estimated by the standard deviation of the mean of the (weighted) climate values associated with the ten best analogues. This procedure probably results in an underestimate of the true uncertainty because it does not include information on the absolute level of the analogues. In other words, a relatively small prediction interval could result if the closest analogues all had similar associated climate values, even if those closest analogues were relatively poor. Further work is needed to improve the estimates of uncertainty in reconstructed climate values. In this chapter and in Whitlock and others (Chapter 17), the fifth percentile (0.205) of the squared-chord distance values calculated among all modern pollen spectra is adopted as a threshold indicator of a good analogue. When this value is exceeded, the climate reconstructions are considered to be less reliable. Because the reconstructions generated by the analogue approach are limited to those climate values present in the modern data set (or weighted averages of them), the analogue approach is unable to extrapolate when applied to fossil-pollen data.

The overall goodness of fit of the analogue approach is judged by estimating modern climate at individual sites from the modern pollen data (excluding the site itself in the search for analogues). The correlation between the observed and estimated modern climate values can be expressed as an R^2 value to allow comparisons among procedures.

Response-surface approach. Response surfaces describe the abundance of a particular pollen type as a function of a small number of climate variables (Bartlein and others, 1986). In a sense they represent the "inverse" of the functions relating pollen and climate data that are established using regression analysis. Response surfaces are fit to individual pollen types with a weighted-averaging technique. A window is moved within the climate space defined (in this case) by mean January and July temperatures and annual precipitation, and a distance-weighted average of the percentages of a particular pollen type for the observations that fall within the window is determined. The window is moved over a regular grid in the climate space formed by these three variables. Further discussion of the fitting procedure is given by Huntley and others (1989), Webb and others (1993), and Prentice and others (1991).

The response surfaces can be extrapolated to some extent by calculating weighted averages of the fitted values for regions just outside the portion of climate space covered by the data. Such mild extrapolation may be necessary in order to extend the surfaces into regions of climate space not represented in the modern data set (see Prentice and others, 1991, for further discussion).

The pollen spectra synthesized in this way generally do have analogues in the fossil data set for eastern North America, and so it is likely that the extrapolation of the surfaces in this manner does not build artifacts into the climate reconstructions. The result of the response surface analysis is a set of "fitted values" for the different pollen types at specific locations in the climate space defined by the three climatic variables.

The response-surface approach differs from the "straight" analogue approach in the nature of the data set that is searched for modern analogues. In the analogue approach, this data set is the modern pollen and climate data set, whereas in the response-surface approach, the data set consists of the fitted values of the individual pollen types and the associated climate values. The response-surface fitting procedure in effect smooths the modern pollen data and removes much of the spatial variability in the modern pollen data that is unrelated to climate.

The overall goodness of fit of the response-surface approach can be evaluated by estimating the observed modern climate from the modern pollen data, in a similar fashion as for the simple analogue approach. Uncertainties in the predicted values can also be estimated as in the simple analogue approach, and again are probably an underestimate of the true uncertainty. Likewise, a threshold level for the squared-chord distance can be determined by calculating the dissimilarities between all modern pollen spectra and the response-surface fitted values.

RESULTS

Specific results of the different approaches

Regression approach. The regression equations pertinent to the reconstruction of climate from the Elk Lake fossil data appear in Table 1. Individual equations were selected on the basis of Mahalanobis distance values, and the particular time range over which each equation was applied is given in Table 2. Figure 1 displays the climate reconstructions obtained by applying the different regression equations to the Elk Lake fossil data. The Mahalanobis distances for individual fossil spectra, expressed as probability values, appear in Figure 2.

Analogue approach. Figure 3 shows the reconstructed climate values obtained as the weighted average of the ten closest analogues to each fossil spectrum, and Figure 2 shows the squared-chord distances between each fossil spectrum and the closest modern analogue, along with the average of the squared-chord distances for the ten closest analogues. Table 3 contains the R^2 values for the estimation of the observed modern climate values with the modern pollen data.

Response-surface approach. The response surfaces generated in the application of this approach describe the continental-scale relationships between pollen abundance and climate. Figure 4 shows the climate reconstructions obtained with this method, and Figure 2 shows the squared-chord distances between each fossil spectrum and the single best and ten best analogues among the response-surface fitted values. A selection of response surfaces for some major pollen types appears in Figure 5. Table 3 lists the R^2 values for the modern climate values inferred from the modern pollen data.

Modern pollen-climate relationships

The response surfaces display graphically the relationship between modern pollen and climate at the continental scale, and provide some insight into the manner in which the different pollen types may have responded to past climatic variations. Each panel in Figure 5 shows the surface plotted on the July temperature–January temperature plane for a slice through climate space at a particular value of annual precipitation.

The surfaces for individual pollen types are unique, and each have distinct optima or locations in climate space where the maximum abundance of each pollen type is reached. The surfaces are generally unimodal, or bimodal, as for *Betula,* which has two abundance maxima: one in the mixed forest generated by tree birch and one in the boreal forest and tundra generated by shrub birch.

The shapes of the surfaces suggest that a rich variety of changes in pollen abundance could occur in response to a particular change in climate. For example, consider a site with annual precipitation of about 650 mm, mean July temperature of 15 °C, and mean January temperature of –20 °C. An increase in July temperature to 21 °C and January temperature to –10 °C with no change in precipitation would give that site a trajectory in climate space that would (1) move from the abundance maximum (>40%) of spruce to a location with less than 10% spruce; (2) move from a location with around 20% pine, to one with around 40% pine, and then back down to about 30% pine; (3) move from a location with very little oak pollen to one with greater than 10% oak pollen; and (4) result in very little net change in the abundance of birch or prairie forb pollen. The implication of this example for the interpretation of the fossil-pollen data from Elk Lake (and for the interpretation of fossil pollen data in general) is that whereas some climatic changes may produce quite sharp changes in the abundance of particular pollen types, other climatic changes may produce only subtle changes in pollen abundance, and still other changes result in a mixture of responses.

Comparison of reconstruction approaches

The reconstructions are broadly similar among methods. In general, both January and July temperatures increase from the beginning of the record to about 6000 varve yr, and annual precipitation is lowest between about 8000 and 4000 varve yr. The reliability of all three approaches (Fig. 3) is lowest during the *Picea* assemblage, and again during the *Quercus-Ostrya-Carpinus* assemblage, when the fossil spectra do not have good modern analogues (see Whitlock and others, Chapter 17). The general agreement among the approaches becomes evident when the individual reconstructions are superimposed (Fig. 6).

Comparison of the individual reconstructions here in light of

TABLE 1. REGRESSION EQUATIONS

Regression Equations for January Mean Temperature (°C)

Calibration Region: 45–55°N, 95–105°W

Pollen Sum = *Alnus* + *Betula* + Cyperaceae + Forbs + Gramineae + *Picea* + *Pinus* + *Quercus*

Jan. Temp = -13.116 - 0.913 *Alnus*$^{0.5}$ - 0.304 Cyperaceae$^{0.5}$ + 0.283 Forbs$^{0.5}$ + 0.379 Gramineae$^{0.5}$ - 0.266 *Picea*$^{0.25}$

$R^2 = 0.842$; $F = 114.21$; $Pr = 0.000$

Calibration Region: 40–50°N, 85–95°W

Pollen Sum = *Abies* + *Acer* + *Alnus* + *Betula* + *Carya* + *Fagus* + *Fraxinus* + Herbs + *Juglans* + *Juniperus* + *Picea* + *Pinus* + *Quercus* + *Tsuga* + *Ulmus*
where Herbs = Gramineae + *Artemisia* + Compositeae + (Chenopodiaceae-Amaranthaceae)

Jan. Temp = -11.981 + 0.279 *Acer* - 0.316 *Betula*$^{0.5}$ + 0.284 *Fagus* + 0.335 *Fraxinus* - 0.893 *Picea*$^{0.25}$ - 0.350 *Pinus*$^{0.5}$ + 2.520 *Quercus*$^{0.25}$ - 0.194 *Ulmus* + 0.751 *Tsuga*$^{0.5}$

$R^2 = 0.729$; $F = 29.29$; $Pr = 0.000$

Calibration Region: 45–55°N, 85–105°W

Pollen Sum = *Alnus* + *Betula* + Forbs + Gramineae + *Picea* + *Pinus* + *Quercus* + *Tsuga*
where Forbs = *Artemisia* + Compositeae + (Chenopodiacese-Amaranthaceae)

Jan. Temp = -12.943 - 0.217 *Alnus* + 0.044 *Betula* - 1.039 Forbs$^{0.25}$ - 0.461 Gramineae$^{0.5}$ + 2.581 *Quercus*$^{0.25}$ - 0.987 *Picea*$^{0.5}$ + 0.192 *Tsuga*

$R^2 = 0.876$; $F = 86.43$; $Pr = 0.000$

Calibration Region: 40–55°N, 75–87°W

Pollen Sum = *Abies* + *Acer* + *Alnus* + *Betula* + *Carya* + *Fagus* + *Fraxinus* + Gramineae + *Picea* + *Pinus* + *Quercus* + *Tsuga* + *Ulmus*

Jan. Temp = - 5.348 + 0.564 *Acer*$^{0.5}$ - 0.581 *Alnus*$^{0.5}$ - 0.256 *Betula*$^{0.5}$ + 1.042 *Carya*$^{0.25}$ - 1.282 *Picea*$^{0.5}$ - 1.740 *Pinus*$^{0.25}$ + 1.979 *Quercus*$^{0.25}$ + 0.776 *Tsuga*$^{0.50}$ - 0.646 *Ulmus*$^{0.5}$

$R^2 = 0.949$; $F = 236.92$; $Pr = 0.000$

Regression Equations for July Mean Temperature (°C)

Calibration Region: 45–55°N, 95–105°W

Pollen Sum = *Alnus* + *Betula* + Cyperaceae + Forbs + Gramineae + *Picea* + *Pinus* + *Quercus*

July Temp. = 21.796 - 0.272 *Alnus*$^{0.5}$ - 0.015 Forbs + 0.297 Gramineae$^{0.5}$ - 1.902 *Picea*$^{0.25}$

$R^2 = 0.701$; $F = 57.33$; $Pr = 0.000$

Calibration Region: 40–50°N, 85–95°W

Pollen Sum - *Abies* + *Acer* + *Alnus* + *Betula* + *Carya* + *Fagus* + *Fraxinus* + Herbs + *Juglans* + *Juniperus* + *Picea* + *Pinus* + *Quercus* + *Tsuga* + *Ulmus*
where Herbs = Gramineae + *Artemisia* + Compositeae + (Chenopodiaceae-Amaranthaceae)

July Temp. = 18.686 - 0.334 *Abies* - 0.218 *Betula*$^{0.5}$ - 0.097 *Fagus* + 0.106 *Fraxinus* + 0.704 Herbs$^{0.25}$ - 0.364 *Picea*$^{0.25}$ + 1.006 *Quercus*$^{0.25}$

$R^2 = 0.799$; $F = 56.79$; $Pr = 0.000$

Calibration Region: 45–55°N, 85–105°W

Pollen Sum = *Alnus* + *Betula* + Forbs + Gramineae + *Picea* + *Pinus* + *Quercus* + *Tsuga*
where Forbs = *Artemisia* + Compositeae + (Chenopodiaceae-Amaranthaceae)

July Temp. = 19.199 - 0.384 *Betula*$^{0.5}$ - 0.285 *Picea*$^{0.5}$ + 0.079 *Pinus*$^{0.5}$ + 1.307 *Quercus*$^{0.25}$ + 0.040 *Tsuga*

$R^2 = 0.786$; $F = 64.13$; $Pr = 0.000$

Calibration Region: 40–55°N, 75–87°W

Pollen Sum = *Abies* + *Acer* + *Alnus* + *Betula* + *Carya* + *Fagus* + *Fraxinus* + Gramineae + *Picea* + *Pinus* + *Quercus* + *Tsuga* + *Ulmus*

July Temp. = 20.767 - 0.253 *Alnus*$^{0.5}$ - 0.160 *Betula*$^{0.5}$ + 0.372 *Carya*$^{0.25}$ + 0.170 *Fraxinus*$^{0.5}$ - 0.294 *Picea*$^{0.5}$ - 0.811 *Pinus*$^{0.25}$ + 1.097 *Quercus*$^{0.25}$

$R^2 = 0.903$; $F = 154.92$; $Pr = 0.000$

Regression Equations for Annual Precipitation (cm)

Calibration Region: 45–55°N, 95–105°W

Pollen Sum = *Alnus* + *Betula* + Cyperaceae + Forbs + Gramineae + *Picea* + *Pinus* + *Quercus*

Ann. Precip. = 57.180 - 2.365 Forbs$^{0.5}$ -0.170 Gramineae - 3.580 *Picea*$^{0.25}$ + 4.941 *Quercus*$^{0.5}$

$R^2 = 0.578$; $F = 33.18$; $Pr = 0.000$

Calibration Region: 45–55°N, 85–105°W

Pollen Sum = *Alnus* + *Betula* + Forbs + Gramineae + *Picea* + *Pinus* + *Quercus* + *Tsuga*
where Forbs = *Artemisia* + Compositeae + (Chenopodiaceae-Amaranthaceae)

Ann. Precip. = 88.283 - 4.208 *Alnus*$^{0.5}$ + 1.993 *Betula*$^{0.5}$ - 15.340 Forbs$^{0.25}$ - 1.788 Gramineae$^{0.5}$ + 2.950 *Quercus*$^{0.5}$ + 0.684 *Tsuga*

$R^2 = 0.874$; $F = 47.29$; $Pr = 0.000$

Calibration Region: 40–55°N, 75–87°W

Pollen Sum = *Abies* + *Acer* + *Alnus* + *Betula* + *Carya* + *Fagus* + *Fraxinus* + Gramineae + *Picea* + *Pinus* + *Quercus* + *Tsuga* + *Ulmus*

Ann. Precip. = 65.552 + 1.624 *Abies* + 0.708 *Betula* + 0.589 *Carya* + 0.298 Gramineae + 0.132 *Picea* + 0.402 *Quercus* + 0.865 *Tsuga* -3.693 *Ulmus*$^{0.5}$

$R^2 = 0.680$; $F = 12.33$; $Pr = 0.000$

Calibration Region: 40–50°N, 85–105°W

Pollen Sum = *Abies* + *Acer* + *Alnus* + *Betula* + *Carya* + Forbs + *Fagus* + *Fraxinus* + Gramineae + *Juniperus* + *Picea* + *Pinus* + *Quercus* + *Tsuga* + *Ulmus*
where Forbs = *Artemisia* + Compositeae + (Chenopodiaceae-Amaranthaceae)

Ann. Precip. = 95.445 + 4.335 *Abies*$^{0.5}$ - 3.564 *Alnus*$^{0.25}$ - 2.243 *Betula*$^{0.25}$ - 2.526 *Fraxinus*$^{0.25}$ - 6.145 Forbs$^{0.5}$ + 2.07 *Juniperus*$^{0.5}$ - 1.682 *Pinus*$^{0.5}$ + 6.108 *Quercus*$^{0.25}$ + 3.967 *Tsuga*$^{0.25}$ - 4.206 *Ulmus*$^{0.25}$

$R^2 = 0.940$; $F = 94.64$; $Pr = 0.000$

*The equations include those reported by Bartlein and Webb (1985) for the reconstruction of mean July temperature, along with additional ones for mean January temperature and annual precipitation not reported, but constructed as part of the same work. Note that the equations for mean July temperature for the calibration region 40–50°N, 85–95°W and for annual precipitation for the calibration region 40–50°N, 85–105°W were also reported in Bartlein and others (1984).

TABLE 2. REGRESSION EQUATION APPLICATIONS TO THE ELK LAKE POLLEN DATA

	Calibration Region	Age Range (varve yrs)	R^2
Mean January Temperature	45–55°N, 85–105°W	320–10,084	0.876
	45–55°N, 95–105°W	10,134–11,638	0.842
Mean July Temperature	40–50°N, 85–95°W	320–6,562	0.799
	40–55°N, 85–105°W	6,746–10,084	0.786
	45–55°N, 95–105°W	10,134–11,638	0.701
Annual Precipitation	40–55°N, 85–105°W	320–3,692	0.578
	40–50°N, 85–105°W	3,794–7,662	0.940
	45–55°N, 85–105°W	7,862–11,638	0.578

the assumptions that underlie each approach allows us to make some observations on the relative merits of each approach. In general, when the assumptions that underlie each approach are not violated, and the various measures of extrapolation or unreliability do not signal that such problems exist, there is little difference between reconstructions, as is evident for the past 10,000 varve yr. (The largest difference among reconstructions during this interval is the slightly higher January temperature reconstructed by the response-surface approach in the period prior to 4000 varve yr.) In contrast, prior to 10,000 varve yr, when problems with the reconstructions are apparent (Fig. 2), the amplitude of the variations of the individual reconstructions differ markedly, while the overall patterns of the reconstructions are still generally similar.

The regression approach has the intrinsic merit of being statistically optimal when the assumptions underlying the approach are not violated. Because the prediction confidence intervals arise from a formal statistical model, whenever the uncertainties in the reconstructed values are an issue, the regression approach might be preferred. Because the regression approach is particularly prone to "hidden extrapolations" (Weisberg, 1985), the application of the approach must be carefully monitored.

The analogue approach might be considered to have the distinct advantage of being the least statistical of the three approaches, because it involves the straightforward comparison of individual spectra (which could be done subjectively). In contrast to the regression and response surface approaches, the analogue approach does not involve the construction of a generalized relationship between pollen and climate, and therefore reconstructions derived using the analogue approach may depend overmuch on the nature of the particular spectra in the analogue data set. The analogue reconstructions (Fig. 3) generally show more short-

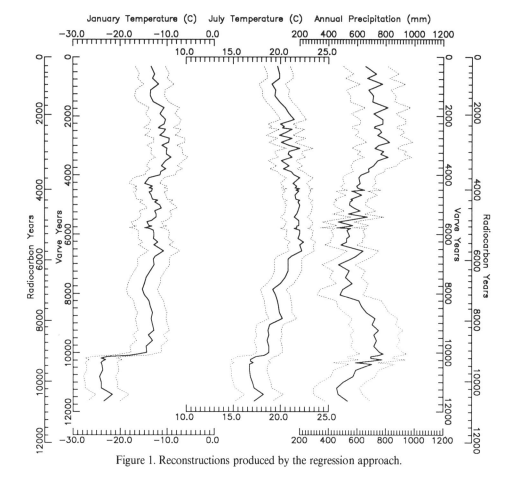

Figure 1. Reconstructions produced by the regression approach.

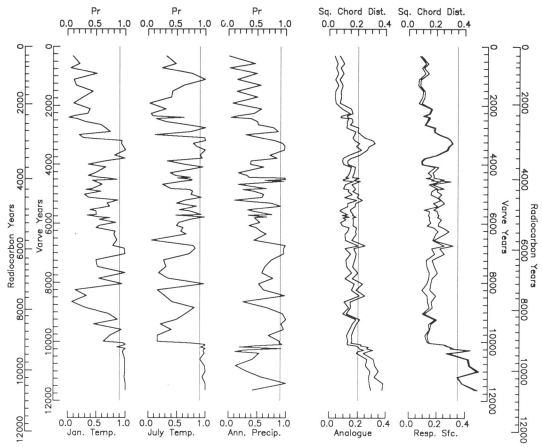

Figure 2. Extrapolation measures. The three series on the left show the series of probability levels of the Mahalanobis distances for the fossil spectra as calculated using the calibration data sets for each climate variable. The fourth series shows the squared-chord distance values for the single closest (on the left), and the average of the ten closest (on the right) modern analogues for each fossil spectrum. The fifth series shows the squared-chord distance values for the single closest (on the left), and the average of the ten closest (on the right) analogues between each fossil spectrum and the response surface fitted values.

term variability than those derived using the other two approaches, particularly during the interval 7000 to 4000 varve yr, which may be a manifestation of that dependence.

The response-surface approach is the only approach that confronts the extrapolation problem in an explicit fashion. In the regression approach, extrapolations are possible but may be unwarranted. Reconstructions produced by the analogue approach are restricted to lie within the domain of the analogue data set. By including an explicit extrapolation procedure, the response-surface approach is arguably the better approach when it is likely that climates prevailed without modern analogues.

Overall, when the assumptions that underlie the individual approaches are not violated, there is little to distinguish among the reconstructions, as for the interval at Elk Lake from 10,000 varve yr to present. Prior to 10,000 varve yr at Elk Lake there are clearly problems with all three approaches (Fig. 2), and there are greater differences among approaches. For this interval, and in consideration of the above, we are therefore inclined to have somewhat greater confidence in the response-surface and analogue reconstructions, and lesser confidence in the regression reconstructions. As noted above, however, the overall pattern of the reconstructions is still quite consistent among approaches for this interval.

General results: Elk Lake climate history

The longer term trends in the climate reconstructions divide the Elk Lake record into the following climate intervals.

Late glacial (11,600–11,000 varve yr). The bottom four samples of the Picea *assemblage are characterized by high percentages of* Picea, Gramineae, Artemisia, *and Cyperaceae pollen. These values imply a cold dry climate (about 5 °C lower in January than present, 2.5 °C lower in July, with precipitation about 200 mm less annually). In general, the individual methods produce similar results, although the response surfaces yield reconstructions of annual precipitation slightly lower than those produced by the other approaches. The climate reconstructions are probably less reliable than later (Fig. 2), because significant

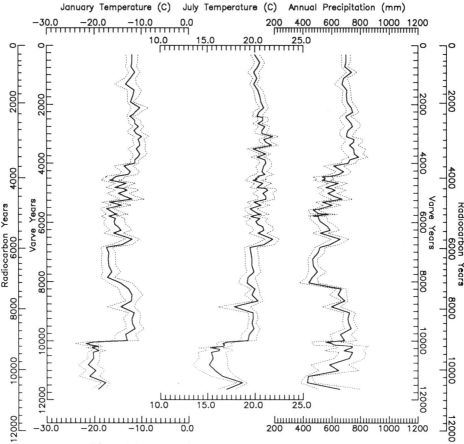

Figure 3. Reconstructions produced by the analogue approach.

TABLE 3. R² VALUES FOR ANALOGUE AND RESPONSE SURFACE/ANALOGUE RECONSTRUCTION APPROACHES

	Analogue	Response Surface/Analogue
Mean January temperature	0.928	0.730
Mean July temperature	0.928	0.856
Annual precipitation	0.870	0.751

percentages of indeterminate pollen, which are excluded from the pollen sum, occur in the unvarved portion of the core. The exclusion of the indeterminate pollen may inflate the percentages of *Picea,* thereby producing biased (systematically too cold) reconstructions. The specific values of temperature and precipitation may be in error, owing to extrapolation and bias, but the overall interpretation of conditions colder and drier than present is consistent with the reconstructions of parkland vegetation.

Late glacial–early Holocene transition (11,000–10,000 varve yr). This interval includes the transition from the *Picea* assemblage to the *Pinus banksiana-resinosa–Pteridium* assemblage. The transition begins with high percentages of *Picea, Larix,* and *Betula* percentages, followed by high percentages of *Picea* and *Betula* and various hardwood taxa, including *Fraxinus, Quercus,* and *Ulmus,* and terminates with a dramatic shift from an assemblage dominated by high percentages of *Picea* to one dominated by *Pinus.* Several marked changes in climate values are inferred through the course of this transition. Reconstructed July temperature decreases about 2.5 °C with the increases of *Picea, Larix,* and *Betula* pollen. It subsequently rises with increasing percentages of *Betula* and hardwood taxa, as does annual precipitation. The shift from the *Picea* to *Pinus* assemblage generates equally sharp increases in January and July temperature to values only a few degrees cooler than present. Within the transition, measures of extrapolation decrease from fairly high to insignificant values, and the reconstructions from the three methods are comparable after 10,000 varve yr (Fig. 2).

Early Holocene (10,000–8500 varve yr). This interval encompasses the *Pinus banksiana-resinosa–Pteridium* assemblage and is characterized by relatively cool and moist conditions (warmer than earlier and slightly cooler than present, moister than the late glacial interval, and about as moist as present). A slight warming and drying trend occurs within the interval.

Early Holocene–middle Holocene transition (8500–7800 varve yr). This interval marks the transition from forest to prairie during the early part of the *Quercus*-Gramineae-*Artemisia* as-

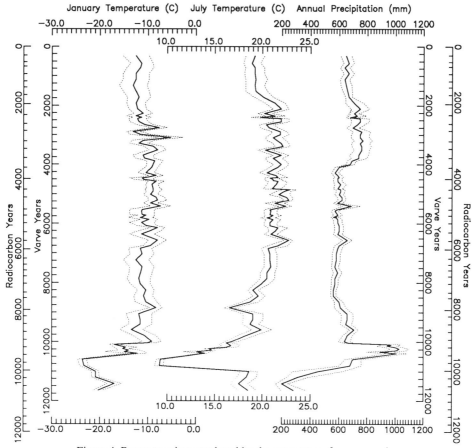

Figure 4. Reconstructions produced by the response surface approach.

semblage, and is characterized by a continuation and acceleration of the decrease in annual precipitation. January temperature also decreases slightly, while July temperature continues the increasing trend started in the previous interval.

Middle Holocene (7800–4500 varve yr). Most of the prairie period (*Quercus*-Gramineae-*Artemisia* assemblage) is included in this warm, dry interval. Annual precipitation is about 100 mm lower than present, and July temperature is about 2.0 °C warmer near the end of this interval. All three variables show relatively low amplitude short-term variations, including two episodes of warmer, wetter conditions, one between about 6750 and 6250 varve yr, and another, more pronounced one between about 5400 and 4800 varve yr. These fluctuations are inferred from subtle increases in *Quercus* percentages and concomitant decreases in herbaceous taxa. Similar episodes have been identified in other paleoecological and sedimentological records from Elk Lake (see discussion).

Middle Holocene–late Holocene transition (4500–3500 varve yr). This transition embraces the shift from the *Quercus*-Gramineae-*Artemisia* assemblage to the *Quercus–Ostrya-Carpinus* assemblage. Notable are the decreases in herbaceous taxa and *Quercus*, and the increases in *Pinus, Betula,* and *Ostrya-Carpinus*. The reconstructions of annual precipitation increase to values greater than present. January temperature increases, while July temperature remains slightly greater than present.

Late Holocene (3500 varve yr to present). The last parts of the *Quercus–Ostrya-Carpinus* assemblage and all the *Pinus strobus* assemblage are characterized by a warm, moist climate. At the beginning of this interval, July temperature was about 1.5 °C higher than present, January temperature about 2.0 °C higher, and annual precipitation was about 100 mm greater. All three climate variables gradually decrease toward their modern value, with a slight acceleration toward cooler, drier conditions after 1750 varve yr. These reconstructed trends can be inferred from the increased abundances of *Pinus* at the expense of *Betula, Quercus,* and *Ostrya-Carpinus*.

To summarize, the sequence of climatic changes at Elk Lake reconstructed from the pollen evidence shows a transition from relatively cold and dry conditions during the first 1000 yr of the record to relatively cool and moist conditions during the early Holocene. The vegetation changed from parkland to spruce forest and finally to pine forest. In the middle Holocene, prairie became established, with the introduction of warm and dry conditions. Modern vegetation and climate prevailed after 3500 varve yr. Superimposed on these broad climatic trends were low-amplitude short-term fluctuations.

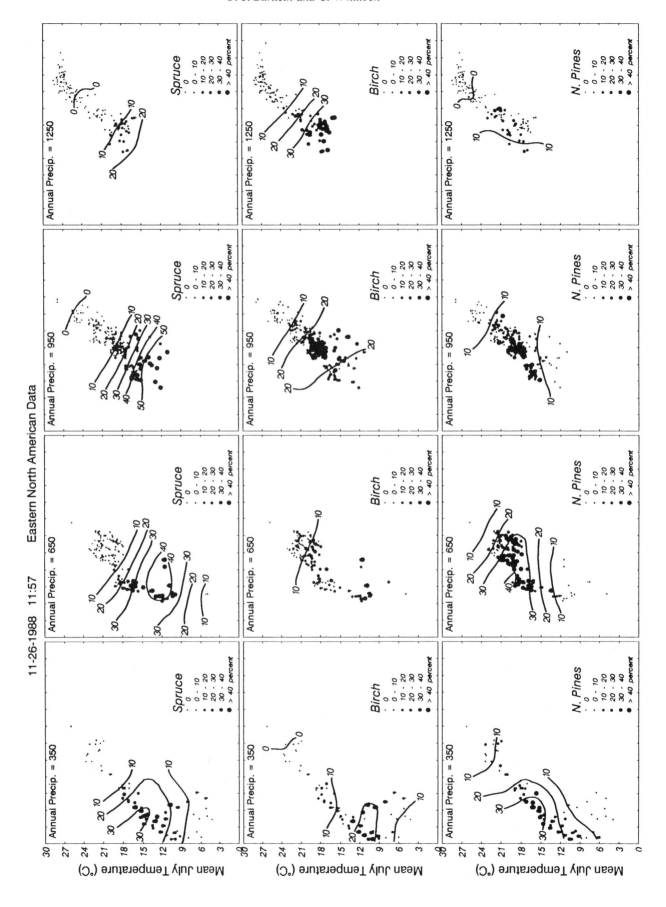

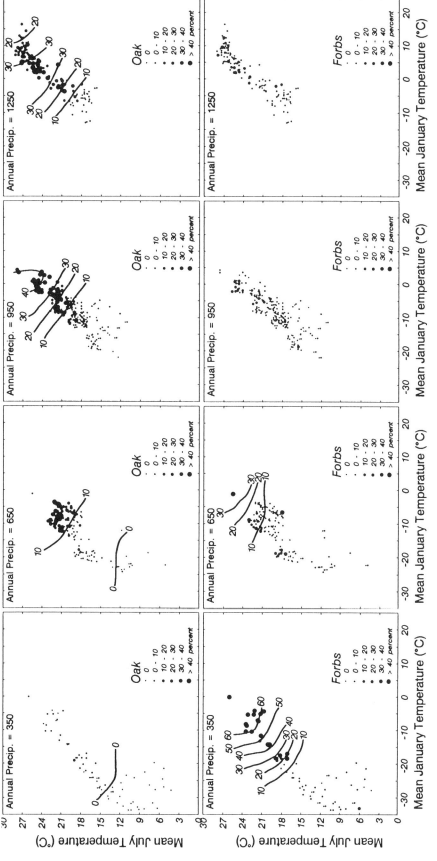

Figure 5. Response surfaces for some of the major pollen types.

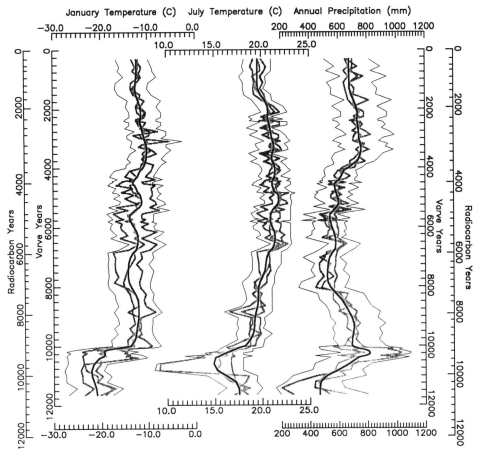

Figure 6. Elk Lake climate reconstruction summary. The three series plotted with stippled lines show the reconstructions produced by the individual approaches, the series plotted with the thin line show the envelope of the prediction intervals, and the series plotted with a thick black line represents the stacked and smoothed reconstruction of each variable (constructed by simple averaging of the individual reconstructions for each level, followed by smoothing [Velleman, 1980]). The modern observed values (1978-1984) for Itasca Park are also shown.

Comparisons with other paleoenvironmental indicators

Every paleoenvironmental indicator responds to environmental changes in a unique way. Each has its own, usually nonlinear, multivariate response function of the kind typified for pollen types (Fig. 5). The multivariate nature of most response functions thus implies that the record of an individual indicator contains the influence of several controls, and, moreover, different indicators will respond independently to a given change in controls. Where multiple paleoenvironmental indicators are available, as they are at Elk Lake, there is no reason to expect that the indicators would all produce identical paleoenvironmental reconstructions. On the contrary, different reconstructions from various indicators are a natural result of the different response functions. Where multiple and diverse paleoenvironmental indicators yield similar results, the reconstructions are probably relatively robust.

Overall, the various indicators at Elk Lake suggest similar millenial-scale variations in the environment. The most extensive agreement occurs for the dry middle Holocene climate interval (7800-4500 varve yr), and the transition into and out of it. Diatom evidence for the interval between 8200 and 4000 varve yr registers a greater influx of diatoms than before or after, and dominance by taxa that imply generally dry conditions and lower lake level (Bradbury and Dieterich-Rurup, Chapter 15). The ostracode record has a similar interpretation (Forester and others, 1987): the interval from 7800 to 4000 varve yr is interpreted as a time of generally drier conditions and increasing temperature. The influx of detrital clastic material to the lake was higher from 8200 to 4000 varve yr (Dean, Chapter 10), again an indication of drier conditions during that interval. A similar level of general agreement exists for middle Holocene temperature changes at Elk Lake, as inferred from different paleoenvironmental indicators.

What is more remarkable, however, is the consistent evidence among the different records for an interval of generally wetter conditions between 5400 and 4800 varve yr that occurred within the dry middle Holocene climate interval (Dean and others, 1984). This interval of wetter conditions has been inferred independently from a number of indicators, including pollen, diatoms, ostracodes, and the physical and chemical characteristics of the sediments. The varve chronology of Elk Lake permits the timing of its occurrence to be closely dated. That the different indicators register the climatic interval synchronously implies that little lag (i.e., <200 yr) occurs in their response to climatic change.

DISCUSSION

The sequence of climatic variations recorded in the paleoecological record of a particular location reflects the superimposition of several different controlling factors that operate at different scales. At the global or hemispheric scale the boundary conditions of the climate system include the seasonal and latitudinal distribution of solar radiation; the area, height, and reflectivity of the ice sheets; the temperature of the oceans; and the composition of the atmosphere. Together these factors act to control the nature of the large-scale circulation of the atmosphere. Atmospheric circulation governs the location and activity of storm tracks and the duration at which different air masses dominate particular regions. Solar radiation has an additional direct effect by controlling in part the net radiation at a particular location, which in turn governs both sensible heating and evapotranspiration. At the regional scale or mesoscale, local geography exerts some control on climate. The influence of the North American Great Lakes on the climate of adjacent regions is an example. As the spatial scale decreases to the size of the local region that contributed fossils to the record (e.g., the "pollen catchment"), landscape factors assume increasing importance. All three scales of control must be considered with respect to the paleoclimatic record at Elk Lake, even though it is difficult to treat each scale independently.

Large-scale controls: The ice sheet and insolation

During the past 18,000 yr, the regional climatic variations during the transition from glacial to interglacial conditions have been controlled by the size of the Laurentide ice sheet and the latitudinal and seasonal distribution of solar radiation (COHMAP Members, 1988). The ice sheet exerted control over regional climates by acting both globally (on atmospheric circulation) and locally, through proximal or periglacial influences (Wright, 1987). At the global scale, the ice sheet at its largest extent had a major influence on atmospheric circulation in the Northern Hemisphere: it split the jet stream and created a glacial anticyclone over eastern North America (COHMAP Members, 1988). The great elevation and high reflectivity of the ice sheet also contributed to the great cooling of the mid-latitude regions. The areas along the southern ice margin experienced both these global influences, as well as the more direct influence of the glacial anticyclone. When the ice sheet was large, the anticyclone was probably well developed, and adiabatic warming of the winds descending from the broad dome of the ice sheet resulted in climates that were milder and drier than might be anticipated along the southern margin of the ice sheet (Bryson and Wendland, 1967). As the ice sheet decreased in size, this effect probably became attenuated.

During the interval spanned by the Elk Lake record, considerable changes in the size of the ice sheet occurred (Dyke and Prest, 1987; Fig. 7). In terms of the oxygen isotope changes from glacial to interglacial conditions, 60% of the full-glacial isotopic level remained at 12 ka, decreasing to 25% at 9 ka, to 10% at 6 ka, and to approximately present values by 3 ka (Mix, 1987). The ice sheet probably exerted considerable influence on the Elk Lake record before 6 ka, both locally, due to its periglacial influence, and globally, through its influence on atmospheric circulation.

Insolation, the second major control at the global scale, also varied considerably during the interval spanned by the Elk Lake record (Berger, 1978). At 12 ka, summer (July) insolation at 47°N was greater than present by 7.7%, and winter insolation was less than present by 9.8%. This difference was caused by the slightly greater tilt of the Earth's axis then and the occurrence of perihelion in summer, rather than winter, as at present. By 9 ka, the insolation anomaly had increased to 8.6% greater than present in July, and 10.6% less in January. From 9 ka to present the amplification of the seasonal cycle of insolation attenuated gradually, but it was still important at 6 ka. As for the ice sheet, insolation probably had both global and local effects. On a global scale, the greater summer insolation about 10,000 yr ago probably produced heating of the centers of the continents relative to coastal areas, consequently modifying atmospheric circulation (Kutzbach, 1987). On a local scale, the greater summer insolation probably increased net radiation. The additional net radiation potentially served to increase both sensible heating (heating of the ground surface and the air above it) and latent heating (evapotranspiration).

The specific influence of these large-scale controls on the Itasca region can be evaluated with the help of paleoclimatic model simulations. Kutzbach and Guetter (1986) and Kutzbach (1987) described a series of experiments with the NCAR CCM wherein they simulated past climates at 3000 yr intervals from 18 ka to present. For eastern North America, the main responses of the simulated climate to the changing controls included (Webb and others, 1987) (1) strong control of atmospheric circulation by the large ice sheet at 18 and 15 ka in both summer and winter; (2) diminished control of atmospheric circulation by the smaller ice sheet at 12 and 9 ka, the main influence being confined to summer; and (3) earlier response to the increasing summer insolation in regions farther from the ice sheet, and delayed response in regions close to the ice. Several secondary responses of the simulated climate are also important, including a shift in the location of the largest temperature differences in July (relative to present) toward the center of the continent from 9 to 6 ka and consequent shift in the surface wind patterns, and a tendency for January temperatures in the simulations to remain lower than the model's control (modern) simulations until 6 ka, while July temperatures increased to values higher than the modern values by 9 ka.

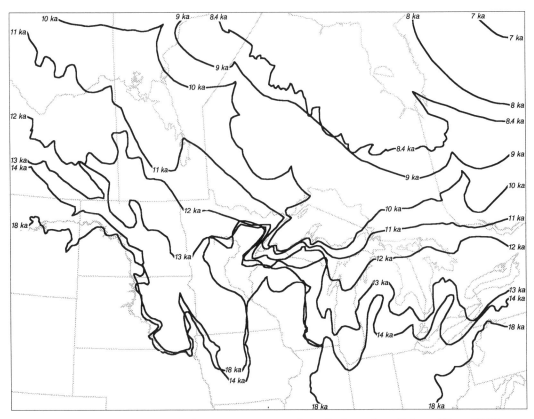

Figure 7. Retreat of the Laurentide ice sheet. Contours are in ka. Redrawn from Dyke and Prest (1987).

The detailed results of these simulations that are relevant to the Elk Lake record are obtained by examining areal averages of specific climate variables. Several limitations of the simulation results, however, must be kept in mind when interpreting such information. The resolution of the model is coarse (4.4° of latitude by 7.5° of longitude), and mesoscale features that have important influences on climate, such as the Great Lakes and the Appalachians, are not represented in the model. Similarly, the western Cordillera is crudely represented in the model, and its downstream influences on the climate of eastern North America may also be crudely represented. The simulations of precipitation and precipitation minus evaporation by the model are also poor. Simulations are described as "snapshots" of climate taken at different specific times, but in reality the boundary conditions change slowly and individual simulations therefore represent fairly broad intervals within which the controls were at generally similar levels.

The glacial anticyclone in the model simulations creates stronger (than present) northerly and northeasterly winds in the Itasca region from 18 to 12 ka (Fig. 8). During the interval from 9 to 3 ka, the shift in the heat low and its attendant circulation toward the center of the continent is evident in the simulation of stronger westerlies (than present) in July. Although both July and annual net radiation were greater than present at 12 ka (Fig. 9), July and January temperatures remained lower than present in the simulations. At 9 ka in the simulations, July temperature increased to levels greater than present and January temperature remained close to modern values in response to the seasonal distribution of insolation. The difference in the temperature response to similar radiation anomalies at 12 and 9 ka can be ascribed to the stronger influence of the ice sheet at 12 ka.

In summary, the modeling results suggest that as the northeasterly winds generated by the glacial anticyclone weakened (they were still quite strong at 12 ka), westerly winds strengthened as a result of a shift in the "heat low" generated by the stronger summer insolation (Kutzbach, 1987). The westerlies reached a maximum around 6 ka in both seasons, and decreased thereafter. The proximal or periglacial effects of the ice, still large at 12 ka, also decreased, probably becoming insignificant by 7.5 ka. As the influence of the ice was waning, the influence of the insolation anomaly increased (see also Wright, 1992).

This sequence of changes in the large-scale controls has several implications for the Itasca region. When the ice sheet was large, generally cool conditions should have prevailed, consistent with the generally cool conditions inferred from the pollen data from Elk Lake for the interval 11,600–6000 varve yr. With the replacement of the glacial anticyclonic wind regime by stronger westerlies during the interval between 9000 and 6000 varve yr,

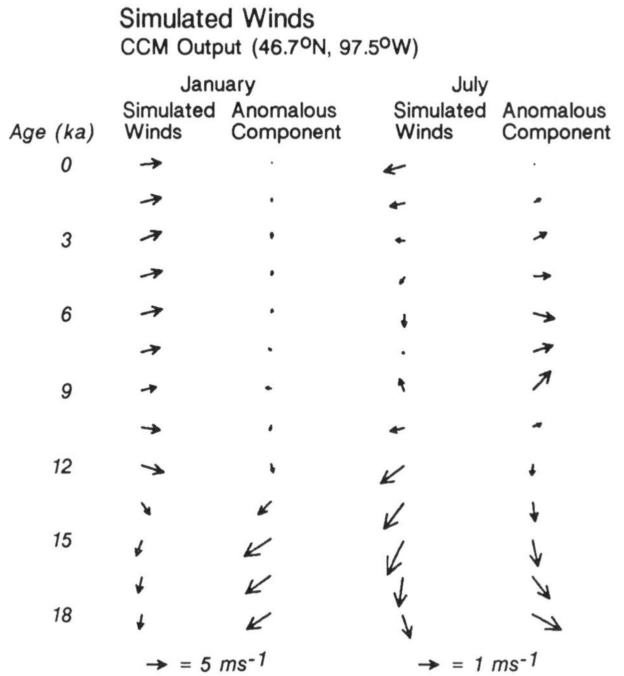

Figure 8. Winds simulated by the CCM. Simulated winds are displayed as vectors, where the length and orientation of the vectors represent the strength and direction of winds. The vectors were constructed by taking a distance-weighted average of the wind components at the model's grid points surrounding the grid point at 46.7°N, 97.5°W (the closest one to Elk Lake), and including that point. The anomalous components show the sense in which the simulated winds differ from those simulated for the model's control experiments.

precipitation should have decreased to its lowest levels during the Holocene, again consistent with the Elk Lake evidence. Finally, during the past 6000 yr, modern conditions should have developed as the seasonal distribution of insolation gradually approached present-day levels.

Mesoscale controls: Lake Agassiz

During the early part of the Elk Lake record, the presence of Lake Agassiz (Teller, 1985, 1987) probably modified the regional climate, much the same way that the North American Great Lakes generate a lake effect today (cooler summers, warmer winters, greater annual precipitation on the leeward shore). The record from Elk Lake provides an opportunity to examine the significance of Lake Agassiz on the regional climate. This discussion of the Lake Agassiz chronology follows Teller (1985, 1987; Fig. 10), but the ages cited there have been adjusted to calendar years with the relationship described by Stuiver and others (1986) (Fig. 11).

Lake Agassiz formed around 12,800 yr ago as the Laurentide ice retreated (Figs. 6 and 10). For the first 1000 yr of its existence (the Cass and Lockhart phases), the lake occupied much of the Red River Valley to the west of Elk Lake. The size of this lake suggests that the lake effect was large. As the ice retreated, outlets to the east opened and Lake Agassiz receded into Canada during the Moorhead phase (between about 11,900 and 10,750 yr ago). During this interval the lake effect in the Itasca region was probably diminished. During the Emerson phase, between 10,750 and about 10,250 yr ago, ice readvanced into the Lake Superior basin, closing the eastern outlets and allowing the lake to reoccupy its former position south of the International Boundary. The lake effect on the Itasca region thus probably increased. After 10,000 yr ago, during the Nipigon, Ojibway, and Terrell Sea phases, the ice retreat again caused the lake to recede into Canada and the local lake effect to diminish (Fig. 10).

The readvance of the lake into the Red River Valley and probable reestablishment of the lake effect on surrounding areas provide a possible explanation for variations in the reconstructed climate between 11,000 and 10,000 varve yr. Between 11,000 and 10,250 varve yr, inferred July temperatures were low. Annual precipitation increased, reaching levels greater than present near the end of the interval. Both cooler and wetter conditions implied by these reconstructions are consistent with a lake effect that would occur as the shore of Lake Agassiz moved within 75 km of Elk Lake during the Emerson phase.

This explanation has some weaknesses. First, the increased precipitation about 10,250 varve yr is inferred from elevated abundances of hardwood taxa in a few fossil spectra. Throughout the Midwest there is an episode of higher hardwood abundances between 12,000 and 10,000 varve yr ago (Amundson and Wright, 1979; Webb and others, 1983), and it is within this broader interval that the hardwood peak at Elk Lake occurs. If this minor hardwood oscillation was fostered by the lake effect during the Emerson phase, its expression should have become attenuated with increasing distance from Lake Agassiz. Unfortunately most fossil records from the Midwest are not analyzed in sufficient detail to identify this oscillation nor are they adequately dated to make site to site comparisons. Second, there may be some uncertainties in dating. The beginning of the oscillation lies in the unvarved part of the core, and is thus poorly dated. The chronology of Lake Agassiz comes from radiocarbon dating of material within the terraces and deposits of the lake, and indirectly by radiocarbon dating of glacial deposits in the Midwest. The chronology is therefore subject to the uncertainties attached to such a heterogeneous suite of dates. The Elk Lake record does not include the older Cass and Lockhart phases, when the lake effect should have been equally as pronounced. The tentative correlation between the reconstructed climate and Lake Agassiz stages is based on but a single example, the Emerson phase, and it could be fortuitous. Some of these problems could be resolved with additional detailed regional analyses of the late glacial–Holocene transition (see Webb and others, 1983).

The Elk Lake record provides an interesting perspective on the Younger Dryas climatic fluctuation. The drainage of Lake Agassiz to the east during the Moorhead phase redirected meltwater flow from the Mississippi drainage into the North Atlantic. The introduction of this meltwater has been implicated as a trig-

North-Central U.S.

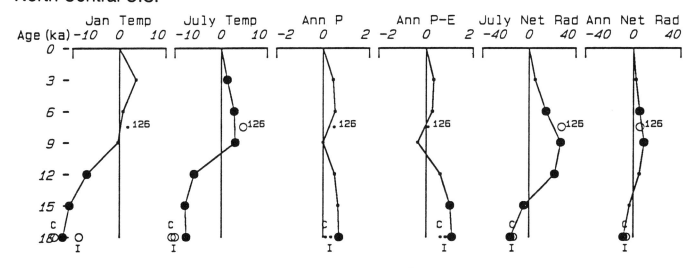
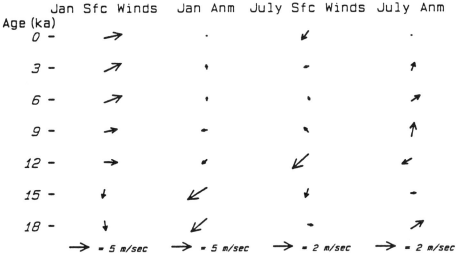

Figure 9. Averages of selected climate variables for a region centered on the upper Midwest. See Webb and others (1993) for details. The variables are all expressed as anomalies, or differences from the model's control simulations. Large dots indicate differences that are "significant," i.e., that exceed the model's natural variability.

gering mechanism for the Younger Dryas-age cooling in the North Atlantic region (see Wright, 1987, for discussion), but whether this climatic event is hemispheric or global in extent is controversial (Rind and others, 1986). At Elk Lake, reconstructions of July temperature and annual precipitation do show a pronounced change at the *end* of Younger Dryas time (10,600 to 10,000 ^{14}C yr; Cwynar and Watts, 1989), or 11,600 to 11,000 calendar years before present (applying the calibration of Stuiver and others, 1986). The change in temperature, however, is toward cooling with the increased lake effect, rather than the warming observed elsewhere. Even though the change at Elk Lake has the same ultimate cause and is contemporaneous with the end of the Younger Dryas reversal elsewhere, its proximal cause is something quite different.

Whereas Lake Agassiz may thus have been the ultimate cause of the Younger Dryas cooling, the mesoscale climatic changes in the Itasca region were in an opposite direction. Should the Elk Lake record then be cited as evidence of a global Younger Dryas-age cilmatic event? On one hand, it is contemporaneous with and related to the same ultimate cause as the Younger Dryas reversal in other parts of the world. On the other hand, the climate interval at Elk Lake was not controlled by changes in the North Atlantic region that might have driven the climate system on a global scale.

Short-term paleoclimatic variations

A striking feature of the Elk Lake record is the similarity in registration by different paleoenvironmental indicators of shorter term (i.e., century scale) climatic variations, such as those that occurred within the dry middle Holocene interval. The ultimate controls of these short-term climatic variations cannot be as easily identified as those that caused the millenial-scale variations. There are a number of potential controls that may operate on the century scale, and their individual effects may be impossible to separate in a single record. These controls include solar variability, volcanism, and internal variations in the climate system (Bartlein, 1988). In order to identify the specific controls responsible for a particular short-term climatic fluctuation (such as that at Elk Lake between 5400 and 4800 varve yr) it will be necessary to develop spatial networks of paleoecological data that have the appropriate temporal resolution (e.g., Gajewski, 1988).

SUMMARY AND CONCLUSIONS

The past climatic variations inferred from the Elk Lake pollen record disclose a sequence of changes that are consistent with what is known about the large-scale controls of climate over the past 12,000 yr, and the record inferred from other paleoenvironmental indicators. The results of three different numerical approaches used to reconstruct climate differ only in detail.

The late glacial climate interval (11,600–11,000 varve yr) was characterized by conditions much colder and drier than present, when *Picea* dominated the pollen record. This interval was followed by a transitional interval, lasting about 1000 yr, characterized by marked increases in temperature and precipitation. The early Holocene interval (10,000–8500 varve yr) is characterized by relatively cool and moist conditions when *Pinus* dominated (warmer and moister than earlier, and a little cooler than and about as moist as present). Another transition at the time of change from forest to prairie led to the middle Holocene interval (7800–4500 varve yr), when the reconstructions show low annual precipitation, and a general increase in July temperatures. The warm dry prairie interval terminated with a transition in

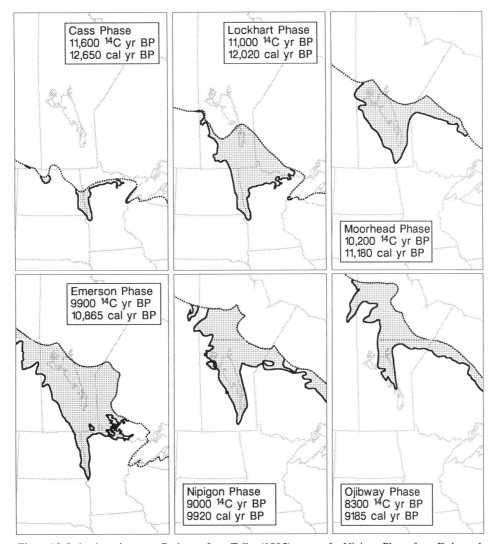

Figure 10. Lake Agassiz stages. Redrawn from Teller (1985), except for Nipigon Phase, from Dyke and Prest (1987). The dotted line shows the ice margin at different times, and the heavy line indicates proglacial lake shorelines.

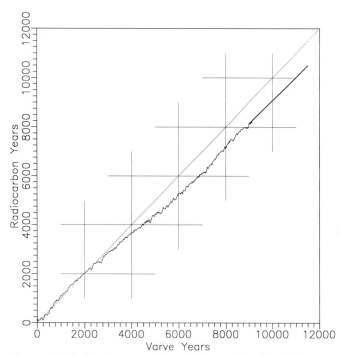

Figure 11. Radiocarbon years/varve years relationship. The curve plotted here illustrates the relationship between absolute ages expressed in calendar or varve years and those derived from radiocarbon analyses. The data were obtained from the file "atm20.14c" that accompanies the program of Stuiver and Reimer (1986). The straight line segment plotted between 9000 and 11,500 varve yr is that given by Stuiver and others (1986).

which the pollen of forest taxa increased. The late Holocene interval (3500 varve yr to present) is characterized by modern vegetation and climate.

Superimposed on these millenial-scale variations are shorter term variations in climate, such as those related to Lake Agassiz and those that occurred within the middle Holocene interval. The synchroneity among the different indicators for the latter implies that the pollen record, and hence the vegetation, is able to respond rapidly to environmental changes.

The ultimate cause of all of these climatic variations is difficult to determine precisely, because the pollen record reflects the combination of climatic changes that have taken place on a hierarchy of spatial scales, ranging from global to local. Nevertheless, the particular sequence of climatic changes at Elk Lake seems to be consistent with the effects of a gradually diminishing Laurentide ice sheet, superimposed on the amplification of the seasonal cycle of insolation as produced by orbital variations. In particular, as long as the proximal effects of the ice sheet on circulation and temperature prevailed, the response to the higher summer insolation during the early Holocene was attenuated.

ACKNOWLEDGMENTS

This research was supported by National Science Foundation grants to the University of Oregon (ATM-8713980) and to the Carnegie Museum of Natural History and the University of Oregon (ATM-8815170). We thank Bev Lipsitz for technical assistance, and Bob Thompson and Vera Markgraf for reviews. J. Platt Bradbury provided subtle, yet effective, encouragement throughout.

REFERENCES CITED

Amundson, D. C., and Wright, H. E., Jr., 1979, Forest changes in Minnesota at the end of the Pleistocene: Ecological Monographs, v. 49, p. 1–16.

Anderson, P. M., Bartlein, P. J., Brubaker, L. B., Gajewski, K., Ritchie, J. C., 1989, Modern analogues of late-Quaternary pollen spectra from the western interior of North America: Journal of Biogeography, v. 16, p. 573–596.

Barnosky, C. W., Anderson, P. M., and Bartlein, P. J., 1987, The Northwestern U.S. during deglaciation: Vegetational history and paleoclimatic interpretations, in Ruddiman, W. F., and Wright, H. E., Jr., eds., North America and adjacent oceans during the last deglaciation: Boulder, Colorado, Geological Society of America, The Geology of North America, v. K-3, p. 289–321.

Bartlein, P. J., 1988, Late-Tertiary and Quaternary paleoenvironments, in Huntley, B., and Webb, T., III, eds., Vegetation history: Dordrecht, Kluwer, p. 113–152.

Bartlein, P. J., and Webb, T., III, 1985, Mean July temperature at 6000 yr B.P. in Eastern North America: Regression equations for estimates from fossil-pollen data, in Harrington, C. R., ed., Climatic change in Canada 5: Critical periods in the Quaternary climatic history of northern North America: Syllogeus, v. 55, p. 301–342.

Bartlein, P. J., Webb, T., III, and Fleri, E. C., 1984, Holocene climatic change in the northern Midwest: Pollen-derived estimates: Quaternary Research, v. 22, p. 361–374.

Bartlein, P. J., Prentice, I. C., and Webb, T., III, 1986, Climatic response surfaces from pollen data for some eastern North American taxa: Journal of Biogeography, v. 13, p. 35–57.

Berger, A., 1978, Long-term variation of caloric insolation resulting from the Earth's orbital elements: Quaternary Research, v. 9, p. 139–167.

Bryson, R. A., and Wendland, W. M., 1967, Tentative climatic patterns for some late-glacial and postglacial episodes in central North America, in Mayer-Oakes, W. J., ed., Life, land, and water: Winnipeg, University of Manitoba Press, p. 271–298.

COHMAP Members, 1988, Climatic changes of the last 18,000 years: Observations and model simulations: Science, v. 241, p. 1043–1052.

Cwynar, L. C., and Watts, W. A., 1989, Accelerator-mass spectrometer ages for late-glacial events at Balybetagh, Ireland: Quaternary Research, v. 31, p. 377–380.

Dean, W. E., Bradbury, J. P., Anderson, R. Y., and Barnosky, C. W., 1984, The variability of Holocene climate change: Evidence from varved lake sediments: Science, v. 226, p. 1191–1194.

Dyke, A. S., and Prest, V. K., 1987, Late Wisconsinan and Holocene history of the Laurentide ice sheet: Geographie Physique et Quaternaire, v. 41, p. 237–263.

Forester, R. M., Delorme, L. D., and Bradbury, J. P., 1987, Mid-Holocene climate in northern Minnesota: Quaternary Research, v. 28, p. 263–273.

Gajewski, K., 1988, Late Holocene climatic changes in eastern North America estimated from pollen data: Quaternary Research, v. 29, p. 255–262.

Guiot, J., 1987, Late Quaternary climatic change in France estimated from multivariate pollen time series: Quaternary Research, v. 28, p. 100–118.

Howe, S., and Webb, T., III, 1983, Calibrating pollen data in climatic terms: Improving the methods: Quaternary Science Reviews, v. 2, p. 17–51.

Huntley, B., Bartlein, P. J., and Prentice, I. C., 1989, Climatic control of the distribution and abundance of beech (Fagus L.) in Europe and North America: Journal of Biogeography, v. 16, p. 551–560.

Klimanov, V. A., 1984, Paleoclimatic reconstructions based on the information statistic method, in Velichko, A. A., ed., Late Quaternary environments of the Soviet Union: Minneapolis, University Minnesota Press, p. 297–303.

Kutzbach, J. E., 1987, Model simulations of the climatic patterns during the deglaciation of North America, *in* Ruddiman, W. F., and Wright, H. E., Jr., eds., North America and adjacent oceans during the last deglaciation: Boulder, Colorado, Geological Society of America, The Geology of North America, v. K-3, p. 425–446.

Kutzbach, J. E., and Guetter, P. J., 1986, The influence of changing orbital patterns and surface boundary conditions on climate simulations for the past 18,000 years: Journal of the Atmospheric Sciences, v. 43, p. 1726–1759.

Lichti-Federovich, S., and Ritchie, J. C., 1965, Contemporary pollen spectra in central Canada, II. The forest-grassland transition in Manitoba: Pollen et Spores, v. 7, p. 63–87.

—— , 1968, Recent pollen assemblages from the western interior of Canada: Review of Paleobotany and Palynology, v. 7, p. 297–344.

Macdonald, G. M., and Ritchie, J. C., 1986, Modern pollen spectra from the western interior of Canada and the interpretation of late Quaternary vegetation development: New Phytologist, v. 103, p. 245–268.

Mix, A. C., 1987, The oxygen-isotope record of glaciation, *in* Ruddiman, W. F., and Wright, H. E., Jr., eds., North America and adjacent oceans during the last deglaciation: Boulder, Colorado, Geological Society of America, The Geology of North America, v. K-3, p. 111–135.

Neter, J., Wasserman, W., and Kutner, M. H., 1983, Applied linear regression models: Homewood, Illinois, R. D. Irwin, Inc., 547 p.

Overpeck, J. T., Prentice, I. C., and Webb, T., III, 1985, Quantitative interpretation of fossil pollen spectra: Dissimilarity coefficients and the method of modern analogs: Quaternary Research, v. 23, p. 87–108.

Prentice, I. C., 1983, Postglacial climatic change: Vegetation dynamics and the pollen record: Progress in Physical Geography, v. 7, p. 273–286.

—— , 1986, Vegetation responses to past climatic changes: Vegetatio, v. 67, p. 131–141.

Prentice, I. C., Bartlein, P. J., and Webb, T., III, 1991, Vegetation and climate change in eastern North America since the Last Glacial Maximum: Ecology, v. 72, p. 2038–2056.

Rind, D., Peteet, D., Broecker, W., McIntyre, A., and Ruddiman, W., 1986, The impact of cold North Atlantic sea surface temperatures on climate: Implications for the Younger Dryas cooling (11–10k): Climate Dynamics, v. 1, p. 3–33.

Stuiver, M., and Reimer, P. J., 1986, A computer program for radiocarbon age calibration: Radiocarbon, v. 29, p. 1022–1030.

Stuiver, M., Kromer, B., Becker, B., and Ferguson, C. W., 1986, Radiocarbon age calibration back to 13,300 years BP and the ^{14}C age matching of the German oak and US bristlecone pine chronologies: Radiocarbon, v. 29, p. 969–979.

Teller, J. T., 1985, Glacial Lake Agassiz and its influence on the Great Lakes, *in* Karrow, P. F., and Calkin, P. E., eds., Quaternary evolution of the Great Lakes: Geological Association of Canada Special Paper 30, p. 1–16.

—— , 1987, Proglacial lakes and the southern margin of the Laurentide Ice Sheet, *in* Ruddiman, W. F., and Wright, H. E., Jr., eds., North America and adjacent oceans during the last deglaciation: Boulder, Colorado, Geological Society of America, The Geology of North America, v. K-3, p. 39–69.

Velleman, P., 1980, Definition and comparison of robust nonlinear data smoothers: Journal of the American Statistical Association, v. 75, p. 609–615.

Webb, T., III, 1980, The reconstruction of climatic sequences from pollen data: Journal of Interdisciplinary History, v. 10, p. 749–772.

—— , 1986, Is the vegetation in equilibrium with climate? How to interpret late-Quaternary pollen data: Vegetatio, v. 67, p. 75–91.

—— , 1993, Vegetation, lake levels, and climate in eastern North America since 18,000 yr B.P., *in* Wright, H. E., Jr., Kutzbach, J. E., Webb, T., III, Ruddiman, W. F., Street-Perrott, F. A., and Bartlein, P. J., eds., Global climates since the Last Glacial Maximum: Minneapolis, Minnesota, University of Minnesota Press (in press).

Webb, T., III, Cushing, E. J., and Wright, H. E., Jr., 1983, Holocene changes in the vegetation of the Midwest, *in* Wright, H. E., Jr., ed., Late-Quaternary environments of the United States, Volume 2: The Holocene: Minneapolis, University of Minnesota Press, p. 142–165.

Webb, T., III, Bartlein, P. J., and Kutzbach, J. E., 1987, Post-glacial climatic and vegetational changes in eastern North America since 18 ka: Comparison of the pollen record and climate model simulations, *in* Ruddiman, W. F., and Wright, H. E., Jr., eds., North America and adjacent oceans during the last deglaciation: Boulder, Colorado, Geological Society of America, The Geology of North America, v. K-3, p. 447–462.

Weisberg, S., 1985, Applied linear regression: New York, Wiley, 324 p.

Wright, H. E., Jr., 1987, Synthesis; the land south of the ice sheets, *in* Ruddiman, W. F., and Wright, H. E., Jr., eds., North America and adjacent oceans during the last deglaciation: Boulder, Colorado, Geological Society of America, The Geology of North America, v. K-3, p. 479–488.

—— , Jr., 1992, Patterns of Holocene climatic change in the Midwestern United States: Quaternary Research, v. 38, p. 129–134.

MANUSCRIPT ACCEPTED BY THE SOCIETY JULY 27, 1992

Fire, climate change, and forest processes during the past 2000 years

James S. Clark*
Limnological Research Center and Department of Ecology, Evolution, and Behavior, University of Minnesota, Minneapolis, Minnesota 55455

ABSTRACT

Lake sediment records of vegetation, climate, and fire history indicate dynamic responses to climate changes of the past 2000 years. Studies of nitrogen (N) mineralization and forest structure within catchments of the same lakes suggest that responses may represent complex direct and indirect effects of climate on vegetation. Existing patterns in plant composition, mineralization rates, and topography indicate that spatial variability is influenced by microclimate and drainage. Xeric southwest aspects have low mineralization rates, high fire frequency, and support species that tightly cycle N. During the past 2000 years, vegetation, fire, and sediment indicators of drought have changed several times. High fire frequencies during more xeric intervals before A.D. 1200 prevailed during periods dominated by fire-tolerant hardwoods and then white pine. Decade and century scale climate fluctuations since A.D. 1200 resulted in altered fire frequencies may in turn explain the recent expansion of red pine in the region. Taken together with patterns of N cycling in the existing catchments, these changes in climate, fire, and vegetation may have had important consequences for terrestrial nutrient budgets. Such terrestrial effects may not have influenced nutrient loading to lakes.

INTRODUCTION

Few forested landscapes have undergone the degree of spatial and temporal heterogeneity over the past two millennia observed at Itasca State Park. The patchy character of existing stands attests to centuries of climate change, expansion and demise of forest-tree populations at local and regional scales, and fires that burned in an irregular fashion. Here, within 80 km of tallgrass prairie, are boreal spruce-fir, pine, northern hardwoods, and mixed oak forests. Interrupted groves of red and white pine emerge above sugar maple–basswood or birch-aspen canopies, dense hazel thickets, or thick balsam fir–white spruce stands. This patchwork includes some of the more important forest types that occur in eastern North America, and it has long attracted the attention of plant ecologists, including students of Cowles (Lee, 1924), Clements (Bergman and Stallard, 1916), and Cooper (Grant, 1934). Understanding the causes for these vegetation patterns has been a recurring research theme.

Paleoecologists have likewise found the region of great interest (McAndrews, 1966; Janssen, 1967b; Jacobson, 1979). McAndrews's (1966) classic study of space-time patterns in vegetation was fascinating, at least in part because of the magnitude and frequency of forest transitions revealed by a transect of pollen cores extending from prairie eastward across Itasca. His results suggested times when vegetation types similar to many of those now on the landscape may have dominated the local scene. For example, spruce was widespread in the region early in the Holocene, and mixed oak forests dominated from 4500 to 3000 yr B.P. (Whitlock and others, Chapter 17). The pine assemblages, as they exist today, are a recent phenomenon, whereas northern hardwoods, which include *Acer saccharum* and *Tilia americana*, may have been restricted to enclaves on mesic sites since their reappearance or modest expansion in the region after the middle Holocene. The composition and structure of these prehistoric forests certainly differed in important ways from the suggested

---

*Present address: Department of Botany, Duke University, Durham, North Carolina 27706.

Clark, J. S., 1993, Fire, climate change, and forest processes during the past 2000 years, *in* Bradbury, J. P., and Dean, W. E., eds., Elk Lake, Minnesota: Evidence for Rapid Climate Change in the North-Central United States: Boulder, Colorado, Geological Society of America Special Paper 276.

modern analogues. However, today these different forest types show some consistent patterns in the cycling of nutrients and the occurrence of fire that suggest the usefulness of analogy as an avenue for interpreting past environments in light of existing patterns.

One of the first steps for understanding past vegetation patterns involves the need to explain the occurrence of these forest types on existing landscapes. Despite the broad, subcontinental patterns in water balance (Transeau, 1905; Thornwaite, 1948; Thornthwaite and Mathew, 1957; Willmott and others, 1985), seasonality (Woodward, 1987), and fire (Chandler and others, 1983) that correlate with transitions at the same scale in forest pattern, there has been no generally accepted explanation for the heterogeneity in forest types within Itasca Park. One might suppose, for example, that some of these large-scale patterns are reproduced at a finer scale across the modern park landscape. Yet the case for spatial pattern within the park includes many explanations that span the range of biotic and abiotic interactions recognized by ecologists in forests, including local climatic variability (Buell and Gordon, 1945), edaphic heterogeneity (Grant, 1934; Kell, 1938; Buell and Martin, 1961; Hanson and others, 1974), herbivory (Ross and others, 1970; Peet, 1984), fire (Spurr, 1954; Frissell, 1973), and treefall (Webb, 1988). Most of these investigations of forest pattern involved analysis of a single variable, and consequently the interactive effects of these factors are poorly understood. Species composition, for example, may depend on the availability of nitrogen (N), which depends in turn on local water balance, litter chemistry, and possibly time since the last disturbance. Some potentially important factors have also been overlooked. For example, few efforts have been undertaken to understand nutrient cycling within the park, and no systematic analyses have been carried out concerning effects of light availability on forest composition.

Because of the large number of factors that may be important for understanding existing stands and explaining vegetation change, I initiated a multidisciplinary investigation to determine how forest processes vary across the existing landscape, how patterns have changed through time, and how those changes in vegetation might relate to contemporaneous shifts in the physical environment. In this chapter I consider results of this study that deal specifically with processes most likely to influence lakes and thus that are relevant to the question of paleolimnology of Elk Lake. I focus on the effects of fire and topographic influences on nutrient cycling and species composition, because it is possible that past changes in species composition or fire regime could have important consequences for lakes.

Because I am interested in long-term dynamics, I focus here on forest assemblages that occur in old-growth forests. The forests of Itasca have been the subject of many investigations, notably those of Janssen (1967a), Buell and Gordon (1945), Peet (1984), Webb (1988), and Hansen and others (1974), that will not be reviewed here. Logging throughout much of the park in the late nineteenth and early twentieth centuries is responsible for some of the modern diversity of plant assemblages. Thus, many of the existing patterns are transient or successional, and they are therefore difficult to relate to local edaphic variability and microclimate. For example, the clear relations between species composition and topography in uncut forests discussed below have not previously been described in the park. These spatial patterns that develop with time are probably obscured in the disturbed forests, where most analyses of vegetation have been conducted.

Usefulness of analogies to address processes operative in existing stands is limited, because fire has been suppressed since 1910. Thus, it is not possible to establish precisely how forest processes during and following fire changed through the past. Rather, the data provide insights into how processes might have changed, and they suggest hypotheses regarding nutrient supply to lakes that can be explored with paleolimnological data.

METHODS

Interpretations presented here draw on results of a forest-dynamics study within 1 km^2 of mostly old-growth forests not on map (Fig. 1). The data include stratigraphic analyses of varved sediments from the lakes Deming, Arco, and Buddy (Fig. 1), and analyses of compositional patterns and ecosystem processes in the surrounding forests. Stratigraphic analyses encompass the past 2000 varve yr, but they focus in more detail on the past 350 yr to permit comparison of sediment results with stand reconstructions that correspond to the ages of living trees and climate records that begin in A.D. 1840 (Clark, 1989a). Sediments were obtained using a freeze corer (Swain, 1973). Past forest composition was reconstructed in part by pollen analysis. Sediments were dated using varve counts. Accuracy of dating methods was confirmed by comparisons with cultural horizons in pollen data (Clark, 1988a). Fossil pollen was analyzed in contiguous 1 cm samples, dated A.D. 1640 to 1985 from each of the lakes and at larger intervals back to A.D. 100 at Deming Lake (Clark, 1988b).

Fire history was reconstructed from fire scars on *Pinus resinosa* trees and from fossil charcoal in varved sediment (Clark, 1990a). *Pinus resinosa* trees up to 400 yr old are scarred by surface fires that can be dated from tree rings. 150 fire scars on 21 trees were dated to provide an independent record of fire through the past and to determine spatial patterns of burns (Clark, 1989a). Fire scars were dated by ring counts.

Charcoal analysis was completed on petrographic thin sections (Clark, 1988a). A continuous record is available from A.D. 1240 to 1985 in the Deming Lake core. Only short segments spanning 40–80 yr have been completed from older sections of that core. Sediments were embedded in epoxy resin, sectioned, and mounted on glass slides. Charcoal fragments are clearly visible on these preparations, and year-by-year counts of charcoal yield annual charcoal accumulation rates C, in units of cm^2 charcoal/cm^2 sediment/yr. To permit evaluation of changes in fire regimes through time and among lake catchments, I identified an accumulation rate value $C' = 1$ cm$^2 \cdot$ cm$^{-2} \cdot$ yr^{-1} representing the charcoal accumulation rate that was observed in years when fire scars occurred within a lake's catchment (Clark, 1990a). In years

Figure 1. Location of 1 km² study area in Itasca State Park and study plots within this area. Plot numbers match those in Figure 3. Black areas are Lakes Budd, Arco, and Deming.

when a fire did not occur within a lake's catchment, accumulation rates were usually below this value. The average time interval between charcoal peaks $> C'$ (i.e., >1 cm^2 charcoal/cm^2 of lake bottom/yr) is used as an index to compare how frequency of fire may have changed through the past. This index contains error that results when fires occur that do not result in charcoal peaks $> C'$ and when charcoal values $> C'$ are estimated when no fire occurs within a catchment. Comparisons with fire-scar data indicate that the average time interval between the peaks can be underestimates of the fire interval on any given location on the landscape, because a charcoal peak can occur when only a portion of a catchment burns. For example, a charcoal peak could represent a fire that burned half of the lake catchment, and a second peak could represent a fire burning the other half. In this case, the time interval between charcoal peaks underestimates the average interval between fires for any given piece of ground. Where different parts of catchments burn at different times, each with some average interval, the charcoal record in the sediments describes the shorter intervals represented by this composite process (Clark, 1990a).

Species composition, standing crop, and nitrogen (N) mineralization were analyzed in each of the different kinds of stands within the study area (Clark, 1990c). Above-ground biomass of each plant species on each of 18 study plots was determined by harvesting all shrubs and herbs on four 1 m^2 subplots and by allometric equations applied to diameter measurements on all trees within 400 m^2 plots (Ohmann and Grigal, 1985a, 1985b). N mineralization was estimated from in situ incubations (Pastor and others, 1984; Vitousek and Matson, 1985; Clark, 1990c) that determine the accumulation of nitrate (NO$_3^-$) and ammonium (NH$_4^+$) in loosely capped pipes containing the upper 10 cm of surface litter and soil. Plant uptake and solution transport do not occur in the pipes, because roots are severed, and there is no perculation of water. These data are used to estimate rates of N mineralization (Clark, 1990c).

RECENT CLIMATE CHANGE

Recent climate changes in the region are interpreted from several sources. Gajewski (1988) predicted relatively moist conditions 1400–1600 yr B.P. at three sites in western Wisconsin on the basis of fossil pollen evidence (his Fig. 4; curves for Dark, Ruby, and Little Pine lakes). Temperature predictions from the same data set (Gajewski, 1988, Fig. 3) and other sources (Leonard, 1986) suggest that cooler conditions commenced about A.D. 1600, following a period of higher temperatures from A.D. 1400 to 1600. Grimm's (1983) pollen and stratigraphic evidence show expansion of Bigwoods vegetation at 1600, suggesting a more positive water balance. Charcoal deposition rate in the three Wisconsin lakes mentioned above also decreased at this time (Gajewski and others, 1985, Fig. 5). Climatic changes of that time may also explain increased varve thickness at Deming Lake at A.D. 1600 (Clark, 1988b).

Loss on ignition (LOI) for the three lakes studied here suggests that the generally cool and moist times since A.D. 1600 were interrupted by warm and dry interludes from A.D. 1770 to 1810 and 1810 to 1860 (Clark, 1990a). Increased accumulation of clastic material in the sediments of all of the lakes at these times suggests erosion of exposed littoral silts as a result of low lake levels. Inorganic accumulation rates, determined from LOI and varve counts, increased in cores from each of the three lakes from 1770 to 1810 (Clark, 1990a). Petrographic thin sections of these sediments show silt-size clasts (presumably quartz) that probably were eroded from exposed littoral zones at times of low lake levels. Similar sedimentology characterizes sediments deposited during the drought years of the 1930s, when lake levels were about 1 m lower than at present (Clark, 1989a).

EXISTING FORESTS

Forest vegetation

Within the old-growth stands there exists a decided topographic effect on forest composition. South and west aspects receive the largest amount of solar radiation late in the day when moisture levels are lowest, they are exposed to the prevailing westerlies, and so appear to be most xeric (Stoeckler and Curtis, 1960). Here *Corylus* thickets are most abundant, as are *Betula papyrifera, Quercus macrocarpa,* and, to a lesser degree, *Pinus resinosa* (Fig. 2). *Aster macrophyllus,* graminoids, and *Pteridium aquilinum* are common in the herb layer. At the other extreme are the mesic northeast aspects that support *Populus grandidentata, P. tremuloides, Acer saccharum, Tilia americana,* and *Quercus borealis. Acer spicatum* and *Ostrya virginiana* are important understory species on northeast aspects. *Picea glauca* and *Pinus banksiana* are found mostly on coarse-textured soils, *P. banksiana* being recruited almost exclusively following fire (Heinselman, 1973). *Abies balsamea* is most abundant on steep north aspects or other areas with coarse soils. Occurrence of *Pinus strobus* is not explained by site conditions in studies conducted to date, although it is less abundant on coarse sandy soils (e.g., Kell, 1938). Neither *P. strobus* nor *P. resinosa* have a topographic dependency, and so they are not plotted in Figure 2.

Fire

Frissell (1971, 1973) used fire scars from throughout the park to show that forests burned with an average interval of 20 to 25 yr during the two centuries preceding the beginning of fire suppression in A.D. 1910. A more intensive analysis of fire scars on trees in my study area showed that the spatial pattern of burns

Figure 2. Topographic effects on the standing crop of some of the important canopy and shrub species in the study area. Slope units are height/distance (m/m). Standing crop is ln (1 + t ha^{-1}). There are moisture and N availability gradients from mesic, high-N sites on northeast aspects to xeric, low-N sites on southwest aspects.

Fire, climate change, and forest processes

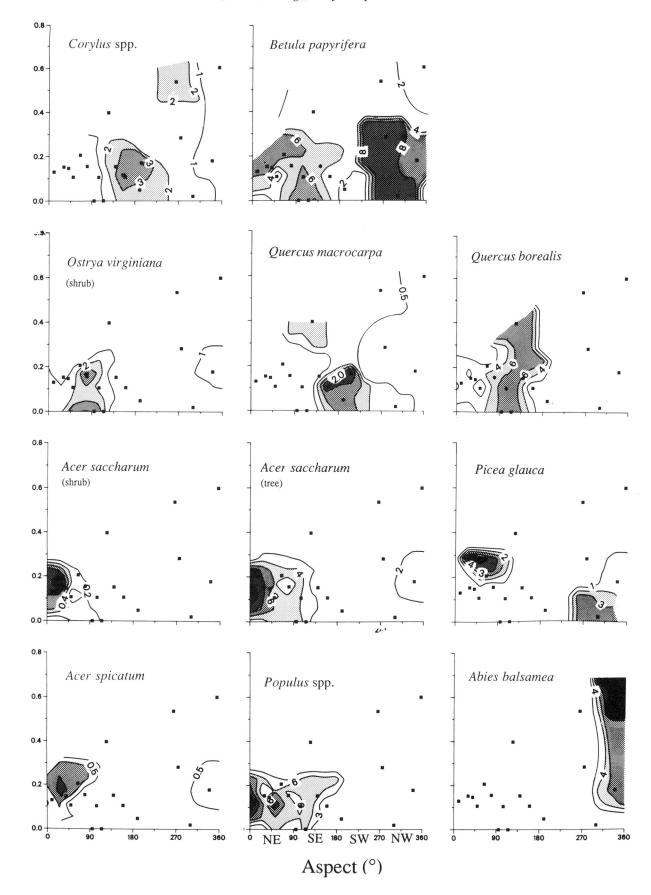

depended on topography, with southwest aspects burning most frequently in the past (Clark, 1990a, Fig. 3). The average interval between fires over the past 350 yr was 50–60 yr on northeast aspects and 20–30 yr on southwest aspects. Fire frequency also changed together with climate (see below).

N cycling

N mineralization is the microbially mediated transformation of organic N to NH_4^+. Mineralization of organic matter represents one of the most important steps in the turnover of N, because it determines the rate at which N returned in annual litterfall is made available for uptake by growing plants. It is important to understand N turnover, because N is the element most often limiting to plant growth in temperate forests (Keeney, 1980; Pastor and others, 1984). The rate at which N turns over can influence net primary production, cycling of other important elements such as carbon, phosphorus, and labile cations, and even the rate at which water is lost from the soil profile by evapotranspiration. N may also determine species composition (Pastor and others, 1984; Tilman, 1988). Once mineralized, N is available for oxidation to NO_3^- by nitrifiers. NO_3^- is highly mobile in the soil profile, and leaching losses can elevate cationic denudation. In the extreme case, NO_3^- losses through solution transport may adversely affect water quality (Bormann and Likens, 1979; Vitousek, 1983), and, in some lakes, N may limit primary production or influence algal species composition during some part of the year (Reynolds, 1984). Leaching losses of N depend in part on how N is cycled. Thus, some limnological effects of forest compositional transitions at Lake Itasca may have resulted from the rates at which different forest types cycle N.

Mineralization rates in the existing forests are tied to the same topographic patterns important for explaining forest composition, and nitrification is also significant only on northeast aspects. N is turned over most rapidly on northeast aspects (Fig. 3), probably for several reasons. Quantity and quality (i.e., lignin:N) of litter is greatest on northeast aspects (Fig. 3). Comparatively low lignin:N in the litter of *Acer saccharum* and *Tilia americana* (Gosz and others, 1973; Mellilo and others, 1982; McClaugherty and others, 1985) on northeast aspects compared to the higher lignin:N litter of *Pinus resinosa* on southwest aspects results in less immobilization of N by the microbial community. I am not aware of lignin:N data for *Corylus,* but other *Betulaceae* (e.g., *Betula papyrifera,* which is also most abundant on southwest aspects) have lignin:N values substantially higher than northern hardwood species such as *Acer saccharum* (Gosz and others, 1973; McClaugherty and others, 1985). Thus, microbial populations on northeast aspects are probably strongly limited by carbon and they compete less with autotrophic plants for mineral N.

It is possible that mineralization is also limited by water more often on the xeric southwest aspects, as has been suggested at sites having a more positive water balance than does Itasca (Vitousek and Matson, 1985). Although the bulk of the annual

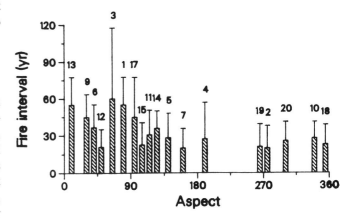

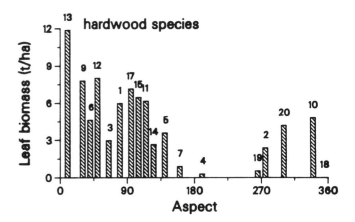

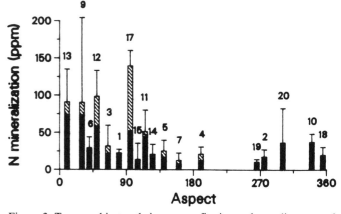

Figure 3. Topographic trends in average fire interval, standing crop of leaf biomass, and N mineralization. Bars are one standard deviation. Letters above bars are plot numbers located in Figure 1. N mineralization bars include an upper hatched portion, indicating net nitrification, and a lower portion, indicating net NH_4^+ accumulation. The total height of the bar is total net mineralization.

precipitation falls during the growing season, summer droughts are common (Clark, 1989b), and southwest aspects are expected to be most severely affected. Indeed, soil moisture may provide the overriding control on forest composition here. Moisture limitation of N mineralization may lead to a competitive advantage for species that efficiently use N, which in turn exaggerates these differences in rates at which N is cycled. This type of feedback between local water balance, N cycling, and species composition is the basis for Pastor and Post's (1988) argument that future decreases in soil moisture may result in increased spatial heterogeneity in some midwestern forests. Dry sites, where water limits mineralization, become still less productive as species more conservative with N dominate, further tightening the N cycle. Moist sites, where moisture does not limit mineralization rates, would display increased mineralization rates with increasing temperatures, promoting species that cycle more N (Pastor and Post, 1988; Clark, 1990d). Mesic northeast aspects in my study area allow for higher mineralization, plants are less N limited (Clark, 1990c), and, thus, less N is translocated from leaves prior to abscission in autumn (Vitousek, 1983).

An alternative explanation for spatial patterns in mineralization and species composition is that more frequent fire on xeric southwest aspects selected for species more conservative in their use of N. If so, higher mineralization rates on northeast aspects could be viewed more as a result of vegetation patterns than a cause. Abrupt transitions in forest composition and structure that parallel ridgetops suggest that past fires on south or west exposures stopped at the tops of ridges and did not move down into northeast-facing slopes (Clark, 1990a). Therefore, possible explanations for vegetation patterns can be based either on fire history or microclimatic effects on water and N cycling.

Evidence to date suggests that topographic patterns in vegetation structure and composition are best explained as a consequence of topography (i.e., drainage, parent material, and microclimate) rather than of fire frequency directly. This explanation is supported by the fact that mineralization rates and species composition show stronger dependencies on topography than does fire. Although the fire-topography dependency exists, that dependency is modest (a model including slope, aspect, and position on slope yields an $r^2 = 0.44$, $P = 0.09$). Moreover, the relations between species composition and fire frequency (Clark, 1990c) and between mineralization rates and fire frequency ($r = 0.38$; not significant) are also less clear than are the relations between these variables and topography; aspect, slope, and position-on-slope explains 79% of the variance in 1987 mineralization rates ($P = 0.0003$) in my study area. The evidence is consistent with the interpretation that microclimate has important effects on species composition and potential mineralization rates. Fire also depends on microclimate (primarily fuel moisture content), in addition to characteristics of the vegetation that result from microclimate (vertical and horizontal continuity of fuels and fuel moisture). Thus mineralization shows a weaker dependency on fire frequency. However, fire serves to exaggerate the topographic effects on mineralization and composition, because more frequent fire on southwest aspects favors efficient N users (Clark, 1990c).

The patterns of establishment following fire may depend in part on these topographic effects on mineralization rates. *Populus* species occur mostly on the northeast aspects (Fig. 2), where rapid turnover of N is possible (Fig. 3), and moisture conditions may be more favorable for seeds that remain viable only for a few days (Fowells, 1965). *Betula papyrifera* also becomes established following fire, but it does so mostly on south and west aspects (Fig. 2), where mineralization rates are lower (Fig. 3). *Quercus macrocarpa* regenerates almost exclusively on the driest southwest aspects (Fig. 2). Thus, fire together with mineralization rates and water balance may be necessary to explain coexistence of many different species in the park.

These processes associated with vegetation patterns in the existing stands provide a basis for interpreting factors that influenced limnology through the past. In the next section I discuss changes in compositional patterns suggested by pollen evidence from Deming Lake, and I apply results from modern stands in an analogue fashion to speculate on how environments may have changed during the past 2000 yr.

PHYSICAL ENVIRONMENTS AND FORESTS OF THE PAST

Hardwoods at A.D. 0–400 (2000–1550 varve yr B.P.)

Forests of 2000 yr ago differed most prominently from existing ones in having less *Pinus* (Fig. 4). Pollen of *Quercus, Ostrya, Populus,* and *Betula* is most abundant at this time. Using the value $C' = 1$, the limited charcoal evidence spanning a period of 76 yr (Fig. 5) suggests an average fire interval of 11 yr. On the existing landscape, similar vegetation assemblages are most common on well-drained ridgetops or southwest aspects that burned on average every 18–20 yr during the past 200 yr. Charcoal data tend to underestimate the interval between fires, so the actual interval may have been 10–20 yr at this time.

Existing hardwood assemblages that contain large amounts of *Betula* are characterized by open canopies, high understory light levels, and often dense *Corylus* understories. These assemblages are conservative with N, so leaching losses may have been low between fires. There is still the possibility that solution transport to lakes occurred immediately following fire. Mineralization of N by fire can produce a flush of available N following fire (Raison, 1979), and depletion of organic matter with repeated burning can lead to chronic decreases in N. Presence of *Corylus*, which could sprout rapidly following fire, suggests that any increases in mineral N that might have occurred following fire would probably have been short lived. It is also uncertain whether much of the soluble nutrients and particulates deposited on the soil surface would have been transported in solution to lakes, after subsurface runoff had been filtered by soil (Richter and others, 1982). It is clear that much ash and char was transported to lakes in this and subsequent times, probably to a large

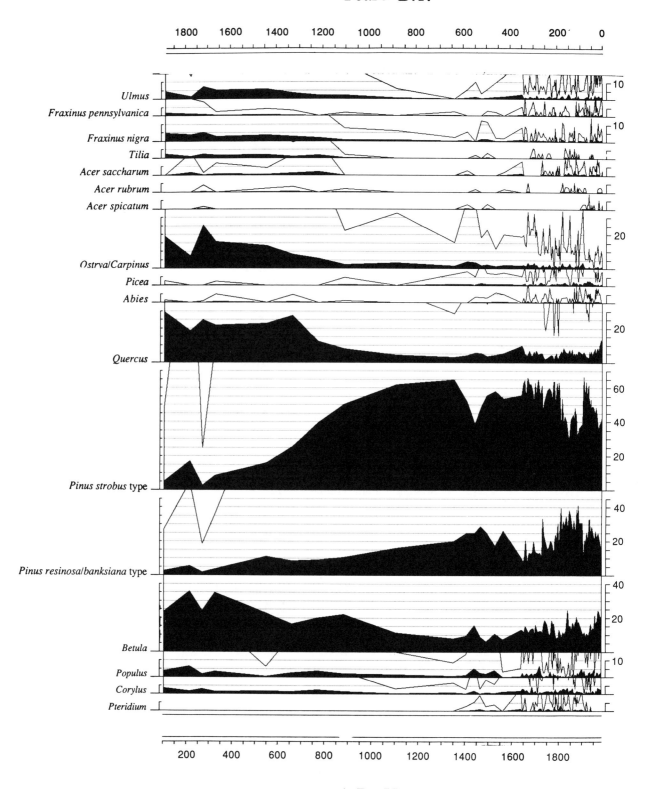

Figure 4. Important pollen types expressed as percent of total terrestrial forest pollen at Deming Lake.

burns on southwest aspects. Many of the intervals between fires at this time may have been too short to permit reproductive maturity, so much regeneration may have been vegetative. Both species resprout from root collars or roots. The *Quercus* species now present with *Betula* and *Corylus* are *Q. borealis* and *Q. macrocarpa* on the driest sites. If *Q. macrocarpa* were the dominant *Quercus* species then, as suggested by McAndrews (1966) and Jacobson (1979), mature individuals may have survived frequent surface fires in the understory. *Quercus borealis* is more commonly associated with *Populus* spp. and *Ostya. Ostrya* is an important understory species on well-drained ridgetops (Fig. 2), and it is also a prolific sprouter. An open *Quercus* canopy at the time may have allowed more prolific flowering and more efficient transport of pollen from *Ostrya* than occur at present (Jacobson, 1979).

Pollen percentages of northern hardwoods species are more abundant at this time than they are following the subsequent increase of *Pinus strobus*–type pollen. Northern hardwoods were probably rare, and they may have been limited to moist sites (McAndrews, 1966) protected from fire. Average N losses from these northern hardwood assemblages are expected to have been higher than those from more xeric sites, because mineralization and nitrification rates are highest in these stands (Fig. 3). Less frequent fire, however, suggests that the nutrient losses averaged over time may not have been higher. The relative contributions of N from these different kinds of assemblages may have depended on the degree to which solution losses increased following the frequent fires on southwest aspects.

Taken together, the evidence suggests that the assemblages of this time were somewhat different in composition from any of those now on the landscape, largely because of the decreased role of *Pinus*. Each of the dominant species of that time might be explained, however, on the basis of a warmer and/or drier climate and more frequent fire. The fire regime may have been similar to that which characterized forests on ridgetops and southwest slopes during the past 200 yr. In general, nutrient losses to lakes via solution transport may have been lower than occurs in the more mesic assemblages of existing forests as a result of more conservative cycling of N by species dominant at that time. There was, however, frequent redistribution of C, N, and P as a result of volitalization and convection of ash during fires.

Expansion of Pinus strobus (A.D. 400–1200; 1550–750 varve yr B.P.)

Percentages of *Pinus strobus*–type pollen begin an increase lasting 800 yr about A.D. 400 (Fig. 4). The pollen diagram of Whitlock and others (Chapter 17) shows a similar increase beginning somewhat earlier. It is possible that (1) the *Pinus strobus* increase occurred earlier in Elk Lake than it did in Deming Lake, or (2) the increases were contemporaneous within the two lakes, and one or both of the chronologies contain error. Percentages of most other pollen types decline as *Pinus strobus* reaches values of 50%–70% from A.D. 1000 to 1750. Limited charcoal evidence

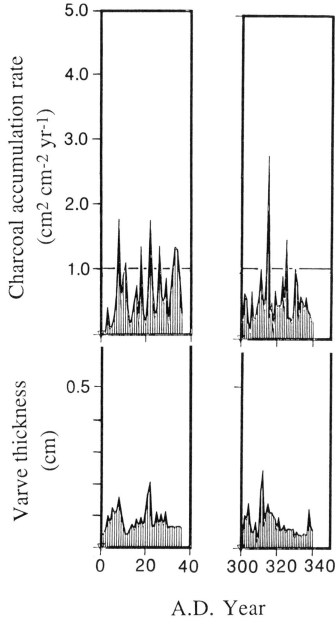

Figure 5. Stratigraphic charcoal accumulation and varve thickness spanning two intervals of 40 yr each from A.D. 0 (left) and A.D. 300 (right), when hardwoods dominated the landscape.

extent by wind (Clark, 1988c), and that ash might have represented a source of alkalinity additional to ground-water carbonates and mineralized N and P (Raison, 1979).

The "early successional" tree species *Betula papyrifera* and *Populus* spp. may have coexisted by virtue of differences in competitive ability for N or moisture. If northeast aspects allowed for higher mineralization rates, as they do now (Fig. 3), *Populus* may have been the more important early successional tree on northeast slopes, while *Betula papyrifera* tended to colonize recent

(Fig. 6) suggests an average interval of 12 yr between fires, but the charcoal peaks are more clustered in time than are those in the previous zone. As a result, several longer intervals of 20 yr are suggested. Although *Pinus strobus* now occurs on areas that burned frequently in the past, it depended on these longer intervals to become established and reach maturity before the next fire (Clark, unpublished). Once mature, *Pinus strobus* can withstand low-intensity surface fires, and a longevity of up to 450 yr (Loehle, 1988) ensures that mature individuals will survive long enough to supply seed to burned areas when these long interval between fires do occur.

The expansion of *Pinus strobus* suggests a water balance more positive than that of the previous zone. Cooler or moister conditions (Jacobson, 1979) may have commenced between A.D. 0 to 300 (1950–1650 varve yr B.P.), and *Pinus strobus* subsequently expanded over the next several generations of trees. A more positive water balance would have resulted in more water moving through the system and thus the potential for greater leaching losses. The establishment of *P. strobus* within mixed-oak assemblages may have followed fires (McAndrews, 1966), and pine litter could have inhibited further establishment of oaks (Jacobson, 1979).

If soil moisture did increase with this transition, it is possible that mineralization rates would also have increased, albeit somewhat modestly. *Pinus strobus* is conservative with N, and stands dominated by this species (e.g., plot 14, Fig. 1) exhibit low mineralization rates (Fig. 3). Like the previous zone, leaching losses to lakes may have been minimal, except following fires. If fires tended to be more clustered in time when *Pinus strobus* was dominant, then these periods of short intervals may have been times when N losses were greatest as a result of extended periods of low plant cover and reduced plant uptake. More charcoal data are needed, however, to better characterize the fire regime and thus to assess the degree to which fire might have influenced expansion of *Pinus strobus* and nutrient cycling.

Fluctuations after A.D. 1200 (750 varve yr B.P.)

Since A.D. 1200 there appear to have been several climatic fluctuations and attendant changes in fire regime and vegetation. As mentioned above, the fifteenth and sixteenth centuries appear to have been warmer or drier than before and after. Charcoal data show that fire was more important during the inferred warm and dry fifteenth and sixteenth centuries than it was before and after (Fig. 7). The average interval between charcoal peaks $>C'$ was 9 yr from A.D. 1440 to 1600, compared to 13 yr from 1240 to 1440, and 14 yr from 1640 to 1920. Significantly more charcoal accumulated per year from 1440 to 1600 than from 1640 to 1920 ($P < 0.01$; Kolmogorov-Smirnov two-sample test). These warm and dry times saw transient increases in percentages of *Populus* and *Pinus resinosa* or *banksiana* pollen and spores of *Pteridium* (Fig. 8). Percentages of each of these pollen types decreased with the onset of cooler and moister times about A.D. 1600.

Fire regimes were responsive both to the more positive water balance after 1600 and the decade-scale fluctuations in climate since that time (Fig. 9). Although mean annual charcoal accumulation decreased significantly following 1600, either annual charcoal accumulation or frequency of large charcoal peaks

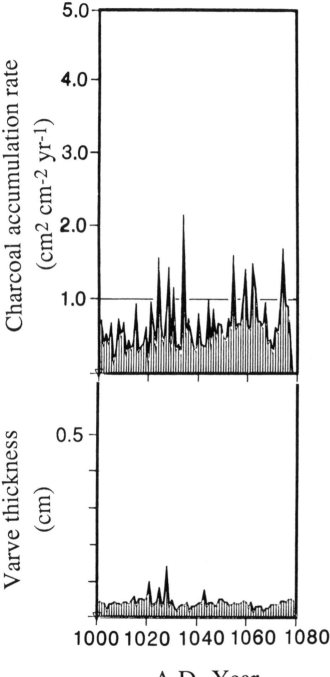

Figure 6. Stratigraphic charcoal accumulation and varve thickness spanning an 80 yr interval from A.D. 1000 when *Pinus strobus* was an important canopy species.

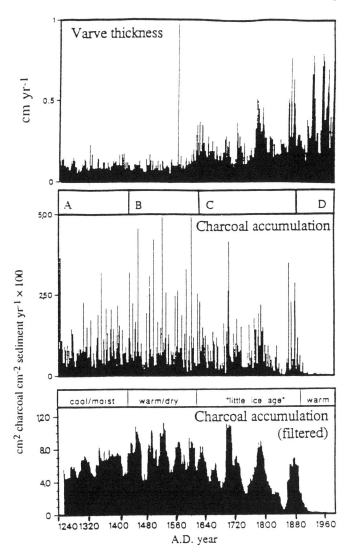

Figure 7. Varve thickness (above), charcoal accumulation (center), and charcoal accumulation filtered with a 15 yr running average (below) from Deming Lake.

was greater in the sediments of Deming and Arco lakes from A.D. 1780 to 1820 than they were during the subsequent mid-nineteenth century ($P < 0.05$; Kolmogorov-Smirnov two-sample tests). The Budd Lake series did not show a significant increase at this time, although fire scars indicate that fires were more frequent in the Budd Lake catchment from 1800 to 1815 than they were from 1815 to 1860. A smoothed charcoal series shows three intervals since 1600 when fire was more important, centered at 1700, 1790, and 1880, or 90 yr apart (Fig. 7) (Clark, 1988b).

Pinus resinosa began to increase in abundance in the seventeenth century (Fig. 8), with most seedling recruitment restricted to intervals between fires that were sufficiently long to allow trees to attain fire resistence (about 20 yr). Thus the distribution of fires in time was important for the *Pinus resinosa* expansion since 1600 (Clark, unpublished). Spores of *Pteridium* increased in sediments of each lake following times of frequent fire (A.D. 1700, 1790, 1880). Thus, species composition was sensitive to frequency of fire, and changes in fire regime over time resulted in shifting species dominance.

Fire frequency increased again late in the nineteenth century (Fig. 9), corresponding to dry conditions in the 1890s. Increases in fire frequency were significant in the Arco series ($P < 0.05$), and significantly more charcoal accumulated in Deming Lake ($P < 0.05$). All three series clearly indicate the onset of fire suppression in 1910.

The expansion of early successional species with the increased fire frequency beginning A.D. 1440 may have increased or decreased nutrient losses. Although these early successional species are generally conservative with N, they are less conservative than *Pinus strobus*. Thus changes in species composition might have led to increased solution transport of nutrients. However, *Pinus strobus* and *P. resinosa* were still abundant on the landscape. An abrupt increase in varve thickness with the onset of cooler and moister conditions in A.D. 1600–1640 (Fig. 7) could reflect higher primary production in the lake as a result of increased nutrient loading that would be predicted by more moisture moving through the soil profile. Although species composition also changed at this time, the ways in which these different tree species cycle N may have less importance for nutrient loadings to lakes than does the more positive water balance.

CONCLUSIONS

The dramatic transitions in vegetation and fire regimes during the past 2000 yr appear to have been closely linked to climate change (Table 1). Together with evidence from existing forests on N mineralization, those changes suggest possible effects on nutrient cycling and therefore, perhaps, the nutrient balance of lakes. Conclusive statements concerning the loading of N to lakes is difficult, because the increases that result from fire occurrence are likely offset to some degree, as a consequence of tighter cycling of N by species that dominated when fire was frequent (Clark, 1990d). I did not analyze phosphorus, which generally limits primary production in lakes, because it is less commonly limiting to forest growth in temperate regions. But nutrient cycles interact as a consequence of the ratios of nutrients required by plants and decomposers, so phosphorus and other nutrients are expected to have changed through time as well. The overriding influence of microclimate on species composition, N mineralization, and fire frequency in existing forests indicates that forest processes are sensitive to the degree of variability in water balance that occurs at this landscape scale. The changes in forest patterns that have occurred through the past provide evidence for sensitivity to climate changes that have occurred over this interval.

It is nonetheless difficult to speculate from these data concerning how nutrient supplies to lakes changed over time. The

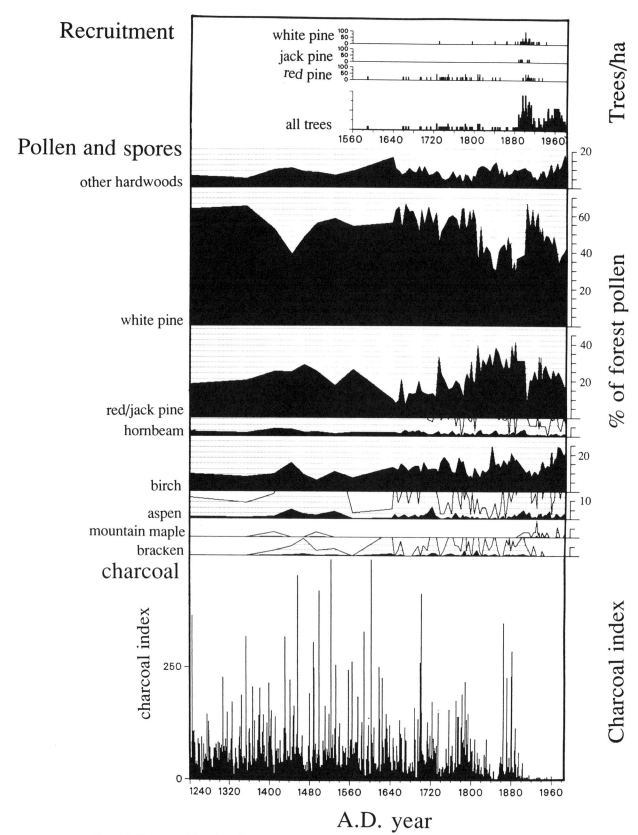

Figure 8. Tree ages (above), pollen percentages (center), and charcoal accumulation (below) for the interval A.D. 1240 to 1985 from Deming Lake. Tree ages were determined from increment cores.

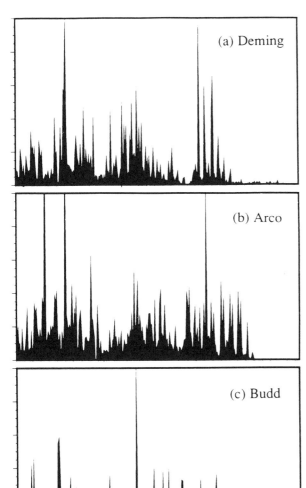

Figure 9. Charcoal accumulation in the three lakes shown in Figure 1 since A.D. 1640.

TABLE 1. POSSIBLE CHANGES IN WATER BALANCE, FIRE FREQUENCY, AND N CYCLING SUGGESTED BY DATA DISCUSSED IN THE TEXT

Time interval (A.D. year)					
from	0	400	1400	1600	1920
to	400	1400	1600	1920	Present
Water balance*	−	− or 0	−	+	0
Fire frequency†	5 to 10	5 to 10	12	4 to 8	None
N conservation§	+	+	+	+	0
N losses					
between fire**		−	−	−	−
total losses‡	±	±	±	±	+

* Relative to today: − = more negative in the past and + = more positive.
† Number of fires per 100 yr.
§ + = Dominance by species more conservative in the cycling of N than those of today.
** − = Smaller losses than today at times when fires were not occurring.
‡ Because amounts of nutrients transported to lakes during and immediately following fires are unknown and potentially large, it is uncertain whether nutrient supply to lakes at times of fire could have offset smaller of amounts of nutrients reaching lakes between fires as a result of high nutrient-use efficiency.

relative contributions of nutrient inputs immediately following fires versus those that occurred between fires, and the degree to which nutrients suspended or dissolved in the soil solution were filtered out before reaching lakes are unknown. Because most of the plant assemblages were more conservative with N than are those of today (Table 1), it is likely that the supply of N to lakes before fire suppression was lower than occurs in modern forests. The time-averaged inputs could still have been higher in the past, however, if fires were an important vector for transport in the past of particulates containing mineralized N to lakes. Paleolimnological investigations of these sediments would provide an independent means for assessing the extent to which nutrient regimes may have changed in the past.

ACKNOWLEDGMENTS

This chapter benefited from efforts of individuals acknowledged in each of the original papers cited here. I especially thank my thesis advisor H. E. Wright for supporting the project, E. J. Cushing, Dave Grigal, and Don Zak for discussion, and Platt Bradbury for inviting me to prepare this chapter. I also benefited from comments of Platt Bradbury, Eric Grimm, and Tom Swetnam. Financial support was provided by National Science Foundation grants BSR-8715251 and BSR-8818355 and grants from the Dayton-Wilkie Foundation and the Whitehall Foundation. This is contribution number 624 of the Education Department of the State of New York.

REFERENCES CITED

Bergman, H. F., and Stallard, H., 1916, The development of climax formations in northern Minnesota: Minnesota Botanical Studies, v. 4, p. 333–378.

Bormann, F. H., and Likens, G. E., 1979, Pattern and process in a forested ecosystem: New York, Springer-Verlag, 253 p.

Buell, M. F., and Gordon, W. H., 1945, Hardwood-conifer forest contact zone in Itasca Park, Minnesota: American Midland Naturalist, v. 34, p. 433–439.

Buell, M. F., and Martin, W. E., 1961, Competition between maple-basswood and fir-spruce communities in Itasca Park, Minnesota: Ecology, v. 42, p. 428–429.

Chandler, C., Cheney, P., Thomas, P., Trabaud, L., and Williams, D., 1983, Fire in Forestry, Volume I, Forest fire behavior and effects: New York, John Wiley, 450 p.

Clark, J. S., 1988a, Charcoal-stratigraphic analysis on petrographic thin sections:

Recent fire history in northwest Minnesota: Quaternary Research, v. 30, p. 81–91.

——, 1988b, Effects of climate change on fire regime in northwestern Minnesota: Nature, v. 334, p. 233–235.

——, 1988c, Particle motion and the theory of charcoal analysis: Source area, transport, deposition, and sampling: Quaternary Research, v. 30, p. 67–80.

——, 1989a, Effects of long-term water balance on fire regime, northwestern Minnesota: Journal of Ecology, v. 77, p. 989–1004.

——, 1989b, Ecological disturbance as a renewal process: Theory and application to fire history: Oikos, v. 56, p. 17–30.

——, 1990a, Fire occurrence during the last 750 years in northwestern Minnesota: Ecological Monographs, v. 60, p. 139–155.

——, 1990b, Twentieth century climate change, fire suppression and forest production and decomposition in northwestern Minnesota: Canadian Journal of Forest Research, v. 20, p. 219–232.

——, 1990c, Long-term interactions among nitrogen mineralization, species composition, and fire frequency: Biogeochemistry, v. 11, p. 1–22.

——, 1990d, Ecosystem sensitivity to climate change and complex responses, *in* Wyman, R., ed., Global climate change and life on earth: New York, Chapman and Hall, p. 65–98.

Fowells, H. A., 1965, Silvics of forest trees of the United States: Washington, D.C., USDA Forest Service Agricultural Handbook, v. 271, 762 p.

Frissell, S. S., 1971, An analysis of the maintenance of presettlement biotic communities as an objective of management in Itasca State Park, Minnesota [Ph.D. thesis]: St. Paul, University of Minnesota, 240 p.

——, 1973, The importance of fire as a natural ecological factor in Itasca State Park, Minnesota: Quaternary Research, v. 3, p. 397–407.

Gajewski, K., 1988, Late Holocene climate changes in eastern North America estimated from pollen data: Quaternary Research, v. 29, p. 255–262.

Gajewski, K., Winkler, M. G., and Swain, A. M., 1985, Vegetation and fire history from three lakes with varved sediments in northwestern Wisconsin: Review of Palaeobotany and Palynology, v. 44, p. 277–292.

Gosz, J. R., Likens, G. E., and Bormann, F. H., 1973, Nutrient release from decomposing leaf and branch litter in the Hubbard Brook Forest, New Hampshire: Ecological Monographs, v. 47, p. 173–191.

Grant, M. L., 1934, The climax forest community in Itasca County, Minnesota, and its bearing upon the successional status of the pine community: Ecology, v. 15, p. 243–257.

Grimm, E. C., 1983, Chronology and dynamics of vegetation change in the prairie-woodland region of southern Minnesota, U.S.A.: New Phytologist, v. 93, p. 311–350.

Hansen, H. L., Kurmis, V., and Ness, D. D., 1974, The ecology of upland forest communities and implications for management in Itasca State Park, Minnesota: University of Minnesota Agricultural Experiment Station Technical Publication, v. 298, 43 p.

Heinselman, M. L., 1973, Fire in the virgin forest of the Boundary Waters Canoe Area, Minnesota: Quaternary Research, v. 3, p. 329–382.

Jacobson, G. L., 1979, The palaeoecology of white pine (*Pinus strobus*) in Minnesota: Journal of Ecology, v. 67, p. 697–726.

Janssen, C. R., 1967a, A floristic study of forests and bog vegetation, northwestern Minnesota: Ecology, v. 48, p. 751–765.

——, 1967b, A postglacial pollen diagram from a small *Typha* swamp in northwestern Minnesota, interpreted from pollen indicators and surface samples: Ecological Monographs, v. 37, p. 145–172.

Keeney, D. R., 1980, Prediction of soil nitrogen availability in forest ecosystems: A literature review: Forest Science, v. 26, p. 159–171.

Kell, L. L., 1938, The effect of the moisture-retaining capacity of soils on forest succession in Itasca Park, Minnesota: American Midland Naturalist, v. 20, p. 682–694.

Lee, S. C., 1924, Factors controlling forest successions at Lake Itasca, Minnesota: Botanical Gazette, v. 78, p. 129–174.

Leonard, E. M., 1986, Varve studies at Hector Lake, Alberta, Canada, and the relationship between glacial activity and sedimentation: Quaternary Research, v. 25, p. 199–214.

Loehle, C., 1988, Tree life history strategies: The role of defenses: Canadian Journal of Forest Research, v. 18, p. 209–222.

McAndrews, J. H., 1966, Postglacial history of prairie, savanna, and forest in northwestern Minnesota: Torrey Botanical Club Memoirs, v. 22, p. 1–72.

McClaugherty, C. A., Pastor, J., Aber, J. D., and Melillo, J. M., 1985, Forest litter decomposition in relation to soil N dynamics and litter quality: Ecology, v. 66, p. 266–275.

Melillo, J. M., Aber, J. D., and Muratore, J. F., 1982, N and lignin control of hardwood leaf litter decomposition dynamics: Ecology, v. 63, p. 621–626.

Ohmann, L. F., and Grigal, D. F., 1985a, Biomass distribution of unmanaged upland forests in Minnesota: Forest Ecology and Management, v. 13, p. 205–222.

——, 1985b, Plant species biomass estimates for 13 upland community types of northeastern Minnesota: Washington, D.C., USDA Forest Service Research Bulletin NC-88, 52 p.

Pastor, J., and Post, W. M., 1986, Influence of climate, soil moisture, and succession on forest carbon and N cycles: Biogeochemistry, v. 2, p. 3–27.

Pastor, J., and Post, W. M., 1988, Response of northern forests to CO_2-induced climate change: Nature, v. 334, p. 55–58.

Pastor, J. J., Aber, J. D., McClaugherty, C. A., and Melillo, J. M., 1984, Aboveground production and N and P cycling along a N mineralization gradient on Blackhawk Island, Wisconsin: Ecology, v. 65, p. 256–268.

Peet, R. K., 1984, Twenty-six years of change in a *Pinus strobus, Acer saccharum* forest, Lake Itasca, Minnesota: Torrey Botanical Club Bulletin, v. 111, p. 61–68.

Raison, R. J., 1979, Modification of the soil environment by vegetation fires, with particular reference to nitrogen transformations: A review: Plant and Soil, v. 51, p. 73–108.

Reynolds, C. S., 1984, The ecology of freshwater phytoplankton: Cambridge, England, Cambridge University Press, 384 p.

Richter, D. D., Ralston, C. W., and Harms, W. R., 1982, Prescribed fire: Effects on water quality and forest nutrient cycling: Science, v. 215, p. 661–663.

Ross, B. A., Bray, J. R., and Mitchell, W. H., 1970, Effects of long-term deer exclusion on a *Pinus resinosa* forest in north-central Minnesota: Ecology, v. 51, p. 1088–1093.

Spurr, S. H., 1954, The forests of Itasca in the nineteenth century as related to fire: Ecology, v. 35, p. 21–25.

Stoeckeler, J. H., and Curtis, W. R., 1960, Soil moisture regime in southwestern Wisconsin as affected by aspect and forest type: Journal of Forestry, v. 58, p. 892–896.

Swain, A. M., 1973, A history of fire and vegetation in northeastern Minnesota as recorded in lake sediment: Quaternary Research, v. 3, p. 383–396.

Thornthwaite, C. W., 1948, An approach toward a rational classification of climate: Geographical Review, v. 38, p. 55–94.

Thornthwaite, C. W., and Mather, J. R., 1957, Instructions and tables for computing potential evapotranspiration and the water balance: Publications in Climatology, v. 10, 185 p.

Tilman, D., 1988, Plant strategies and the dynamics and structure of plant communities: Princeton, New Jersey, Princeton University Press, 360 p.

Transeau, E. N., 1905, Forest centers of eastern America: American Naturalist, v. 39, p. 875–889.

Vitousek, P. M., 1983, The effects of deforestation on air, soil, and water, *in* Bolin, B., and Cook, R. B., eds., The major biogeochemical cycles and their interactions: Scientific Committee on Problems of the Environment, p. 223–225.

Vitousek, P. M., and Matson, P. A., 1985, Disturbance, N availability, and N losses in an intensively managed loblolly pine plantation: Ecology, v. 66, p. 1360–1376.

Webb, S. L., 1988, Windstorm damage and microsite colonization in two Minnesota forests: Canadian Journal of Forest Research, v. 18, p. 1186–1195.

Willmott, C. J., Rowe, C., and Mintz, Y., 1985, Climatology of the terrestrial seasonal water cycle: Journal of Climatology, v. 5, p. 589–606.

Woodward, F. I., 1987, Climate and plant distribution: Cambridge, England, Cambridge University Press, 174 p.

MANUSCRIPT ACCEPTED BY THE SOCIETY JULY 27, 1992

Printed in U.S.A.

Geological Society of America
Special Paper 276
1993

Holocene climatic and limnologic history of the north-central United States as recorded in the varved sediments of Elk Lake, Minnesota: A synthesis

J. Platt Bradbury and Walter E. Dean
U.S. Geological Survey, Box 25046, Federal Center, Denver, Colorado 80225
Roger Y. Anderson
Department of Earth and Planetary Sciences, University of New Mexico, Albuquerque, New Mexico 87131

ABSTRACT

Integration of the results and interpretations of geochemical, paleoecological, and sedimentological analyses of a varved sediment record provides a detailed chronicle of limnological and climatic changes for the past 10 ka at Elk Lake, west-central Minnesota. The early Holocene record at Elk Lake was controlled by circumstances of glacial history (e.g., basin morphometry and surrounding till lithology) in combination with global warming at the end of the Pleistocene. Later, the interplay of climate change and a disintegrating ice sheet determined the character of local environments that were affected by reduction of precipitation and increased windiness during the middle Holocene. Elk Lake became more productive and clastic sediment increased to dominate the record as the forests thinned and prairie vegetation characterized the region. During this prairie period, winters may have been cold because disintegrating northern ice sheets ceased to block winter outbreaks of Arctic air. Correlations of wind-deposited materials throughout much of the eastern two-thirds of the United States suggest that drought conditions and strong winds were widespread between 8 and 4 ka. The mid-Holocene was climatically variable, however, with strong fluctuations in varve thickness at decadal, centennial, and millennial scales testifying to rapid climatic changes. Although the cause of such climatic cycles is not yet clear, correlations between ^{14}C anomalies and varve thickness suggest that variations in solar flux and resulting magnetic storms and zonal winds may have induced strong climatic changes. A particularly strong 600 yr fluctuation to cool and wet climates that may document neoglacial conditions around 5 ka interrupted the prairie period. After 4 ka, the climate at Elk Lake was dominated by a tropical airstream during the summer, and dry arctic and Pacific airstreams during the winter. Large-scale variations ceased, although decadal and multidecadal variations in varve thickness chronicle changes similar, but not clearly correlative, to historically documented climatic episodes such as the Medieval Warm Period and the Little Ice Age.

INTRODUCTION

This volume reports on the scientific efforts of many individuals, all brought to bear on Elk Lake, a small lake in Minnesota with a varved Holocene sediment record. The first part of this final chapter combines the findings of the various investigations into a summary of changes that occurred in and around Elk Lake. In this synthesis we also examine how the new data from

Bradbury, J. P., Dean, W. E., and Anderson, R. Y., 1993, Holocene climatic and limnologic history of the north-central United States as recorded in the varved sediments of Elk Lake, Minnesota: A synthesis, *in* Bradbury, J. P., and Dean, W. E., eds., Elk Lake, Minnesota: Evidence for Rapid Climate Change in the North-Central United States: Boulder, Colorado, Geological Society of America Special Paper 276.

Elk Lake fit into previously assembled information for the region. The objective is a step-by-step historical survey of important findings, with emphasis on the local response during and after the disintegration of the Laurentide ice sheet. In tracking these responses, it is helpful to recognize the dominant influence of the ice sheet and easterly anticyclonic winds until about 9 ka. Thereafter, local and regional responses can be related to the shifting influence of the three major airstreams (Fig. 1) and changes in net radiation. Anderson and others (Chapter 1) point out that Elk Lake is located at a "climatic triple junction" where three regional airstreams join to form sharp climatic gradients across Minnesota. A warm, moist, tropical Atlantic airstream dominates during the summer months, whereas the winter months experience competition between a cold, dry Arctic airstream and a dry Pacific airstream. Periods of drought in the Great Plains occur when the dry Pacific airstream dominates over the other two airstreams as a result of a stronger westerly wind (Bryson, 1966; Muhs, 1985). The summary in the first part of this synthesis examines local changes at Elk Lake in terms of climatic variables associated with these airstreams, as well as the progressive or evolutionary changes peculiar to Elk Lake and its drainage.

Now that we have many details about past changes at Elk Lake, we can examine the value of that record in resolving larger questions related to climate and climate change. Most investigations of cores from a single site, such as Elk Lake, concentrate on only a few parameters suspected to be important. One simply does not know in advance, however, which of many possible parameters will contain critical information on climate and the environment. Here we test the assumption that comprehensive investigation of a single site contains elements of serendipity as

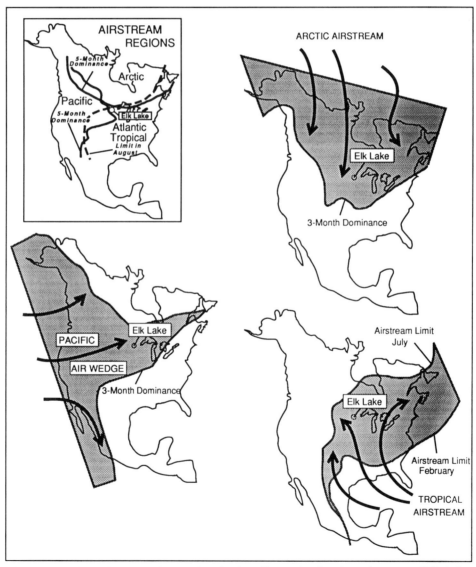

Figure 1. Dominant airstreams of North America that affected the climatic environment of Elk Lake. Modified from Bryson and Hare (1974).

well as synergy, and the second part of this final chapter explores the regional changes in climate that are defined at Elk Lake, including information that may bear on their ultimate cause.

One piece of information that we have obtained from examining evidence from several disciplines is the recognition of the magnitude, regional extent, and potential importance of arid conditions in the mid-Holocene. This mid-Holocene dry interval often is interpreted as having been warmer as well as drier (e.g., altithermal or hypsithermal period), but several proxy indicators in Elk Lake sediments suggest that this period was colder than present, at least until about 6.8 ka. The timing of maximum aridity, at about 6 ka, reflects a lag of about 3 ka from maximum insolation predicted by the Milankovitch climate forcing mechanism. This lag, not peculiar to Elk Lake, is found at enough sites to be considered a regional effect in the northern mid-latitudes (Webb, 1985). In addition, the aridity maximum appears to represent the low point in the "sawtooth" pattern of changes in ice volume (marine $\delta^{18}O$) that has been repeated at the point of reversal of each major cycle in glaciation. The record from Elk Lake is probably the most detailed compiled for that critical interval, and in the second part of this chapter we bring together the evidence for this event at Elk Lake and regionally, and consider what this evidence suggests for Milankovitch orbital forcing and other causal mechanisms.

Unless otherwise stated, age values for the varved Elk Lake core are given in thousands of varve years (ka) for the age of the midpoint of a sample interval relative to the varve time scale (T_o = A.D. 1927; Anderson and others, Chapter 4). Events outside of Elk Lake that are dated by radiocarbon, or presumed to be dated by radiocarbon, are given the designation ^{14}C yr B.P. (Anderson and others, Chapter 4). These radiocarbon ages can be converted to calendar ages (ka) using Figure 4 of Anderson and others (Chapter 4) based on data from Stuiver and Reimer (1986).

PALEOLIMNOLOGICAL SYNTHESIS OF ELK LAKE DATA

Lake origins: Changes before 8 ka

The limnological history of Elk Lake began when an ice block was deposited and buried by outwash from the Wadena ice lobe sometime before 14 ka. The block was subsequently covered by insulating drift from the Des Moines–St. Louis sublobe and began to melt (Wright, Chapter 2). A sag pond formed over the melting ice, probably about 11 ka, and abundant pollen and needles, primarily of spruce (*Picea*; Fig. 2), were deposited along with sand, silt, and woody trash in the shallow water (Wright, Chapter 2; Whitlock and others, Chapter 17). The pond supported cold-water ostracodes, especially *Cytherissa lacustris*, that are partial to boreal forest habitats (A. J. Smith, 1991), a calcareous green alga, *Phacotus*, mollusks, and charophytes indicating that the hydrochemistry of the pond was comparatively alkaline from the start. Some Cretaceous pollen (Stark, 1976) and diatoms (Bradbury and Dieterich-Rurup, Chapter 15) near the base of the Holocene sequence in Elk Lake indicate that erosion of surrounding till provided some of the sediment to Elk Lake.

The limnological characteristics of this initial Elk Lake stage were closely tied to local conditions and circumstances. For example, the alkaline pond chemistry was controlled by the lithology of the Des Moines–St. Louis till, which, in turn, owed its origin to the route of that glacier over Paleozoic and Cretaceous rocks to the northwest (Wright, Chapter 2). Even variations in the pollen deposition rate during this time probably resulted from locally variable sedimentation patterns in the small pond, and occurrences of minor hardwood and herbaceous taxa were related to local, nearby habitats within the open spruce forest (Whitlock and others, Chapter 17). Although these local influences were driving the limnology of this early pond above the ice block, the ultimate driving force responsible for the melting of the ice block and the growth of spruce forests was the regional climatic change forced by insolation and the gradual demise of the Laurentide ice sheet (Bartlein and Whitlock, Chapter 18). Certainly the great expansion of Lake Agassiz between about 10,000 and 9500 ^{14}C yr B.P. (about 11.4 to 10.4 ka in calendar years; Teller, 1985, 1987) just to the north of Elk Lake must have had a marked effect on temperature and precipitation in the entire region, in addition to hastening the melting of the Laurentide ice sheet.

By 10.4 ka the ice block had largely disappeared, and Elk Lake was about 50 m deep where the ice block was thickest. Seasonal sediment production was preserved as couplets of fine laminations or varves that could be counted to establish a high-resolution (annual) chronology for the lacustrine deposits of this lake. The ratios of chlorophyll derivatives plus lutein to procaryote pigments (C+L/P) in sediments from 10.4 ka to about 9.2 ka were about twice as high as in those deposited 9.0 ka to the present (Fig. 3), reflecting a rapid and significant influx of terrestrial plant detritus (Sanger and Hay, Chapter 13). Spruce continued to dominate the vegetation, but between 10.2 and 10.0 ka the spruce forest disappeared and was rapidly replaced by forests of jack and red pines (*Pinus banksiana* and *P. resinosa*; Whitlock and others, Chapter 17; Fig. 2). The replacement of spruce by pine in the north-central United States is well documented and indicates a widespread vegetational response to climate changes accompanying deglaciation (e.g., Bartlein and others, 1984). The spruce-pine transition is time transgressive from south to north (e.g., Webb and others, 1983; Whitlock and others, Chapter 17), and jack pine reached Minnesota from the Great Lakes region about 10.0 ka. The transition from spruce dominance to pine dominance in eastern Minnesota may have occurred in less than a century (Wright, 1984); this was certainly true in the Elk Lake region (Fig. 2). In climatic terms, the vegetational change can be interpreted as representing an increase in summer (July) temperature of about 4 °C and an increase in annual precipitation to near-modern values.

The varves that formed during this interval are characterized by abundant manganese, iron, and $CaCO_3$ (Dean, Chapter 10; Anderson, Chapter 5; Fig. 3) that precipitated from ions leached from the fresh calcareous drift as the climate became wetter and

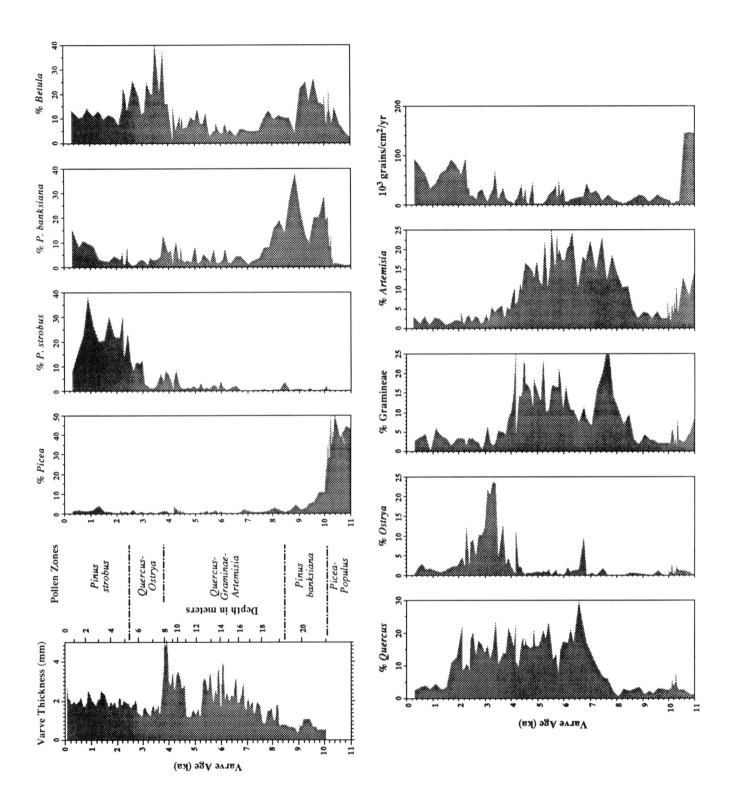

warmer. The thin layers of rhythmically laminated sediment that accumulated in the deep lake reflected the strong seasonality in northern Minnesota, enhanced by increasing distance from the ice front. In summer, without ice cover, calcium and bicarbonate ions exceeded saturation in warmed surface waters and precipitated calcium carbonate under the influence of photosynthesis in relatively pure, white layers. During summer and winter stratification, high concentrations of iron and manganese built up in the hypolimnion, just as they do today, but the hypolimnion of early Elk Lake must have been some 20 m thicker (the thickness of sediment deposited over the past 10 ka). At fall and spring periods of circulation, iron and manganese precipitated mainly as amorphous oxyhydroxides that formed dark-colored winter laminae. The relatively large amounts of manganese, relative to iron, in sediment deposited during the first 2000 yr in Elk Lake (Fig. 3) may reflect both its abundance in the till and its comparatively labile character under reducing conditions in soils (e.g., Mackereth, 1966; Engstrom and others, 1985). Low diatom accumulation rates, small percentages of *Cyclotella comta* and *C. ocellata* (diatoms characteristic of cool forest lakes), ostracodes (*Cytherissa lacustris*), and midges (*Tanytarsus*) between 10.4 and 8.5 ka (Fig. 4; Stark, 1976) suggest an early oligotrophic Elk Lake. Nevertheless, the dominance of *Stephanodiscus minutulus* and *Fragilaria crotonensis* (Fig. 4) indicate that Elk Lake was moderately productive, at least seasonally. The siliceous frustules of these latter diatoms often form discrete, thin, pure laminae in and among the nearly black iron- and manganese-rich layers. The concentration of total plant pigments increases markedly from 9.0 ka to a maximum centered on about 8.0 ka (Fig. 3) indicating a rapid increase in productivity and pigment preservation by about 8.2 ka (Sanger and Hay, Chapter 13).

Portent of drought: 9.5 ka to 9.1 ka

Several paleolimnological indicators suggest a significant environmental change between about 9.5 ka to 9.1 ka. Authigenic calcite, as well as indicators of clastic deposition (e.g., sediment mass accumulation rate, sodium, quartz, and varve thickness; Fig. 3), and birch pollen (*Betula*; Fig. 2) all show peaks in abundance at this time, whereas percentages of pine pollen decrease (Fig. 2). The first turbidites of clay and fine silt also appear among the thin varves in this interval. Although the turbidites could logically be explained by local causes such as oversteepened slopes, synchronous changes in all the related parameters suggests that there may have been a larger climatic control. In modern forests, birch is considered an indicator of xeric conditions and commonly increases in response to fire (Clark, Chapter 19), along with bracken fern (*Pteridium*) (Whitlock and others, Chapter 17). Swain (1973) suggested that the ratio of conifers to sprouters (such as birch) might be a sensitive index of vegetation response to periodic burns because conifers become established by seed reproduction only, whereas sprouters are capable of reproducing from underground roots and stems that escape fire. By 9.0 ka pine was again the dominant arboreal pollen type (Fig. 2). The presence of oxygen-sensitive macrofossils, including larval head capsules of the blood worm (chironomid), the fingernail clam *Pisidium*, and ostracodes, in cores from the margin of the northwestern basin of Elk Lake (Stark, 1976) all indicate high bottom-water oxygen concentrations (= shallow water) within the *P. banksiana-resinosa* zone (10.0 to 8.5 ka; Fig. 2) with peak values in the middle of this pollen zone.

A warmer, drier than modern climate during the period 9.5–9.0 ka could also be reflected in the increases of the clastic indicators, especially if forest cover were thinned by increases in fire frequency. Greater slope instability and sheet-flood underflows may be reflected by turbidite deposition. A thinner forest cover might also result in greater transportation of labile ions such as manganese, calcium, and bicarbonate from soils and increased deposition in the lake. A minimum in iron concentration (Fig. 3) at the same time as the manganese maximum suggests that soil redox conditions may have been slightly less reducing, thereby retaining oxidized iron in the soils but with continued reduction of manganese. The diatoms do not respond as dramatically to the environmental changes at this time, perhaps because nutrients released from vegetation by fire do not always enter lacustrine ecosystems (e.g., McColl and Grigal, 1977; Bradbury and others, 1975). On the other hand, the minor increases in *Fragilaria crotonensis* and reduction of *Stephanodiscus minutulus* (Fig. 4) could be due to hot summers and lake stratification early in the year (Bradbury, 1988). This would be consistent with increased fire frequency.

The relation between inferred warm and dry climates (high fire frequency) at 9.5 ka to the ultimate effects of increasing seasonality of solar insolation and to the rate of deglaciation is speculative. Models suggest that the anticyclonic northeasterly glacial winds would have been decreasing with the retreat and reduction of the ice sheet, while cyclonic winds would have been increasing in strength as a low-pressure system that formed in the continental interior gradually moved toward the northeast as the ice sheet shrank (Bartlein and Whitlock, Chapter 18; Webb and others, 1987). The interplay of these two wind systems could have resulted in alternating periods of calm and storminess and/or in positive and negative water balances as the major climatic driving mechanisms gradually changed in location and magnitude during the early Holocene. Such short-term climatic variations are implied by charcoal fluctuations in the late Holocene record of fire in Itasca State Park (Clark, Chapter 19).

Figure 2. Plots of percentages of spruce (*Picea*), white pine (*Pinus strobus*), jack pine (*P. banksiana*), birch (*Betula*), oak (*Quercus*), hophornbeam (*Ostrya*), grasses (Gramineae), and sagebrush (*Artemisia*), and pollen flux in 10^3 pollen grains per cm^2 per year, all vs. varve age (ka) for 0.5 cm^3 samples from the varved Elk Lake core. A plot of varve thickness vs. varve years and the major pollen zones are shown for reference. Data are from Whitlock and others (Chapter 17).

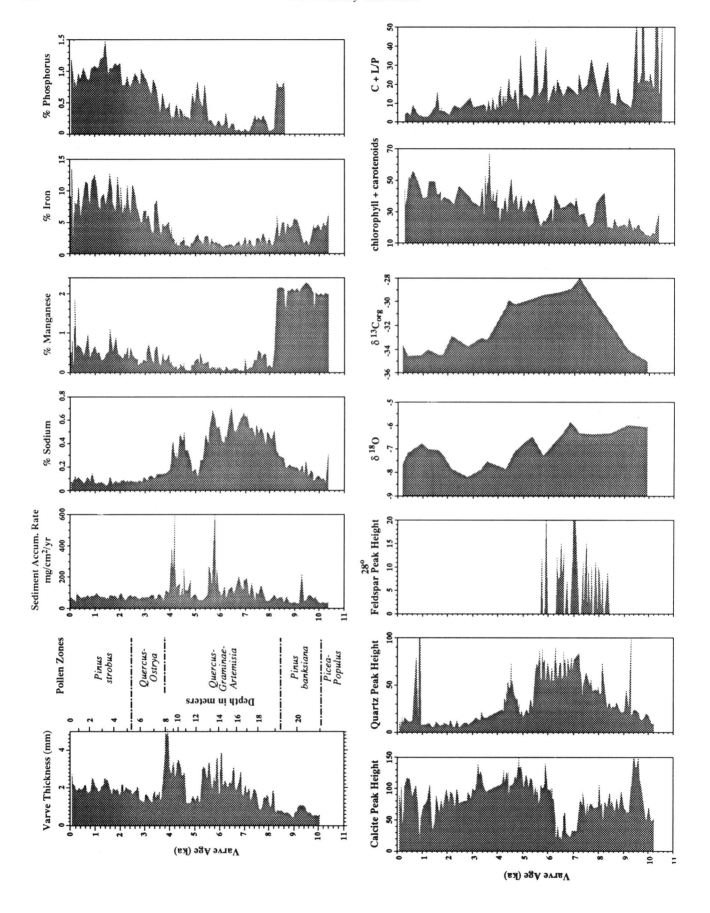

Eastward expansion of the prairie: 9.0 ka to 7 ka

In the millennium between 9.0 and 8.0 ka, pine forests became greatly reduced around Elk Lake and first sagebrush (*Artemisia*) and then grass (Gramineae) began to dominate the pollen assemblage (Fig. 2). *Fragilaria crotonensis* was the dominant planktonic diatom during most of this millennium (Fig. 4). Because this diatom seems to prosper when spring circulation is weak and/or during the summer when the lake is stratified (Bradbury, 1988), its abundance may relate to warm summers (high insolation) and low summer wind stress that are modeled between 9.0 and 8.0 ka (e.g., Bartlein and Whitlock, Chapter 18). Cyanobacterial pigments dominated (low C+L/P ratio due to higher concentrations of procaryote pigments; Fig. 3) between 9.2 and 8.3 ka (Sanger and Hay, Chapter 13), and high phosphorus concentrations in the sediment (Fig. 3) add additional support to the interpretation of warm, calm summers and strongly stratified limnological conditions that would promote blooms of blue-green algae (cyanobacteria) and burial of phosphorus.

Today sagebrush typifies the prairie environments to the west of Elk Lake, and appearance of sagebrush pollen in abundance at about 8.7 ka marks the initial expansion of prairie environments eastward. Varve thickness and concentrations of sodium and plagioclase feldspar (Fig. 3), which are interpreted as indicating increased wind stress, erosion, and deposition of relatively undecomposed, windborne clastic material (Dean, Chapter 10; Anderson, Chapter 5), begin to increase markedly at 8.2 ka. Pollen influx also increases after 8.2 ka (Fig. 2), reflecting both the amount of vegetation and wind-driven lake turbulence that increases pollen redeposition from littoral areas (e.g., Davis, 1973). Most pollen profiles from Elk Lake suggest that the transition into the prairie period was gradual, in contrast to the rapid spruce-pine transition at about 10 ka. McAndrews (1966), however, on the basis of pollen in a core from Bog D just east of Elk Lake in Itasca State Park, found that the forest-prairie transition occurred very abruptly just above the level of a sediment sample dated to 8560 ± 120 ^{14}C yr B.P. Some geochemical variables suggest that the beginning of the prairie period occurred suddenly at 8.2 ka (Dean, Chapter 10; see profiles for manganese and phosphorus in Fig. 3). Q-mode factor analyses of the geochemical data tend to "sharpen" geochemical zones in the Elk Lake sediments, and show that the forest-prairie transition occurred within about 100 yr (Dean, Chapter 10). In addition, during the main prairie period, most indicators document extreme cyclic fluctuations in local and regional environmental conditions with periodicities of 100 to 200 yr, presumably linked to regional climatic cycles (Anderson, Chapter 5).

At about 8.2 ka abundance peaks of *Stephanodiscus minutulus* and diatom influx (Fig. 4) are coincident with a zone of thick, quartz-rich varves (Fig. 3). These indicators suggest maximal conditions of wind stress and associated storminess, but it is not clear how this significant climatic event relates to the hypothesized gradual changes in the ultimate controlling mechanisms of insolation and ice-sheet extent. Perhaps they indicate the crossing of threshold conditions, and may mark the onset of strong northwesterly winds of the dry Pacific air wedge that became dominant as anticyclonic circulation related to the Laurentide ice sheet collapsed (Anderson and others, Chapter 1). Suffice it to say that 8.2 ka marks the beginning of the mid-Holocene arid prairie period during which the forest-prairie border ultimately migrated more than 100 km eastward (Wright, 1976).

Grass dominates the pollen assemblage between 8.0 and 7.0 ka during the progressive increase of oak in the savanna vegetation (Fig. 2). Phosphorus accumulation in Elk Lake was moderately high during this time (Fig. 3), and *Fragilaria crotonensis* again dominated the planktonic diatom flora (Fig. 4), but total diatom productivity (flux) was comparatively low (Fig. 4). Lack of vigorous lake circulation could promote preservation of phosphorus in the sediments and keep organic productivity fairly low (Bradbury and Dieterich-Rurup, Chapter 15). Lower productivity is also suggested by concentrations of plant pigments, which decrease markedly from high values between 8.2 and 7.8 ka to low values at about 7.5 ka, then increase again by 7.0 ka (Fig. 3; Sanger and Hay, Chapter 13). These indicators may be interpreted as representing generally dry conditions, but with sufficient spring precipitation for abundant grass, followed by hot summers that favored early and strong stratification of the lake, which, in turn, limited nutrient recycling and productivity in the epilimnion. Modern pollen analogs for this period are located in prairies along the United States–Canadian border near long. 100°W (Whitlock and others, Chapter 17).

By 8.0 ka the influx of oak (*Quercus*; Fig. 2) pollen began to increase, signaling the initial development of an oak-grassland with scattered sagebrush that characterized most of northwestern and north-central Minnesota during the prairie period (McAndrews, 1966; Wright, 1976, 1992). The lag of oak behind the first increase of sagebrush (8.7 ka) and grass (8.5 ka) may relate to cooler summer temperatures during the early part of the prairie period (e.g., Prentice and others, 1991).

Ostracode profiles from Elk Lake (Forester and others, 1987) show a mid-Holocene (7.8–6.8 ka) water chemistry of higher and more variable conductivity than at present, suggesting that northwestern Minnesota was indeed drier during this time. However, analogous ostracode populations live today in regions of long and cold winters with conditions similar to those

Figure 3. Plots of sediment mass accumulation rate (MAR in mg dry sediment/cm^2/yr); percentages of sodium, manganese, iron, and phosphorus; X-ray diffraction relative peak heights for calcite, quartz, and 28° 2Θ feldspar; values of δ^{18}O in bulk carbonate and δ^{13}C in organic carbon, both in ‰ relative to PDB, the marine carbonate standard; and concentrations of total pigments (chlorophyll derivatives plus carotenoids) and the ratio of chlorophyll derivatives plus lutein to procaryote pigments (C+L/P), all vs. varve age (ka) for samples from the varved Elk Lake core. A plot of varve thickness vs. varve years, and the major pollen zones are shown for reference. Data are from Dean (Chapter 10), Dean and Stuiver (Chapter 12), and Sanger and Hay (Chapter 13).

316   J. P. Bradbury and Others

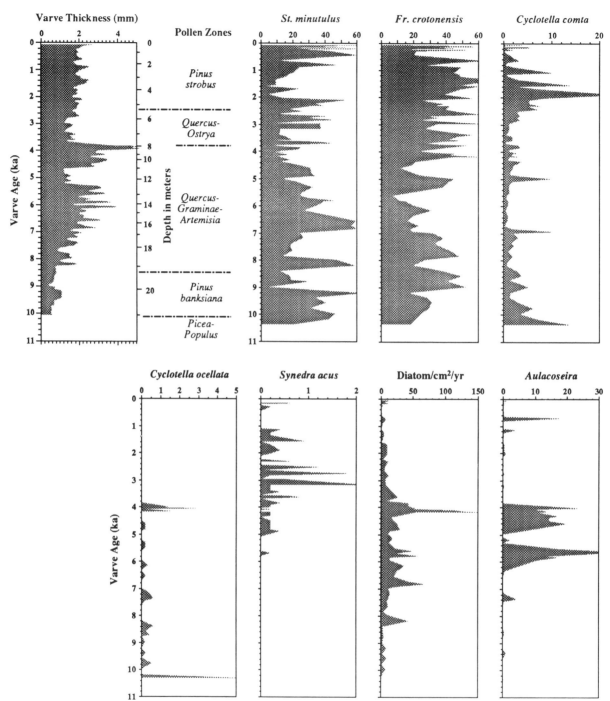

Figure 4. Plots of percentages of *Stephanodiscus minutulus, Fragilaria crotonensis, Cyclotella comta, C. ocellata, Synedra acus,* and *Aulacoseira* spp., and total diatom flux in diatoms per cm² per year, all vs. varve years for samples from the varved Elk Lake core. A plot of varve thickness vs. varve age (ka) and the major pollen zones are shown for reference. Data are from Bradbury and Dieterich-Rurup (Chapter 15).

found today in Canadian prairie lakes. Generally cool climates are also suggested by oxygen isotope analyses from marl in the deep-basin core, which show that values of $\delta^{18}O$ in carbonate were high (= cold water) and varied very little until about 6.8 ka (Fig. 3). In addition, analogous pollen assemblages during the interval 7.8–6.6 ka are found in southern Saskatchewan (Whitlock and others, Chapter 17). This cold and dry climate may have been partly responsible for the late arrival of oak in the prairie vegetation, according to pollen-climatic response surfaces for oak (Bartlein and Whitlock, Chapter 18).

Oxygen and hydrogen isotope data from cellulose in fossil wood from an esker in eastern Ontario (Edwards and Fritz, 1986) indicate that cold, dry conditions persisted here until 7400 ^{14}C yr B.P. (ca. 8.4 ka), when the climate became somewhat warmer but remained dry. The main hypsithermal period with distinctly warm and wet conditions did not begin in eastern Ontario until about 6000 ^{14}C yr B.P. (ca. 6.8 ka). The differences between the pollen and isotope data of Edwards and Fritz for eastern North America and between the pollen and ostracode data for northwestern Minnesota may reflect large-scale climatic gradients. Perhaps cold winters in Minnesota between 7 and 8 ka reflect an increase in south-penetrating Arctic fronts formerly blocked by the large mass of glacial ice to the north (e.g., Wright, 1992).

On the basis of pollen-climatic response surfaces (Webb, 1985; Bartlein and others, 1986), Bartlein and Whitlock (Chapter 18) estimate that the annual precipitation in northwestern Minnesota was about 100 mm lower than present in the interval between 7.8 and 4.5 ka, and July temperature was about 2 °C warmer than present near the end of that interval. Pollen, charcoal, and sediment data from Lake Mendota show that southern Wisconsin was about 0.5 °C warmer in July and as much as 10% drier between 6500 and 3500 ^{14}C yr B.P. (7.3–3.8 ka) relative to present (Winkler and others, 1986). The water level of Lake Mendota also was lower during this mid-Holocene interval. There is some suggestion that the forest-prairie transition may have been time transgressive, beginning earlier (ca. 8000 ^{14}C yr B.P. = 8.8 ka) and ending earlier (ca. 5000–4000 ^{14}C yr B.P. = 5.8–4.4 ka) in Minnesota, northern Wisconsin, and northern Michigan than in southern Wisconsin (7000–6000 ^{14}C yr B.P. = 7.5–6.8 ka to ca. 3000 ^{14}C yr B.P. = 3.1 ka) (Webb and others, 1983; Bartlein and others, 1984; Winkler and others, 1986).

Warm-dry mid-Holocene: 7 ka to 4 ka

Following the initial eastward expansion of the prairie, the Pacific airstream with its dry, westerly winds became increasingly dominant, and sediments in Elk Lake began to reflect the reduced moisture and other effects related to the zonal winds. Concentrations of sodium, quartz, and plagioclase feldspar all show peak values at about 7.0 ka, remain at high values for the next 1200 yr, and reach peak values again at about 5.8 ka (Fig. 3). These high concentrations suggest an increase in influx of relatively unweathered clastic material, probably moved by wind, forming thicker varves (Dean, Chapter 10; Anderson, Chapter 5). The magnetic susceptibility (Fig. 5) also was high during this period, reaching a maximum between 6.0 and 5.4 ka. This suggests that the increased influx of clastic material also resulted in an increase in detrital magnetite (Sprowl and Banerjee, 1989, and Chapter 11).

Small amounts of *Aulacoseira* began to appear at about 7.3 ka (Fig. 4), indicating windy summer conditions (Bradbury and Dieterich-Rurup, Chapter 15). The effects of increased temperature and evaporation between 7.2 ka and 6.5 ka also produced the highest salinity in the history of Elk Lake, as evidenced by ostracode assemblages (Forester and others, 1987) and the presence of aragonite needles in the sediments (Dean, Chapter 10; Anderson, Chapter 5). The water level in the lake probably was at an all-time low during this interval (Stark, 1976; Forester and others, 1987). Diatom productivity during this interval also was high and dominated by *Stephanodiscus minutulus* (Fig. 4), which requires high phosphorus loading rates. Extended periods of circulation and recycling of phosphorus resulted in lower accumulation of phosphorus in the sediments (Fig. 3). Concentrations of algal pigments were high and fairly stable during this interval (Fig. 3), indicating high overall productivity (Sanger and Hay, Chapter 13). Carbon isotope ratios (Fig. 3) and pyrolysis hydrogen and oxygen indices lend additional support for increased productivity and burial of well-preserved, hydrogen-rich, oxygen-poor, ^{13}C-enriched algal organic matter beginning about 7.2 ka (Dean and Stuiver, Chapter 12). The climate during this turbulent and productive lake period is interpreted as having been relatively warm and dry; mild winters and windy springs produced long and intense periods of spring circulation in the lake that promoted nutrient cycling and phytoplankton productivity.

The critical date for increased temperature within the already dry prairie period appears to be about 6.7 ka. After this date the ostracode species-composition data indicate that Elk Lake became warmer and less saline (Forester and others, 1987). In response to higher temperature and lower salinity, $CaCO_3$ in sublittoral (10 m water depth) and deep-basin cores from Elk Lake became depleted in ^{18}O by about 2‰ by the end of the prairie period (Fig. 3; Dean and Stuiver, Chapter 12). Climate reconstructions from pollen data suggest an air temperature increase of about 1 °C in both summer and winter at this time (Bartlein and Whitlock, Chapter 18).

Short-term mid-Holocene climatic perturbation: 5.4 ka to 4.8 ka

During the remainder of the mid-Holocene prairie period (between 6 and 4 ka), oak pollen remains comparatively high while grass and sagebrush pollen decrease (Fig. 2). A slight increase in birch pollen and other arboreal taxa and a decrease in nonarboreal taxa after 5.5 ka suggests the development of somewhat more mesic environments in the area (Whitlock and others, Chapter 17). Many paleolimnological parameters such as *Aulacoseira* (Fig. 4), varve thickness (Fig. 3), magnetic susceptibility (Fig. 5), quartz, sodium, and sediment mass accumulation rate (Fig. 3), indicate that wind stress, turbulence, and lake productivity reached maxima at about 5.8 ka. Massive development of *Aulacoseira* is related to mid-late summer turbulent mixing, presumably resulting from strong convective thunder storms during this season. Pollen percent and influx (Fig. 2) mimic the *Aulacoseira* peaks and imply increased transport and deposition (or redeposition?) of pollen in the lake by wind or by increased lake turbulence. The parameters indicating clastic deposition (e.g., sodium, quartz, and mass accumulation rate, Fig. 3; magnetic susceptibility, Fig. 5) also can be best explained by increased eolian

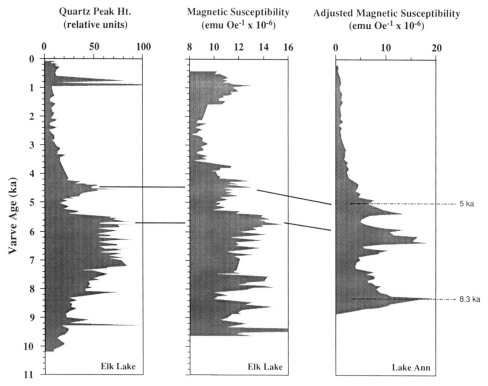

Figure 5. Plots of X-ray diffraction peak height for quartz and magnetic susceptibility in varved sediments from Elk Lake vs. varve years. Data are from Dean (Chapter 10) and Sprowl and Banerjee (Chapter 11). Also shown is a plot of magnetic susceptibility from Lake Ann, 230 km southeast of Elk Lake (modified from Keen and Shane, 1990). The Lake Ann plot is tied to the Elk Lake varve time scale (in ka) by two dates, 4390 ± 80 ^{14}C yr B.P. and 7400 ± 80 ^{14}C yr B.P. shown by the two dashed lines at the calibrated ages of 5.0 and 8.3 ka, respectively. Suggested correlations of two major peaks in the two magnetic susceptibility plots are shown by solid lines.

input, turbulence, and storms. The covariation in diatom abundance, pollen abundance, and clastic influx are most certainly related, with wind velocity as the common denominator. *Aulacoseira* only occurs in Elk Lake in the interval 6.4–3.8 ka, but during this interval there is an exact correspondence between its abundance and concentration of the clastic variables. *Physocypria globulosa* characterizes the interval where *Aulacoseira* species are common. This ostracode is a low-alkalinity, warm-water species (A. J. Smith, 1991), and suggests that Elk Lake was somewhat fresher between 6.0 and 4.0 ka than before 6.0 ka.

The influx of windborne clastic material ended suddenly at 5.4 ka (see quartz profile in Fig. 3). At this time the lake entered a brief phase (ca. 600 yr) that appears to have been a precursor to lake conditions characteristic of the latest stage in Elk Lake's development beginning at 3.8 ka, but differs from the more permanent change becaus it was not accompanied by the same changes in vegetation. Burial of phosphorus in the sediments (Fig. 3), concentrations of cyanobacterial pigments (Sanger and Hay, Chapter 13), and abundance of *Fragilaria crotonensis* (Fig. 4) all increased, whereas varve thickness and other clastic variables (e.g., sodium and quartz, Fig. 3; magnetic susceptibility, Fig. 5), and pollen influx (Fig. 2) all decreased from 5.4 ka to 4.8 ka. Within the same interval, assemblages of silicified resting stages (stomatocysts) of chrysophycean algae exhibit a striking return to an assemblage that existed in Elk Lake between 10.4 and 8.5 ka and during the Little Ice Age (A.D. 1450–1850) (Zeeb and Smol, Chapter 16). Reduced cycling of nutrients and diatom production during this same interval can be interpreted as reduced turbulence and weak circulation in the lake, possibly related to reduced strength of regional winds.

Varve thickness data (Fig. 6) indicate that this climatic and paleolimnologic transition occurred within a decade and ended about as rapidly. We interpret this 600 yr period as a time when wind-driven late summer lake turbulence was minor, and phosphorus was trapped in the anoxic sediments and not returned to the epilimnion during periods of circulation. The concomitant reduction of clastic indicators suggests that climate during this time was unusually calm. By about 4.8 ka the winds suddenly returned and clastic indicators returned nearly to the previous maximum levels at 5.8 ka. These values remain high until about 4.0 ka, when they suddenly fall to low levels to terminate the prairie period. Because suspension and transport of dust are

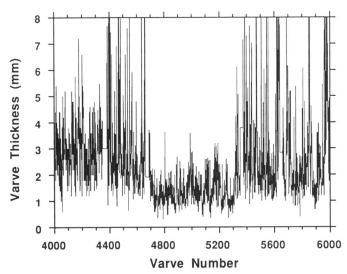

Figure 6. Plot of varve thickness vs. varve years for the time interval 4000 to 6000 varve yr showing abrupt cessation and restart of thick varves and accumulation of eolian materials in Elk Lake.

closely related to antecedent precipitation and soil moisture (drought; Brazel, 1989), the rapid response is consistent with the triggering of dust suspension at a shear threshold that is determined by soil moisture. The long-term changes in soil moisture that at first increased and then lowered the threshold, shutting off and then turning on dust suspension, probably were more gradual than indicated by the event itself, occurring in an interval that was longer than 600 yr. This suspected longer interval must have been accompanied by a significant increase and later decrease in moisture in order to account for the event. This long-term change in moisture may be reflected in the *Quercus* pollen profile that rises and falls over a period of 1200 yr between 5.6 and 4.4 ka (Fig. 2). The pollen data suggest that July temperatures were about 1° higher and annual precipitation about 50 mm greater during this interval than before or after.

Return of eolian clastic material after 4.8 ka was accompanied by a return of *Aulacoseira* nearly to its previous maximum level at 5.8 ka. Conditions of high clastic influx and *Aulacoseira* production continue until about 3.8 ka, when they suddenly fall to low levels to mark the termination of the prairie period perhaps related to increased frequencies of maritime tropical airstreams. Abundances of pollen from prairie plants (sagebrush and grass) had already begun to decline about 6.0 ka, and by 4.0 ka were nearly at background levels as a mixed mesic forest of pine, birch, oak, and hophornbeam (*Pinus, Betula, Quercus,* and *Ostrya*; Fig. 2) succeeded the oak savanna. This horizon was dated by radiocarbon at 3930 ± 100 yr B.P. (ca. 4.4 ka) in a core from Bog D about 3 km east of Elk Lake (McAndrews, 1966).

Increasing dominance of the tropical airstream: the past 4 ky

From about 4 ka to the present, many paleolimnological parameters of the Elk Lake record remain comparatively constant; for example, varve thickness, sediment mass accumulation rate, and sodium (Fig. 3), and diatom flux (Fig. 4). These reductions in variance can be explained most simply by the cessation of a variable clastic influx. Other indicators, such as *Cyclotella comta, Synedra acus,* quartz, oxygen isotope composition, and magnetic susceptibility (Figs. 3, 4, and 5) continue to show significant excursions that probably are related to both local environmental and/or climatic events. A few geochemical parameters, such as iron and phosphorus (Fig. 3) increase dramatically.

Pollen stratigraphy divides the last 4.5 ka into an interval of hardwood forest with high percentages of oak, birch and hophornbeam (*Quercus, Betula,* and *Ostrya*; Fig. 2) between 4.5 and about 3.5 ka, and a subsequent cooler interval of mixed conifer-hardwood forests characterized by white pine (*Pinus strobus*). Both forest types suggest more mesic conditions than the preceding prairie period (Whitlock and others, Chapter 17). Paleoclimatic interpretations of Elk Lake pollen data by Bartlein and Whitlock (Chapter 18) suggest that at 3.5 ka the July temperature was 1.5 °C higher than the present, January temperature was about 2 °C higher, and annual precipitation was about 100 mm greater. The correspondence between hardwood forest elements and *Synedra acus* between 4.0 and 3.0 may relate to warm climatic conditions that were manifested in Elk Lake by long and stable periods of summer stratification (Bradbury and Dieterich-Rurup, Chapter 15).

The increase in *Cyclotella comta* (Fig. 4) and values of $\delta^{18}O$ coincident with the rise in white pine pollen probably relate to the increase of forest cover caused by a cooler climate (Whitlock and others, Chapter 17). *Cyclotella comta* correlates with deeper, fresher, and nutrient poor lakes in Minnesota today (Brugam, Chapter 14) and such lacustrine environments are common in the cooler and wetter regions of northeastern Minnesota (Gorham and others, 1983). The importance of cyanobacteria in the accumulated organic matter is indicated by minimal C+L/P ratios (Fig. 3). The $1.5^0/_{00}$ increase in values of $\delta^{18}O$ from 2.8 ka to 0.9 ka (Fig. 3) may relate to a decrease in temperature, but other factors such as surface-water salinity and isotopic composition of ground water and precipitation also affect values of $\delta^{18}O$. However, most variables in the Elk Lake core indicate that by about 3.0 ka environmental conditions in the lake had reached an equilibrium level with respect to water chemistry, and, therefore, salinity variation is unlikely. If we assume that the $1.5^0/_{00}$ increase in values of $\delta^{18}O$ from 2.8 ka to 0.9 ka is due entirely to temperature, this would equate to about a 6 °C decrease in late summer water temperature during the main period of calcite precipitation (Dean and Megard, Chapter 8). However, a similar isotopic signature could be derived from precipitation originating in a warmer air mass such as the tropical airstream (K. Kelts, 1992, personal commun.).

As in the early history of Elk Lake (e.g., 10.4–8.2 ka), the high values of iron in the sediments for the past 4.0 ka (Fig. 3) probably relate to increased moisture, higher lake level, anoxic soils, and leaching of iron from surrounding till substrates (Mackereth, 1966). The lower concentrations of manganese (relative to those in early Holocene sediments) could result from the prior removal of this more labile, redox-sensitive till component in the early Holocene. In this case, the manganese profile may record progressive, local effects of weathering rather than distinctive climatic conditions in the late Holocene. Dean (Chapter 10) described the geochemical evolution of Elk Lake sediments within the three component system Fe-Mn-Al, with an early Holocene lake rich in Mn and Fe, a mid-Holocene lake rich in Fe and Al, and a late Holocene lake rich in Fe and Mn. Transitions between these three lake stages were accomplished within one or, at most, two centuries. Another way of viewing the chemical evolution of Elk Lake is that influx of iron, presumably from ground water, remained constant or increased somewhat, whereas the influx of manganese decreased as manganese was progressively removed from the surrounding glacial till. Superimposed on this iron and manganese evolution was a mid-Holocene eolian event that is recorded by aluminum (and other clastic variables).

Some of the late Holocene increase in iron, particularly in sediments that accumulated over the past 2 ka, is associated with an increase in fine-grained magnetite. Bulk magnetic measurements of Elk Lake sediments (Sprowl and Banerjee, 1989, and Chapter 11) show that both magnetic susceptibility (Fig. 5) and anhysteretic remnant magnetization (ARM) increase beginning about 2 ka. Magnetic susceptibility (interpreted as a proxy for amount of magnetite) increases about three fold from 3 ka to 1 ka, but the most dramatic increase is in ARM (interpreted as a proxy for magnetite grain size), which increases eight fold over the same interval. Sprowl and Banerjee suggest that this dramatic increase in fine-grained magnetite in the upper few meters of Elk Lake sediment may be biogenic, suggesting that 2 ka marks the advent of magnetotactic bacteria in Elk Lake.

Overall, the past 4.0 ka of the Elk Lake record represents the establishment of modern climatic and environmental regimes. By this time the Laurentide ice sheet had disappeared and summer insolation had approached modern values. The climatic changes suggested by the fluctuations of some paleolimnological parameters in Elk Lake may be associated with the decadal to centennial variability in modern climates, such as the Little Ice Age and the Medieval Warm Period (Anderson, Chapter 5). A combination of pollen, diatom, and oxygen isotope data indicate that climatic conditions in northwestern Minnesota became cooler and perhaps windier beginning about 2.8 ka. If the observed $1.5^0/_{00}$ increase in $\delta^{18}O$ beginning about 2.8 ka and peaking at about 0.9 ka (Fig. 3) is due entirely to temperature, a decrease of about 6 °C in late summer epilimnetic water temperatures is indicated. The abundance of white pine (*Pinus strobus*) pollen, indicating cool, moist conditions (Jacobson, 1979) in the drainage of Elk Lake, also reached a maximum at about 0.9 ka (Fig. 2). Assemblages of siliceous chrysophyte cysts within this time interval are remarkably similar to an assemblage that existed in Elk Lake between 10.4 and 8.5 ka (Zeeb and Smol, Chapter 16). However, the marked increases in *Aulacoseira* species (Fig. 4), quartz, and magnetic susceptibility (Fig. 5) between 1.0 ka and 0.7 ka suggest that influx of windborne clastic material, and associated turbu-

lence in the lake, characterized part of this interval, perhaps relating to the Medieval Warm Period in some way. Varve-thickness data show cyclic variations with periodicities of about 25, 40–50, and 200 yr, similar to periodicities in thickness of clastic-rich varves during the mid-Holocene that may be related to atmospheric ^{14}C production (Anderson, Chapter 5), but the late Holocene varves are a more complex mixture of iron, manganese, organic matter, $CaCO_3$, biogenic silica, and only minor clastic material. Close-interval paleolimnological time series for the past 3.0 ka in the Elk lake varved-sediment record, now under investigation, will be required to properly define the more subtle variations in a number of parameters that may be forced by high-frequency climatic changes during the latest part of the Holocene.

REGIONAL COMPARISONS AND CAUSES OF CHANGE

Many of the paleolimnological and paleoecological changes at Elk Lake identified in the first part of this synthesis can be correlated with changes in records from other sites and with changes in past climate inferred for the region. It is helpful to separate those regional changes that can be attributed to climatic effects related to the collapse of the Laurentide ice sheet and Northern Hemisphere insolation from subsequent changes that occurred under a different set of boundary conditions. The deglaciation history in the Itasca region of northwestern Minnesota is a reflection of changes on a larger scale that involved the entire Laurentide ice sheet. The ultimate cause of these large changes that occurred over several thousands of years is probably related to Milankovitch orbital forcing. Shorter events that occurred over decades to centuries, such as the Younger Dryas cold event, are either related to still-unestablished mechanisms that forced climate at higher frequencies, or to feedback between the ice sheet, atmosphere, oceans, and other water masses such as proglacial lakes and meltwater streams (Ruddiman and Wright, 1987). The first task, before suggesting possible causes, is to compare changes at different temporal and spatial scales and use the Elk Lake record to help identify regional effects. In the following comparisons, ^{14}C dates (^{14}C yr B.P.) of paleoenvironmental events in other records have been recalculated to approximate calendar (varve) years from the Elk Lake record according to Anderson and others (Chapter 4, Fig. 4) and Bartlein and Whitlock (Chapter 18, Fig. 11).

Time-transgressive changes during deglaciation

The clearest effects of deglaciation in Elk Lake sediments are found in palynological data. For example, Whitlock and others (Chapter 17) examined major episodes in the vegetation history of the upper midwest at sites along two south-north transects from Iowa to Manitoba, and one west-east transect from South Dakota to northeastern Minnesota. For investigation of major pollen events during deglaciation, Whitlock and others chose the spruce (*Picea*) decline/pine (*Pinus*) rise, and the late-glacial elm (*Ulmus*) maximum. Correlations of these pollen events have provisionally established the south (11.2 ka) to north (10.0 ka) time-transgressive nature of the spruce decline as a result of a northward shift of the dominance of the polar airstream during deglaciation. The spread of pine following the spruce decline was apparently much more variable spatially, ranging from 10,800 to 9000 ^{14}C yr B.P. (11.8–10.0 ka). The late-glacial elm maximum possibly represents a synchronous vegetation response to warm climates associated with the insolation maximum at 9.0 ka. Wright (1992) suggested that the presence of the ice sheet only 600 km from central Minnesota at 9.0 ka may have prevented winter outbreaks of extremely cold arctic air, such as occurs today, that freeze temperate deciduous trees such as elm. Therefore, elm could have survived in the northern Great Plains whereas it cannot today.

The possible dry interval at Elk Lake about 9.5 to 9.1 ka, indicated by increase in several clastic variables (sodium, quartz, and varve thickness; Fig. 3) and birch pollen (*Betula*; Fig. 2), corresponds approximately in time with the calculated calendar date of the Cochrane (= Cockburn) advances or surges in southern Hudson Bay (8600–8200 ^{14}C yr B.P. = 9.5–9.2 ka; Andrews, 1987; Miller and others, 1988) that produced a prominent oxygen isotope meltwater spike in cores from the Hudson Strait. This well-documented glacial readvance presumably also correlates with glacial advances or surges in Ontario dated between 8500 and 8000 ^{14}C yr B.P. (9.5–9.0 ka) (Wright, 1987). These events may all relate to climatic changes associated with the insolation maximum.

Some regional changes during glaciation in the range of hundreds to several thousand years can be attributed to changes in the ice sheet itself and to possible feedback mechanisms operating on a regional scale. For example, Lake Agassiz, just to the north and west of Elk Lake (Fig. 7), must have had a profound influence on the climate and vegetation of the region. Bartlein and Whitlock (Chapter 18) suggest that the lake effect during the Emerson phase of Lake Agassiz, when the lake shore was within 75 km of Elk Lake, may explain the cooler and wetter conditions inferred from the Elk Lake pollen record prior to 10.2 ka, which is the end of the Younger Dryas. As Bartlein and Whitlock point out, this evidence has implications for differentiating between local versus regional causes of climatic events such as the Younger Dryas event, the origin of which has been a controversial topic (e.g., Broecker and others, 1988, 1989; Lewis and Anderson, 1989; Teller, 1987, 1990; see review by Kennett, 1990). The draining of Lake Agassiz into the North Atlantic and subsequent readvance of the Greenland ice sheet may have been the ultimate cause of Younger Dryas cooling regionally or even globally, but the main effect on climate change in the Itasca region of Minnesota was decreased lake-effect cooling and increased precipitation during the Moorhead phase of Lake Agassiz.

Advances and retreats in margin the Laurentide ice sheet, on the approximate time scale of the Younger Dryas event, occurred in oscillations of about 2500 yr (Denton and Karlén, 1973). Such

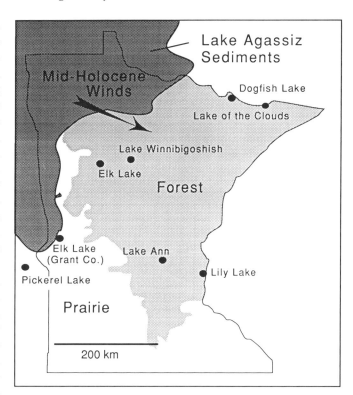

Figure 7. Map of Minnesota showing location of Elk Lake and other lakes discussed in this paper, relative to Lake Agassiz, and trajectory of regional winds during the mid-Holocene as determined from dune fields southeast of Lake Winnibigoshish and northwest of Lake Ann.

oscillations may be global in extent (Lindsley and Thunnel, 1990; DeDeckker and others, 1991) and are manifested in such diverse paleoclimate proxy records as Greenland ice (Oeschger and others, 1988), lake levels (Allen and Anderson, unpublished), and Antarctic dust (Pedersen and others, 1991). The origin of forcing for these millennial-scale oscillations is highly speculative, and includes changes in volcanicity (Bryson, 1989), solar activity (Anderson and others, 1990; Anderson, 1991), deep-ocean dynamics, and high-frequency components of Milankovitch orbital cycles. At Elk Lake, climatic oscillations in this frequency band, if they were regionally important, were obscured by much stronger decadal- to centennial-scale responses to deglaciation and developmental changes in the lake and its drainage. After the episode of deglaciation, the best chance for finding an expression of regional millennial-scale climatic oscillations is during the prolonged interval of aridity that occurred from 8.2 ka to 3.8 ka.

Regional aridity during the mid-Holocene

Warmer than modern reconstructed summer temperatures occurred over most of Europe about 6000 ^{14}C yr B.P. (6.8 ka) with greatest heating (~2 °C) in middle and northern Europe (Huntley and Prentice, 1988). Regional aridity during the same approximate time interval in midwestern United States is well documented by the eastward expansion of oak-savanna and prairie vegetation as shown by lines of equal pollen percentages in isopoll maps (Webb and others, 1983; Jacobson and others, 1987). These studies suggest that maximum eastward expansion or prairie in the midwest occurred between 7000 and 6000 ^{14}C yr B.P. (ca. 7.8 to 6.8 ka). Bartlein and others (1984) concluded that the mid-Holocene temperature maximum in the midwest was time transgressive with maximum warmth occurring about 6000 ^{14}C yr B.P. (6.8 ka) in the northern midwest and about 4000 ^{14}C yr B.P. (4.4 ka) in the southern midwest.

Whitlock and others (Chapter 17) note that the increase in prairie herbs in the western part of their transect (at about 9000 ^{14}C yr B.P. [10.0 ka] in South Dakota and western Minnesota) begins about 1000 ^{14}C yr earlier than at correlative sites in eastern Minnesota (at about 8000 ^{14}C yr B.P. = 8.8 ka). The exception is Elk Lake, where the increase in prairie herbs also occurs at 8.8 ka as in eastern Minnesota. This observation led Whitlock and others (Chapter 17) to conclude that the radiocarbon dates from carbonate-bearing lake sediments in the west were too old because of a carbonate error (see Anderson and others, Chapter 4, for discussion of this error), and that the transition from forest to prairie was a synchronous event across Minnesota. The only other site with a varve chronology that documents increases in prairie vegetation is Lake of the Clouds (Fig. 7), about 325 km east northeast of Elk Lake in the Boundary Waters Canoe Area (Craig, 1972). Here Chenopodiineae pollen blown from prairie vegetation in the west became significant about 8.2 ka, according to varve counts, the same time as at Elk Lake, when diatom flux, quartz, and other clastic indicators suggest that winds became stronger.

Paleolimnological data from ostracodes and diatoms (Forester and others, 1987; Bradbury and Dieterich-Rurup, Chapter 15) and mineralogy and geochemistry of sediments from Elk Lake (Dean, Chapter 10) indicate that the lowest lake levels and most alkaline conditions existed between 7.2 and 6.8 ka. This interpretation conforms with an assessment of the driest conditions throughout the upper midwest at the same time derived from pollen evidence of many sites (Bartlein and others, 1984). Individual lacustrine records of dry conditions are difficult to evaluate, however, because of poor age control. For example, the clear paleolimnological evidence of low water levels at Pickerel Lake, northeastern South Dakota, is only generally dated between 8000 and 4000 ^{14}C yr B.P. (ca. 8.9–4.5 ka) (Watts and Bright, 1968). At Elk Lake, Grant County, Minnesota, a low-water, alkaline lake stand is recorded by the ostracode *Limnocythere staplini* in an interval dated at about 7200 ^{14}C yr B.P. (ca. 8 ka) that may be somewhat too old (A. J. Smith, 1991).

Increased aridity and wind intensity on a regional scale (at least the eastern two-thirds of the United States and eastern Canada) during the period between 8 and 5 ka is indicated by the development of active dunes in several places in Minnesota (Grigal and others, 1976; Keen and Shane, 1990), on the Great Plains and Rocky Mountain basins (Ahlbrandt and others, 1983), in the upper Ohio River valley (Simard, 1989), on the Carolina coastal plain (Soller, 1988), and in Quebec (Filion, 1984, 1987).

Grigal and others (1976) defined a large (580 km^2) dune field that resulted from the desiccation and deflation of Lake Winnibigoshish in north-central Minnesota (Fig. 7) by strong and persistent winds from the northwest (Fig. 7). This dune field was active between about 8000 and 5000 ^{14}C yr B.P. (ca. 8.8–6.0 ka), based on radiocarbon dates of charcoal in soil above and below the dunes. The Anoka sand plain in east-central Minnesota is a large (2200 km^2) late Wisconsin glacial outwash plain (Cooper, 1935). However, a parabolic dune field began to develop on the sand plain about 8000 ^{14}C yr B.P. (8.8 ka); dominant winds were from the northwest, and arboreal vegetation on the sand plain was severely reduced by drought conditions (Keen and Shane, 1990). Using magnetic susceptibility as a proxy for eolian sand and silt grains, Keen and Shane postulated three Holocene pulses in the influx of eolian sediment to Lake Ann on the western part of the Anoka sand plain (Fig. 7).

Dune fields of the northern Great Plains, epitomized by the Nebraska Sand Hills, had been thought to be late Pleistocene in age (e.g., Wright, 1970). Although there may have been some late Pleistocene dune migration, it is now known that the most recent dune migration occurred during the Holocene (see reviews by Ahlbrandt and others, 1983; Muhs, 1985; Swinehart, 1990). The primary paleowind direction in most of these dune fields was from the northwest (Ahlbrandt and Fryberger, 1980; Gaylord, 1982; Ahlbrandt and others, 1983; Muhs, 1985; Swinehart, 1990), except in the North Park and Great Sand Dunes fields of Colorado (Andrews, 1981; Ahlbrandt and others, 1983). Ahlbrandt and others (1983) postulated four phases of aridity and active dune migration during the Holocene in the Nebraska Sand Hills and intermontane basins in Colorado and Wyoming. The main (second) phase of dune activity in both the Sand Hills and Rocky Mountain basins corresponds to an extended period of drought (about 7000 to 5000 ^{14}C yr B.P. = 7.8–5.7 ka) in western North America based on archaeological evidence (Benedict, 1979; Holen, 1990) and dwarfing of bison (Wilson, 1974). Gaylord (1982) was able to subdivide the main phase of dune migration in the Ferris dune field of south-central Wyoming into two separate phases from about 7660 to 6460 ^{14}C yr B.P. and from 6460 to 5500 ^{14}C yr B.P. (ca. 8.3–7.4 ka and 7.4–6.4 ka, respectively). We suggest that this main mid-Holocene period of dune activity in the Nebraska Sand Hills and Rocky Mountain basins corresponds to the main period of aridity and eolian influx documented in Elk Lake and Lake Ann (Fig. 5) that reached a maximum 6.0 ka. This is when the strong westerlies reached a maximum in both winter and summer (Bartlein and Whitlock, Chapter 18).

Episodes of Holocene eolian deposition are widespread in eastern North America that may correlate in part with aridity interpreted from the Elk Lake record. For example, parabolic dunes were active in the St. Lawrence lowland in southern Quebec between 10,000 and 7500 ^{14}C yr B.P. (ca. 9–8.4 ka) (Filion, 1984, 1987), and dune migrations in the lower Cape Fear River valley on the Carolina coastal plain were dated at between 7700 and 5720 ^{14}C yr B.P. (ca. 8.5–6.5 ka) (Soller, 1988). Presumed

Holocene eolian sand caps Quaternary fluvial terraces in the upper Ohio River valley in West Virginia (Simard, 1989).

Although the climatic mechanisms for dune formation may vary from place to place, it would appear that the period between about 8000 and 5000 ^{14}C yr B.P. (ca. 9–6 ka) was a period of widespread aridity and eolian activity on North America, with large expanses of dune fields most extensively developed in the Rocky Mountains and Great Plains, but possibly extending to the east coast. At least several other periods of increased aridity and eolian activity during the Holocene are also suggested. From the many different lines of evidence, we conclude that the mid-Holocene was a dynamic period of climatic change, punctuated by rapid transitions between dry and moist intervals (Bryson and others, 1970; Bryson, 1974; Ahlbrandt and others, 1983).

Rapid climatic oscillations in the mid-Holocene

Evidence that rapid changes in climate during the mid-Holocene were regional in extent can be seen by comparing changes in Elk Lake with changes in Lake Ann, about 230 km to the southeast of Elk Lake (Fig. 7). Variations in magnetic susceptibility in sediments from Lake Ann (Fig. 5) show three distinct episodes of eolian activity with maxima in eolian flux at about 7400, 5800, and 4900 ^{14}C yr B.P. (ca. 8.2, 6.6, and 5.7 ka, respectively; Keen and Shane, 1990). The Lake Ann core had only two ^{14}C dates, 7400 ± 80 yr B.P. and 4390 ± 80 yr B.P. (ca. 8.3 and 5.0 ka, respectively). These two calibrated dates are shown in Figure 5 in their correct stratigraphic position relative to the magnetic susceptibility profile. These dates are anchored to the Elk Lake varve record, represented in Figure 5 by quartz concentration and magnetic susceptibility, at 8.3 ka and 5.0 ka. The rest of the Lake Ann magnetic susceptibility profile has been scaled proportionally.

We suggest that there are radiocarbon errors in the Lake Ann dates and that the upper two peaks in magnetic susceptibility in this record correlate with peaks in magnetic susceptibility and quartz in the Elk Lake core at about 5.7 and 4.4 ka, as indicated by the correlation lines in Figure 5. This would make the magnetic susceptibility minimum between these peaks correspond to the minimum in eolian influx, centered on about 5 ka and as shown by many parameters in the Elk Lake core (Figs. 3 and 6). These same two peaks also occur in the influx of grass and *Artemisia* pollen in Elk Lake and Lake Ann. The spacing of the two maxima define an episode of 1300 yr; changes in precipitation during this time probably responsible for exceeding the soil moisture and shear threshold that led to the 600 yr cessation of clastic influx into Elk Lake.

Possible solar origin of fine-scale mid-Holocene climatic Oscillations

The thermal and eolian maximum in the Elk Lake record at about 6 ka lags behind the insolation maximum at about 9 ka during the last Milankovitch cycle (Kutzbach, 1987). A lag of about 3 ka does not preclude a role for large-scale Milankovitch forcing, and some of the changes in vegetation at the maximum of insolation at 9 ka can be interpreted as a result of warming effects associated with deglaciation (Bartlein and Whitlock, Chapter 18).

Modulation of solar radiation by volcanic aerosols in the atmosphere has been suggested as a likely control on global climatic changes at scales of several years to several millennia (e.g., Kennett and Thunell, 1977; Bryson and Goodman, 1980; Bryson, 1989). Bryson (1989) constructed a time series of ^{14}C-dated volcanic eruptions, which he then used as input, along with Milankovitch orbital variations, to model global ice volume and Northern Hemisphere surface temperatures. The model predictions produce fluctuations in temperature on scales of centuries to millennia that Bryson considers to be driven by atmospheric volcanic dust input because there are no fluctuations at those scales in the Milankovich parameters. These fluctuations are particularly well developed in the interval between 8000 and 5000 yr B.P. However, Rampino and others (1979) suggested that although positive volcanic-dust feedback occurs for short-term (<10 yr) global cooling, it is unlikely to produce even minor climate fluctuations in the decadal to centennial range. They further suggest that the inverse may be true, i.e., variations in global climate can lead to changes in stresses in the Earth's crust that might affect volcanicity.

Solar flux, through its control on ^{14}C production, magnetic storms, and zonal winds, may be one way of forcing rapid, short-term climate change. For example, Stuiver and others (1991) observed that the termination of the Younger Dryas cold event is associated with the beginning of a marked decrease in ^{14}C production. Anderson (Chapter 5) presents graphical and statistical evidence to show that mid-Holocene oscillations in varve thickness with periods of about 200 yr appear to correlate with similar oscillations in ^{14}C production, as determined by ^{14}C:^{12}C ratio in tree rings over the same time interval. Time series analyses of Elk Lake varve thickness for the interval 8.0–3.8 ka show increased spectral density at periodicities of 200 yr, 40–50 yr, and 22 yr that are best developed in the interval 7.3–5.3 ka and a similar spectral structure was found for secular changes in ^{14}C production (Fig. 8). Varves that formed in Elk Lake between 8.0 and 3.8 ka have a high content of eolian clastic material (Anderson, 1992, and Chapter 5), and the increase in varve thickness in the mid-Holocene probably reflects an increase in the surface wind field. Changes in the production of ^{14}C are known to be directly linked to an increase in cosmic ray flux. The cosmic ray flux is in turn affected inversely by changes in solar corpuscular emissions (i.e., solar flares), the solar wind, and increases in the interplanetary magnetic field (E. J. Smith, 1991), as well as by changes in the geomagnetic field (Damon and others, 1989).

If the apparent association between ^{14}C production and varve thickness at Elk Lake can be confirmed, it will follow that regional surface winds, as they were recorded at Elk Lake, were affected by changes in solar activity. That this ^{14}C–wind–varve-thickness association is plausible is indicated by the fact that daily changes in coronal mass ejections (solar flares) are followed a few

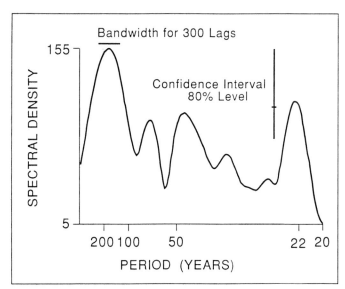

Figure 8. Evolutionary spectrum for Elk Lake varve-thickness time series. The spectrum was obtained by computing sequential and overlapping power spectra for annual data in 600 yr segments of the varve time series (Anderson, Chapter 5).

days later by an increase in tropospheric zonal winds at 700 mbar, and by the fact that Elk Lake lies near the center of the region between the Great Lakes and Hudson Bay where the strongest response to zonal winds has been reported (Stolov and Shapiro, 1974; see discussion by Anderson, Chapter 5). Another interesting aspect of the association is that the strongest oscillations in both varve thickness (regional surface winds) and radiocarbon (solar activity) occur at the time scale of the Maunder deviations, during which there appears to be a maximum effect on the solar wind (E. J. Smith, 1991; Jokipii, 1991). In addition, the 200 yr deviations in radiocarbon over the past 9000 yr are associated with oscillations of about 2100 to 2400 yr, suggesting that century-scale oscillation in effects related to winds at Elk Lake may be related to the millennial-scale oscillations. The changes in varve thickness at Elk Lake that occur just before, during, and after the 600 yr event centered on about 5 ka (Fig. 6) are concident with exceptionally large deviations in ^{14}C that are part of a 2000 to 2500 yr oscillation (Damon and others, 1989).

This 600 yr interruption of clastic accumulation is the most distinctive and abrupt change in the mid-Holocene paleolimnological and paleoenvironmental record in Elk Lake. This event is recorded by magnetic susceptibility in both Elk Lake and Lake Ann (Fig. 5), but the lack of well-dated, high-resolution paleolimnological records from other lakes makes wider correlation difficult. At Pickerel Lake (Fig. 7), Haworth (1972) documented a probable correlative of this event by the stratigraphic distribution of the diatom *Cyclotella comta* that interrupts a large peak of *Melosira* (*Aulacoseira*) *granulata* in the mid-Holocene. The event is not dated, but its relation to the pollen stratigraphy (Watts and Bright, 1968) suggests that it may correlate with the 600-yr interruption of eolian activity at Elk Lake. Both records would reflect the same environmental change, cooler and perhaps fresher water conditions, and a significant reduction in wind stress. A similar fluctuation occurs in Lily Lake, about 320 km southeast of Elk Lake (Fig. 7) in the deciduous forests of east-central Minnesota (Brugam and others, 1988). Here, *Fragilaria crotonensis* interrupts a continuous dominance of *Fragilaria construens,* again suggesting increase of water levels without significant wind-driven summer turbulence. A date on the event of 4.3 ka (corrected) makes it too young to correlate with the Elk Lake event, although the accompanying pollen stratigraphy places the Lily Lake event near the end of the prairie period, as at Elk Lake.

This same climatic event, between 5.4 and 4.8 ka, also may relate to mid-Holocene glacier advances of similar temporal scale in the Rocky Mountains (Benedict, 1983), in several other areas in the world (Denton and Karlén, 1973), and to colder temperatures in the Canadian Arctic deduced from palynological determinations of tree-line position (Nichols, 1974), although dating uncertainties prevent direct correlation. The dramatic expression of this oscillation at Elk Lake is in contrast to changes in other records, which show a gradual increase in aridity and/or temperature to a mid-Holocene maximum (hypsithermal) followed by a gradual increase in moisture and a decrease in temperature (e.g., Wright, 1976). Possibly because of threshold effects, sedimentation in Elk Lake was especially responsive to this particular climatic event. These examples suggest that a strong, but brief, late mid-Holocene (ca. 5 ka) climate perturbation may have a wide distribution in north-central United States.

Although a relation between solar activity, wind intensity, and varve thickness in Elk Lake is speculative, the timing and periodicities of events suggest this possibility. Perhaps the most intriguing association between the time-series for varve thickness and secular changes in radiocarbon is that the relation is restricted to the same 2000-yr window in time when the Earth's dipole moment reached its minimum value in the Holocene (Merrill and McElhinny, 1983). This implies that both changes in solar activity and zonal winds were subject to greater modulation at 6 ka, when the geomagnetic field was at its weakest, and for which there is a logical explanation in the greater flux of low-energy galactic cosmic rays. The expression of the association is restricted to the dry mid-Holocene, when varve thickness was related to wind activity. Evaluation of records at other sites are required to confirm the relation. If this can be done, and if regional winds are related to cosmic ray flux at all frequencies of change, it leaves the possibility that changes in the geomagnetic field, independent of changes from solar activity, may have an effect on winds and climate and thereby account for widespread aridity at 6 ka.

Late Holocene and neoglacial environments

The sudden end of the prairie period and beginning of modern climate regimes at Elk Lake at 3.8 ka can be correlated to several pollen sites in the north-central United States that have recorded the onset of cooler climates at about 4.0 ka (Whitlock and others, Chapter 17). Cooler climates and increased moisture also are documented by the initiation of bog growth in many parts of north-central Minnesota (Webb and others, 1983), and suggested by paleolimnological changes shown by ostracodes in Elk Lake, Grant County, Minnesota (Fig. 7) (A. J. Smith, 1991). The migration of white pine (*Pinus strobus*) westward across northern Minnesota documents a progressive replacement of dry, oak-dominated forests by mixed coniferous vegetation as cooler and wetter climates extended westward (Jacobson, 1979). White pine largely had replaced the oak savanna around Elk Lake about 2.7 ka and peaked about 0.9 ka (Fig. 2). This corresponds to an increase in values of $\delta^{18}O$ that also peak at about 0.9 ka (Fig. 3).

This inferred cooler period occurred long before the historical European Little Ice Age (ca. 0.5–0.1 ka or A.D. 1450–1850). However, neoglacial advances are recorded in many parts of the world beginning about 3.0 and 2.0 ka (Denton and Karlén, 1973), and may correlate generally with the cool and moist climates in the Elk Lake area. The dating of glacial fluctuations and the paleoclimatic interpretations of the Elk Lake record are loose enough, however, to prevent detailed comparisons at this time.

Some climatic characteristics of the Medieval Warm Period and the Little Ice Age are well known because of their historical documentation in Europe (e.g., Grove, 1988; Lamb, 1982), but it is clear that the climatic expression of such episodes varies from place to place. The Medieval Warm Period (= Neo-Atlantic, A.D. 700–1200) in the north-central United States was characterized by a northerly position of the Pacific airstream and consequently mild winters and wet summer conditions in the northern Plains (Bryson, unpublished). Such climates apparently allowed nonirrigated corn cultivation as far west as central Colorado (Baerreis and Bryson, 1967), and may have been partly responsible for the dominance of *Pinus strobus* in the Elk and Deming lakes pollen records about 0.9 ka (Fig. 2) (Whitlock and others, Chapter 17; Clark, Chapter 19). The abrupt end of mild and wet conditions in the northern plains was caused by extended drought as the Pacific airstream (the westerlies) moved southward into Minnesota and neighboring states by A.D. 1200 under the influence of an expanding circumpolar vortex (Bryson, unpublished). At Elk Lake, increased windiness suggested by peaks of *Aulacoseira* and quartz (Figs. 3 and 4) at 0.7 ka (A.D. 1200) may relate to this change.

The Little Ice Age (= Neo-Boreal, ca. A.D. 1450–1850) may reflect temperate climates under the effect of a greatly expanded circumpolar vortex and a dominance of meridional circulation of the Arctic airstream. Such climatic patterns were established between A.D. 1350 and 1600, but apparently continued to the end of this climatic episode by A.D. 1850 (Bryson, unpublished). Lack of relevant historical climatic records in North America has prompted correlations of proxy paleoclimatic evidence with the European events. For example, Grimm (1983) and Swain (1973) found evidence for reduced fire frequency (cooler and wetter climates) in pollen records from south-central Minnesota (A.D. 1550–1750) and Lake of the Clouds (A.D. 1400–1700). However, forest-fire frequencies determined from charcoal flux in varved sediments of Deming Lake (4 km southeast of Elk Lake) are highest between A.D. 1400 and 1600 and less abundant before and after (Clark, Chapter 19; Fig. 9). Diatom profiles that span the Little Ice Age at Dogfish Lake, about 90 km west of Lake of the Clouds, also suggest drier conditions (Bradbury, 1986; Bradbury and Dieterich-Rurup, Chapter 15), although this record is not precisely dated.

It is clear that neither the Medieval Warm Period nor the Little Ice Age can be characterized uniformly from paleoecological and paleolimnological records in the north-centeral United States. Records sensitive to interannual variability, such as diatoms, varve thickness, charcoal fluxes, geochemistry, and high resolution pollen records (e.g., Gajewski, 1989) all show considerable fluctuations during the last millennium. The close-interval sampling of pollen and charcoal flux in Deming Lake (Clark, Chapter 19; Fig. 9) illustrate extreme cyclicity at decadal time

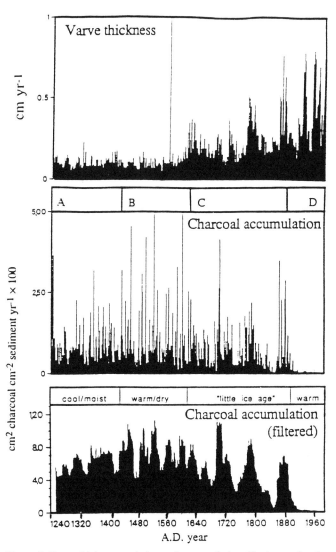

Figure 9. Varve thickness and charcoal accumulation (flux) rates for the past 720 yr at Deming Lake, Itasca State Park, Clearwater County, Minnesota (Clark, Chapter 19).

scales and broader climatic changes affecting fire frequencies at 40 to 100 yr intervals, similar to those found in varve-thickness variations at Elk Lake (Anderson, Chapter 5). Denton and Karlén (1973) acknowledged that many glacial records of the Little Ice Age and the Medieval Warm Period elsewhere suggest variable climates. Climatic variability would seem a logical consequence of increased meridional circulation (Bryson, unpublished), and the strong variations in proxy records of climate change during the past 500 yr can be taken as evidence of the prevalence of this circulation pattern. Detailed analyses of geochemical, sedimentological, and paleontological parameters for the past 3.0 ka at Elk lake are now under investigation and may provide insights into the climatic character of those times.

EPILOGUE

The integrated geochemical, biological, and sedimentological studies at Elk Lake have reinforced and clarified paleolimnological and paleoclimatic interpretations. In many cases the addition of independent data sets has removed interpretational ambiguities that characterize paleoenvironmental studies. There still exist ambiguities, however, that can only be resolved by increased knowledge of "process" limnology. The use of sediment traps (Nuhfer and others, Chapter 7) in the Elk Lake study was a first step in an attempt to relate how this lake's interannual and seasonal limnological dynamics were ultimately expressed in sediment accumulating on the lake bottom. In this endeavor we were only partly successful. The traps were experimental and some annual or biennial sediment records were not recovered. In addition, during much of the trapping period Elk Lake was not monitored chemically, physically, or biologically, resulting in an unfortunate loss of information that could have been correlated to the deposits in the traps. The single most important lesson that the Elk Lake study contributes to future studies is that a lake monitoring and trapping program must begin as early as possible in the study and be continued throughout the investigation and into the future to provide the scientific community a continuous flow of data on how the lake, a sensitive climate monitor, is responding to natural and anthropogenic climate change.

REFERENCES CITED

Ahlbrandt, T. S., and Fryberger, S. G., 1980, Eolian deposits in the Nebraska Sand Hills: U.S. Geological Survey Professional Paper 1120A, p. A1–A24.

Ahlbrandt, T. S., Swinehart, J. B., and Maroney, D. G., 1983, The dynamic Holocene dune fields of the Great Plains and Rocky Mountain basins, U.S.A., in Brookfield, M. E., and Ahlbrandt, T. S., eds., Eolian sediments and processes: Amsterdam, Elsevier Science Publishers, p. 379–406.

Anderson, R. Y., 1991, Solar variability captured in climatic and high-resolution paleoclimatic records: A geologic perspective, in Sonett, C. P., Giampapa, M. S., and Mathews, M. S., eds., The Sun in time: Tucson, University of Arizona Press, p. 543–561.

——, 1992, Possible connection between changes in climate, solar activity, and earth's magnetic field: Evidence in varved sediments from a Minnesota lake: Nature, v. 358, p. 51–53.

Anderson, R. Y., Linsley, B. K., and Gardner, J. V., 1990, Expression of seasonal and ENSO forcing in climatic variability at lower than ENSO frequencies: Evidence from marine varves off California, in Meyers, P. A., and Benson, L. V., eds., Paleoclimates: The record from lakes, ocean, and land: Palaeogeography, Palaeoclimatology, Palaeoecology, v. 78, p. 287–300.

Andrews, J. T., 1987, The late Wisconsin glaciation and deglaciation of the Laurentide ice sheet, in Ruddiman, W. F., and Wright, H. E., Jr., eds., North America and adjacent oceans during the last deglaciation: Boulder, Colorado, Geological Society of America, The Geology of North America, v. K-3, p. 13–37.

Andrews, S. G., 1981, Sedimentology of Great Sand Dunes, Colorado, in Ethridge, F. G., and Flores, R. M., eds., Non-marine depositional environments: Models for exploration: Society of Economic Paleontologists and Mineralogists Special Publication 31, p. 270–291.

Baerreis, D. A., and Bryson, R. A., eds., 1967, Climate change and the Mill Creek culture of Iowa: Archives of Archaeology, v. 29, 673 p.

Bartlein, P. J., Webb, T., III, and Fleri, E., 1984, Holocene climate change in the northern Midwest: Pollen derived estimates: Quaternary Research, v. 22, p. 361–374.

Bartlein, P. J., Prentice, I. C., and Webb, T., III, 1986, Climatic response surfaces from pollen data for some eastern North American taxa: Journal of Biogeography, v. 13, p. 35–57.

Benedict, J. B., 1979, Getting away from it all: A study of man, mountains, and the two-drought altithermal: Southwestern Lore, v. 456, no. 3, p. 1–12.

——, 1983, Chronology of cirque glaciation, Colorado Front Range: Quaternary Research, v. 3, p. 584–599.

Bradbury, J. P., 1986, Effects of forest fire and other disturbances on wilderness lakes in northeastern Minnesota II. Paleolimnology: Archiv für Hydrobiologie, v. 106, p. 203–217.

——, 1988, A climatic-limnologic model of diatom succession for paleolimnological interpretation of varved sediments at Elk Lake, Minnesota: Journal of Paleolimnology, v. 1, p. 115–131.

Bradbury, J. P., Tarapchak, S. J., Waddington, J.C.B., and Wright, R. F., 1975, The impact of a forest fire on a wilderness lake in northeastern Minnesota: Verhandlung der Internationalen Vereinigung für Theoretische und Angewandte Limnologie, v. 19, p. 875–883.

Brazel, A. J., 1989, Dust and climate in the American Southwest, in Leinen, M., and Sarnthein, M., eds., Paleoclimatology and paleometeorology: Modern and past patterns of global atmospheric transport: Dordrecht, Kluwer Academic Publishers, p. 65–96.

Broecker, W. S., Andree, M., Wolfli, W., Oeschger, H., Bonani, G., Jennett, J., and Peteet, D., 1988, The chronology of the last deglaciation: Implications for the cause of the Younger Dryas event: Paleoceanography, v. 3, p. 1–19.

Broecker, W. S., Jennett, J. P., Flower, B. P., Teller, J. T., Trumbore, S., Bonani, G., and Wolfli, W. G., 1989, Routing of meltwater from the Laurentide ice sheet during the Younger Dryas cold episode: Nature, v. 341, p. 318–321.

Brugam, R. B., Grimm, E. C., and Eyster-Smith, N. M., 1988, Holocene environmental changes in Lily Lake, Minnesota, inferred from fossil diatom and pollen assemblages: Quaternary Research, v. 30, p. 53–66.

Bryson, R. A., 1966, Air masses, streamlines, and the boreal forest: Geographical Bulletin, v. 8, p. 228–269.

——, 1974, A perspective on climatic change: Science, v. 184, p. 753–760.

——, 1989, Late Quaternary volcanic modulation of Milankovitch climate forcing: Theoretical and Applied Climatology, v. 39,, p. 115–125.

Bryson, R. A., and Goodman, B. M., 1980, Volcanic activity and climatic changes: Science, v. 207, p. 1041–1044.

Bryson, R. A., and Hare, F. K., 1974, Climates of North America, in Landsberg, H. E., ed., World survey of climatology, Volume 11: New York, Elsevier, 420 p.

Bryson, R. A., Baerreis, D. A., and Wendland, W. M., 1970, The character of late-glacial and post-glacial climatic changes, in Dort, W., Jr., and Jones, J. K., Jr., eds., Pleistocene and recent environments of the central Great Plains: Lawrence, Kansas, University of Kansas Press, p. 53–74.

Cooper, W. S., 1935, The history of the upper Mississippi River in late Wisconsin and post-glacial time: Minneapolis, University of Minnesota Press, 116 p.

Craig, A. J., 1972, Pollen influx to laminated sediments, a pollen diagram from northeastern Minnesota: Ecology, v. 53, p. 46–57.

Damon, P. E., Cheng, S., and Linick, T. W., 1989, Fine and hyperfine structure in the spectrum of secular variations in atmospheric ^{14}C: Radiocarbon, v. 31, p. 697–703.

Davis, M. B., 1973, Differential sedimentation of pollen grains in lakes: Limnology and Oceanography, v. 18, p. 635–646.

DeDeckker, P., Correge, T., and Head, J., 1991, Late Pleistocene record of cyclic eolian activity from tropical Australia suggesting the Younger Dryas is not an unusual climatic event: Geology, v. 19, p. 602–605.

Denton, G. W., and Karlén, W., 1973, Holocene climatic variations—Their pattern and cause: Quaternary Research, v. 3, p. 155–205.

Edwards, T.W.D., and Fritz, P., 1986, Assessing meteoric water composition and relative humidity from ^{18}O and ^{2}H in wood cellulose—Paleoclimatic implications for southern Ontario, Canada: Applied Geochemistry, v. 1, p. 715–723.

Engstrom, D. R., Swain, E. B., and Kingston, J. C., 1985, A paleolimnological record of human disturbance from Harvey's Lake, Vermont: Geochemistry, pigments, and diatoms: Freshwater Biology, v. 15, p. 261–288.

Filion, L., 1984, A relationship between dunes, fire, and climate recorded in the Holocene deposits of Quebec: Nature, v. 309,, p. 543–546.

—— , 1987, Holocene development of parabolic dunes in the central St. Lawrence lowland: Quaternary Research, v. 28, p. 196–209.

Forester, R. M., Delorme, L. D., and Bradbury, J. P., 1987, Mid-Holocene climate in northern Minnesota: Quaternary Research, v. 28, p. 263–273.

Gajewski, K., 1989, Late Holocene climate changes in eastern North America estimated from pollen data: Quaternary Research, v. 29, p. 255–262.

Gaylord, D. R., 1982, Geologic history of the Ferris dune field, south-central Wyoming, in Marrs, R. W., and Kolm, K. E., eds., Interpretation of wind-flow characteristics from eolian landforms: Geological Society of America Special Paper 192, p. 65–82.

Gorham, E., Dean, W. E., and Sanger, J. E., 1983, The chemical composition of lakes in the north-central United States: Limnology and Oceanography, v. 28, p. 287–301.

Grigal, D. F., Severson, R. C., and Goltz, G. E., 1976, Evidence of eolian activity in north-central Minnesota 8,000 to 5,000 yr ago: Geological Society of America Bulletin, v. 87, p. 1251–1254.

Grimm, E. C., 1983, Chronology and dynamics of vegetation change in the prairie-woodland region of southern Minnesota, U.S.A.: New Phytologist, v. 93, p. 311–350.

Grove, J. M., 1988, The little ice age: London, New York, Routledge, 498 p.

Haworth, E. Y., 1972, Diatom succession in a core from Pickerel Lake, northeastern South Dakota: Geological Society of America Buleltin, v. 83, p. 157–172.

Holen, S. R., 1990, Anthropology: Native American occupation of the Sand Hills, in Bleed, A., and Flowerday, C., eds., An atlas of the Sand Hills: Lincoln, Nebraska, Conservation and Survey Division, Institute of Agriculture and Natural Resources, Resource Atlas no. 5a, p. 189–205.

Huntley, B., and Prentice, I. C., 1988, July temperatures in Europe from pollen data 6,000 years before present: Science, v. 241, p. 687–690.

Jacobson, G. L., Jr., 1979, The paleoecology of white pine (*Pinus strobus*) in Minnesota: Journal of Ecology, v. 67, p. 697–726.

Jacobson, G. L., Jr., Webb, T., III, and Grimm, E. C., 1987, Patterns and rates of vegetation change during the deglaciation of eastern North America, in Ruddiman, W. F., and Wright, H. E., Jr., eds., North America and adjacent oceans during the last deglaciation: Boulder, Colorado, Geological Society of America, The Geology of North America, v. K-3, p. 277–288.

Jokipii, J. R., 1991, Variations of the cosmic ray flux with time, in Sonnett, C. P., Giampapa, M. S., and Mathews, M. S., eds., The Sun in time: Tucson, University of Arizona Press, p. 205–220.

Keen, K. L., and Shane, L.C.K., 1990, A continuous record of eolian activity and vegetation change at Lake Ann, east-central Minnesota: Geological Society of America Bulletin, v. 102, p. 1646–1657.

Kennett, J. P., 1990, The Younger Dryas cooling event: An introduction: Paleoceanography, v. 5, p. 891–895.

Kennett, J. P., and Thunell, R. C., 1977, On explosive Cenozoic volcanism and climatic implications: Science, v. 196, p. 1231–1234.

Kutzbach, J. E., 1987, Model simulations of the climatic patterns during the deglaciation of North America, in Ruddiman, W. F., and Wright, H. E., Jr., eds., North America and adjacent oceans during the last deglaciation: Boulder, Colorado, Geological Society of America, The Geology of North America, v. K-3, p. 425–446.

Lamb, H. H., 1982, Climate history and the modern world: London, Methuen, 387 p.

Lewis, C.F.M., and Anderson, T. W., 1989, Oscillations of levels and cool pulses of the Laurentide Great Lakes caused by inflows from glacial Lakes Agassiz and Barlow-Ojibway: Journal of Paleolimnology, v. 2, p. 99–146.

Lindsley, B. K., and Thunnel, R. C., 1990, The record of deglaciation in the Sulu Sea: Evidence for the Younger Dryas event in the western tropical Pacific: Paleoceanography, v. 5, p. 1025–1039.

Mackereth, F.J.H., 1966, Some chemical observations on post-glacial lake sediments: Royal Society of London Philosophical Transactions, ser. B, v. 250, p. 165–213.

McAndrews, J. H., 1966, Postglacial history of prairie, savanna, and forest in northwestern Minnesota: The Torrey Botanical Club Memoirs, v. 22, p. 1–72.

McColl, J. G., and Grigal, D. F., 1977, Nutrient changes following a forest wildfire in Minnesota: Effects in watersheds with differing soils: Oikos, v. 28, p. 105–112.

Merrill, R. T., and McElhinny, M. W., 1983, The Earth's magnetic field, its history, origin, and planetary perspective: New York, Academic Press, p. 95–133.

Miller, G. H., Hearty, P. J., and Stravers, J. A., 1988, Ice-sheet dynamics and glacial history of southeasternmost Baffin Island and outermost Hudson Strait: Quaternary Research, v. 30, p. 116–136.

Muhs, D. R., 1985, Age and paleoclimatic significance of Holocene sand dunes in northeastern Colorado: Annals of the Association of American Geographers, v. 75, p. 566–582.

Nichols, H., 1974, Arctic North American paleoecology: The recent history of vegetation and climate deduced from pollen analysis, in Ives, J. D., and Barry, R. J., eds., Arctic and alpine environments: London, Meuthen, p. 637–667.

Oeschger, H., Neftel, A., Staffelbach, T., and Stauffer, B., 1988, The dilemma of rapid variations of CO_2 in Greenland ice cores [abs.]: Annals of Glaciology, v. 10, p. 215–216.

Pederson, T. F., Nielsen, B., and Pickering, M., 1991, Timing of late Quaternary productivity pulses in the Panama Basin and implications for atmospheric CO_2: Paleoceanography, v. 6, p. 657–677.

Prentice, I. C., Bartlein, P. J., and Webb, T., III, 1991, Vegetation and climate change in eastern North America since the last glacial maximum: Ecology, v. 72, no. 6, p. 2038–2056.

Rampino, M. R., Self, S., and Fairbridge, R. W., 1979, Can rapid climatic change cause volcanic eruptions?: Science, v. 206, p. 826–828.

Ruddiman, W. F., and Wright, H. E., Jr., 1987, Introduction, in Ruddiman, W. F., and Wright, H. E., Jr., eds., North America and adjacent oceans during the last deglaciation: Boulder, Colorado, Geological Society of America, The Geology of North America, v. K-3, p. 1–12.

Simard, C. M., 1989, Geologic history of the lower terraces and floodplains of the upper Ohio River valley: West Virginia Geological and Economic Survey Open-File Report 8903, 159 p.

Smith, A. J., 1991, Lacustrine ostracodes as paleohydrological indicators in Holocene lake records of the north-central United States [Ph.D. thesis]: Providence, Rhode Island, Brown University, 306 p.

Smith, E. J., 1991, The Sun and interplanetary magnetic field, in Sonett, C. P., Giampapa, M. S., and Mathews, M. S., eds., The Sun in time: Tucson, University of Arizona Press, p. 175–201.

Soller, D. R., 1988, Geology and tectonic history of the lower Cape Fear River valley, southeastern North Carolina: U.S. Geological Survey Professional

Paper 1466-A, 60 p.

Sprowl, D. R., and Banerjee, S. K., 1989, The Holocene paleosecular variation record from Elk Lake, Minnesota: Journal of Geophysical Research, v. 94, p. 9369-9388.

Stark, D. M., 1976, Paleolimnology of Elk Lake, Itasca State Park, northwestern Minnesota: Archiv für Hydrobiologie, Supplement 50 (Monographische Beitrage), p. 208-274.

Stolov, H. L., and Shapiro, R., 1974, Investigation of the responses of the general circulation at 700 Mbar to solar-geomagnetic disturbance: Journal of Geophysical Research, v. 79, p. 2161-2170.

Stuiver, M., and Reimer, P. J., 1986, A computer program for radiocarbon age calibration: Radiocarbon, v. 28, p. 1022-1030.

Stuiver, M., Braziunas, T. F., Becker, B., and Kromer, B., 1991, Climatic, solar, oceanic, and geomagnetic influences on late-glacial and Holocene atmospheric $^{14}C/^{12}C$ change: Quaternary Research, v. 35, p. 1-24.

Swain, A. M., 1973, A history of fire and vegetation in northeastern Minnesota as recorded in lake sediments: Quaternary Research, v. 3, p. 383-396.

Swinehart, J. B., 1990, Wind-blown deposits, *in* Bleed, A., and Flowerday, C., eds., An atlas of the Sand Hills: Lincoln, Nebraska, Conservation and Survey Division, Institute of Agriculture and Natural Resources, Resource Atlas no. 5a, p. 43-56.

Teller, J. T., 1985, Glacial Lake Agassiz and its influence on the Great Lakes, *in* Karrow, P. F., and Calkin, P. E., eds., Quaternary evolution of the Great Lakes: Geological Association of Canada Special Paper 30, p. 1-16.

—— , 1987, Proglacial lakes and the southern margin of the Laurentide ice sheet, *in* Ruddiman, W. F., and Wright, H. E., Jr., eds., North America and adjacent oceans during the last deglaciation: Boulder, Colorado, Geological Society of America, The Geology of North America, v. K-3, p. 39-70.

—— , 1990, Meltwater and precipitation runoff to the North Atlantic, Arctic, and Gulf of Mexico from the Laurentide ice sheet and adjacent regions during the Younger Dryas: Paleoceanography, v. 5, p. 897-906.

Watts, W. A., and Bright, R. C., 1968, Pollen, seed and mollusk analysis of a sediment core from Pickerel Lake, northeastern South Dakota: Geological Society of America Bulletin, v. 79, p. 855-876.

Webb, T., III, 1985, Holocene palynology and climate, *in* Hecht, A. D., ed., Paleoclimate analysis and modeling: New York, John Wiley and Sons, Inc., p. 163-195.

Webb, T., III, Cushing, E. J., and Wright, H. E., Jr., 1983, Holocene changes in the vegetation of the midwest, *in* Wright, H. E., Jr., ed., Late Quaternary environments of the United States, Volume 2, The Holocene: Minneapolis, University of Minnesota Press, p. 142-165.

Webb, T., III, Bartlein, P. J., and Kutzbach, J. E., 1987, Climatic change in eastern North America during the last 18,000 years; comparisons of pollen data with model results, *in* Ruddiman, W. F., and Wright, H. E., Jr., eds., North America and adjacent oceans during the last deglaciation: Boulder, Colorado, Geological Society of America, The Geology of North America, v. K-3,, p. 447-462.

Wilson, M., 1974, History of the bison in Wyoming, with particular reference to early Holocene forms, *in* Wilson, M., ed., The Holocene history of Wyoming: Laramie, Wyoming Geological Survey Report of Investigations 10, p. 91-99.

Winkler, M. G., Swain, A. M., and Kutzbach, J. E., 1986, The middle Holocene dry period in the northern midwestern United States: Lake levels and pollen stratigraphy: Quaternary Research, v. 25, p. 235-250.

Wright, H. E., Jr., 1970, Vegetational history of the central plains, *in* Dort, W., Jr., and Jones, J. K., Jr., eds., Pleistocene and recent environments of the central Great Plains: Lawrence, University of Kansas Press, p. 157-172.

—— , 1976, The dynamic nature of Holocene vegetation: Quaternary Research, v. 6, p. 581-596.

—— , 1984, Sensitivity and response time of natural systems to climatic change in the late Quaternary: Quaternary science reviews, v. 3: p. 91-131.

—— , 1987, Synthesis; The land south of the ice sheets, *in* Ruddiman, W. F., and Wright, H. E., Jr., eds., North America and adjacent oceans during the last deglaciation: Boulder, Colorado, Geological Society of America, The Geology of North America, v. K-3, p. 479-488.

—— , 1992, Patterns of Holocene climatic change in the midwestern United States: Quaternary Research, v. 38, p. 129-134.

MANUSCRIPT ACCEPTED BY THE SOCIETY JULY 27, 1992

Index

[Italic page numbers indicate major references]

A

Abies, 255, 265, 271
  balsamea, 298
  balsamifera, 252
Acer
  rabrum, 252
  saccharum, 252, 265, 295, 298, 300
  spicatum, 198
acidophil, 207
acoustic backscattering, 22
Acraperus harpae, 34
aeration, 24
aerosols, volcanic, 323
aggregates, 49, 51, 57
airstreams, *1*, 5, *20*, 46, 172, 252, 259, 310, 317, *319*, 321, 325
albite, 140
Alexandria moraine complex, 8, 11
algae, 311
  debris, 175
  flagellated, 239
  planktonic, 235
alkalinity, 24, 25, 27, 191, 203
allochthonous components, 136
Alnus, 256
  crispa, 252
  rugosa, 252, 257
Alona quadrangulata, 35
Alonella excisa, 34
aluminosilicate minerals, 154
aluminum, 5, 64, 80, 84, 88, 145
Ambrosia, 38, 40, 218, 224, 233, 254, 256, 259
Amnicola, 35
Amphora
  ovalis, 218, 221, 228
  perpusilla, 218, 228, 229
  type pollen, 259
Anabaena, 34
analogue method, 275, 276, *277*, 280
  surface sample, *189*
Anchnanthes, 229, 230
anhysteretic remanent magnetization (ARM), *159*, 320
Anoka sandplain, 11, 52, 322
anoxic conditions, 23, 24, 29, 31, 98, 136, 149, 181, 185, 216, 226, 234
Antarctic dust, 321
anticyclone, glacial, 287, *288*
Appalachian region, 255
aragonite, 46, 48, 54, 98, 102, 111, 112, 139, 140, 317
Arco Lake, 24, 127, 128, 129, 130, 131, 132, 192, 296, 305
aridity, 311
  regional, *321*
Artemisia, 42, 46, 57, 186, 228, 254, 255, 257, 258, 271, 281, 315, 323

Artemisia (continued)
  frigida, 257
  ludoriciana, 257
assemblages
  fossil diatom, *189*
  hardwood, 303
  Picea, *254*, 278, 282
  Picea-Populus, 255
  Pinus banksiana–Pinus resinosa-pteridium, *254*, 255, 282
  Pinus strobus, *254*, 265, 271, 283
  Quercus-Gramineae-*Artemisia*, 254, *256*, 282
  Quercus-Ostrya-Carpinus, 278, 283
Aster macrophyllus, 298
Asterionella, 218, 231
  formosa, 80, 82, 85, 155, 208, 218, 224, 226, 230, 231, 232, 233
atmosphere, *287*
Aulacoseira, 210, 211, 218, 221, 227, 230, 317, *319*, 320, 325
  ambigua, 155, 192, 203, 208, 210, 211, 224, 227, 229, 230, 232, 234, 235
  distans, 211
  granulata, 192, 203, 210, 227, 230, 232, 234
  italica, 210, 211, 227
autochthonous components, 136
autrophication, 177

B

Bacillariophyta, 32
backscattering, acoustic, 22
Ball Club Lake, 127, 128, 129, 130, 131, 132
barium, 84, 89, 149
bathymetry, *50*, *55*
Beaver Lake, 127, 128, 129, 130, 131, 132
bedrock, 120
Betula, 42, 230, 254, 255, 257, 259, 265, 271, 282, 283, 301, 303, 313, 319, 321
  lutea, 252
  papyrifera, 252, 298, 300, 301, 303
Betulaceae, 300
bicarbonate ions, 313
Big Kandiyohi Lake, 127, 128, 129, 130, 131, 132
Big Lake, 127, 128, 129, 130, 131, 132
Bog D, 16, 40, 252, 255, 270, 315
bogs, 16
  growth, 325
Boxmina longirostris, 34
Browns Lake, Ohio, 186
Budd Lake, 296, 305
Burntside Lake, 231

C

calcite, 54, 55, 85, 93, 97, 98, 100, 101, 102, 107, 108, 109, 110, 138, 139, 140, 149, 153, 168, 171, 172, 177, 228, 313
  crystalline, 106
  nucleation, 109
  precipitates, 101, 102, 109
calcium, 2, 4, 16, 25, 27, 55, 80, 85, 88, 89, 97, 99, 100, 102, 148, 313
  depletion, 109, 111
  ions, 313
calcium bicarbonate, 25
calcium carbonate, 5, 6, 16, 25, 27, 33, 46, 49, 54, 64, 84, *97*, 116, 139, 145, 154, 163, 169, 184, 185, 311, 320
  deposition environment, *97*
Campbell beach, 13
Campbell level, 13
Camptocercus rectirostis, 34
Candona, 35, 171, 172, 174
  decora, 35
  ohioensis, 35, 166, 171, 172, 173
canthaxanthin, 183
Cape Fear River valley, 322
carbon, 152, 300, 313
  burial, 174
  dissolved inorganic, 172
  isotopes, *45*, *60*, *62*, *172*, 317, 323
  isotope studies, *163*
  organic, 138, 149, 151, 152, 155, 166, *174*, 178
carbonate, 9, 11, 25, 48, 50, 91, 98, 145, 149, 154, 156, 171, 178, 185
  bulk-sediment, 166
  carbon, *172*
  deposition, *98*
  lacustrine materials, 164
  marine materials, 164
  minerals, 5, 98, 140, 228
  precipitation, 99, 100
  saturation, *106*
carotenes, 183
carotenoids, 181, *184*
Carpinus, 252
Cass phase, 289
catchment, 25, 51, 296, 305
cations, 98, 116, 300
Cedar Bog Lake, 119, 121, 127, 128, 129, 130, 131
Ceratophyllum deversum, 35
Chambers Creek, 19
Chaoborus, 22, 33, 34
Chara, 34, 98
charcoal, 296, 304

329

charophytes, 168, 311
chemical analysis, Minnesota lakes, *127*
chemistry, *25*
chenopodiineae, 256, 257, 322
chironomids, *35*, 85, 228, 313
Chironomus, 35
  plumosus, 35
chlorophyll, 24, 25, 33, 311
  derivatives, *182, 184*
  undecomposed, 181
Christmas Lake, 127, 128, 129, 130, 131, 132
Chrysophyceae, 239
chrysophytes, 80, *239, 245*
Chydorus sphaericus, 34, 234
circulation events, seasonal events, *78*
Cladocera, 22, 33, *34*
clastics, 5, 51, 53, 57, 64, 85, 92, 121, 125, 149, 161, 230, 313, 317
  debris, 25
  minerals, 153
clay, 5, 13, 39, 46, 48, 50, 51, 54, 57, 62, 313
Clear Lake, 127, 128, 129, 130, 131, 132
Clearwater Lake, 121, 127, 128, 129, 130, 131, 132
Clearwater County, Minnesota, *75*
climate, 15, *19, 63*, 172, 178, 310
  arid, 311
  carbon isotope connection, *45, 60*
  change, *6*, 9, 46, 234, *295, 298*, 305, 310, 311, 320
  data, *275*
  fluctuations, 304
  glacial, 184
  glacial intervention, 291
  gradients, 115, 169, 189, 317
  history, *281, 309*
  oscillations, 1, *6*, 46, 52, 57, 59, *63*, 160, 289, 321, *323*
  perturbation, *317*
  reconstruction, 278
  record, *1*
  reversal, 6, 16
  solar-geomagnetic connection, *45, 60, 63*
  solar origin oscillations, *323*
  succession, *6*
  values, 278
  variability, 1, *45, 57, 64*, 186, 187, *326*
  variables, 288
  variations, *275, 287, 291*, 313
  warming, 168
Cocconeis placentula, 224
colloids, 153
cooling, global, 323
Copepoda, 22, 33, 91
copper, 122
cores, 255, 295, 310
  deep-basin, 136, 166, 168, 170, 171, 172, 178
  frozen, 221, 241, 296

cores *(continued)*
  long, *242*
  short, *244*
  splicing, *38*
  sublittoral, 166, 168, 170, 171, 172, 178
coring, *37*, 50, 69, *70, 207*, 217, 227, 228
Cornus, 252
Corylus, 252, 256, 257, 298, 300, 301, 303
cosmic particle flux, *45, 60, 62*
cosmic rays, 62
Coteau des Prairies, 11
Cowdry Lake, 127, 128, 129, 130, 131, 132
Crane Lake, 119, 121, 122, 127, 128, 129, 130, 131, 132
crustaceans, 231
Crystal Lake, Wisconsin, 32
crystalline rocks, 115, 120
crystallites, 49
crystals, 16
currents
  lake bottom, 90
  turbidity, 98, 186
cyanobacteria, 32, 33, 84, 85, 183, 185, 187, 228, 315, 319
cyanophyceans, 186
Cyclocypris sp., 35
Cyclops, 33
Cyclotella, 85, 192, 218, 227
  comta, 192, 203, 218, 227, 231, 232, 233, 313, 319, 324
  kuetzingiana, 218, 227, 228, 231
  michiganiana, 84, 192, 208, 211, 218, 223, 227, 228, 231
  ocellata, 192, 218, 228, 313
  stelligera, 81, 211
Cymbella, 218
Cyperaceae, 254, 255, 281
Cypridopsis, 35
  vidua, 35
cysts
  abundance, 242
  assemblage, 242, 245, 247
  chrysophycean, 6
  chrysophycean (postglacial) record, *239*
  chrysophyte, 241, 320
Cytherissa, 35
  lacustris, 228, 311, 313

D

Daphnia, 231
  catawba, 34
  galeata, 34
  longispina, 34
Dark Lake, 298
Darwinula, 35
  stevensoni, 35
Dead Coon Lake, 121, 127, 128, 129, 130, 131, 132
debris
  algal, 175
  clastic, 25

debris *(continued)*
  plant, 13, 255
deglaciation, 311, 313, 320
  time-transgressive changes, 320
deltas, 13
Deming core, 296
Deming Lake, 24, 127, 128, 129, 130, 131, 132, 296, 301, 303, 305, 325
dentrification, 31
depression, 48, 53, 55, 91, 93, 104, 185, 217
Des Moines glacier, 190
Des Moines lobe, 8, 11, 19, 252, 311
Des Moines River valley, 11
Des Moines–St. Louis till, 311
De Sota Lake, 11
detritus, 13, 16, 48, 98, 311
  phytoplankton, 181
diagenesis, *48*, 182
Diaphanosoma, 34
diatoms, 6, 16, 31, 32, 46, 51, 53, 54, 80, 91, 109, 151, 217, 247, 286, 311, 313, 322
  benthic, 224, 229, 232, 234, 247
  concentration, *221*
  debris, 25
  ecology, *224*
  flux, 5, 319
  fossil assemblages, *189*
  growth, 226
  influx, 229, 315
  littoral, 229
  marine, 227
  planktonic, 224, 228, 234
  production, 177
  productivity, 226
  relative frequency, *218*
Dinobryan, 245
  cylindrieum, 245
discharge, 13
Dogfish Lake, 232, 325
dolomite, 8, 98, 100, 102, 107, 108, 109, 110, 111, 138, 139, 140, 149, 154, 171
drainage, 2, 5, 46, 320
drainage basin, 136, 153
drainage channels, 216
drift
  calcareous, 116
  glacial, 8, 11, 13, 19, 110, 151, 155
drought, 63, 65, 187, 232, 301, 310, *313*, 322
  cycles, 5
drumlins, 7, *8*
Duluth Gabbro, 121
dunes, 52, *322, 323*
  formation, 323
  migration, 322
  parabolic, 322

E

echinenone, 183
ecotone, 136
electrical conductance, 27

Elk-Grant Co. Lake, 127, 128, 129, 130, 131, 132
Elk Spring, 29, 30, 31
Emerson phase, 289, 321
Endochironomus, 35
English Lake District, 122, 124, 210
Entomoneis ornata, 229
eolian activity, 258, 323
epilimnion, 22, 23, 31, 32, 79, 84, 87, 88, 91, 104, 105, 111, 124, 152, 172, 226, 230, 234, 235
erosion, wind, 184
eskers, 11
Eucalyptus, 253
Euglenophyta, 183
eukaryotes, 181
Eunotia, 224
European Little Ice Age, 324
European settlement, *233*
evaporation, 168, 317
evapotranspiration, 300
evolution, brime (closed-basin), 100

F

factor variance data, 119
Faquar Lake, 127, 128, 129, 130, 131, 132
feldspar, 138, 228, 229
    orthoclase, 139
    plagioclase, 139, 153, 166, 315, 317
Ferris dune field, 322
fire, 15, 57, 232, 255, *295, 298*
    frequency, 301, 305, 325
    regime, 304, 305
    scars, 296, 298
    topography dependency, 301
Fish Lake, 127, 128, 129, 130, 131, 132, 192
flumes, 91
flux
    cosmic particle, *45, 60, 62*
    neutron (secondary), 62
    solar, 323
forbs, 265, 271
forest-prairie border, 136, 252
forests
    aspen, 184
    assemblages, 296
    birch, 4, 5, 51
    boreal, *227*, 252, 255, 311
    coniferous, 15, 45, 190, 216, *227, 228, 230*
    deciduous, 15, 46, 190, 216, *230*, 231, 232, 265, 324
    environments, *227, 230*
    existing, *298*
    hardwoods, 5, 46, 190, 252, 289, 295, *301*, 319
    mesic, 2, 5, 55
    oak, 295
    old-growth, 296
    past, *301*
    pine, 2, 5, 16, 46, 184, 252, 255, 283, 295

forests *(continued)*
    prehistoric, 295
    processes, *295*
    spruce, 2, 4, 15, 16, 51, 184, 255, 270, 283, 295, 311
    transitions, 295
    types, 295
    vegetation, *298*
fossil pigments, *181*
fossil-pollen record, 275
Fragilaria, 218, 224, 227, 231
    brevistriata, 228
    construens, 324
    crotonensis, 32, 80, 81, 82, 84, 85, 155, 192, 203, 208, 209, 210, 211, 218, 221, 224, 226, 227, 228, 229, 230, 231, 232, 234, 235, 313, 315, 318, 324
    vaucheriae, 228
Francis Lake, 127, 128, 129, 130, 131, 132
Fraxinus, 252, 254, 255, 256, 257, 265, 271, 282
freeze dates, 22
frustules, 33, 53, 241, 313
Frustulia rhomboides, 207

G

gabbro sills, 126
geochemical history, *154*
geochemistry, *115*, *135*, *140*
geology, *159*, *215*
geomagnetic disturbances, 62
geomagnetic field, *63*
geomagnetic-solar events, *45*
George Lake, 127, 128, 129, 130, 131, 132
glacial, late, *181*
glacial advance, 324
glacial anticyclone, 287, *288*
glacial debris, 13
glacial deposits, 215
glacial history, *7, 252*
glacial retreat, 19
glacial till, 2, 5, 7, 9, 13, 25, 45, 46, 190, 270, 320
glaciation, 7, 13
glaciers, 190, 321
Gladstone Lake, 127, 128, 129, 130, 131, 132
Glenboro, 270
Gomphonema, 221, 232
Gramineae, 186, 254, 255, 256, 257, 271, 281, 315
gramminerals, 298
Grand Traverse Bay, Lake Michigan, 124
granite, 122
Grantsburg sublobe, 10, 11
Grantsburg till, 190
Graptolebris testudinaria, 35
grass, 5, 315
gravel, 11, 12, 13
Great Basin, 178
Great Lakes region, 122, 172, 265, 270, 271

Green Bay, Lake Michigan, 102, 140
Green Lake, Fayetteville, New York, 27, 183
Green Lake, Minnesota, 127, 128, 129, 130, 131, 132
ground water
    inflows, 32, 168
    seepage, 4, 5, 55
Grove Lake, 127, 128, 129, 130, 131, 132
gypsum, 116, 151, 155
Gyralus, 35
Gyrosigma attenuatum, 228

H

Hale solar cycle, 63
Ham Lake, 127, 128, 129, 130, 131, 132
Harvey's Lake, Vermont, 149, 231
heat budget, 78
hematite, 88
Henderson Lake, Maine, 170
Hennepin County, 192
Herman beach, 13
Herman level, 13
Holocene, *135*, *181*, *215*, *309*
    early, *282*
    late, *283*, *324*
    middle, *247*, *283*, *311*, *317*, *321*
Hudson Bay, 13
hydrogen isotope data, 317
hydrogen sulfide, 151
hydroxides, 124, 152
    ferric, 55, 64, 91, 103, 120, 124
hypolimnion, 22, 23, 24, 29, 31, 32, 79, 80, 82, 87, 89, 92, 104, 105, 108, 122, 140, 149, 152, 153, 172, 226, 313
Hypsithermal, 247

I

ice
    advance, 19, 289
    block, 3, 7, 11, 13, 19, 48, 215, 217, 311
    buried, 15
    cover, 22, 78, 79, 85, 95, 104
    flowing, 11
    glacial, 11
    lobes, *8*
    melt, 209
    movement, 7, 8, 9
    out, 78, 79, 226
    southern margin, 287
    stagnant, 13, 19, 215, 252
ice sheet, 7, 11, 13, 45, *287*, 310, 321
    retreat, 46, 289
illite, 138
insolation, 287, 313, 320, 321, 323
invertebrates, benthic, *34*
ions, major, *25*

Iron Lake, 127, 128, 129, 130, 131, 132
iron, 2, 4, 6, 25, *29*, 46, 49, 51, 55, 64, 80, 82, 84, 88, 89, 92, 122, 124, 126, *149*, 153, 155, 156, 230, 311, 313, 319, 320
  dissolved, 104
  spring, 88
iron oxides, 88
iron phosphate, 124, 177
iron sulfide, 124, 151
Itasca Bison Kill Site, 12, 15, 42
Itasca Lake, 27, 127, 128, 129, 130, 131, 132
Itasca moraine, 7, 9, 10, 12, 19, 215
Itasca Park, 11, 19, 21, 216, 221, 232, 233, 252
Itasca region, 1, 13, 287
  landscape history, 7
Itasca transect, 190
Iva, 254
  ciliata-type taxa, 256

J

Josephine Lake, 24, 127, 128, 129, 130, 131, 132
Juniperus, 254, 255, 271
  communis, 252

K

kaolinite, 138
Kimball Lake, 127, 128, 129, 130, 131, 132
Kirchner Marsh, 187
Kylen Lake, 42, *161*, 270

L

Lake Agassiz, 7, *13*, *289*, 311, 321
Lake Aitkin, 12
Lake Ann, 52, 322, *323*
Lake Ann core, 323
Lake Balaton, Hungary, 101
Lake Constance, 109
Lake E, 270
Lake Greifen, Switzerland, 172, 174, 177
Lake Itasca, 7, 11, 12, 19, 106, 216, 234
Lake Koochiching, 12
Lake Manitoba, 111
Lake Mendota, Wisconsin, 27, 33, 317
Lake Michigan, 231
  southern, 124
Lake Mille, 11
Lake of the Clouds, 42, 63, 232, 265, 322, 325
Lake Ontario, 229, 231
Lake St. Croix, *161*
Lake Sallie, 233
Lake Superior, 121
Lake Superior basin, 8
Lake Superior lowland, 11
Lake Traverse, 13
Lake Upham, 12
Lake Washington, 172
Lake West Okoboji, 270
Lake Winnibigoshish, 322
Lake Zurich, 149
lakes, *16*
  acid, 189, 207
  alkaline, 189, 210, 232
  alkalinity, 208, 209
  bottom, 88, 93, 104
  calcareous, 192
  carbonate, 118
  carbonate-bearing sediments, 322
  central Minnesota, 115
  circulation, 315
  classification, *118*
  clastic, 118
  dilute, 227
  dimictic, 23, 24, 90, 226, 234
  effects, 289
  end-member, 119
  eutrophic, 153, 186, 187, 190
  group III, 27
  hard-water, 98, 140, 185
  holomictic, 183, 185
  kettle, 190
  loco-carbonate, 190
  marl, 97, 98, 102
  meromictic, 24, 183, 184, 185
  mesic-forest, *5*, *46*, *55*, 93, 154
  modern, *46*, *55*, 94, 154, 192, 208
  muds, 15
  northeastern Minnesota, 115, 125
  oligotrophic, 186, 228
  organic, 118
  origins, *311*
  phases, 2
  postglacial, *3*, *45*, *50*, *57*, 92, 154, 155
  prairie, *4*, *46*, *51*, *57*, 92, 98, 100, 111, 116, 122, 125, 140, 153, 154, 190, 192, 207, 210
  proglacial, 12
  saline, 5, 102
  seepage, 216, 234
  shallow, 194, 210
  string, 11
  transparent, 25
  varved, 232
  western Minnesota, 116
laminae, 4, 33, 37, 38, 40, 47, 48, 50, 51, 53, 54, 91
  carbonate, *57*, 64
  concentrated diatom, *56*, 64
  ferric hydroxide, *56*, 64
  organic, *56*, 64
laminations, 5, 16, 25, 37, 64, 98, 159, 217, 241, 311
  seasonal processes, 93
landscape history, 7
Larix, 42, 254, 255, 265, 271, 282
  laricina, 252
LaSalle Lake, Lower, 11
Late Holocene, *247*
Laurentide ice sheet, 287, 310, 311, 321
Lawrence Lake, Michigan, 27
leaching, 2, 46, 300, 319
light
  intensities, 23
  transmission, *25*
Lily Lake, 324
limestone, 8, 110
Limnocythere, 35
  herricki, 229
  staplini, 322
limnology, *19*, 311
  history, *309*
limonite, 88
Linwood Lake, 127, 128, 129, 130, 131, 132
lipscom bite, 85
Little Ice Age, 65, *232*, 248, 318, *325*
Little Pine Lake, 119, 121, 127, 128, 129, 130, 131, 132, 298
loadings, 119, *146*, 154, 227, 305, 317
Lockhart phase, 289
loess, 46
logging, *233*
Long Lake, 127, 128, 129, 130, 131, 132
Long Prairie River, 10
loss on ignition (LOI), 298
Lotus Lake, 127, 128, 129, 130, 131, 132
lowlands, 8

M

macrophytes, 109
  aquatic, 175, 181
magnesite, 102
magnesium, 4, 25, 97, 99, 102, 111, 116, 149
magnesium bicarbonate, 25
magnetic measurements, *159*
magnetic properties, *159*
magnetic storms, 323
magnetic susceptibility, 69, 319, 320
magnetite, 6, 69, 159, 160, 320
Magnolia, 254
Mallomonas, 245
  allomonas caudata, 247
  allomonas crassisquama, 245
  crassisquama, 244, 245
  pseudocoronata, 232
manganese, 2, 4, 6, 25, *29*, 49, 51, 55, 80, 82, 84, 85, 88, 89, 92, 93, 103, 104, 122, 124, 140, 148, *149*, 152, 155, 156, 230, 311, 313, 319, 320
manganese carbonate, 49, 122
manganese oxide, 49, 91, 104, 124, 140
Mantrap Lake, 11
Maple Lake, 127, 128, 129, 130, 131, 132
marl, 16, 27, 97, 168, 169, 171, 172, 173, 178
marshes, 216

Marstonia, 35
Martin Pond, 271
Massomonas caudata, 244, 245, 247
Mastogloia, 218, 228
Maunder deviations, 324
Maunder solar cycle, 63
McIntyre Lake, 192
Meander Lake, 233
Medicine Lake, 270
Medieval Warm Epoch, 65, *325*
Melosira (Aulacoseira) granulata, 324
melt, 7, 311
melting, 11, *13*, 48, 215, 311
meltwater, 12, 19, 215, 289
Mendota Lake, 27
meromixis, *24*
Mesabi iron range, 11
metalimnion, 22, 23, 24, 124, 226, 230
microfossils, Siliceous, *221*
microlaminae, 91
microzone, 120, 122
Milankovich cycle, 323
Milankovich orbital forcing, 320
Mina Lake, 127, 128, 129, 130, 131, 132
mineralization rates, 300
mineralogy, *100*, *109*, *135*, *139*
Minneapolis area, 8
Minneapolis Valley, 10
Minnesota lakes, *98*, *115*, 184
　　chemical analysis, *127*
　　sediment samples, *127*
Minnesota River Valley, 11, 13
Mirror Lake, 232
Mississippi Lake, 11
Mississippi River, 12, 19
models
　　brine evolution, 101
　　climate change, 247
　　climatic, 155
　　five-factor, *120*
　　four-factor, *119*
　　paleoclimate, *287*
molluscs, 168, 311
monimolinmion, 24, 92
Moorhead phase, 289
Moose Lake, 127, 128, 129, 130, 131, 132
Morgan Lake, New Mexico, 89
morphometry, *21*, *90*, 217
morphotypes, 239
　　stomato cyst, 242
Morrison Lake, 11
Mountain Lake, 121, 122, 127, 128, 129, 130, 131, 132
myxoxanthophyll, 183

N

N cycling, *300*
Nebraska Sand Hills, 322
neoglacial environments, *324*
Nicollet Creek, 42
Nipigon phase, 289
nitrogen, 31, 226

Nitzschia
　amphibia, 228
　romana, 228
Nokay Lake, 127, 128, 129, 130, 131, 132
nutrients, *31*, 80, *224*, 229, 234, 296, 301, 305
　cycles, 305
　fluxes, 21, 31, 226, 228, 235
　loading, 229

O

oak pollen, 57
oak savanna, 4, 5, 15, 186, 265, 321
Ojibway phase, 289
O'Leary Lake, 121, 127, 128, 129, 130, 131, 132
ooze, 13
opal, biogenic, 85
organic matter, 25, 48, 64, 80, 84, 85, 118, 149, 174, *175*, 177, 186, 187
organisms, burrowing, 217
oscillations, climatic, 1, *6*, 46, 52, 57, 59, *63*, 160, 289, 321, *323*
Oscillatoria, 228
oscillaxanthin, 183
ostracodes, 3, 6, 35, 57, 166, 168, 171, 178, 228, 229, 234, 247, 255, 286, 311, 313, 315, 317, 322
Ostrya, 230, 301, 303, 319
　virginiana, 252, 298
Ostrya-Carpinus, 265, 283
Ottawa River sediment, 124
outflow, 13
outwash, 7
oversaturation, 106, 108, 109
oxidation, 185
oxides, 89, 152
　manganese, 49, 91, 104, 124, 140
oxygen, 24, 48, 51, 88, 106, 108, 226
　depletion, *23*
　dissolved, *104*
　hypolimnetic, 23
　isotopes, *163*, *168*, 170, 171, 287, 317, 319, 320
oxyhydroxides, 89, 93, 103, 104, 122, 149, 152, 156, 313

P

paleoclimate, *275*
　variations, *290*
paleoenvironment
　indicators, 275, *286*
　reconstructions, 286
paleolimnology, *227*
　diatom, *215*
　Holocene, *215*
　synthesis, *311*
paleomagnetism, *159*
Park Rapids, 10
Park Rapids outwash plain, 11

Park Rapids sand plain, 7
particulates, 301
peat, 175
pelagial zone, 98
Pentaneura, 35
perihelion, 287
Petalostemum, 259
　candidum-type pollen, 259
　purpureum-type pollen, 259
Phacotus, 57, 227, 311
phosphate, 149, 153
　iron, 149, 152, 153
　minerals, 122
phosphorus, 30, 31, 32, 49, 51, 82, 84, 88, 89, 92, 149, 152, 153, 155, 191, 203, 224, 226, 227, 228, 230, 235, 300, 315, 318
　minerals, 25
photosynthesis, *32*, 109, 174, 177, 230
　oxygenic, 23
　phytoplankton, 24, 106, 111, 112
　plankton, 109
Physocypria, 35
　globulosa, 318
　pustulosa, 35
phytoliths, 231, 232, 247
phytoplankton, 31, *32*, 85, 109, 233
Picea, 40, 42, 227, 228, 252, 254, 255, 265, 270, 271, 281, 282, 291, 321
　decline, 265, *270*
　glauca, 209, 252
　mariana, 252
Pickerel Lake, South Dakota, *170*, 173, 270, 322, 324
Pierre Shale, 151
pigments
　algal, 317
　concentrations, 187
　cyanobacterial, 315, 318
　diversity, 185
　plant, 313
　prokaryote, 183, 185, 187, 311
　sedimentary, *183*
pine pollen, 278
Pinnularia biceps, 203
Pinus, 252, 253, 254, 255, 259, 270, 271, 282, 283, 291, 303, 319, 321
　banksiana, 228, 232, 252, 255, 265, 298, 304, 311
　banksiana-resinosa pollen, 255
　banksiana-resinosa zone, 313
　resinosa, 232, 252, 255, 265, 296, 298, 300, 304, 305, 311
　rise, 265, *270*
　strobus, 42, 65, 172, 231, 232, 252, 265, 298, 303, 304, 305, 319, 320, 324
　strobus-type pollen, 265, 303
Pisidium, 35, 313
plankton, 88, 187
　lake, 175
plant debris, 13, 255
plant pigments, 175, 176

Pleuroxus
  denticulatus, 34
  procurvus, 34, 35
pollen, 217
  abundance, 278
  modern data, *275*
  pine, 278
  rain, 15
  record, *275*
  stratigraphy, *265*
pond, sag, 311
Populus, 255, 301, 303
  balsamifera, 252, 255
  grandidentata, 298
  tremuloides, 252, 255, 298
porosity, 88
postglacial, early, *245*
potash feldspars, 122, 125
potassium, 88
prairie, 252
  expansion, *315*
  forb pollen, 278
  herbs, 322
  vegetation, 2, 322
prairie-forest boundary, 136, 252
prairie lakes, *4, 46, 51, 57*, 92, 98, 100, 111, 116, 122, 125, 140, 153, 154, 190, 192, 207, 210
prairie period, 168, 173, 176, 177, 184, 186, *228*, 248, 257, 258, 265, *271, 315*
  See also prairie lakes
prairie-savannah vegetation, 4, 229
precipitation, *21*, 98, 99, 111, 152, 168, 226, 227, 228, 232, 258, 278, 283, 289, 301, 311, 317
  simulations, 288
  spring, 230
  summer, 230
precipitation-evaporation gradient, 99, *100*, 101
Pretty Lake, Indiana, 170
Procladius, 35
productivity, *106*, 152
profundal basin, 37
profundal zone, 178
prokaryotes, 183, 185, 187, 311
Prunus, 252
Pseudochydorus globosus, 34
Ptamogeton, 35
Pteridium, 255, 304, 305, 313
  aquilinum, 298
pyrite, 88, 122, 149, 151
pyrite framboids, 49

Q

Q-mode factor analysis, 118, 125, *145*, 149
quartz, 85, 138, 139, 140, 228, 229, 313, 315, 317, 318, 319, 320, 321, 325
Quercus, 186, 228, 230, 254, 255, 256, 258, 265, 271, 282, 301, 303, 315, 319
  alba, 252
  borealis, 252, 298, 303

Quercus (continued)
  ellipsoidalis, 252, 256
  macrocarpa, 252, 256, 265, 298, 301, 303
  rubra, 265
Quercus-Gramineae-*Artemesia* pollen zone, 153
Quercus-Ostrya subassemblage, 254, *265*, 271

R

radiation, 287, 288, 310, 323
radiocarbon chronology, 38, *40*, 50, 74
rainfall, 78, 80, 88, 92, 189
Rainy lobe, *8*, 11
Ramsey County, 192
rates
  accumulation, *221*
  evaporation, 111
  influx accumulation, 221
  mass-accumulation, 80, 87, *89*, 145, 149, 166, 177, 313, 317, 319
  pollen-accumulation, 255, 256, 259, 265
  sedimentation, 184
Recent, *247*
Red Lake peatlands, 16
Red Lakes lowland, 8, 11, 12, 13
Red River, 8
Red River Valley, 11, 289
redox association elements, 122
redundancy, 70
Reeds Lake, 127, 128, 129, 130, 131, 132
regression method, 275, 276, *277, 278*, 280
respiration, 23, 109
response-surface analyses method, 2, 275, 276, *277*, 280
resuspension, 92, 93
Rhizosolenia, 38, 53
  eriensis, 32
rhodochrosite, 49, 85, 87, *102*, 111, 112, 122, 139, 140, 149, 152, 154, 156, 171
Rhopalodia gibba, 203, 218
Ribes, 252
Rideau River sediment, 124
River Warren, glacial, 13
rockbridgeite, 85, 87, 177
Rocky Mountain basins, 322
Ruby Lake, 298
runoff, 52, 82, 181, 186, 233

S

sagebrush, 5, 247, 315
St. Croix moraine, 7, 8, 9, 11
St. Lawrence lowland, 322
St. Louis River, 12
St. Louis sublobe, *11*, 19, 252, 311
St. Olaf lake, 127, 128, 129, 130, 131, 132

salinity, 25, 27, 54, 99, *100*, 111, 116, 140, 166, *168*, 171, 178, 229, 317
  gradients, 169
Salix, 256, 257
Sallie Lake, 127, 128, 129, 130, 131, 132
Salt Lake, 121, 127, 128, 129, 130, 131, 132
sand, 11, 12, 13, 48, 215
  dunes, 13
  eolian, 52
Sand Point Lake, 119, 120, 121, 122, 127, 128, 129, 130, 131, 132
Sarah Lake, 192
saturation, *107*
seasons, lake (defined), 76
sedimentation, 16, 25, 27, 47, 59, 91
  accumulation, 87, 90
  apparent rate (defined), 76
  clastic, 230
  modern, *75*
  rate, 76, 82, 85, *87*, 89, 90, 91, 92, 93
  seasonal, 78
sediments, 7, *122*
  accumulation, 234
  anoxic, 186, 227, 318. See also anoxic conditions
  bottom, *88*
  carbonate-bearing, 322
  chronology, *37*
  clastic, 54, 151
  density, 160
  flux (defined), 76
  glacial, 7
  Holocene, *135*
  isotopic studies, *163*
  littoral, 149
  Minnesota lakes samples, *127*
  organic-carbon context, 177
  postglacial, 245
  prairie-period, 177
  redox-sensitive, 122, 124
  resuspended, 90
  stratigraphy, *69*
  surface, 25, *115*, 191, *224*
  transect, *224*
  traps, 76, *80, 84,* 88*, 91*, 93, 181, 226, 326
  types, *125*
  varved, 4, *48*, 98, *135, 181*, 296, *309*
seepage, ground-water, 4, 5, 55
seston mineralogy, *85*
Shagawa Lake, 153
shale, 11, 12, 19, 116, 120, 125, 151, 155
Shetek Lake, 127, 128, 129, 130, 131, 132
Sida crystallina, 35
siderite, 153
silica, 25, 31, 32, 55, 64, 80, 82, 84, 88, 91, 93, 154, 155, 226, 230, 239, 320
Siliceous, *221*
silicon, 32, 224, 226, 230, 235
  fluxes, 233

silt, 5, 13, 46, 48, 50, 51, 57, 62, 215, 313
slumping, 48, 98
Soap Lake, Washington, 89
sodium, 100, 116, 145, *153*, 166, 229, 313, 315, 317, 319, 321
soils
  anoxic, 315
  erosion, 233
  moisture, 301, 319
solar activity, 5, *62*, 65, 324
solar-geomagmetic activity, *45, 60, 62*, 65
solar radiation, *287*
Spectacle Lake, 127, 128, 129, 130, 131, 132
spectral density, 60, 62, 323
springs, iron-rich, 149
sprouters, 313
spruce pollen, 278
stagnation, 226
statospores, 239
Stephanodiscus, 51, 57, 191, 192, 194, 209, 218, 226, 227, 229
  alpinus, 191, 218, 229, 230, 231, 232
  dubius, 191
  hantzschii, 191, 227
  minutula, 191, 209, 211
  minutulus, 32, 38, 85, 152, 155, 176, 218, 221, 224, 226, 227, 228, 229, 230, 231, 232, 233, 234, 235, 313, 315, 317
  niagarae, 192, 194, 203, 208, 211, 221, 229, 230, 231
  parvus, 191, 229
  subtilus, 191
  spp., 208, 211, 226, 230
Stewart's Dark Lake, Wisconsin, 183, 187
stillstand, 11
stomatocysts, 239, *241*, 318
strandlines, 13
stratification, 108, 152, 153, 226
  seasonal, *78*
  summer, 107, 108, 149, 152
  wintcr, 107
stratigraphy, 91
streams, subglacial, 11
strontium, 84, 85, 149
sublittoral zone, 178
succession
  climatic, *6*
  vegetational, 15
sulfate, 100, 101, 116, 124, 151
sulfur, 30, 138, 149, 151
Sunfish Lake, Ontario, 183
sunspot, 63
Superior lobe, 8, 9, 11, 190
Superior lowland, 8
Surirella, 221, 229, 232
Synedra, 155, 218, 226, 231, 233
  acus, 218, 221, 226, 230, 319
  cyclopum, 231
  nana, 227
  parasitica, 229
  spp., 227
Synurophyceae, 239

T

Tabellaria, 218
  fenestrata, 218, 224, 226, 227, 233
  flocculosa, 80, 85, 221
Tanytarsus, 35, 228, 313
Taxus canadensis, 252
temperature, *21, 104*, 172, 174, 178, 226, 247, 252, 311, 317
  oceans, 287
tephra, 91
Terrell Sea phase, 289
thaw dates, 22
thermal events, 80
thermal stratification, 105
thermocline, 33, 79, 104, 226
Tilia, 265
  americana, 252, 295, 298, 300
till, glacial, 2, 5, 7, 9, 13, 25, 45, 46, 190, 270, 320
titanium, 88
transitions
  early Holocene–middle Holocene, *282*
  forest-prairie, 315
  late glacial–early Holocene, *282*
  middle Holocene–late Holocene, *283*
transparency, 25
transport, 49, *51*, 59, 62, 90
trap
  accumulations, 90
  rates, *90*
  results, *91*
tree rings, 62, 63, 233
triple junction, climatic, 136, 172, 310
Trout Lake, 121, 127, 128, 129, 130, 131, 132
Tubulliflorae, 256
tunnel valleys, *11*, 12, 19, 215, 252
turbidites, 48, 155, 228, 232, 313
turbidity, optical, 24

U

Ulmus, 230, 254, 255, 265, 270, 271, 282, 321
  maximum, 265, *270*
  spp., 252
undersaturation, 106, 108, 109

V

Valvata, 35
varved sediments, diagenesis, *48*
varves, 1, 5, 6, 33, 37, *46*, 87, 152, 217, 241, 311, 315, 320
  chronology, 37, 38, *40*, 60, *69, 74*, 161, 227, 252, 275
  chronometer, 1, *45, 49*
  chronostratigraphy, 50
  climate variability, *57*
  counts, *69, 70*, 161, 217
  interpretation, *91*

varves *(continued)*
  origin, *46*
  petrology, 38
  preservation, *48*
  sequential changes, *50*
  thickness, 37, 46, 54, 57, *60, 62, 64*, 160, 161, 228, 313, 315, 317, 318, 319, 321, 323, 324
  time series, *39*, 46, 59, 60
  types, *46*
  years, 160, 161, 166, 171, 172, 173, 178, 184, 185, 186, 187, 252, 255, 257, 271, 278, 280, 281, 286, 301, 304, 311
vegetation
  aquatic, 168
  gradients, 275
  history, *15, 40, 251*
  patterns, past, 296
  prairie, 2, 322
  prairie-savanna, 4, 229
  terrestrial, 175
  transitions, 305
  types, 295
  wetlands, 16
Viburnum, 252
vivanite, 49, 51
volcanic aerosols, 323

W

Wadena drumlin field, 7, *8*
Wadena ice lobe, *8, 11*, 19, 215, 252, 311
Wadena till, 252
waters
  anoxic conditions, 23, 24, 29, 31, 98, 136, 149, 181, 185, 216, 226, 234
  balance, 172
  bottom, 23, 24, 29, 31, 51, 98, 136, 174, 181, 182, 185, 227, 313
  calcium-magnesium, 116
  chemistry, *98*
  dilute bicarbonate, 115
  fluxes, *25*
  ground, 122, 149, 320
  metalimnetic, 230
  stratification, *23*
  surface, 174
weathering, 47, 320
  indices, *153*
Weber Lake, 42, 271
wetlands, 7, 19, 25, 216
  vegetation, 16
Williams Bay, Jack's Lake, Ontario, 30
Williams Lake, Hubbard County, Minnesota, 32
Williams Lake, Washington, 89
Wilson Lake, *149*
winds
  anticyclonic, 310
  energy, 226
  erosion, 184
  intensity, 324

winds *(continued)*
    troposphere, 63, 65
    zonal, 323, 324
Wolf Creek, 42
Wolf Creek Lake, 270
wood, fossil, 171

X, Y, Z

xanthophylls, 183
xeric aspects, 301, 313
xeric sites, 303
x-ray diffractograms, 137

Younger Dryas climate fluctuation, 289
Younger Dryas cold event, 320, 321

zooplankton, 22, *33*, 85, 91
    mobile, *91*
    pellets, 85

Typeset by WESType Publishing Services, Inc., Boulder, Colorado
Printed in U.S.A. by Malloy Lithographing, Inc., Ann Arbor, Michigan